ADVANCES IN ARTHROPOD REPELLENTS

ADVANCES IN ARTHROPOD REPELLENTS

Edited by

JOEL COATS
Iowa State University, Department of Entomology, Pesticide Toxicology Laboratory, Ames, IA, United States

CALEB CORONA
Iowa State University, Department of Entomology, Pesticide Toxicology Laboratory, Ames, IA, United States

MUSTAPHA DEBBOUN
Delta Mosquito & Vector Control District, Visalia, CA, United States

ELSEVIER

ACADEMIC PRESS
An imprint of Elsevier
elsevier.com/books-and-journals

Academic Press is an imprint of Elsevier
125 London Wall, London EC2Y 5AS, United Kingdom
525 B Street, Suite 1650, San Diego, CA 92101, United States
50 Hampshire Street, 5th Floor, Cambridge, MA 02139, United States
The Boulevard, Langford Lane, Kidlington, Oxford OX5 1GB, United Kingdom

British Library Cataloguing-in-Publication Data
A catalogue record for this book is available from the British Library

Library of Congress Cataloging-in-Publication Data
A catalog record for this book is available from the Library of Congress

ISBN: 978-0-323-85411-5

For Information on all Academic Press publications visit our website at https://www.elsevier.com/books-and-journals

Publisher: Charlotte Cockle
Acquisitions Editor: Anna Valutkevich
Editorial Project Manager: Lindsay Lawrence
Production Project Manager: Maria Bernard
Cover Designer: Mark Rogers

Typeset by Aptara, New Delhi, India

Contents

6. Repelling mosquitoes with electric fields

Ulla Gordon, Farooq Tanveer, Andreas Rose, Krijn Paaijmans

7. Multimodal mechanisms of repellency in arthropods

Fredis Mappin and Matthew DeGennaro

8. Finding a repellent against ticks: neurophysiological and behavioral approaches

Zainulabeuddin Syed, Kenneth L. O'Dell Jr

9. Arthropod repellents and chemosensory reception

Robert Renthal

10. Semifield system and experimental huts bioassays for the evaluation of spatial (and topical) repellents for indoor and outdoor use

Mgeni Mohamed Tambwe, Johnson Kyeba Swai, Sarah Jane Moore

11. Semi-field evaluation of arthropod repellents: emphasis on spatial repellents

Daniel L. Kline, Karen McKenzie, Adam Bowman

12. Human subject studies of arthropod repellent efficacy, at the interface of science, ethics, and regulatory oversight

Shawn B. King, Cassandre H. Kaplinsky, Ralph Washington, Jr., Scott P. Carroll

13. Arthropod repellent research in Northwest Florida, United States

John P. Smith

14. Current status of spatial repellents in the global vector control community

Nicole L. Achee, John P. Grieco

15. Repellent semiochemical solutions to mitigate the impacts of global climate change on arthropod pests

Agenor Mafra-Neto, Mark Wright, Christopher Fettig, Robert Progar, Steve Munson, Darren Blackford, Jason Moan, Elizabeth Graham, Gabe Foote, Rafael Borges, Rodrigo Silva, Revilee Lake, Carmem Bernardi, Jesse Saroli, Stephen Clarke, James Meeker, John Nowak, Arthur Agnello, Xavier Martini, Monique J. Rivera, Lukasz L. Stelinski

16. The role of arthropod repellents in the control of vector-borne diseases

Stephen P. Frances, Mustapha Debboun

Contributors

Nicole L. Achee Department of Biological Sciences, Eck Institute for Global Health, University of Notre Dame, Notre Dame, IN, United States

Arthur Agnello Cornell University, College of Agriculture and Life Sciences, Ithaca, NY, United States

Natasha M. Agramonte DeKalb County Board of Health, Decatur, GA, United States

Carmem Bernardi ISCA, Inc., Riverside, CA, United States

Ulrich R. Bernier United States Department of Agriculture, Center for Medical, Agricultural, and Veterinary Entomology, Gainesville, FL, United States

Darren Blackford USDA Forest Service, Forest Health Protection, Ogden, UT, United States

Jeffrey Bloomquist Emerging Pathogens Institute, Entomology and Nematology Department, University of Florida, Gainesville, FL, United States

Rafael Borges ISCA, Inc., Riverside, CA, United States

Adam Bowman USDA-ARS, Center for Medical and Veterinary Entomology, Gainesville, FL, United States

Scott P. Carroll Carroll-Loye Biological Research, Davis, CA, United States; Department of Entomology and Nematology, University of California, Davis, CA, United States

Stephen Clarke USDA Forest Service, Forest Health Protection, Lufkin, TX, United States

Joel Coats Iowa State University, Department of Entomology, Pesticide Toxicology Laboratory, Ames, IA, United States

Caleb Corona Iowa State University, Department of Entomology, Pesticide Toxicology Laboratory, Ames, IA, United States

Mustapha Debboun Delta Mosquito & Vector Control District, Visalia, CA, United States

Matthew DeGennaro Department of Biological Sciences, Biomolecular Sciences Institute, Florida International University, Miami, FL, United States

Muhammad Farooq Anastasia Mosquito Control District, St. Augustine, Florida, United States

Christopher Fettig USDA Forest Service, Pacific Southwest Research Station, Davis, CA, United States

Gabe Foote University of California, Department of Entomology and Nematology, Davis, CA, United States

Stephen P. Frances Australian Defence Force Malaria and Infectious Disease Institute, Gallipoli Barracks, Enoggera, Qld, Australia

Larry I. Goodyer Leicester School of Pharmacy, De Montfort University, Gateway, United Kingdom

Ulla Gordon BioGents AG, Regensburg, Germany

Elizabeth Graham USDA Forest Service, Forest Health Protection, Juneau, AK, United States

John P. Grieco Department of Biological Sciences, Eck Institute for Global Health, University of Notre Dame, Notre Dame, IN, United States

Shiyao Jiang Emerging Pathogens Institute, Entomology and Nematology Department, University of Florida, Gainesville, FL, United States

Amy L. Johnson U.S. Army Combat Capabilities Development Command Soldier Center, Natick, MA, United States

Cassandre H. Kaplinsky Carroll-Loye Biological Research, Davis, CA, United States

Shawn B. King Carroll-Loye Biological Research, Davis, CA, United States

Daniel L. Kline USDA-ARS, Center for Medical and Veterinary Entomology, Gainesville, FL, United States

Revilee Lake ISCA, Inc., Riverside, CA, United States

Kenneth J. Linthicum United States Department of Agriculture, Center for Medical, Agricultural, and Veterinary Entomology, Gainesville, FL, United States

Agenor Mafra-Neto ISCA, Inc., Riverside, CA, United States

Fredis Mappin Department of Biological Sciences, Biomolecular Sciences Institute, Florida International University, Miami, FL, United States

Xavier Martini University of Florida, Institute of Food and Agricultural Sciences, Gainesville, FL, United States

Karen McKenzie Woodstream, Melbourne, FL, United States

James Meeker USDA Forest Service, Forest Health Protection, Pineville, LA, United States

Jason Moan Alaska Department of Natural Resources, Division of Forestry, Anchorage, AK, United States

Sarah Jane Moore Vector Control Product Testing Unit, Ifakara Health Institute, Environmental Health, and Ecological Sciences, Bagamoyo, Tanzania; Vector Biology Unit, Swiss Tropical and Public Health Institute, Basel, Switzerland; University of Basel, Basel, Switzerland; Nelson Mandela African Institute of Science and Technology (NM-AIST), Tengeru, Tanzania

Steve Munson USDA Forest Service, Forest Health Protection, Ogden, UT, United States

Edmund Norris Emerging Pathogens Institute, Entomology and Nematology Department, University of Florida, Gainesville, FL, United States; United States Department of Agriculture, Center for Medical, Agricultural, and Veterinary Entomology, Gainesville, FL, United States

John Nowak USDA Forest Service, Forest Health Protection, Asheville, NC, United States

Kenneth L. O'Dell Jr. Department of Entomology, University of Kentucky, Lexington, KY, United States

Krijn Paaijmans Center for Evolution and Medicine, School of Life Sciences, Arizona State University, Tempe, Arizona, United States; The Biodesign Center for Immunotherapy, Vaccines and Virotherapy, Arizona State University, Tempe, Arizona, United States; ISGlobal, Barcelona, Spain

Steven T. Peper Anastasia Mosquito Control District, St. Augustine, Florida, United States

Melynda K. Perry U.S. Army Combat Capabilities Development Command Soldier Center, Natick, MA, United States

Robert Progar USDA Forest Service, Sustainable Forest Management Research (SFMR), Washington, DC, United States

Whitney A. Qualls Anastasia Mosquito Control District, St. Augustine, Florida, United States

Robert Renthal University of Texas at San Antonio, Department of Biology, San Antonio, TX, United States

Gary Richoux Emerging Pathogens Institute, Entomology and Nematology Department, University of Florida, Gainesville, FL, United States

Monique J. Rivera University of California at Riverside, Department of Entomology, Riverside, CA, United States

Andreas Rose BioGents AG, Regensburg, Germany

Jesse Saroli ISCA, Inc., Riverside, CA, United States

Rodrigo Silva ISCA, Inc., Riverside, CA, United States

John P. Smith Public Health Entomology Services, LLC, Panama City Beach, FL, United States

Lukasz L. Stelinski University of Florida, Institute of Food and Agricultural Sciences, Gainesville, FL, United States

Johnson Kyeba Swai Vector Control Product Testing Unit, Ifakara Health Institute, Environmental Health, and Ecological Sciences, Bagamoyo, Tanzania

Zainulabeuddin Syed Department of Entomology, University of Kentucky, Lexington, KY, United States

Mgeni Mohamed Tambwe Vector Control Product Testing Unit, Ifakara Health Institute, Environmental Health, and Ecological Sciences, Bagamoyo, Tanzania; Vector Biology Unit, Swiss Tropical and Public Health Institute, Basel, Switzerland; University of Basel, Basel, Switzerland

Farooq Tanveer BioGents AG, Regensburg, Germany

Ralph Washington, Jr. Carroll-Loye Biological Research, Davis, CA, United States

Colin Wong Iowa State University, Department of Entomology, Pesticide Toxicology Laboratory, Ames, IA, United States

Mark Wright University of Hawaii at Manoa, College of Tropical Agriculture and Human Resources, Honolulu, HI, United States

Rui-De Xue Anastasia Mosquito Control District, St. Augustine, Florida, United States

Liu Yang Emerging Pathogens Institute, Entomology and Nematology Department, University of Florida, Gainesville, FL, United States

Preface

Arthropod repellents are a key component of the first line of defense and protection against biting arthropods and the pathogens they transmit to humans and animals. They do not require large equipment, no organized effort of community vector control, and distribute the responsibility for personal protection to the individual. The public throughout the world is concerned about the current worldwide emphasis on the prevention of infection from arthropod-borne pathogens, such as malaria, dengue, chikungunya, Zika, West Nile virus, Lyme disease, leishmaniasis, Chagas disease, tick-borne encephalitis, etc. In addition, public health professionals are currently interested in the development and use of arthropod repellents due to the widespread of insecticide resistance by arthropods, high cost of developing effective prophylactic vaccines and drugs, and the recent global increase in incidences of arthropod-borne diseases. Currently, research and development of arthropod repellents is moving fast enough to meet the public health demand for effective topical and spatial arthropod repellents for human personal protective measures.

Advances in Arthropod Repellents is organized in four parts exploring recent advances and growth of new knowledge in the field of repellent research, particularly in the understanding of the molecular basis of repellency of new repellent compounds. The first part provides new information on novel arthropod repellent molecule discovery and assay development. The second part covers the recent mechanisms of arthropod repellent research and development discussing the exciting and cutting-edge multimodal mechanisms of repellency in arthropods, chemosensory reception, and neurophysiological and behavioral approaches. The third part concentrates on topical and spatial repellent studies in field and semifield trials. The fourth and final part of the book discusses for the first time arthropod repellent semiochemical solutions to mitigate the impacts of global climate change on agricultural pests and concludes with an update on the current status and future uses of arthropod repellents.

As the editors of this book, we greatly appreciate and thank the contributing chapter authors and coauthors who donated their time and decades of accumulated professional experience in the field of arthropod repellents. We also thank Lindsay Lawrence, Praveen Anand, Maria Bernard, and Anna Valutkevich of Elsevier who guided us throughout the publication process. We hope this book will be useful and interesting to professional arthropod repellent researchers, public health professionals, vector-control professionals, environmentalists, wildlife professionals, and other members of the scientific community in academia, industry, and government. It is our goal that this book will play an important role in the personal protection of arthropod bites and arthropod-borne diseases.

Mustapha Debboun, Ph.D.
Delta Mosquito & Vector Control District
Caleb Corona, Ph.D.
Department of Entomology, Iowa State University
Joel R. Coats, Ph.D.
Department of Entomology, Iowa State University

About the Editors

Joel Coats, Charles F. Curtiss Distinguished Professor of Entomology in the Department of Entomology at Iowa State University, received his B.S. degree from Arizona State University and did his graduate work at the University of Illinois at Urbana-Champaign, receiving his M.S. and Ph.D. in Entomology, with specialization in Insecticide Toxicology and Environmental Toxicology. His research program focuses on insect toxicology and environmental toxicology and chemistry. Current studies are focused on natural products as insecticides and repellents, as well as biorational analogs of them. Dr. Coats received the International Award for Research in Agrochemicals from the American Chemical Society; he is a Fellow of the American Association for the Advancement of Science, Fellow of the American Chemical Society, and a Fellow of the Entomological Society of America. He has trained 53 graduate students and published over 200 peer-reviewed publications, including 15 books.

Caleb Corona received his Ph.D. degree in Entomology and Toxicology from the lab of Dr. Joel Coats, Iowa State University, where his research focused on the development of novel insecticides and spatial repellents derived from natural products. He received his B.Sc. degree in Biology with a Chemistry minor from King University in Bristol, Tennessee, where he also served as a Teaching Assistant in Chemistry and later an undergraduate research assistant for federally funded Toxicology laboratory under Dr. Vanessa Fitsanakis . He has been invited as a symposium speaker at the Society for Vector Ecology, the Entomological Society of America, and the American Chemical society multiple times over the course of his career.

Mustapha Debboun received the Ph.D. degree in Medical and Veterinary Entomology from the University of Missouri-Columbia, is a Board-Certified Entomologist with the Entomological Society of America (ESA), and confirmed with the title of Fellow of the ESA. He has more than 27 years of experience in public health entomology, integrated vector management, personal protective measures, and mosquito/vector-borne diseases where this work has taken him to over 35 countries in Africa, Asia, Australia, Europe, and South America. He is currently the General Manager of Mosquito & Vector Control District in Visalia, California. He has published over 120 peer-reviewed scientific articles and co-edited five books. Dr. Debboun organizes national and international symposia, serves on 10 journal editorial review boards, and a reviewer for eight scientific peer-reviewed journals. He is also nationally and internationally recognized for his extensive work on arthropod repellent research and development.

Acknowledgments

Dr. Joel Coats dedicates this book to the many students, postdocs, and interns who have worked in his research group over the decades.

Dr. Caleb Corona dedicates this book to his wife, Emily who has been his rock throughout the publication process, completion of his Doctorate degree, and through all the days in between.

Dr. Mustapha Debboun dedicates this book to his beloved parents, three brothers, four sisters, his beautiful wife, Natalie, and their three extraordinary children, Ameena, Adam, and David, who keep him humble and remind him daily of how important they are in his life.

In Memoriam
Dr. Daniel Strickman

The editors are publishing this book in memory of Dr. Daniel A. Strickman, a revered colleague and friend to all of us involved with vector biology and medical entomology. During his stellar career, Dan's contributions were myriad, at numerous different levels and in service to many institutions and organizations.

After growing up in San Diego, CA, Dan studied at Dartmouth College, and then received the Bachelor of Science degree in Biology from the University of California at Riverside. Dr. Joel Coats first met him when he came to the University of Illinois at Urbana-Champaign for his graduate work; there he received the Ph.D. degree in Entomology in 1978, with a specialization in Medical Entomology. Dan's professional career was heavily devoted to public service, as well as international service: Dan served in the Peace Corps in Paraguay; as a medical entomologist in the US Air Force and the US Army, including a stint as Chief of the Department of Entomology at Walter Reed Army Institute of Research. Next, he served as the National Program Leader for the US Department of Agriculture's Veterinary, Medical, and Urban Entomology Program. Later in his career, Dan became a Senior Program Officer at the Bill and Melinda Gates Foundation.

During his remarkable career, Dan has been the winner of numerous prestigious awards. He has published over a hundred journal articles, as well as book chapters and four books. His latest book, *Mosquitoes of the World*, will stand as a capstone of the rich legacy he leaves us. Dan Strickman will be missed by his many colleagues, collaborators, and friends, but he will also be long remembered.

Joel R. Coats, Ph.D.
Iowa State University, Ames, Iowa

Caleb Corona, Ph.D.
Iowa State University, Ames, Iowa

Mustapha Debboun, Ph.D., BCE
Delta Mosquito and Vector Control District

CHAPTER

1

Arthropod repellents in public health

Mustapha Debboun[a], *Larry I. Goodyer*[b]

[a]Delta Mosquito & Vector Control District, Visalia, CA, United States, [b]Leicester School of Pharmacy, De Montfort University, Gateway, United Kingdom

1.1 Arthropod repellents

Arthropod repellents are compounds applied to skin, clothing, or other surfaces that discourage arthropods from landing or climbing on those surfaces. Dethier et al. (1960) defined a repellent as "any stimulus which elicits an avoiding reaction" by an arthropod. They also made a further distinction of terms of the physical state of the chemical by recognizing contact repellents and vapor repellents, i.e., those that are touched by an arthropod or detected in the air. Later, Roberts (1993) used the term excitorepellency to include all chemically induced irritant and repellent behaviors by classifying chemicals as irritants when tarsal contact is required and as repellents when avoidance is elicited through the vapor phase. Thus, a repellent product, whether topical, clothing, or spatial is one that is used to interrupt damage from biting arthropods such as mosquitoes, bed bugs, fleas, ticks, etc., by disrupting their normal behavior during their host-seeking and blood-feeding process. For example, when mosquito repellents are applied to sources such as human skin, clothing, or other surfaces, they discourage and disrupt mosquitoes from landing or climbing on those surfaces and serve as the first line of defense against them. Commercially available arthropod repellents are divided into synthetic arthropod repellents or natural arthropod repellents. The US Environmental Protection Agency (EPA) regulates repellent products in the United States, and the Centers for Disease Control and Prevention (CDC) recommends that consumers use only those repellent products that are approved and registered by the EPA and used in accordance with the label instructions. In addition, CDC, National Institute of Health (NIH), Bill & Melinda Gates Foundation, industry and academia continue to support ongoing research efforts for use of novel arthropod repellents to prevent and protect individuals from vectors and vector-borne pathogens. Many countries and regions have approval and registration agencies for arthropod repellents, for instance, the EU Biocide regulatory system.

Advances in Arthropod Repellents.
DOI: https://doi.org/10.1016/B978-0-323-85411-5.00016-9

1.2 Role of arthropod repellents

The use of arthropod repellents is a unique individual measure of personal protection and the first line of defense against biting hematophagous arthropods on humans and animals that are capable of transmitting human and animal pathogens throughout the world. When it comes to biting arthropods, all users have one thing in common, and that is none of them want to be bitten by mosquitoes and other biting arthropods such as fleas, sand flies, black flies, bed bugs, kissing bugs, mites, ticks, etc. Most or all of users believe that the prevention of arthropod bites is the goal of any kind of arthropod repellent. In addition, if it is not possible to control biting arthropods with conventional pesticides and since arthropod repellents are generally used as personal protection from arthropod bites as the first line of defense, and do not require large and complicated control equipment or organized community vector control, it is the users who decide whether or not to use an arthropod repellent, what kind to use, how much to apply, and the control over exposure to biting arthropods. What appeals to the users is that a repellent product applied to their skin will stop the biting arthropods from reaching their skin to cause any damage to the skin or deny them a blood meal to transmit any pathogen. Therefore, arthropod repellents have been widely used to avoid and lessen the chance of getting vector-borne diseases (Barnard et al., 1998; Barnard, 2000; Barnard and Xue, 2004; Rowland et al., 2004; Hill et al., 2007; Debboun and Strickman, 2013; Onyango and Moore, 2015). They are a multimillion-dollar global industry with an estimated global repellent market greater than $3 billion (BMGF and BCG, 2007). Arthropod repellents are formulated in many different formulations including pump sprays, aerosols, lotions, roll on, wipes, spatial (coils and sticks), laundry fabric treatments, and combinations with sunscreen and other cosmetic products. In addition, formulation of arthropod repellents whether topical, clothing, or spatial will provide opportunities for better improvement in new and novel repellent products. For example, Salafsky et al. (2007) developed a system of encapsulating N,N-diethyl-3-methylbenzamide (DEET) in liposomes that produced remarkable duration in a cosmetically acceptable form. Eventually, technology will result in more effective and accepted active ingredients that last longer, easy to use, and become important additions to the vector control tools available to prevent the transmission of arthropod-borne pathogens.

Due to overreliance on pyrethroids in public health and vector control, development of biting arthropod resistance to conventional synthetic pesticides, their toxicity to nontarget organisms, adverse environmental effects for insecticide toxicity, high operational cost, and community acceptance have prompted the use of new and effective arthropod repellents as means of biting arthropod control. Thus, arthropod repellents have become an important augmented and effective tool in integrated vector management (IVM) to reduce the impact of biting arthropods on humans and animals. This is so because their use has a strong effect on the overall vectorial capacity by reducing the probability of infecting or being infected by an arthropod vector (Smith et al., 2012).

Currently, there is a great public concern throughout the world about vector-borne pathogens that result in outbreaks of a wide range of vector-borne diseases, i.e., West Nile fever and neuroinvasive disease, St. Louis encephalitis, eastern equine encephalitis, Japanese encephalitis, dengue, chikungunya, Zika, malaria, leishmaniasis, Lyme disease, Chagas disease, scrub typhus, and onchocerciasis. Therefore, medical entomologists and public health professionals are interested in the development and use of arthropod repellents given the increase in incidences of arthropod-borne diseases, insecticide resistance, and the high cost of developing new effective prophylactic vaccines and drugs. In addition, an ideal arthropod repellent is one that is effective against multiple species of biting arthropods, long-lasting, i.e., 8 or more hours, safe, resistant to wetting and abrasion, nonirritating, nonallergenic, greaseless, odorless, nondamaging to clothing or other material, feels good on human skin, has a good shelf-life, inexpensive, and accepted for use by regulators.

To date, a range of scientific evidence is available on concerning the efficacy of arthropod repellents in reducing vector-borne diseases (Philip et al., 1944; McCulloch, 1946; Schwartz and Golstein, 1990; Soto et al., 1995; Lwin et al., 1997; McGready et al., 2001; Durrheim and Govere, 2002; Moore et al., 2002; Asilian et al., 2003; Rowland et al., 2004; Kimani et al., 2006; Hill et al., 2007; Dutta et al., 2011; Vaughn and Meshnick, 2011; Deressa et al., 2014; Onyango and Moore, 2015). Thus, we believe that the time has come for the World Health Organization (WHO) to support and implement the use of arthropod repellents in public health to reduce the incidence of vector-borne diseases. In addition, arthropod repellents (topical and spatial) have become an important part of IVM, particularly when used strategically to prevent infective bites that are not prevented by other vector control methods.

1.3 Brief history of arthropod repellents

For thousands and even millions of years, prehistoric humans and indigenous tribes attempted to use various forms of repellents. Examples include smoke in the form of burning plants and leaves, covering the skin with mud and soil, animal fats and greases, various plant oils, hanging bruised plant parts in homes, applying essential oils to the skin, and wearing various herbs and poultices around the neck or clothing to prevent and protect from arthropod bites (Peterson and Coats, 2001; Gerberg and Novak, 2007; Dolan and Panella, 2011; Maia and Moore, 2011). Smoke is still the most widely used method of repelling mosquitoes and other biting arthropods throughout the world (Moore and Debboun, 2007). For example, Torr et al. (2011) showed that smoke from burning wood or dried cow dung was effective in preventing bites from tsetse flies and demonstrated true repellency causing the tsetse flies to orient away from the unbaited traps.

The first recorded use of arthropod repellents was the reference by Homer circa 900-800 BCE to "pest-averting sulfur" (Keatinge, 1949) and among the writings of Herodotus circa 484-425 BCE, who observed Egyptian fishermen using oil extracted from the castor oil plant (Herodotus, 1996). Pyrethrum was first used in China and later introduced to Middle East and Europe during the 19th century (Eisner, 1991; Casida and Quistad, 1995). In 1890, the Japanese businessman, Eiichiro Ueyama improved the pyrethrum powder and developed a spiral-shaped mosquito stick burning repellent which was marketed in 1902 and in 1957 mass-produced as a mosquito coil (Uemura, 2004). In North America, native cultures relied on plants and used them to repel biting arthropods (Moerman 1998). For example, the Colville Indians used leaves and stems of common yarrow, *Achillea millefolium* as a smudge to repel mosquitoes. Before World War II, the active ingredients in arthropod repellents were mostly natural plant oils, such as oil of citronella. In 1940, Granett (1940) evaluated 1000 compounds and developed a commercial product, "Sta-Way Insect Repellent Lotion" containing diethylene glycol monobutyl ether and its acetate. Then, in 1942 as a direct response to the war effort, the US Department of Agriculture (USDA) screened over 20,000 chemicals for repellent activity and prioritized indalone, ethyl hexanediol (EH), dimethyl phthalate (DMP), and DEET as the most promising active ingredients and two military mixtures, M-250 which consisted of six parts DMP, two parts Indalone, and two parts EH (6-2-2) and M-2020 consisting of four parts

DMP, three parts EH, and three parts dimethyl carbate (4-3-3) as the most promising mixtures (Strickman, 2007). Later, Christophers (1947) and McCulloch and Waterhouse (1947) reported on and made a distinction between "contact" repellents and "vapor" repellents from the action of pyrethrins which led to the search for "spatial" repellents by the USDA in 1948. Bernier et al. (2007) provided a brief early history of research on spatial repellents.

After World War II, basic research on arthropod repellents in the United States continued to be conducted by the US Department of Defense and the USDA, which led to the discovery of DEET in the late 1940s. Since its introduction and availability in 1956, DEET has become the standard and most widely used active ingredient against a wide variety of biting flies, especially mosquitoes and other biting arthropods, such as chiggers, ticks, fleas, gnats, biting midges, and leeches throughout the world.

1.4 Types of arthropod repellents

1.4.1 Topical (skin) repellents

In addition to DEET, active ingredients of topical arthropod repellents registered by the EPA and endorsed by CDC include 2-(2-hydroxyethyl)-1-piperidinecarboxylic acid 1-methylpropyl ester (Picaridin), ethyl butylacetyl-aminopropionate (IR3535), para-menthane 3,8 diol (PMD), 2-undecanone, (BioUD), and the most current active ingredient, 4-α, 5-Dimethyl-1,2,3,4,4α,5,6,7-octahydro-7-keto-3-isopropenylnaphthalene (nootkatone) that will be discussed in chapter 16 of this book, i.e. the role of repellents in the control of vector-borne diseases. This list will vary in different countries and regions of the world.

1.4.2 Clothing and fabric repellents

In addition to the use of topical (skin) arthropod repellents, another system for personal protection from biting arthropods is the use of repellents in clothing and fabric. In the early 1940s, the two mixture repellents, M-250, also known as 6-2-2, and M-1960, containing 30% each of 2-butyl-2-ethyl-1,3-propanediol, benzyl benzoate, N-butylacetanilide, and 10% Tween were developed and found to be highly effective clothing arthropod repellents (Travis and Morton, 1946). M-1960 was applied to US uniform soldiers in the Pacific Theater against mites to prevent scrub typhus and continued to be used as a clothing arthropod repellent throughout the Korean and Vietnam wars. In 1990, a new clothing arthropod repellent, i.e., permethrin [3-(phenoxyphenyl) methyl (+)-cis, trans-3-(2,2-dichlorothenyl)-2,2-dimethyl-cyclopropanecarboxylate], a synthetic pyrethroid derived from crushed dried flowers of the plant, *Chrysanthemum cinerarifolium* was developed and following EPA registration and approval of four impregnation methods, the US Army implemented it as the standard military clothing repellent (U.S. Environmental Protection Agency, 1990; Casida and Quistad, 1995). Permethrin is unique in that it serves as a contact pesticide and as an arthropod repellent. It has been proven to be a very effective clothing, bed net, and fabric repellent against a wide variety of biting arthropods (Schreck et al., 1978, 1980; Breeden et al., 1982; Gupta et al., 1989; Sholdt et al., 1989; Eamsila et al., 1994; Rowland et al., 1999; Miller et al., 2004; Debboun et al., 2005). Some pyrethroids such as lambda-cyhalothrin, beta-cyfluthrin, etofenprox, and alpha-cypermethrin were also recommended by WHO to treat bed nets and were found to be effective (World Health Organization, 2004). Other EPA-registered repellents that consumers can spray on clothing to repel arthropods include DEET and Picaridin. With the improved technology, factory permethrin-treated clothing, including children's clothing have been EPA registered and marketed to the general public. For example, Insect Shield technology uses permethrin-treated clothing and carries a diverse line of arthropod repellent gear and clothing for men, women, and children.

1.4.3 Spatial (area) repellents

Spatial or area repellents have become important in preventing biting arthropods from reaching their human hosts. They are repellent products that are applied between the human and the source of biting arthropods by preventing biting within a volume of air occupied by humans. They release an active ingredient into the area either actively by volatilizing it with heat and aerosolization or passively without the heat or electricity. They elicit spatial repellency due to a range of arthropod behaviors such as movement away from a chemical, attraction-inhibition, or feeding inhibition induced by airborne chemicals that result in a reduction in human-arthropod contact. Examples of commercial active spatial repellents are mosquito coils, lamps, candles, Thermacell and Off-Clip On devices, and passive spatial repellents include mosquito beaters, Transfluthrin, Metofluthrin strips, and No-Pest Dichlorvos strips. Of all the spatial repellent systems, there is little evidence to support the use of "mosquito buzzer" claimed to emit a sonic frequency that deters mosquitoes (Cabrini et al., 2006) and should generally not be recommended. Good published reviews of spatial repellents are provided by Strickman (2007), Achee et al. (2012), Kline and Strickman (2014), and Norris and Coats (2017).

Recently and currently, repellent workers have been researching the use of highly volatile chemicals as spatial repellents to be an important part of IVM, particularly when used to prevent infective arthropod bites that are not prevented by other methods. They have also been shown to be effective against insecticide-resistant biting arthropods and provide long-lasting repellency without the need for their continual reapplication to human skin. Thus, as mentioned earlier, spatial repellents are becoming one of the effective tools in the fight against vector-borne transmission, and hopes are that the WHO incorporates and endorses them in disease vector control programs.

1.4.4 Plant-based arthropod repellents

Plant-based arthropod repellents have been used by humans for generations in traditional community practices as personal protective measures against a wide variety of biting arthropods. Some of the earlier ones included the use of a variety of plants and flowers that were burnt, hung in homes or on porches, or rubbed on the skin, including chrysanthemum, geranium, and lantana (Brown and Hebert, 1997). Most plant-based arthropod repellents contain essential oils from plants such as citronella, eucalyptus, cedar, geranium, lemongrass, Osage orange, beautyberry, peppermint, soybean, etc. Currently, increased interest in the use of commercial arthropod repellent products containing plant-based active ingredients has gained popularity and demand by consumers, particularly due to their belief that they are "safer" than other synthetic repellents which is a fallacy and not completely true. Thus, the growing demand for natural arthropod repellents by consumers illustrates the continual need to develop and evaluate new plant-based arthropod repellents for personal protection against biting arthropods (Khater, 2012). In addition, since communities tend to accept and favor plant-based arthropod repellent products, their use can serve as a supplementary or alternative personal protective measures or tools in IVM of vectors and vector-borne diseases.

Some of the current popular and effective plant-based arthropod repellents include oil of lemon eucalyptus (PMD), 2-undecanone, citronella, permethrin, neem, nepetalactone, and other essential oils. However, with the exception of those containing PMD, most of these plant-based arthropod repellents are not advised by public health agencies for reducing the risk of contracting arthropod-borne diseases. Good published reviews on plant-based arthropod repellents are provided by Moore et al. (2007), Maia and Moore (2011), Moore (2014), Diaz (2016) and Bekele (2018). In the future, we believe that plant-based arthropod repellents and essential oils will continue to serve as alternatives to synthetic arthropod repellent products as they are readily available throughout the world and could play an important role in new and novel repellent technology.

1.4.5 Arthropod repellent mixtures

One of the earliest arthropod repellents developed from a mixture or combination of chemical compounds was "Sta-Way Insect Repellent Lotion" (Granett, 1940). The idea of using mixtures and combinations of arthropod repellents was developed to get a broader range of efficacy (Travis et al., 1949) which resulted in the development of the combined arthropod repellent known as 6-2-2 or M-250 that consisted of six parts of DMP, two parts of indalone, and two parts of EH. M-250 became the standard US military topical arthropod repellent in the latter part of World War II and provided good protection for 4–6 hours from mosquitoes (Travis and Morton, 1946). In 1951, another repellent mixture, M-2020 or 4-3-3 which consisted of four parts DMP, three parts EH, and three parts DMC was adopted as the US military standard topical arthropod repellent and another arthropod repellent mixture, M-1960 as the US standard clothing arthropod repellent (Gilbert and Gouck, 1953). M-1960 or 3-3-3 consisted of 30% 2-butyl-2-ethyl-1,3-propanediol for protection against mosquitoes and other biting flies, 30% N-butyl-acetamide for ticks, 30% benzyl benzoate for chigger mites and fleas, and the remainder being Tween 80 as an emulsifier. It was used by the soldiers in the Pacific Theater and was successful in stopping the devastating effects of scrub typhus.

Other examples of successful use of arthropod repellent mixtures included the use of the combination of Thanaka and DEET in Burma (McGready et al., 2001), a repellent soap from Australia containing 20% DEET and 0.5% permethrin was used successfully in Thai-Myanmar (Lindsay et al., 1998), Malaysia (Yap, 1986), Papua New Guinea (Charlwood and Dagoro, 1987), Australia (Frances, 1987), India (Mani et al., 1991), Ecuador and Peru (Kroeger et al., 1997), Pakistan (Rowland et al., 2004), and 20% DEET and 15% EH in Senegal (Izri, 2001). Due to the development of pyrethroid resistance in mosquitoes, the use of a combination of an arthropod repellent and insecticide-treated bed nets or mixtures of arthropod repellents and nonpyrethroid -treated fabrics was used as a tool for disease vector control (Pennetier et al., 2005, 2007, 2008).

Recently, plant essential oil mixtures have also been used and shown enhanced protection against a wide range of biting arthropods (Debboun et al., 2014; Gross et al., 2017; Norris et al., 2018). Arthropod repellent mixtures will continue to be used to improve their efficacy, increase the effective sensitivity of the target arthropod by combining more than one mode of action, and extend duration, or acquire better application characteristics (Strickman, 2007). In addition, the use of arthropod repellent mixtures will remain an integral and important part of improving the limited tools available for protection against vectors of disease pathogens, their annoyance, and bites.

1.5 Personal protection from vector-borne diseases

Vector control is an important strategy for preventing and potentially eliminating arthropod-borne diseases, which would be instituted on a regional or national basis to control the relevant vector transmitting the endemic or epidemic disease. Personal protection against disease-carrying arthropods can also be promoted as a public health measure to individuals to reduce the risk of contracting the infections. As described earlier and in summary, three modalities may be employed.

1.5.1 Barrier methods

Examples are the wearing of appropriate clothing through which an arthropod cannot bite and the use of bed nets. Protection by such methods is improved if the fabric is treated with a pesticide that discourages arthropods from landing or biting through the material.

1.5.2 Area methods

In this case, a pesticide would be used that would eliminate or discourage arthropods in a contained area. Examples of these are burning mosquito coils or plugging in vaporizers that release a pesticide into an enclosed space.

1.5.3 Applied repellents

These would be used on exposed skin to prevent the arthropod from detecting a host blood meal and discouraging biting.

As has been stated, there is evidence that these modalities of personal protection can all help to control vector-borne diseases, so arguably the implementation of personal protection becomes a public health issue. This is mostly true when the infection relies on human-to-human transmission, but the involvement of major animal intermediary vectors may hinder elimination of the disease, i.e., a zoonotic reservoir. In the case of the tick-transmitted Lyme disease, a zoonotic wildlife reservoir would maintain its presence and transmission despite humans taking precaution to prevent tick bites. A prime example of an effective personal protection method is the outstanding success of the use of bed nets in reducing malaria incidence in sub-Saharan Africa (Lengler, 2004). However, this modality might only be useful against the malaria mosquito transmitting *Plasmodium falciparum* indoors. Such a strategy might not be as successful if outdoor-biting species predominate, such as is the case in many regions of South America (Steinhardt, 2015). Indeed, there is evidence that sub-Saharan species are adapting to outdoor biting and thus reducing the efficacy of the bed net campaign (Russell et al., 2011). The most effective recommended barrier method of covering exposed arms and legs to prevent mosquito bites is often not practical in warm tropical areas.

Therefore, an important question remains whether a skin-applied repellent used by the local population can help reduce the incidence of a vector-borne disease. There is ample evidence that repellents do deter biting arthropods that transmit disease pathogens and reduce the chance of an individual contracting the disease, and also as discussed earlier, studies have demonstrated that by using an arthropod repellent as a public health measure, it is possible to control an arthropod-borne disease. However, it could be argued that currently, the strength of evidence that they do indeed lower incidence of these diseases in a large population is not great. Two meta-analyses have been conducted to examine whether skin-applied repellents do reduce the incidence of malaria when used by the local population (Wilson et al., 2014; Maia et al., 2018). They both concluded that on the strength of available studies, the use of skin repellents did not have an appreciable effect on disease incidence. Although one notable large study in Bolivia (Hill et al., 2007) did show an effect by regular use of a repellent to reduce the incidence of malaria. Such reviews do tend to have very strict inclusion criteria, generally demand large population sample sizes and emphasize possible source of bias. Overall, it seems that further and larger studies need to be conducted to demonstrate and quantify the benefits of regular repellent use by communities. In addition, the benefits of such strategies against dengue, which arguably would be more important when combatting the daytime/outdoor feeding *Aedes* species, have not been studied in large populations.

1.5.4 Bite avoidance behavior

One reason cited for a lack of efficacy in controlling endemic malaria was the potential for poor adherence to the repellent application regimen. In most studies in the Cochrane review (Maia et al., 2018), adherence was measured by self-reports, subject to recall bias, and then followed up by observation, i.e., weighing of bottles or "sniff tests" of a small sample of participants. Results ranged from self-reported high levels of compliance in some studies to relatively poor in others. For instance, one study had a relatively high adherence to repellent use of 70% when assessed by self-reporting, but only 8% when measured through direct observation (Gryseels et al., 2015.) Even if adherence to repellent use by local populations in endemic areas is reasonable over a relatively short study period, an important question that arises is for how long can regular and correct use of the repellent be maintained? This draws out the distinction between promoting a repellent as a public health strategy to reduce disease incidence and occasional use of repellents by travelers on a short-term basis. To deter nuisance biting, then it might be assumed that individuals would apply repellents as and when needed without necessarily considering regular use to protect against disease transmission. Further, many individuals living in environments where there is a high biting pressure of mosquitoes will become desensitized and not develop the characteristic skin reaction, leading them to believe that they do not get bitten by mosquitoes (Peng and Simons, 1998). Again, these factors do demand further systematic study.

A very different population is one traveling to endemic areas for shorter visits and who may have a different behavior regarding the use of repellents or other recommended bite avoidance measures. Two studies by Thrower and Goodyer (2016) and Hasler et al. (2018), simply involved asking travelers to apply their chosen repellent to their arms as normal and then weigh the container before and after administration. By measuring the area of the arms to which it is applied, the dose of repellent was calculated in mg/cm^2 of skin. The studies have been performed both in the home clinics and at the destination of travel. The mean amount of repellent applied by individuals was remarkably similar in both studies and various setting, showing that despite formulation, a mean dose of 1 mg/cm^2 was used by the travelers. As will be described in the next section, this would be suboptimal to achieve the maximum duration of protection by a formulation and does not represent the dosage at which repellents are usually tested. On the other hand, a much higher dose was achieved by application to the neck area (Thrower and Goodyer, 2016) which provided a good length of protection to that part of the body.

There are other aspects of repellent use behavior that demand further study which might include how well and evenly is a repellent dispersed over the applied area, what concentrations are achieved on the ankles and lower legs, and what factors might play a part in poor adherence to personal protection. One study by Goodyer and Song (2014) explored some of the issues related to bite avoidance behavior in travelers by asking them to complete a retrospective survey on their use of personal protection measures on return from a visit to a malaria-endemic area. Among the 132 travelers completing the survey, only 70% used a repellent on a regular basis despite knowing they were visiting a malaria-endemic area and advised by a health professional on the repellent use. Adherence to other recommended modalities was even lower: only 50% stated that they covered arms and legs when going outside in the evening and around 20% used vaporizers or insecticide sprays indoors. The use of bite avoidance methods was lower in those aged under 30, and females covered arms and legs less frequently than men. However, there did seem to be an association of attempting to use arthropod repellents more regularly if bites were experienced.

There is an argument for considering repellents in the same way as a prophylactic medication to prevent a disease. Indeed, for those infections such as dengue where there is no vaccine or prophylactic medication, it is the only method of reducing the risks of disease. Therefore, individual adherence to bite avoidance recommendations needs to be considered and any barriers to adherence minimized. Such factors would include a product that is cosmetically acceptable with few perceived side effects. Just as with a medication fear of side effects may reduce adherence, which has been seen with the largely unfounded fears concerning DEET toxicity (Swale, 2019) and may have deterred many from using this highly effective repellent. A further comparison to drug medication is that it is recognized that the fewer times a day a patient needs to take the medication, the better will be adherence to the regimen. This would also apply to arthropod repellents: applying just once in a 12-hour period to reliably achieve 100% protection would be the holy grail which, for reasons discussed later, has not to date been achieved.

1.6 Assessment of arthropod repellents

Arthropod repellents are assessed for efficacy by application to the skin of human volunteers and then exposing them to mosquitoes either in the field or in more controlled conditions in the laboratory. There are a number of "in vitro" methods more used for screening of potential repellent products, ranging from providing a blood meal behind a membrane upon which a mosquito lands to measuring changes directly of mosquito antennae receptors known to be involved in the detection of a source of blood (Debboun et al., 2014). Repellents can also be tested by exposure of the skin of laboratory animals to arthropods. However, the tests involving humans are the most widely used by regulatory authorities, and the WHO standard protocols (WHOPES, 2009) are accepted or have been adapted by most regulatory authorities for the marketing of repellent products. In summary, the laboratory-based cage tests involve application of measured repellent to the forearm of individuals inserted into a cage of female mosquitoes and observing landing/biting rates. The field tests are similar except that individuals are exposed in the open to the local population of mosquitoes for a set period. Various regulatory agencies do have specific protocols for testing mosquitoes and other arthropods, for instance as in the EU Biocide regulations (ECHA Document, 2012). However, despite the protocols required by regulatory authorities, much of the published and peer-reviewed research is comprised of protocols that can deviate significantly from these standards making comparisons of studies difficult. For example, earlier studies used relatively few insects and exposed a small area of the skin to the mosquitoes (Rutledge, 1985). Different cage sizes, laboratory temperatures and mosquito numbers have been used to assess repellents which all affect the absolute value of the observed protection times (Barnard, 1998). For example, in a study using 50 mosquitoes introduced into a cage, a complete protection time (CPT) of protection time of 5 hours for an applied active ingredient (AI) of 0.2 mg/cm^2 of DEET was found against *Ae. aegypti* (Goodyer et al., 2020), whereas in a very similar study using 200 mosquitoes, this was just 2 hours (Colucci, 2018).

There are essentially three parameters that these assessments will be used to measure.

1.6.1 The effective dose

The ED_{90} or ED_{95} of the repellent, which is the effective dose able to provide 90% or 95%

protection, is a measure of the "potency" of the formulation. This is usually only measured using cage tests by application of incremental doses of the repellent and calculating the effective dose by probit analysis. The measure is important in as much as if the dose required to achieve the ED_{95} is higher than that achievable for the user it would not be practical for use as a repellent. The currently marketed and approved repellent AIs and formulations would all be expected to have sufficiently low effective doses in this respect.

1.6.2 The complete protection time

The CPT is the time taken before the mosquitoes are observed to start feeding after the application of repellent and is probably the most important of the parameters, though its assessment and true relevance are the most controversial. It is assessed in the field or by cage tests by simply exposing the human limb to mosquitoes and measuring the time taken for them to start feeding after application of the repellent, a measure referred to as the time to first bite. The most controversial aspect of this is that there are several variables that will affect the absolute value of the measure, particularly in the field. Some variables might be:

- The mosquito density can be controlled in cage tests, but results will depend on cage size and mosquito numbers present.
- Feeding avidity.
- Environmental conditions such as humidity, wind direction, and temperature. These can all be controlled in cage tests.
- Attractiveness of individual volunteers in the test. This is mitigated by involving a number of volunteers of different genders, but most trials do not involve more than ten subjects and often far fewer.
- Rubbing and sweating off the repellent. In many field trials, volunteers remain in one position.
- Species of mosquitoes. Some species are more sensitive to repellents than others (Van Roey et al., 2014).

All of these variables would come into play for a user outside of any controlled conditions. For example, a user might be walking through a region at various times of day exposed to different mosquito densities and a variety of species giving a great variation in protection. This has been illustrated in one study (Gupta, 1987) where soldiers were asked to roam in a forested area freely after applying a repellent, and the resultant level of protection varied widely throughout the time period. Certainly, it would be unlikely that a repellent can reliably achieve a once-per-day application under all conditions.

Apart from these external variables, the other most important factor will be the amount of AI of the repellent (mg/cm^2 of skin) applied. This has been well demonstrated for DEET as a logarithmic relationship (Buescher, 1983) between achieving greater than 95% protection and applied dose. This is determined by the equation:

CPT = C1 + C2log*X, where C1 and C2 are coefficients and X = dose of AI mg/cm^2

The coefficients will depend upon the variables in the test conditions described earlier. This relationship has only been reasonably shown for DEET, and there is only one field study demonstrating such a relationship for picaridin (Costantini et al., 2004). Another has confirmed the relationship for PMD (Goodyer et al., 2020).

Although a modified release formulation showed a completely different relationship where an S-shaped exponential curve was found as shown in Fig. 1.1 by Goodyer et al. (2020). There is little in the literature regarding the dose/CPT relationship to other repellent ingredients and formulations.

From Fig. 1.1 for 30% DEET, the rise in CPT protection achieved by an increasing application rate, led to smaller incremental increases in protection time around 1.5 mg/cm^2 of 20% DEET, the exact value depended upon the formula

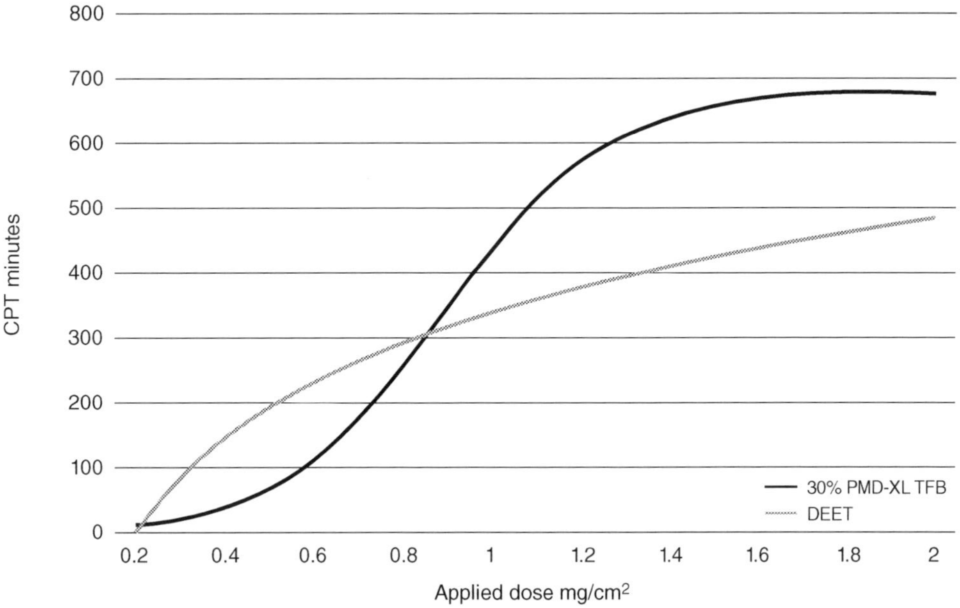

FIG. 1.1 Complete protection times for increasing applied dose of 20% DEET and 30% PMD-XL based on non-linear regression equations (from Goodyer et al., 2020).

coefficients. In practice, an application of greater than 2 mg/cm^2 of a repellent formulation would not be practical as this would simply run off the skin surface. Note if applying a 20% concentration of DEET, the maximum total AI of the formulation achievable in practice would be just 0.5 mg/cm^2 of AI. Therefore, from Fig. 1.1, application of 1 mg/cm^2 20% DEET equates to an AI of 0.2 mg/cm^2 giving a protection time of around 5 hours, whereas application of the same amount of 50% DEET will result in an AI of 0.5 mg/cm^2 and a protection time of around 8 hours. Therefore, the higher the % of the AI in a formulation, the longer the protection time before the repellent needs to be reapplied. Considering the earlier observation that users will tend to always apply a total of 1 mg/cm^2 of repellent, then optimal length of protection is unlikely to be achieved with formulations containing lower concentrations. Further, it should be noted that WHO protocols describe applying repellent formulations at a dose of 1.6 mg/cm^2 and perhaps this should be revised to reflect actual user rates.

With all these variables, it is not surprising that the published data show large variations in reported absolute protection times for repellent formulations. Field trials claimed to be more realistic for user experience compared to cage tests, may show widely varying results. For example, one well-conducted study reported a protection time against *Anopheles* mosquitoes of only 1 hour for DEET (Frances et al., 2004), whereas another study against this genus and using similar concentrations of DEET provided a CPT of 4–5 hours (Costantini et al., 2004). Therefore, it would be expected that user experience of the longevity of a particular repellent would vary greatly, but how to study and quantify fully such a variation would be difficult and has not been attempted meaningfully to date.

Despite this, many manufacturers make claims for longevity of their products based upon the longest achievable protection times, usually with statements such "protection achieved for up to 12 hours." As reapplication times may not be described in the labeling, the user may be misled into believing that this would be the duration before which reapplication is actually required. Manufacturers might allude to differences in protection times with statements referring to sweat or water washing off the repellent, but do not often indicate that a much more frequent reapplication is required. Dose to be applied is also not usually indicated by the manufactures. All of this is further complicated by regulatory requirements for dose application rates from a safety perspective leading to large variations in allowed application rates and AI concentrations between countries depending upon the toxicity risk assessment performed. As stated earlier, DEET has come under the most scrutiny regarding safety concerns, which some would deem unjustified and resulting in using suboptimal use, which would have important health consequences if being used to protect against arthropod-borne diseases.

1.6.3 Half-life

The third important assessment is the half-life of the repellents following the CPT. The is usually measured in a cage test by observing the fall in % protection over time afforded by the repellent from the formula:

Protection = (landings on control arm minus landings on treated arm)/landings on control arm.

The fall in protection with time is an exponential relationship, as shown in Fig. 1.2. Half-lives are not often reported in the literature, but both Goodyer et al. (2020) and Costantini (2004)

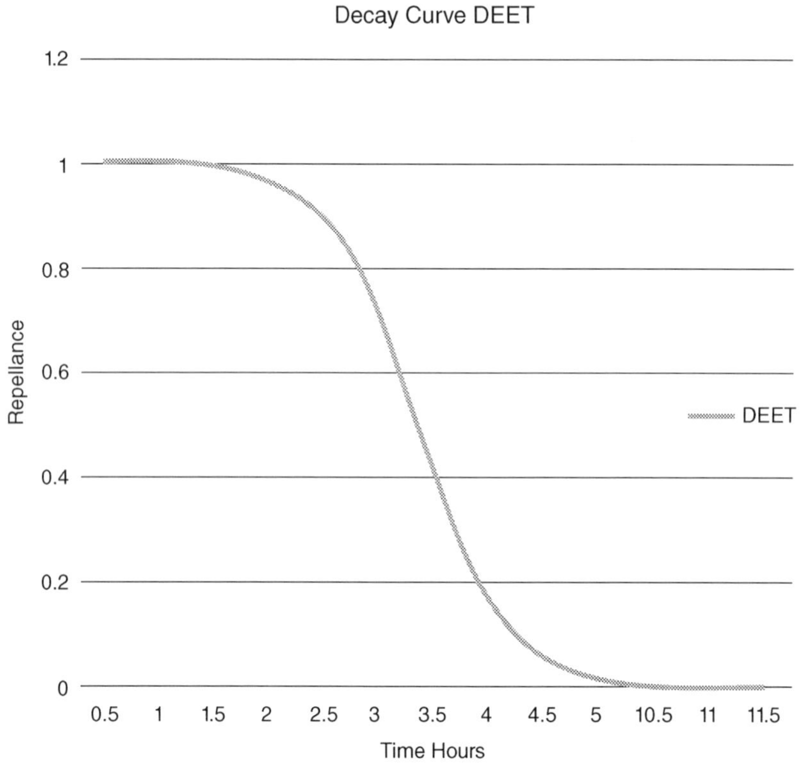

FIG. 1.2 Decay curve of DEET after the complete protection time (from Goodyer et al., 2020).

reported similar half-lives of around 2 hours for DEET. It could be argued that even when repellency has fallen by 50%, some useful protection is still present and could be taken into account in terms of the longevity of action expected. For instance, if in a situation of relatively low biting pressure, then even a large fall in repellency may mean that the user will still not experience any bites.

1.7 Conclusion

In conclusion, there are four widely available and established repellent AIs (PMD, picaridin, DEET, and IR3535) of sufficient potency to reduce the risk of arthropod-borne diseases. The crucial question is which of these if applied in the same formulation and in the same environmental condition gives the longest protection before reapplication is required. In studies where concentrations of different formulations are applied in the same condition, it is reasonable to conclude which formulations would offer the greater protection. Rather than absolute terms of "hours of protection" it would be more meaningful to describe one formulation having "X times the protection of another," i.e., 1.5 or 2 times the length compared to the same dosage of DEET. For users, this would avoid any misconception regarding reapplying the arthropod repellent when necessary if in conditions or undergoing activities that might affect repellent longevity.

References and further readings

Achee, N.L., Bangs, M.J., Farlow, R., Killeen, G.F., Lindsay, S., Logan, J.G., Moore, S.J., Rowland, M., Sweeney, K., Torr, S.J., Zweibel, L.J., Grieco, J.P., 2012. Spatial repellents: from discovery and development to evidence-based validation. Malar. J. 11 (164), 1–9.

Asilian, A., Sadeghinia, A., Shariati, F., Jome, M.I., Ghoddusi, A., 2003. Efficacy of permethrin-impregnated uniforms in the prevention of cutaneous leishmaniasis in Iranian soldiers. J. Clin. Pharm. Ther. 94 (4), 361–366.

Barnard, D.R., Posey, K.H., Smith, D., Schreck, C.E., 1998. Mosquito density, biting rate and cage size effects on repellent tests. Med. Vet. Entomol. 12 (1), 39–45.

Barnard, D.R., 2000. Repellents and Toxicants for Personal Protection: a WHO Position Paper. World Health Organization, Geneva.

Barnard, D.R., Xue, R.D., 2004. Laboratory evaluation of mosquito repellents against *Aedes albopictus*, *Culex nigripalpus*, and *Ochlerotatus triseriatus* (Diptera:Culicidae). J. Med. Entomol. 41, 726–730.

Bekele, D., 2018. Review on insecticidal and repellent activity of plant products for malaria mosquito control. Biom. Res. Rev. 2 (2), 1–7.

Bernier, U.R., Kline, D.L., Posey, K.H., 2007. Human emanations and related natural compounds that inhibit mosquito host-finding abilities. In: Debboun, M., Frances, S.P., Strickman, D. (Eds.), Insect Repellents: Principles, Methods and Uses. CRC Press, Boca Raton, FL, pp. 77–100.

BMGF, BCG, 2007. Market assessment for public health pesticide products. : Bill and Melinda Gates Foundation and Boston Consulting Group, Seattle, WA.

Breeden, G.C., Schreck, C.E., Sorensen, A.L., 1982. Permethrin as a clothing treatment for personal protection against chigger mites (Acarina: Tromiculidae). Am. J. Trop. Med. Hyg. 31, 589–592.

Brown, M., Hebert, A.A., 1997. Insect repellents: an overview. J. Am. Acad. Dermatol. 36, 243–249.

Buescher, M.D., Rutledge, L.C., Wirtz, R.A., Nelson, J.H., 1983. The dose-persistence relationship of DEET against *Aedes aegypti*. Mosquito News 43 (3), 364–366.

Cabrini, I., Andrade, C.F., 2006. Evaluation of seven new electronic mosquito repellers. Entomol. Exp. Appl. 121, 185–188.

Casida, J.E., Quistad, G.B., 1995. Pyrethrum Flowers: Production, Chemistry, Toxicology and Uses. Oxford University Press, Inc, New York, NY.

Charlwood, J.D., Dagoro, H., 1987. Repellent soap for use against malaria vectors in Papua New Guinea. Papua New Guinea Med. J. 30, 301–303.

Christophers, S.R., 1947. Mosquito repellents, being a report of the work of the mosquito repellent inquiry, Cambridge (1943-5). J. Hyg. 45, 176–231.

Colucci, B., Müller, P., 2018. Evaluation of standard field and laboratory methods to compare protection times of the topical repellents PMD and DEET. Sci. Rep. 8 (1), 12578.

Costantini, C., Badolo, A., Iloudo-Sangoro, E., 2004. Field evaluation of the efficacy and persistence of insect repellents DEET, IR3535, and KBR 3023 against *Anopheles gambiae* complex and other Afrotropical vector mosquitoes. Trans. R. Soc. Trop. Med. Hyg. 98 (11), 644–652.

Debboun, M., Strickman, D.A., Klun, J.A., 2005. Repellents and the military: our first line of defense. J. Am. Mosq. Control. Assoc. 21, 4–6.

Debboun, M., Strickman, D., 2013. Insect repellents and associated personal protection for a reduction in human disease. Med. Vet. Entomol. 27, 1–9.

Debboun, M., Paluch, G., Lindsay, D., 2014. Use of chemical mixtures as insecticides and repellents. In: Debboun, M., Frances, S.P., Strickman, D.A. (Eds.), Insect Repellents Handbook. CRC Press, Boca Raton, FL, pp. 283–290 2nd ed.

Debboun, M., Frances, S., Strickman, D., 2014. Insect Repellents Handbook. CRC Press, Boca Raton, FL.

Deressa, W., Yihdego, Y.Y., Kebede, Z., Batisso, E., Tekalegne, A., Dagne, G.A., 2014. Effect of combining mosquito repellent and insecticide treated net on malaria prevalence in southern Ethiopia: a cluster-randomized trial. Parasit. Vectors 7, 132–141.

Dethier, V., Browne, B.L., Smith, C.N., 1960. The designation of chemicals in terms of the responses they elicit from insects. J. Econ. Entomol. 53, 134–136.

Diaz, J.H., 2016. Chemical and plant-based insect repellents: efficacy, safety, and toxicity. Wild. Environ. Med. 27, 153–163.

Dolan, M.C., Panella, N.A., 2011. A review of arthropod repellents. In: Paluch, G., Coats, J.R. (Eds.), Recent Developments in Invertebrate Repellents. CRC Press, Boca Raton, FL, pp. 1–19.

Durrheim, D.N., Govere, J.M., 2002. Malaria outbreak control in an African village by community application of "deet" mosquito repellent to ankles and feet. Med. Vet. Entomol. 180 (1), 112–115.

Dutta, P., Khan, A.M., Khan, S.A., Borah, J., Sharma, C.K., Mahanta, J., 2011. Malaria control in a forest fringe area of Assam, India. Trans. R. Soc. Trop. Med. Hyg. 105 (6), 327–332.

Eamsila, C., Frances, S.P., Strickman, D., 1994. Evaluation of permethrin-treated military uniforms for personal protection against malaria in northeastern Thailand. J. Am. Mosq. Control. Assoc. 10, 515–521.

ECHA Document, 2012. CA-Dec12- Doc.6.2.a—Guidance Efficacy Evaluation of Insecticides PT18 and PT19. ECHA, Helsinki. https://echa.europa.eu/documents/10162/16960215/bpd_guid_tnsg_efficacy_pt18-19_final_en.pdf/9c72241e-0eea-4f23-8e5f-f52d00a83382.

Eisner, R., 1991. "Natural" insecticide research: still working out the bugs. The Scientist 5, 14.

Frances, S.P., 1987. Effectiveness of deet and permethrin, alone, and in a soap formulation as skin and clothing protectants against mosquitoes in Australia. J. Am. Mosq. Control Assoc. 3, 648–650.

Frances, S.P., Waterson, D.G, Beebe, N.W., Cooper, R.D., 2004. Field evaluation of repellent formulations containing DEET and picaridin against mosquitoes in Northern Territory. Australia. J. Med. Entomol. 41, 414–417.

Gerberg, E.J., Novak, R.J., 2007. Considerations on the use of botanically-derived repellent products. In: Debboun, M., Frances, S.P., Strickman, D. (Eds.), Insect Repellents: Principles, Methods and Uses. CRC Press, Boca Raton, FL, pp. 305–309.

Gilbert, I.H., Gouck, H.K., 1953. All-purpose repellent mixtures as clothing treatments against chiggers. Fla. Entomol. 36, 47–51.

Goodyer, L.I., Song, J., 2014. Mosquito bite avoidance attitudes and behaviours on travellers at risk of malaria. J. Travel Med. 21, 33–39.

Goodyer, L., Grootveld, M., Deobhankar, K., Debboun, M., Philip, M., 2020. Characterisation of actions of *p*-menthane-3,8-diol repellent formulations against *Aedes aegypti* mosquitoes. Trans. R. Soc. Trop. Med. Hyg., 1–6. doi:10.1093/trstmh/traa045.

Granett, P., 1940. Studies of mosquito repellents, I. Test procedure and method of evaluating test data. J. Econ. Entomol. 33, 563–565.

Gross, A.D., Norris, E.J., Kimber, M.J., Bartholomay, L.C., Coats, J.R., 2017. Essential oils enhance the toxicity of permethrin against *Aedes aegypti* and *Anopheles gambiae*. Med. Vet. Entomol. 31, 55–62.

Gupta, R.K., Sweeney, A.W., Rutledge, L.C., Cooper, R.D., Francis, S.P., Westrom, D.R., 1987. Effectiveness of controlled-release personal-use arthropod repellents and permethrin-impregnated clothing in the field. J. Am. Mosq. Control Assoc. 3 (4), 556–560.

Gupta, R.K., Rutledge, L.C., Reifenrath, W.G., Gutierrez, G.A., Korte, D.W., 1989. Effects of weathering on fabrics treated with permethrin for protection against mosquitoes. J. Am. Mosq. Control Assoc. 5, 176–179.

Gryseels, C., Uk, S., Sluydt, V., Durnez, L., Phoeuk, P., Sokha, S., Set, S., Heng, S., Siv, S., Gerrets, R., Coosemans, M., Peeters, K., 2015. Factors influencing the use of topical repellents: implications for the effectiveness of malaria elimination strategies. Sci. Rep. 5, 16847.

Hasler, T., Fehr, J., Held, U., Schlagenhauf, P., 2018. Use of repellents by travellers: a randomised, quantitative analysis of applied dosage and an evaluation of knowledge, Attitudes and Practices (KAP). J. Travel Med. Infect. Dis., 2018. https://doi.org/10.1016/j.tmaid.2018.12.007 Accessed December 19, 2019.

Herodotus, 1996. The Histories. Penguin, London reprint.

Hill, N., Lenglet, A., Arnéz, A.M., Carneiro, I., 2007. Plant based insect repellent and insecticide treated bed nets to protect against malaria in areas of early evening biting vectors: double blind randomised placebo controlled clinical trial in the Bolivian Amazon. BMJ 335, 1023–1027. doi:10.1136/bmj.39356.574641.55.

Isri, A., 2001. Efficacy of the combination of DEET (20%) and EHD (15%) against mosquito bites. Results of a study carried out in Senegal. Bull. Soc. Pathol. Ext. 94, 280 Abstract.

Keatinge, A.F.H., 1949. A hundred years of insecticides and repellents in the Army (a historical survey). J. R. Army. Med. Corps. 92, 290–312.

Khater, H.F., 2012. Prospects of botanical biopesticides in insect pest management. Pharmacologia 3 (12), 641–656.

Kimani, E.W., Vulule, J.M., Kuria, I.W., Mugisha, F., 2006. Use of insecticide-treated clothes for personal protection against malaria: a community trial. Malar. J. 5, 63.

Kline, D.A., Strickman, D.A., 2014. Spatial or area repellents. In: Debboun, M., Frances, S.P., Strickman, D.A. (Eds.), Insect Repellents Handbook2nd ed. CRC Press, Bota Raton, FL, pp. 239–251.

Kroeger, A., Gerhardus, A., Kruger, G., Mancheno, M., Pisse, K., 1997. The contribution of repellent soap to malaria control. Am. J. Trop. Med. Hyg. 56, 580–584.

Lengeler, C., 2004. Insecticide-treated bed nets and curtains for preventing malaria 447. Cochrane Database Syst. Rev. https://doi.org/10.1002/14651858.CD000363.pub2. (Accessed 1 March 2020).

Lindsay, S.W., Ewald, J.A., Samung, Y., Apiwathnasorn, Nostein, F., 1998. Thanaka (*Limonia acidissima*) and deet (di-methyl benzamide) mixture as a mosquito repellent for use by Karen women. Med. Vet. Entomol. 12, 295–301.

Lwin, M., Lin, H., Kyaw, M.P., Ohn, M., Maung, SN.S., Soe, K., Oo, T., 1997. The use of personal protective measures in control of malaria in a defined community. South. As. J. Trop. Med. Pub. Health 28 (2), 254–258.

Maia, M.F., Moore, S.J., 2011. Plant-based insect repellents: a review of their efficacy, development and testing. Malar. J. 10 (Suppl 1), S11–S14.

Maia, M.F, Kliner, M., Richardso, M., Lengeler, C., Moore, S.J., 2018. Mosquito repellents for malaria prevention. Cochrane Db. Syst. Rev. doi:10.1002/14651858.CD011595.pub2.

Mani, T.R., Reuben, R., Akiyama, J., 1991. Field efficacy of "Mosbar" repellent soap against vectors of Bancroftian filariasis and Japanese Encephalitis in Southern India. J. Am. Mosq. Control. Assoc. 7, 565–568.

McCulloch, R.N., 1946. Studies in the control of scrub typhus. Med. J. Aust. 1, 717–738.

McGready, R., Simpson, J.A., Htway, M., White, N.J., Nosten, F., Lindsay, S.W., 2001. A double-blind randomized therapeutic trial of insect repellents for the prevention of malaria in pregnancy. Trans. R. Soc. Trop. Med. Hyg. 95, 137–138.

Miller, R.J., Wing, J., Cope, S.E., Klavons, J.A., Kline, D.L., 2004. Repellency of permethrin-treated battle dress uniforms during operation tandem thrust 2001. J. Am. Mosq. Control Assoc. 20, 462–464.

Moerman, D.E., 1998. Native Merican Ethnobotany. Timber Press, Portland, OR.

Moore, S.J., Lenglet, A., Hill, N., 2002. Field evaluation of three plant-based insect repellents against malaria vectors in Vaca Diez Province, the Bolivian Amazon. J. Am. Mosq. Control Assoc. 18, 107–110.

Moore, S.J., Debboun, M., 2007. History of insect repellents. In: Debboun, M., Frances, S.P., Strickman, D. (Eds.), Insect Repellents: Principles, Methods, and Uses. CRC Press, Boca Raton, FL, pp. 3–29.

Moore, S.J., Lenglet, A., Hill, N., 2007. Plant-based insect repellents. In: Debboun, M., Frances, S.P., Strickman, D. (Eds.), Insect Repellents: Principles, Methods, and Uses. CRC Press, Boca Raton, FL, pp. 275–303.

Moore, S.J., 2014. Plant-based insect repellents. In: Debboun, M., Frances, S.P., Strickman, D. (Eds.), Insect Repellents Handbook2nd ed. CRC Press, Boca Raton, FL, pp. 179–211.

McCulloch, R.N., Waterhouse, D.F., 1947. Laboratory and field tests of mosquito repellents. Bull. Council. Sci. Indust. Res. Aust. 213, 28.

Norris, E.J., Coats, J.R., 2017. Current and future repellent technologies: the potential of spatial repellents and their place in mosquito-borne disease control. Int. J. Environ. Res. Public Health. 14, 124. doi:10.3390/ijerph14020124.

Norris, E.J., Johnson, J.B., Gross, A.D., Bartholomay, L.C., Coats, J.R., 2018. Plant essential oils enhance diverse pyrethroids against multiple strains of mosquitoes and inhibit detoxification enzyme processes. Insects 9, 132–142.

Onyango, S.P., Moore, S.J., 2015. Evaluation of repellent efficacy in reducing disease incidence. In: Debboun, M., Frances, S.P., Strickman, D. (Eds.), Insect Repellents Handbook2nd ed. CRC Press, Boca Raton, FL, pp. 117–156.

Peng, Z., Simons, F.E., 1998. A prospective study of naturally acquired sensitization and subsequent desensitisation to mosquito bites and concurrent antibody responses. J. Allergy. Clin. Immun. 101 (2), 284–286.

Pennetier, C., Corbel, V., Hougard, J.M., 2005. Combination of a non-pyrethroid insecticide and a repellent: a new approach for controlling knockdown-resistant mosquitoes. Am. J. Trop. Med. Hyg. 72, 739–744.

Pennetier, C., Costantini, C., Corbell, V., Licciardi, S., Dabire, R.K., Lapied, B., Chandre, F., Hougard, J.M., 2008. Mixture of controlling insecticide-resistant malaria vectors. Emerg. Infect. Dis. 14 (11), 1707–1714.

Pennetier, C., Corbel, V., Boko, P., Odjo, A., N'Guessan, R., Lapied, B., Hougard, J.M., 2007. Synergy between repellents and non-pyrethroid insecticides strongly extends the efficacy of treated nets against *Anopheles gambiae*. Malar. J. 6, 38–44.

Peterson, C., Coats, J., 2001. Insect repellents—past, present and future. Pest Outlook 12, 154–158.

Philip, J.R., Sabin, A.B., 1944. Dimethyl phthalate as a repellent in control of *Phlebotomus* (*pappataci* or sandfly) fever. War Med 6, 27–33.

Roberts, D.R., 1993. Insecticide Repellency in Malaria Vector Control: A Position Paper, VBC Report No. 81131, VBC Project, Tropical Disease Control for Developmnent. Medical Service Corporation International, Arlington, VA.

Rowland, M., Durrani, N., Hewitt, S., Mohammed, N., Bouma, M., Carneiro, I., Rozendaal, J., Schapira, A., 1999. Permethrin treated chaddars and top sheets: appropriate technology for protection against malaria in Afghanistan and other complex emergencies. Trans. R. Soc. Trop. Med. Hyg. 93, 465–472.

Rowland, M., Downey, G., Rab, A., Freeman, T., Mohammad, N., Rehman, H., Durrani, N., Curtis, C., Lines, J., Fayaz, M., 2004. Deet mosquito repellent provides personal protection against malaria: a household randomized trial in an Afghan refugee camp in Pakistan. Med. Vet. Entomol. 9, 335–342.

Russell, T.L., Govella, N.J., Azizi, S., Drakeley, C.J., Kachur, S.P., Killen, F.G., 2011. Increased proportions of outdoor feeding among residual malaria vector populations following increased use of insecticide-treated nets in rural Tanzania. Malar. J 10, 80. doi:10.1186/1475-2875-10-80.

Rutledge, L.C., Wirtz, R.A., Buescher, MD., Mehr, Z.A., 1985. Mathematical models of the effectiveness and persistence of Mosquito repellents. J. Am. Mosq. Control Assoc. 1 (1), 56–61.

Salafsky, B., Shibuya, T., He, Y.X., James, H., Ramaswamy, K., 2007. Lipodeet: an improved formulation for a safe, long-lasting repellent. In: Debboun, M., Frances, S.P., Strickman, D. (Eds.), Insect Repellents: Principles, Methods, and Uses. CRC Press, Boca Raton, FL, pp. 327–336.

Sangoro, O., Turner, E, Simfukwe, E., Miller, J.E., Moore, S.J., 2014. A cluster-randomized controlled trial to assess the effectiveness of using 15% DEET topical repellent with long-lasting insecticidal nets (LLINs) compared to a placebo lotion. Malar. J. 13, 324. https://doi.org/10.1186/1475-2875-13-324.

Schwartz, B.S., Goldstein, M.D., 1990. Lyme disease in outdoor workers:Risk factors, preventive measures, and tick removal methods. Am. J. Epidemiol. 131, 877–885.

Schreck, C.E., Posey, K., Smith, D., 1978. Durability of permethrin as a potential clothing treatment to protect against blood-feeding arthropods. J. Econ. Entomol. 7, 397–400.

Schreck, C.E., Carlson, D.A., Weidhaas, D.E., Posey, K., Smith, D., 1980. Wear and aging tests with permethrin-treated cotton-polyester fabric. J. Econ. Entomol. 73, 451–453.

Sholdt, L.L., Rogers Jr., E.J., Gerberg, E.J., Schreck, C.E., 1989. Effectiveness of permethrin-treated military uniforms fabric against human body lice. Mil. Med. 154, 90–93.

Sluydts, V., Durnez, L., Heng, S., Gryseels, C., Canier, L., Kim, S., Van Roe, Y.K, Kerkhof, K., Khim, N., Mao, S.J., 2016. Efficacy of topical mosquito repellent (picaridin) plus long-lasting insecticidal nets versus long-lasting insecticidal nets alone for control of malaria: a cluster randomised controlled trial. Lancet Infect. Dis. 16, 1169–1177.

Smith, D.L., Battle, K.E., Hay, S.I., Barker, C.M., Scott, T.W., Mckenzie, E., 2012. Ross, Macdonald, and a theory for the dynamics and control of mosquito-transmitted pathogens. PLOS Pathol 8 (4), e1002588 10.02510.1001371/journal.ppat.1002588.

Soto, J., Medina, F., Dember, N., Berman, J., 1995. Efficacy of permethrin-impregnated uniforms in the prevention of malaria and leishmaniasis in Colombian soldiers. Clin. Inf. Dis. 21, 599–602.

Steinhardt, L.C., St Jean, Y., Impoinvil, D., Mace, K.E., Wiegand, R., Huber, C.S., Seme FIls Alexandre, J.S., Fredrick, J., Nkurunziza, E., Jean, S., Wheeler, S., Dotson, E., Slutsker, L., Kachur, S.P., Barnwell, J.W., Lemoine, J.F., Chang, M.A., 2015. Effectiveness of insecticide-treated bednets in malaria prevention in Haiti: a case-control study. Lancet Glob. Health. 5 (1). https://doi.org/10.1016/S2214-109X(16)30238-8.

Strickman, D., 2007. Older synthetic active ingredients and current additives. In: Debboun, M., Frances, S.P., Strickman, D. (Eds.), Insect Repellents: Principles, Methods and Uses. CRC Press, Boca Raton, FL, pp. 361–383.

Swale, D.R., Bloomquist, J.R., 2019. Is DEET a dangerous neurotoxicant? Pest Manag. Sci. 75 (8), 2068–2070.

Syafruddin, D., Asih, P.B.S., Rozi, I.E., Permana, D.H., Nur Hidayati, A.P., Syahrani, L., Alvarez, C., Sidik, D., Bangs, M.J., Bogh, C., Liu, F., Eugenio, E.C., Hendrickson, J., Burton, T., Baird, J.K., Collins, F., Grieco, J.P., Lobo, N.F., Achee, N.L., 2020. Efficacy of a spatial repellent for control of malaria in Indonesia: A cluster-randomized controlled trial. Am. J. Trop. Med. Hyg. 103 (1), 344–358.

Thrower, Y., Goodyer, L.I., 2016. Application of insect repellents by travellers to malaria endemic areas. J. Travel Med 13, 198–203.

Torr, S.J., Mangwiro, T.N.C., Hall, D.R., 2011. Shoofly, don't bother me! Efficacy of traditional methods of protecting cattle from tsetse. Med. Vet. Entomol. 25, 192–201.

Travis, B.V., Morton, F.A., 1946. Treatment of clothing for protection against mosquitoes. Proc. 33rd Ann. Meeting N J. Mosq. Exterm. Assoc 33, 65.

Travis, B.V., Morton, F.A., Smith, C.N., 1949. Use of insect repellents as toxicant. USDA-ARS, E-698, Washington, DC.

Uemura, E.V., 2004. Developing and promoting insecticide together with pyrethrum, Osaka Business Update 4, http://www.ibo.or.jp/e/2004_4/o1_1/1_l.html. (Accessed 1 February 2020).

U.S. Environmental Protection Agency, 1990. Insect/arthropod repellent fabric treatment formulations containing permethrin for military use. Registration Division, Office of Pesticides and Toxic Substances, Washington, DC.

Van Roey, K., Sokny, M., Denis, L., Heng, S., Siv, S., Sluydts, V., Socantha, T., Coosemans, M., Durnez, L., 2014. Field evaluation of picaridin repellents reveals differences in repellent sensitivity between Southeast Asian vectors of malaria and arboviruses. Negl. Trop. Dis. 8, e3326.

Vaughn, M.F., Meshnick, S.R., 2011. Pilot study assessing the effectiveness of long-lasting permethrin-impregnated clothing for the prevention of tick bites. Vect. Born. Zoo. Dis. 11, 869–875.

WHOPES, 2009. Guidelines for Efficacy Testing of Mosquito Repellents for Human Skin. World Health Organization, Geneva.

Wilson, A.L., Chen-Hussey, V., Logan, J.G., Lindsay, S.W., 2014. Are topical insect repellents effective against malaria in endemic populations? A systematic review and meta- analysis. Malar. J. 13 (446), 65.

World Health Organization, 2004. Report of the Fourth Meeting of the Global Collaboration for Development of Pesticides for Public Health, Communicable Disease Control, Prevention and Eradication. WHO Pesticide Evaluation Scheme (WHOPES), Geneva.

Yap, H.H., 1986. Effectiveness of soap formulations containing deet and permethrin as personal protection against outdoor mosquitoes in Malaysia. J. Am. Mosq. Control. Assoc. 2, 63–67.

CHAPTER

2

Novel pyrethroid derivatives as effective mosquito repellents and repellent synergists

Jeffrey Bloomquist[a], *Shiyao Jiang*[a], *Edmund Norris*[a,b], *Gary Richoux*[a], *Liu Yang*[a], *Kenneth J. Linthicum*[b]

[a]Emerging Pathogens Institute, Entomology and Nematology Department, University of Florida, Gainesville, FL, United States, [b]United States Department of Agriculture, Center for Medical, Agricultural, and Veterinary Entomology, Gainesville, FL, United States

2.1 Introduction

Spatial repellents represent one group of an evolving set of tools for interrupting vector insect transmission of disease agents and can provide a bite-free local environment for humans or animals. In contrast to topical repellents applied to clothing or human skin, spatial repellents prevent biting behavior by working in a spatially defined area due to their volatility and behavior-modifying effects. Established spatial repellents include citronella, linalool, and geraniol, natural products used in candles and diffusers (Müller et al., 2009). More recently introduced spatial repellents include volatile pyrethroids, such as metofluthrin (MF), transfluthrin (TF), and prallethrin. These compounds express an initial repelling of insects from a treated area, but their effects can also progress to disorientation and a cessation of host-seeking behavior (Bibbs and Kaufman, 2017). Although vapor-active pyrethroids are quite effective, there are reports that resistance in the field from the kdr mutation can reduce their effectiveness (Wagman et al., 2015), suggesting that continued evaluation of new spatial repellents to circumvent resistance is prudent and appropriate. There is evidence that vapor-active pyrethroids exert their repellent effect via interaction with the olfactory system independent of sodium channel effects, as revealed by electroantennographic (EAG) measurements (Boné et al., 2020). Moreover, TL-I-73, an experimental pyrethroid reported by Chauhan and Bernier (2015), had direct

Advances in Arthropod Repellents.
DOI: https://doi.org/10.1016/B978-0-323-85411-5.00002-9

effects on *Ae. aegypti* odorant receptors expressed in *Xenopus laevis* oocytes (Bohbot et al., 2011). However, an action on the voltage-sensitive sodium channel as a contributing mechanism of repellency is likely, due to the observed kdr cross-resistance (Wagman et al., 2015) and the well-known neuroexcitatory effects of these compounds on that target (Bloomquist, 1996). The studies described here were undertaken to find alternative vapor-active repellents to supplement the pyrethroids, as well as to explore the mode of action of pyrethroid derivatives and structurally related compounds.

2.2 Spatial repellency assay and post-assay behavioral test

An unbiased microassay for testing vapor phase repellency or attractiveness of compounds, especially vapor-active pyrethroids, was described in a recent publication (Jiang et al., 2019) and will be briefly summarized here. When 16 adult female *Ae. aegypti* (2–7 days old) placed in 12.5 × 2.5-cm tubes, they will equally distribute themselves throughout the tube and come to rest with little spatial bias, while showing occasional bouts of walking and flying (Fig. 2.1). Filter papers of 2.5-cm diameter were treated with a 50-μL solution of test compound dissolved in acetone, given 10 min for acetone evaporation, and placed in clear conical polypropylene caps. The end caps were assembled with the glass tubes, and the treated filter papers were approximately 0.5 cm away from the netting to prevent mosquito contact. A rubber band was fixed to the midline (Fig. 2.1) to allow determination of that fraction of mosquitoes attracted or repelled, defined as mosquitoes moving toward or away from the chemical-treated end, respectively. Control experiments were assembled with a filter paper on each end treated with 50 μL of acetone. Repellency was calculated at 15 min, 30 min and 1 h using the formula: number of mosquitoes on the experimental treatment side/16 (Fig. 2.1), where a value of 0 was equal to full

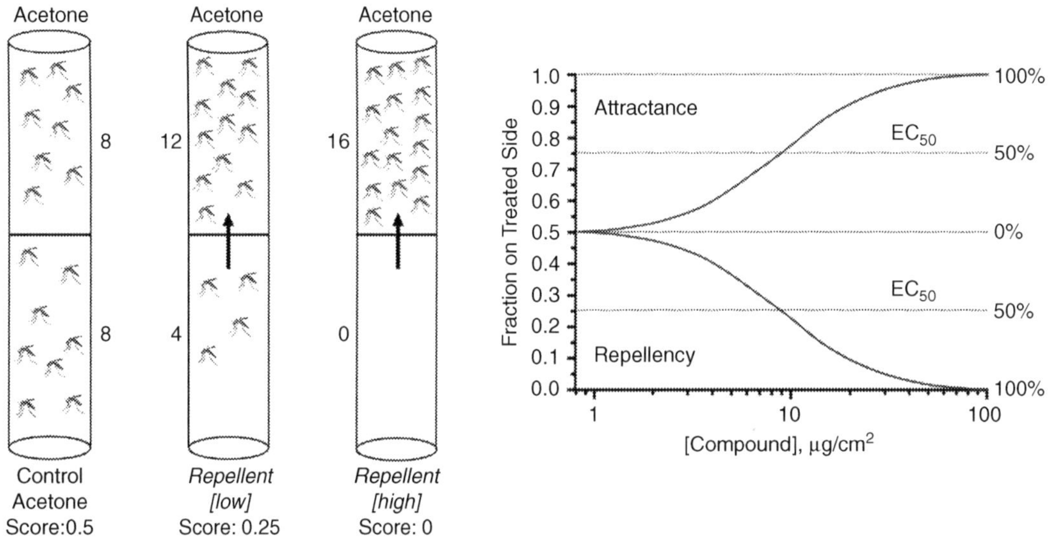

FIG. 2.1 **Nonbiased spatial repellency test apparatus and quantitative behavioral analysis.** (Left) Arrangement of three hypothetical glass tubes showing mosquito distribution, along with control and repellent response scores at low and high concentrations. Numbers indicate mosquitoes on either side of the midline. (Right) Idealized concentration-response curves for attractance and repellency in this assay.

repellency and a value of 0.5 indicated no effect (an even distribution of mosquitoes on either side of the tube midline). Acetone treatment on both ends and a *N,N*-diethyl-3-methyl benzamide (DEET) treatment at 100 μg/cm^2 served as negative and positive controls, respectively. Each concentration was repeated on at least three different batches of mosquitoes to account for cohort variability. Satisfactory results were obtained with tubes held horizontally, but it was found that repellent responses to volatile pyrethroids, such as TF and MF were improved by holding the assay tubes in a vertical orientation, presumably due to more uniform evaporation and spatial diffusion (Jiang et al., 2019).

A post-assay behavioral test was performed when compounds showed incomplete repellency at higher level exposures and little or no obvious knockdown was observed (Yang et al., 2020a). Our working hypothesis is that these insects are disoriented and unable to express a directed movement away from the chemical stimulus. For the post-assay behavioral test, the cap and netting of the control side were removed and the tube held vertically in a cage with the compound-treated side at the bottom. Then, the number of mosquitoes that failed to leave the tube were counted over the ensuing 30 min and compared to matched control tubes. In pilot studies, virtually all mosquitoes left the control tubes in 10–15 min (Yang et al., 2020a). The formula for overall corrected repellency was:

$$\text{Corrected repellency proportion} = \frac{(\#\text{on treated side} - \#\text{failing the PABT})}{16 - \#\text{failing the PABT}}$$

2.3 Pyrethroid fragment screening for vapor phase repellency

It was observed that an older sample of TF exhibited less vapor phase repellency to Orlando (insecticide-susceptible) females than expected, and in particular was less active than a new commercial sample of TF with high purity (99%). Subsequent chemical analysis revealed that the old sample contained no TF, but only the corresponding acid and alcohol. This finding led to fragment screening of transfluthrin acid (TFA) and alcohol (TF-OH), both separately and together (Fig. 2.2). In these studies, the alcohol showed some repellent activity, having an EC_{50} of 80 (69–93) μg/cm^2 (Yang et al., 2020b). This level of activity is 2.5-fold less than DEET, which had an EC_{50} in this assay of 32 (25–39) μg/cm^2 (Jiang et al., 2019). In contrast, it was observed that TFA alone had no statistically significant repellency at levels up to 100 μg/cm^2, and subsequently 50 μg/cm^2 was run with the alcohol to test for any synergistic effects. In the presence of TFA, the EC_{50} of the alcohol declined about 2-fold to a value of 41 (35–54) μg/cm^2, which is similar to that of DEET. This reduction in the EC_{50} is primarily from changing the slope of the concentration-response curve for the alcohol (Fig. 2.2).

2.4 Repellency, synergism, and cross-resistance to pyrethroid acids

Additional pyrethroids and their corresponding acids were screened for repellency in both the insecticide-susceptible Orlando (OR) and pyrethroid-resistant Puerto Rico (PR) *A. aegypti* strains (Table 2.1, Yang et al., 2020b). In contrast to TFA, metofluthrin acid (MFA, Fig. 2.3) and 1*R*-*trans*-chrysanthemic acid (TCA, Fig. 2.3), a component of natural pyrethrins (NP), were active as spatial repellents. However, they showed less potent repellency than the parent pyrethroids MF and TF, which were designed to work in the vapor phase. The spatial activity of NP was similar to that of TCA against OR mosquitoes. Resistance ratios for the pyrethroid acids were < 2, which for MFA was at least 25-fold lower than that of MF, for TCA about 7-fold less than that of NP, and about 3-fold less resistance than TF (Table 2.1). These findings suggested that the acids have a different mode of repellent action than intact pyrethroids,

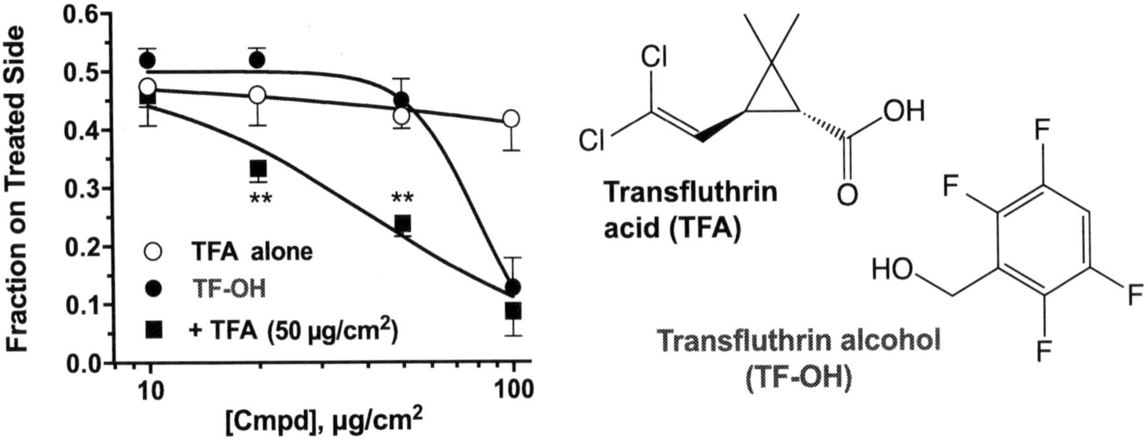

FIG. 2.2 Separate and combined treatments of transfluthrin acid (*TFA*) and the corresponding alcohol (TF-OH) in the spatial repellency assay without a post-assay behavioral test (*PBAT*) correction (plot is modified from Yang et al. 2020b). *Asterisks* indicate statistically significant differences between the mixture and the alcohol alone at 20 and 50 μg/cm² (t-test, $P < .01$).

TABLE 2.1 One-hour repellency of selected pyrethroid acids and their parent pyrethroids in the Orlando and pyrethroid-resistant Puerto Rico strains of *A. aegypti*.

Strain	MFA	MF	TCA	NP	TF
Orlando EC_{50}	14 (10–20)	0.3 (0.1–0.6)	20 (16–24)	31 (16–59)	0.5 (0.3–0.7)
Puerto Rico EC_{50}	18 (9–38)	12 (8–18)	32 (22–46)	343 (170–689)	2.3 (1.5–3.5)
resistance ratio	1.3	40	1.6	11	4.6

EC_{50} values are given in μg/cm² (95% CI) and resistance ratio was calculated by: EC_{50} for PR ÷ EC_{50} for OR. Data taken from Yang et al. (2020a, 2020b).
MF, metofluthrin; *MFA*, metofluthrin acid; *NP*, natural pyrethrins; *TCA*, 1*R*-*trans*-chrysanthemic acid; *TF*, transfluthrin.

whose performance was affected by two kdr mutations (V1016I and F1534C) in the voltage-dependent sodium channel gene expressed in the PR strain of *Ae. aegypti* (Estep et al., 2017). Thus, these lines of evidence are all consistent with the conclusion that pyrethroid acids are interacting primarily with the mosquito olfactory system to elicit repellency and repellent synergism in the vapor phase and not the voltage-sensitive sodium channel.

It was also recently demonstrated that pyrethroid acids can synergize the repellency of pyrethroids and established vapor and contact repellents (Yang 2020b). The synergism ranged from 3.8-fold with citronella to 11.6-fold with 2-undecanone (Table 2.2). Additional repellency and EAG studies by Yang et al. (2020b) showed that TFA worked by two mechanisms; it increased repellent vaporization off the filter and also enhanced their EAG responses. While initial screens indicated the mixtures of TFA and the contact repellent DEET showed no increase in repellency, other experiments demonstrated that synergism of DEET was possible, but only

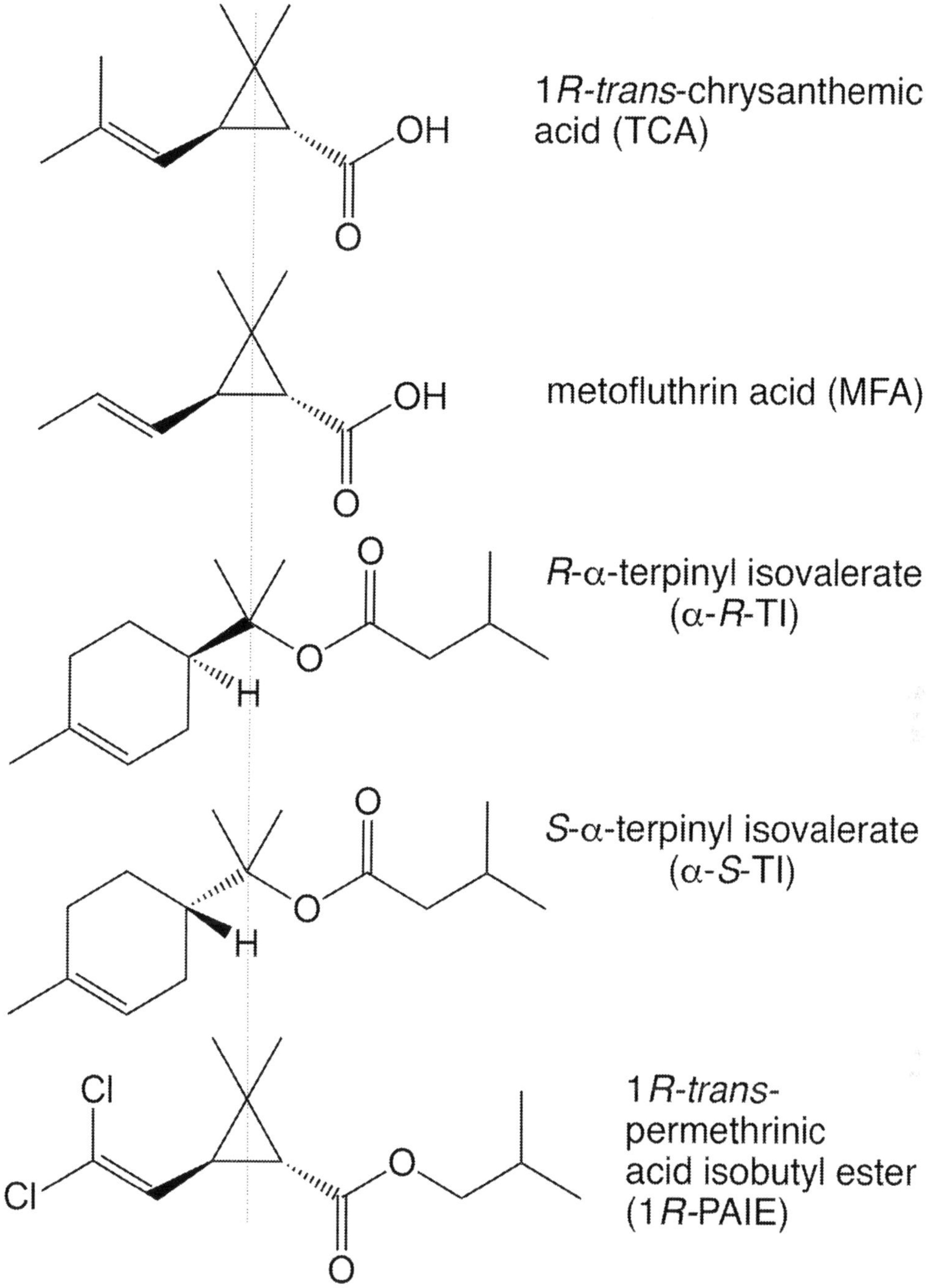

FIG. 2.3 Structural similarity apparent from vertical alignment (*grey dashed line*) of the dimethyl groups of TCA, MFA, both enantiomers of α-TI, and 1*R*-*trans*-permethrinic acid isobutyl ester (1*R*-PIAE). The cognate alcohol moieties for metofluthrin or pyrethroids containing TCA, etc., are readily available from the literature. *MFA*, metofluthrin acid; *TCA*, 1R-*trans*-chrysanthemic acid.

TABLE 2.2 One-hour repellency EC_{50} (μg/cm^2) with (95% Cl) of selected repellents and their mixtures with 50 μg/cm^2 of transfluthrin acid on the Orlando strain of *Ae. aegypti.*

Parameter	Citronella	Methyl jasmonate	2-undecanone	Benzaldehyde
EC_{50}	38 (32–45)	39 (24–64)	59 (48–74)	118 (90–156)
EC_{50} + transfluthrin acid	10 (8–12)	6 (4–13)	5.1 (3–9)	15 (9–26)
Synergistic ratios	3.8	6.5	11.6	8.1

Synergistic ratios were calculated as: EC_{50} compound alone/EC_{50} for compound + transfluthrin acid. Data from Yang et al. (2020b).

occurred when applied side-by-side on the filter paper (Fig. 2.4). Thus, the synergism observed in the side-by-side treatments suggests that TFA was interacting with DEET at the level of sensory detection and that the well-known ability of DEET to trap compounds on surfaces (Zainulabeuddin and Leal, 2008) was suppressing the synergism in mixtures.

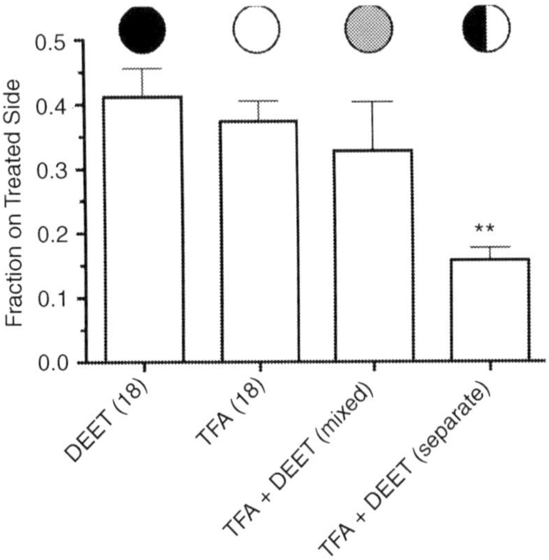

FIG. 2.4 Bar graph showing repellency responses of DEET (filled black circle) and TFA (*filled white circle*) alone, and when mixed (*filled grey circle*) or applied side-by-side to filter papers (*white and black semicircular areas*). Numbers in parentheses are the applied concentration on the filter, in μg/cm^2. *Asterisks* indicate a statistically significant increase in repellency from the side-by-side treatment compared to all other treatments (**$P < .01$ by ANOVA and Newman–Keuls multiple comparisons test). *TFA*, transfluthrin acid.

2.5 Repellency and synergism of transfluthrin acid with experimental anthranilates and pyrazine repellents

In an effort to extend studies of TFA synergism, we assessed its effects of three anthranilates and one pyrazine compound (Fig. 2.5), identified previously by Kain et al. (2013) as active mosquito repellents. This article was subsequently retracted due to problems with experiments that had no impact on the chemical identification of these four repellents (Kain et al., 2016). In our laboratory studies, these compounds showed good activity in the vapor phase assay, and there were no significant differences among the three time points, indicating a rapid spatial repellent effect. Overall, ethyl anthranilate was the most potent, followed by methyl *N,N*-dimethyl anthranilate (MDA), butyl anthranilate (BA), and 2,3-dimethyl-5-isobutyl-pyrazine (DIP), which was comparable in activity to 2-undecanone (Table 2.3). Thus, the increase in size of the alkyl side chain from ethyl to butyl decreased activity nearly 5-fold, suggesting steric hindrance. Similarly, the tertiary amine of MDA gave good activity, within about 2-fold when coupled with a methyl ester (Table 2.3)

In synergism studies with 50 μg/cm^2 TFA, only the treatment group having BA at the EC_{20} level (11 μg/cm^2) coapplied with TFA showed significant synergism (Fig. 2.6A). Further concentration-response experiments showed no statistically significant effect on the potency or efficacy of BA (Fig. 2.6B), where BA+TFA had a

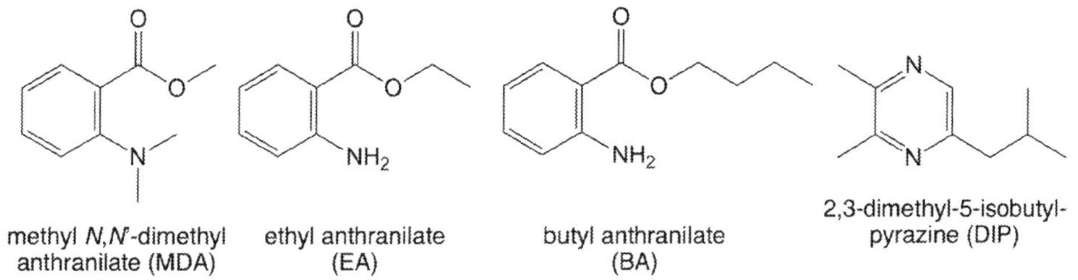

FIG. 2.5 Anthranilate and pyrazine repellents originally described by Kain et al. (2013) evaluated for vapor phase repellency and also for possible synergism by TFA. *TFA*, transfluthrin acid.

TABLE 2.3 EC_{50} (μg/cm^2) values with (95% Cl) and slopes (SEM) for females of the Orlando strain of *Ae. aegypti*.

Compound	15 minutes	Slope	30 minutes	Slope	60 minutes	Slope
Ethyl anthranilate	7 (5–8)	1.6 (0.2)	6 (5–8)	1.4 (0.2)	7 (5–10)	1.3 (0.2)
Butyl anthranilate	32 (20–46)	1.3 (0.4)	35 (22–52)	1.3 (0.4)	22 (15–30)	2.0 (0.5)
Methyl *N,N'*-dimethyl anthranilate	17 (11–25)	1.8 (0.5)	15 (11–20)	1.8 (0.3)	13 (10–17)	2.5 (0.6)
2,3-dimethyl-5-isobutyl-pyrazine	52 (40–68)	1.7 (0.4)	43 (33–56)	1.6 (0.3)	42 (35–50)	1.6 (0.2)

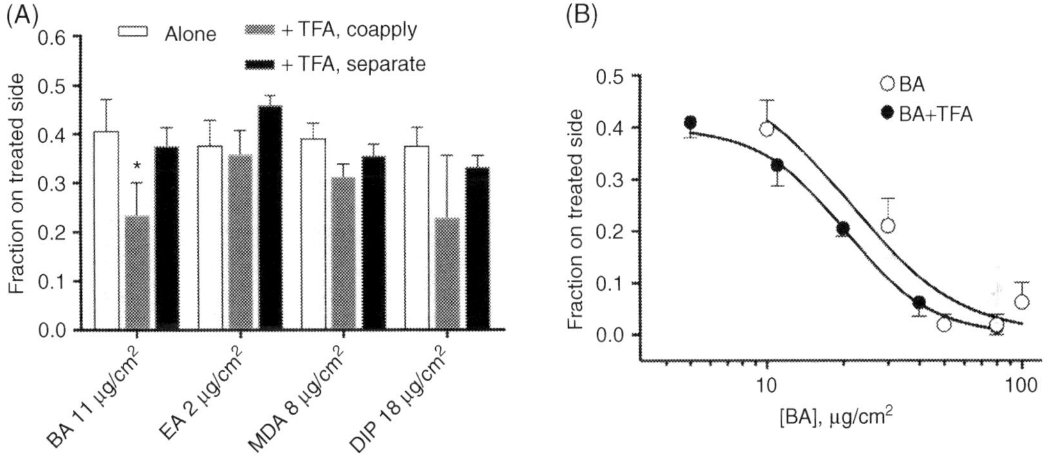

FIG. 2.6 **Spatial repellency synergism experiments with anthranilate and pyrazine compounds tested at the 60-min EC_{20} (fraction on the treated side = 0.4).** (A) Spatial repellency of combination treatments with 50 μg/cm^2 TFA, either mixed (coapply) or separately applied (separate), compared to repellent alone. (B) Concentration-dependent spatial repellency for BA alone and when coapplied with 50 μg/cm^2 TFA. *Asterisks* indicate significant difference (two-tailed Student's t-test, $P < 0.05$). Symbols and bars are means ± SEM. Compound abbreviations are: methyl *N,N'*-dimethyl anthranilate (*MDA*), ethyl anthranilate (*EA*), butyl anthranilate (*BA*), and 2,3-dimethyl-5-isobutyl-pyrazine (*DIP*). *TFA*, transfluthrin acid; BA, butyl anthranilate.

1 h EC_{50} values of 20 (17–24) μg/cm^2, similar to the value for BA alone (Table 2.3). These results extend our observation that TFA synergism is not expressed with all repellent mixtures and not always with a contribution due to evaporative effects (Yang et al., 2020b).

2.6 Repellency and synergism of α-terpinyl isovalerate ester and related compounds

Klimavicz et al. (2018) have reported the spatial repellency of a series of derivatized terpenoids, including a racemic mixture of α-terpinyl isovalerate ester (α-TI). Fig. 2.3 shows a vertical alignment of the two optical isomers of α -terpinyl isovalerate ester, along with the isobutyl ester of TFA (also known as 1*R*-*trans*-permethrinic acid) and TCA. The 1*R*-*trans*-permethrinic acid isobutyl ester (1*R*-PAIE) contains the identical aliphatic *iso*butyl group as terpinyl isovalerate, albeit with reflected attachment to the ester moiety. Both terpinyl isovalerate and 1*R*-PAIE lack ester-conjugated pi-systems. From this similarity, we can infer that the stereoelectronic properties about the *iso*butyl ester of α-TI are mimicked in 1*R*-PAIE. Additionally, the dimethyl-cyclopropyl group of 1*R*-PAIE serves as a possible nonclassical bioisostere for the geminal dimethyl group of α-TI and a similar isostere is seen in TCA. Lastly, while a degree of variability is observed across the terminal ends of the four compounds displayed here, they all exhibit double-bond-containing structures that have similar space-filling steric properties.

In order to test these hypotheses, *R*-α-terpinyl isovalerate, *S*-α-terpinyl isovalerate, and 1*R*-PAIE were synthesized via a Steglich esterification procedure as outlined by Klimavicz et al. (2018). For *R*- and *S*-α-terpinyl isovalerate, the corresponding *R*- or *S*-α-terpineol (1 equivalent) was mixed with isovaleric acid (1.2 equivalents) and 4-dimethylaminopyridine (0.15 equivalents) in methylene chloride (1 mL/mmol) and cooled to 0 °C. A solution of *N*,*N*′-dicyclohexylcarbodiimide (1.05 equivalents) in methylene chloride (2 mL/mmol) was added dropwise, and the mixture was warmed to room temperature and stirred. The reaction progression was monitored by thin-layer chromatography (visualized with KM_nO_4). Upon completion, hexane was added and the solution cooled at 0 °C for 30 min to precipitate unwanted dicyclohexylurea. The dicyclohexylurea was subsequently filtered off, the mother liquor was collected and concentrated via roto-evaporation, and the crude residue was purified via flash chromatography on a silica gel column to afford the desired ester. 1*R*-*trans*-permethrinic acid isobutyl ester was obtained using the same procedure to couple 1*R*-*trans*-permethrinic acid and *iso*butanol.

Biological activity of the enantiomers of α-TI and 1*R*-PAIE was analyzed in the glass tube bioassay for each repellent compound individually. All of the compounds showed increased repellency as the observation period was increased from 15 to 60 min, with about a two-fold decrease in the EC_{50} (Table 2.4). Surprisingly, there was no appreciable difference between the *R* and *S* isomers of α-TI in terms of repellent potency (Fig. 2.7A, Table 2.4). At all time points, the *R* and *S* isomers showed little difference in EC_{50} values, and the extensive overlap in the 95% CL indicate they were not statistically significant (Table 2.4). The racemic α-TI gave a similar EC_{50} value as the two enantiomers, with the biggest differences a large change in slope for α-TI (Fig. 2.7A), along with incomplete repellency at 15 min that gave extremely wide 95% Cl for that EC_{50} (Table 2.4). These results suggest that the two compounds in the racemic mixture were interacting with each other on the filter to change evaporation.

The lack of enantioselectivity for the *R* and *S* isomers of α-TI was surprising, since previous studies have shown such selectivity for both behavioral and odorant receptor response

TABLE 2.4 Time course of repellency by racemic α-TI, its cognate *R* and *S* enantiomer, and 1*R*-*trans*-permethrinic acid isobutyl ester.

Compound	15 minutes	Slope	30 minutes	Slope	60 minutes	Slope
α-TI	113 [38–10,800]	−0.4 (0.16)	61 [15–1350]	−0.68 (0.34)	45 [17–89]	−1.0 (0.38)
α-*R*-TI	93 [66–261]	−2.2 (0.1)	66 [46–99]	−2.1 (0.8)	51 [36–76]	−2.8 (1.1)
α-*S*-TI	87 [56–395]	−1.4 (0.6)	50 [32–82]	−1.6 (0.5)	38 [29–52]	−2.2 (0.6)
1*R*-*trans*-permethrinic acid isobutyl ester	62 [36–142]	−1.3 (0.5)	53 [19–590]	−1.2 (0.7)	34 [12–412]	−0.7 (0.3)

EC_{50} values are given in μg/cm^2 with [95% CI] and slopes (SEM) for females of the Orlando strain of *Ae. aegypti*.

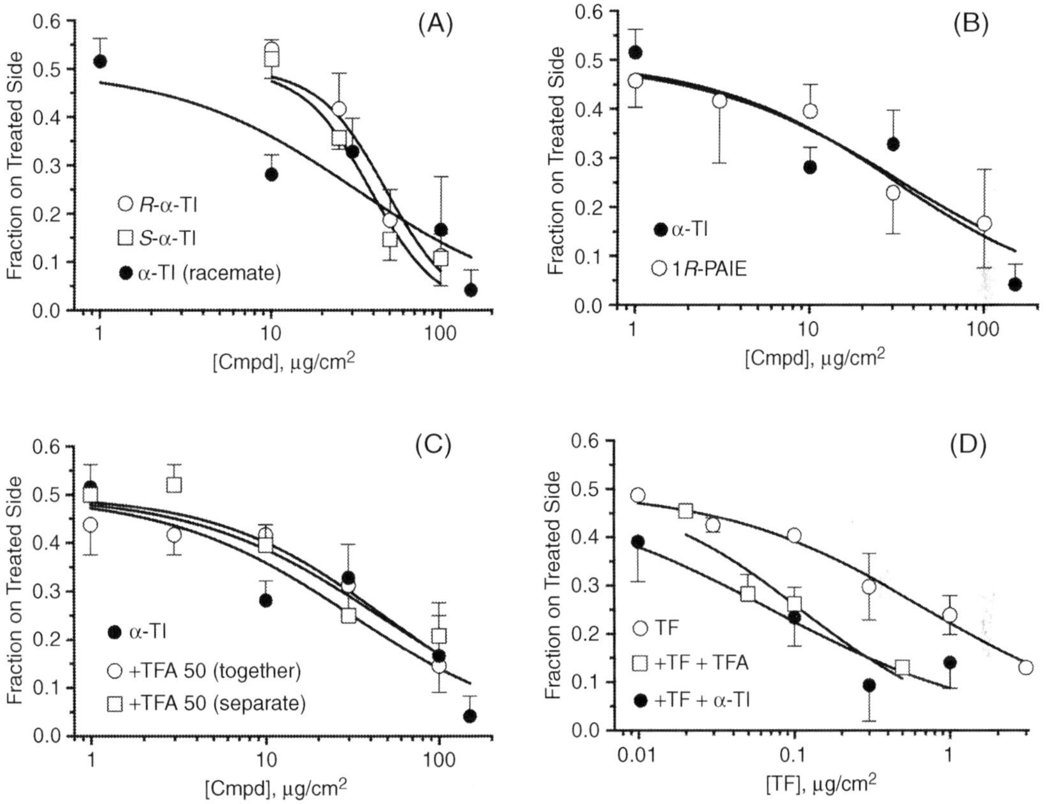

FIG. 2.7 **One-hour repellency experiments with racemic α-terpinyl isovalerate (α-TI), its cognate *R* and *S* enantiomers, 1*R*-*trans*-permethrinic acid isobutyl ester (1*R*-PAIE), and in mixtures with transfluthrin acid (*TFA*) or transfluthrin (*TF*).** (A) Repellency of racemic α-TI and its enatiomers when applied alone. (*B*) Comparative repellent potency of α-TI and 1*R*-PAIE, which are identical in activity. (C) Effect of applying racemic α-TI + TFA (50 μg/cm^2) either together or side by side on the same filter paper. (D) α-TI and TFA evaluated in mixtures as synergists of transfluthrin (*TF*). *Cmpd* = compound.

profiles. For example, the (R)-(—)-enantiomer of 1-octen-3-ol shows about 100-fold more potent activation of *Ae. aegypti* odorant receptor 8 when expressed in *Xenopus laevis* oocytes (Bohbot and Dickens, 2009), and attracts more mosquitoes in field traps than its *S* enantiomer (Kline et al., 2007). Moreover, males of two species of Japanese beetles (*Anomala osakana* and *Popillia japonica*) respond to the enantiomeric pairs of japonilure, but with opposite behavioral effects; the *R* isomer attracting males and the *S* isomer inhibiting this response (Tumlinson et al., 1977). These behavioral responses are mediated by two different chemosensory neurons, located within the same sensillum, and each responds to only one enantiomer (Wotjasek et al., 1998). In contrast, only one pheromone-binding protein was present in these species and associated with both enantiomers to a similar extent. These findings suggest that each enantiomer of α-TI probably interacts with a different receptor protein, but both producing repellency.

1*R*-PAIE had overlapping concentration-response curves compared t α-TI (Fig. 2.7B) and most similar EC_{50} values to α-*S*-TI across all time points (Table 2.4). There was also substantial overlap in the concentration-response curves when α-TI was tested with 50 μg/cm^2 of TFA (Fig. 2.7C), where the mixture $EC_{50} = 45$ (22–157) μg/cm^2 and the side-by-side on the filter paper $EC_{50} = 49$ (25–157) μg/cm^2. Finally, compared to TF alone, both α-TI and TFA synergized the vapor repellency of TF (Fig. 2.7D). The TF $EC_{50} = 0.72$ (0.44–1.17), TF + α-TI $EC_{50} = 0.1$ (0.07–0.17), and the TF + TFA $EC_{50} = 0.07$ (0.02–0.26). The data show a clear lack of overlap in the 95% Cl and give SR = 10 and SR = 7 for α-TI and TFA, respectively. These results indicate that the structural similarity of α-TI and 1*R*-*trans*-permethrinic acid isobutyl ester is reflected in their respective biological activities. Further studies are required to determine the mechanisms underlying this synergism and whether the two compounds act on the same olfactory receptor(s).

2.7 Screening for effects on the central nervous system

A central issue in determining the mechanism underlying TFA-dependent spatial repellency synergism is separating effects on the voltage-sensitive sodium channel from those via the olfactory pathway. In a previous study by Yang et al. (2020b), it was shown that TFA had no effect on nerve firing in larval *Drosophila melanogaster* CNS, even at a concentration of 100 μM. Effects of TFA in this preparation were explored further, as shown in Figs 2.8 and 2.9, and these electrophysiological recordings reveal the following. TFA was confirmed to be inactive on the larval CNS, as expected, and TF-OH and α-TI were similarly without intrinsic activity at 1 μM (data not shown). However, when applied with TF, a 1 μM concentration of TFA was able to enhance the action of TF such that significant nerve block was initiated at 1 nM TF, a concentration that is normally inactive in this assay (Fig. 2.8). No statistically significant effect on TF-induced discharge was observed with TF-OH or α-TI, although the latter did display a mild and prolonged excitatory effect (Fig. 2.8). Such an effect might become significant at a higher concentration of α-TI.

In contrast, 1 μM TFA has no effect on the mild neurophysiological effect of 10 nM imidacloprid, showing enhancement of neither the early hyperexcitation or late block of nerve firing like that caused by 30 nM imidacloprid (Fig. 2.9). These data can be explained via two (or more) possible mechanisms. First, TFA may be occupying an allosteric site on the voltage-sensitive sodium channel to augment the effects of TF. A possible example would be the second putative pyrethroid binding site proposed by Du et al. (2013) for the *Ae. aegypti* sodium channel. A second possible mechanism is that a protein homologous to an odorant receptor is expressed in the fly CNS. This hypothesis was advanced recently to

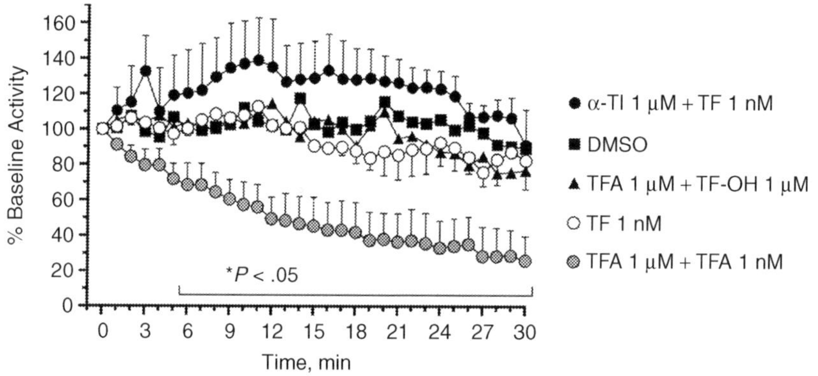

FIG. 2.8 **Evaluation of effects of 1 μM racemic α-terpinyl isovalerate (α-TI), transfluthrin acid (*TFA*), and transfluthrin alcohol (*TF-OH*) in combination with transfluthrin (*TF*) on the larval *D. melanogaster* CNS.** Symbols are means ± SEM. Data for α-TI, TFA, and TF-OH alone are omitted from the graph because they had no statistically significant effect compared to dimethyl sulfoxide (*DMSO*) controls. Likewise, error bars are omitted for the DMSO and TFA + TF-OH groups for clarity of presentation. For the TFA 1 μM + TF 1 nM, all time points from 6 to 30 min (end of the experiment) were significantly different from DMSO control (unpaired t-test). No other treatments were significantly different from DMSO controls at any time point

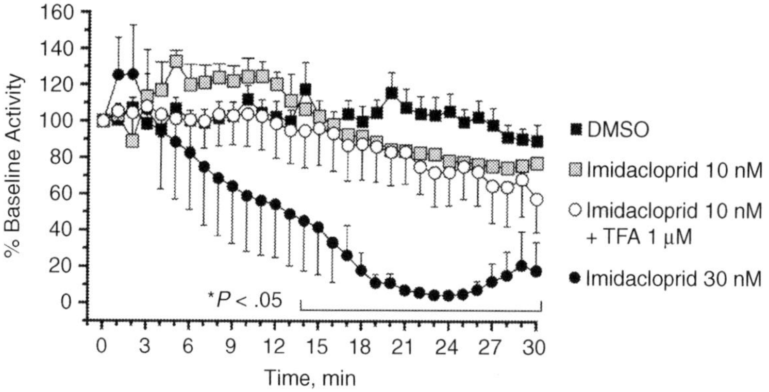

FIG. 2.9 **Evaluation of the effects of transfluthrin acid (*TFA*) and imidacloprid alone or in combination on the larval *D. melanogaster* CNS.** Symbols are means ± SEM, although error bars are omitted for 10 nM imidacloprid from 17 to 30 min to enhance clarity of visualization. For 30 nM imidacloprid, all time points from 14 to 30 min were significantly different from DMSO control (unpaired t-test), as indicated by the bracket. There were no statistically significant effects of the other treatments compared to controls.

explain the neuroexcitation caused by the Orco agonist VUAA1 (Yang et al., 2020c). Further experiments are required to clarify the contributions of these two possible mechanisms, as well as any involvement of α-TI, which in these preliminary studies appears to be less active than TFA.

2.8 Conclusion

As shown previously, fragment screens of the acid and alcohol of TF showed no effect of the acid (TFA) alone, but it did enhance the repellency of the corresponding alcohol, as well as the parent, TF. TFA could also synergize the neurophysiological actions of TF on the larval *D. melanogaster* CNS, but had no effect on imidacloprid-dependent neuroexcitation, perhaps indicating a dual action with TF on the nerve membrane sodium channels. Anthranilate and pyrazine repellents were potent, fast-acting, and not synergized by TFA. Synthesis and repellency evaluation of the pure enantiomers of α-terpinyl isovalerate (α-TI) revealed a more slowly developing effect, but overall little EC_{50} difference from each isomer or the racemic mixture. In addition, the racemate had a lower slope value, perhaps indicating a chemical interaction of the two enantiomers on the filter paper. α-TI synergized the repellency of TF 10-fold, similar to that observed for TFA (7-fold). Taken together, these latter findings and the structural similarity of TFA to α-TI suggests they may work through common mechanisms and receptor(s), although experimental confirmation warrants further study. These studies were performed using a new high-throughput assay system that is simple, inexpensive, and rapid. Thus, it should serve as an ideal testing platform to further improve the efficacy of commercially available mosquito repellents, and also to investigate new cost-effective, low toxicity, and environmentally-friendly repellent compounds.

Acknowledgments

This project was funded by the Department of Defense, Deployed War Fighter Research Program, under USDA Specific Cooperative Agreements 58-6615-4-023 (to J.B.) and 59-6063-8-001 (to K.L.). *Ae. aegypti* mosquitoes were provided by Dr. Dan Kline (Orlando) and Alden Estep (Puerto Rico), of the United States Department of Agriculture Center for Medical and Veterinary Entomology (USDA CMAVE).

References

Bibbs, C., Kaufman, P., 2017. Volatile pyrethroids as a potential mosquito abatement tool: a review of pyrethroid-containing spatial repellents. J. Integr. Pest. Manage. 8 (1), 1–10.

Bloomquist, J.R., 1996. Ion channels as targets for insecticides. Annu. Rev. Entomol. 41, 163–190.

Bohbot, J., Dickens, J., 2009. Characterization of an enantioselective odorant receptor in the yellow fever mosquito *Aedes aegypti*. PLoS One 4 (9), e7032 https://www.ncbi.nlm.nih.gov/pmc/articles/PMC2737144/.

Bohbot, J., Fu, L., Le, T., Chauhan, K., Cantrell, C., Dickens, J., 2011. Multiple activities of insect repellents on odorant receptors in mosquitoes. Med. Vet. Entomol. 25, 436–444.

Bone, E., Gonzales-Audino, P, Sfara, V., 2020. Spatial repellency caused by volatile pyrethroids is olfactory-mediated in the German Cockroach *Blattella germanica*(Dictyoptera: Blattellidae). Neotropical Entomol. 49, 275–283.

Chauhan, K., Bernier, U., 2015. Methods and compositions for repelling and/or killing insects. United States Patent, US 9,101,142 B1. https://www.freepatentsonline.com/9101142.pdf (accessed February 24, 2021).

Du, Y., Nomura, Y., Satara, G., Hua, Z., Nauen, R., Yang He, S., Zhorov, B., Dong, K., 2013. Molecular evidence for dual pyrethroid-receptor sites on a mosquito sodium channel. Proc. Natl. Acad. Sci. USA. 110 (29), 11785–11790.

Estep, A., Sanscrainte, N., Waits, C., Louton, J., Becnel, J., 2017. Resistance status and resistance mechanisms in a strain of Aedes aegypti (Diptera: Culicidae) from Puerto Rico. J. Med. Entomol. 54, 1643–1648.

Jiang, S., Yang, L., Bloomquist, J., 2019. High-throughput screening method for evaluating spatial repellency and vapor toxicity to mosquitoes. Med. Vet. Entomol. 33, 388–396.

Kain, P., Boyle, S., Tharadra, S., Guda, T., Pham, C., Dahanukar, A., Ray, A., 2013. Odour receptors and neurons for DEET and new insect repellents. Nature 502, 507–514.

Kain, P., Boyle, S., Tharadra, S., Guda, T., Pham, C., Dahanukar, A., Ray, A., 2016. Retraction: odour receptors and neurons for DEET and new insect repellents. Nature 536, 488.

Klimavicz, J.S., Corona, C.L., Norris, E.J., Coats, J.R., 2018. Monoterpenoid isovalerate esters as long-lasting spatial mosquito repellentsAdvances in the Biorational Control of Medical and Veterinary Pests. In: ACS Symposium Series, 1289. American Chemical Society, Washington D.C., pp. 205–217.

Kline, D., Allan, S., Bernier, U., Welch, C., 2007. Evaluation of the enantiomers of 1-octen-3-ol and 1-octyn-3-ol as attractants for mosquitoes associated with a freshwater swamp in Florida, U.S.A. Med. Vet. Entomol. 21, 323–331.

Müller, G., Junnila, A., Butler, J., Kravchenko, V., Revay, E., Weiss, R., Schlein, Y., 2009. Efficacy of the botanical repellents geraniol, linalool, and citronella against mosquitoes. J. Vector Ecol. 34 (1), 2–8.

Tumlinson, J., Klein, M., Doolittle, R., Ladd, T., Proveaux, A., 1977. Identification of the female Japanese beetle sex pheromone: inhibition of male response by an enantiomer. Science 197, 789–792.

Wagman, J., Achee, N., Grieco, J., 2015. Insensitivity to the spatial repellent action of transfluthrin in *Aedes aegypti*: A heritable trait associated with decreased insecticide susceptibility. PLoS Neg. Trop. Dis. 9 (4), e0003726. doi:10.1371/journal.pntd.0003726.

Wojtasek, H., Hansson, B., Leal, W., 1998. Attracted or repelled?–a matter of two neurons, one pheromone binding protein, and a chiral center. Biochem Biophys Res Commun 250, 217–222.

Yang, L., Norris, E., Jiang, S., Bernier, U., Linthicum, K., Bloomquist, J., 2020a. Reduced effectiveness of repellents in a pyrethroid-resistant strain of *Aedes aegypti* (Diptera: Culicidae) and its correlation with olfactory sensitivity. Pest Manag. Sci. 76, 118–124.

Yang, L., Richoux, G., Norris, E., Cuba, I., Jiang, S., Coquerel, Q., Démares, F., Linthicum, K., Bloomquist, J., 2020b. Pyrethroid-derived acids and alcohols: Bioactivity and synergistic effects on mosquito repellency and toxicity. J. Agric. Food Chem. 68, 3061–3070.

Yang, L., Demares, F., Norris, E., Jiang, S., Bernier, U., Bloomquist, J., 2020c. Bioactivities and modes of action of VUAA1. Pest Manag. Sci. Epub August 1, 2020. https://onlinelibrary.wiley.com/doi/full/10.1002/ps.6023 (Accessed June 29, 2021).

Zainulabeuddin, S., Leal, W., 2008. Mosquitoes smell and avoid the insect repellent DEET. Proc. Natl. Acad. Sci 105, 13598–13603.

CHAPTER

3

Biorational compounds as effective arthropod repellents against mosquitoes and ticks

Colin Wong, Caleb Corona, Joel Coats

Iowa State University, Department of Entomology, Pesticide Toxicology Laboratory, Ames, IA, United States

3.1 Introduction

Infectious disease outbreaks pose a major risk to both individuals and global institutions (Coibion et al., 2020). The mosquito-borne Zika virus (ZIKV) affected travel, sports, and other economic indicators on a global scale (Jamil et al., 2016). The outbreak also had a horrific human toll, including nearly 1000 cases of ZIKV-related Guillain–Barré syndrome in Brazil alone, with a cost estimate of over ten million US dollars over the lives of the stricken children (Peixoto et al., 2019). Persistent endemic diseases can also present significant threats to life and economic burden. Lyme disease is the most common arthropod-borne disease in the contemporary United States, and infection rates are still rising (Rochlin et al., 2019). Unfortunately, climate change can also be making the environment more habitable for disease-vectoring arthropods. The statewide investigation of *Ixodes scapularis*, the tick that transmits Lyme disease, shows that it is moving into more populated areas over time (Oliver et al., 2017).

This chapter focuses on novel compounds that may be developed into repellents to prevent the spread of these arthropod-borne diseases. Arthropod repellents provide a key resource in protecting individuals, particularly when they must be exposed to the habitats of biting arthropods, whether due to occupation, lack of protective housing, or other personal needs (Bissinger and Roe, 2010, Norris and Coats, 2017). Resistance to insecticides can also cause a resistance to current repellents (Yang et al., 2020). Additionally, consumer demand has sown a desire for additional repellent active ingredients with lower toxicity to humans and the environment (Niesobecki et al., 2019). To fulfill that growing demand, we investigated both natural molecules found in plant essential oils on ticks and biorational derivatives of those compounds on ticks and mosquitoes.

The term biorational has several meanings depending on the context and source, it can

Advances in Arthropod Repellents.
DOI: https://doi.org/10.1016/B978-0-323-85411-5.00012-1

include completely natural biopesticides or plant-incorporated-protectants found within genetically modified organisms (Horowitz et al., 2009; Haddi et al., 2020). In this chapter, we define the term biorational as a molecule consisting of two or more small natural molecules joined to retain the safety and repellency of the parent natural molecules.

Natural molecules have been explored as spatial repellents for mosquitoes, and these often find that the monoterpenoid class of molecules can be effective but short-lived repellents (Paluch et al., 2009; Maia and Moore, 2011). Using these natural molecules, including monoterpenoids commonly found in essential oils, as a basis for the biorational molecules could be a way to maximize the repellent efficacy of monoterpenoids, but improve on the duration of protection as effective repellents. This direction has led to the development of a number of monoterpenoid derivatives for use as repellents (Norris and Coats, 2017; Klimavicz et al., 2018). This research compared some of these biorational molecules with their parent monoterpenoids along with commercially available repellent active ingredients. The testing compared repellency in the short-term with long-term testing, which saw the repellent materials aged for 5 hours prior to testing so that the durability of repellent efficacy could be observed. Additionally, the tests were performed with both mosquitoes and ticks to broaden the scope of a potential arthropod repellent active ingredient found amongst the biorational molecules tested.

3.2 Methods

3.2.1 Tick repellency bioassay

The repellency bioassay, without host cues, was modified slightly from Wong et al., 2021 and uses an open Petri dish to prevent the buildup of volatile compounds in the assay chamber. Within the Petri dish, two filter papers (90-mm) were placed atop each other. The bottom paper was untreated and the top paper had a hole (24-mm radius) cut in the center and was treated with 1 mL of a solution of test compound in acetone. The treatment was dried before assembly of the Petri dish arena. A circular paperweight made of wound wire approximately 90 mm in diameter was used to weigh down the outer edge of the treated paper to reduce any gap between the filter papers. Ticks did not come into contact with the weight. For short-term repellency tests, the arena was constructed, and testing began as soon as the filter papers were dry of acetone (15 minutes). For the long-term repellency tests, the treatment was dried for an additional 5 hours before testing.

An individual tick was placed on the center of the untreated filter paper, in the hole of the treated filter paper. The ticks move toward the edge to climb out of the Petri dish, whether to escape the open area or to climb to higher areas for questing. Ticks were tested individually, and once the tick started moving forward, it was timed until it had crossed all eight legs onto the treated filter paper. The times were recorded and the tick discarded; each tick was only tested once. Times were censored at 300 seconds. Thirty ticks were tested for each treatment (30 replications).

Test solution strengths were initially determined using *N,N*-diethyl-3-methyl benzamide (DEET) to repel *Dermacentor variabilis* ticks in the test arena. A 5% (w/v) solution, which gives a final concentration of 1.098 mg/cm^2 on the treated paper, was chosen for *D. variabilis* ticks. In the preliminary testing with DEET, 5% gave low repellency times, but could still be distinguished from the acetone control when using the model. *I. scapularis* ticks proved to be more sensitive when testing DEET, and a concentration of 0.5% (which gives a final concentration of 0.110 mg/cm^2 on the treated paper) was determined to be closer to the results found with *D. variabilis* ticks. However, using two different concentrations means that a direct comparison between tick species is difficult to interpret, and relative trends between species are more relevant.

3.2.2 Mosquito repellency bioassay

Thirty-centimeter glass chambers are used for this assay. These chambers are enclosed at either end with 90-mm petri dishes. One petri dish contains a 90-mm filter paper treated with 1 mL of a 0.5% solution of a compound in acetone. The other filter paper is left untreated. Twenty unfed female mosquitoes were anesthetized and added to the repellency chamber through a small hole in the middle of the chamber that is sealed.

The distribution of the mosquitoes throughout the chamber is assessed at multiple time points: 15, 30, 60, 90, 120, and 150 minutes after the mosquitoes are added to the chamber. Mosquitoes are observed to be either on the "treated" or "untreated" side of the chamber. These positional observations are used to determine a percentage repellency value for each compound, at each time-point. Ranking was performed by ordering the treatments for each time point and then summing the rank across the different time points. The equation used to determine these repellency values is as follows:

$$\frac{\begin{pmatrix}\text{Number of mosquitoes}\\ \text{in untreated half}\end{pmatrix}-\begin{pmatrix}\text{Number of mosquitoes}\\ \text{in treated half}\end{pmatrix}}{(\text{Total number of mosquitoes})}$$
$$\times 100 = \%\,\textit{Spatial Repellency}$$

The short-term assay type is characterized by the addition of the treated filter paper being added immediately after a drying period (15 minutes). The long-term component of this assay is carried out in the same manner, but instead of the 15-minute drying time, we allow the filter papers to dry for 5 hours before being added to the repellency chamber. During this assay, all compounds were run with four replicates arranged in an alternating pattern to eliminate any preference toward one side of the chambers or the other. The assays are also carried out under consistent lighting and ambient room temperatures.

3.2.3 Ticks

Adult ticks were used and every test used 15 males and 15 females; males and females were visually identified. Both species, *D. variabilis* and *I. scapularis*, were purchased from the Oklahoma State University, National Tick Research and Education Resource, Stillwater, OK. The ticks arrived 1–4 weeks after molting into adults and were tested within one week of receipt. They were maintained in an incubator at 22–24 °C with high humidity (>70% RH) and not fed as adults.

3.2.4 Mosquitoes

The mosquitoes used over the course of this project were provided by the Medical Entomology Laboratory located at Iowa State University. We used one strain of mosquito, a susceptible strain of *Cx. pipiens*; the mosquito colonies were maintained in accordance with standard rearing protocol established by the Medical Entomology Laboratory. Upon obtaining the mosquitoes, they were used 3–5 days after adult emergence. Only adult female mosquitoes were used and not allowed to blood feed prior to their introduction to the repellency chambers.

3.2.5 Chemicals

Perillyl alcohol, citronellol, 2-undecanone, DEET, and *p*-menthane-3,8-diol (PMD) were purchased from Sigma Aldrich (St Louis, Missouri). Nerol and geraniol were purchased from Berjé (Carteret, New Jersey). Esters and carbonates were synthesized in the laboratory and verified using nuclear magnetic resonance spectroscopy.

3.2.6 Tick statistics

All compounds were compared to their respective control treatment using a Cox Proportional Hazard model. The data were divided by test length and tick species because inclusion of all of the data together did not pass the proportionality assumption of the model, as well as because the different tick species used different treatment concentrations. A log-rank test of the data was used to make multiple pairwise comparisons. A Benjamini–Hochberg Procedure was used to adjust for the multiple comparisons, and significance was assigned to paired comparisons with a P value less than 0.05, postprocedure. Within an individual treatment compound, the log-rank test was carried out to determine the effects of species and compound aging within a treatment.

3.3 Results

3.3.1 Mosquito repellent treatments

Nerol and perillyl alcohol performed strongly in the short-term spatial repellency assay eliciting the greatest percentage repellency for most of the time points (Table 3.1). The related monoterpenoid, citronellol, did not show as much spatial repellency in the short-term, with low percentage early in the assay (32 ± 9%), but higher repellency at the end of the assay (88 ± 3%). Citronellol was much worse in the long-term repellency assay with lower percentage repellency at every time point compared to the short-term (Tables 3.1 and 3.2). The long-term tests with citronellol also did not show the increase in repellency toward the end of the assay as observed in the short-term test (Table 3.2). DEET performed similarly between the long-term and short-term tests, starting with lower percentage repellency but achieving an average of 80% repellency by the 90-minute mark and maintaining that level for the duration of the test.

Of the biorational esters and carbonates, ethyl perillyl carbonate was one of the most effective and showed high levels of repellency in both the short-term and long-term spatial repellent tests (Tables 3.1 and 3.2). Menthyl isovalerate, on the other hand, had low percentage repellency in both test types.

3.3.2 Tick repellent control treatments

The acetone vehicle controls were significantly different by species ($P < .05$). Within species, the long-term and short-term treatments had no effect when using the acetone control (*D. variabilis* $P = 0.57$, *I. scapularis* $P = 0.94$). The DEET treatment was used as a positive control and to establish the concentrations used. The response to DEET was very different by species ($P < .005$ for all comparisons); however, comparing within species, the DEET treatments did not have a significant effect, even with differing median repellent times (RT_{50}). *I. scapularis* had an RT_{50} of 56.7 seconds for the short-term DEET test, but only 34.8 seconds for the long-term ($P = 0.19$). *D. variabilis* had an RT_{50} of 14.0 seconds for the short-term DEET test, and the long-term was closer at 15.5 seconds ($P = 0.59$).

3.3.3 Tick repellent experimental treatments

Fig. 3.1 shows all of the experimental treatments grouping together the two species of tick and the short-term and long-term test types for each treatment. The relative ranking of how well these treatments performed in each species-test type grouping is shown in Table 3.3. Ethyl perillyl carbonate had the greatest RT_{50} for 3 of the 4 species-test type combinations. For the long-term tests using *D. variabilis* ticks, ethyl perillyl carbonate had the fourth-highest RT_{50} with only ethyl isopulegyl carbonate and citronellyl cyclopropane carboxylate being significantly better (Table 3.1). The carbonate esters were among the top four compounds for the *D. variabilis* for both long- and short-term tests. The ethyl isopulegyl carbonate was less distinct in the *I. scapularis* tests, with many compounds performing better (Table 3.4).

TABLE 3.1 Mosquito (*Cx. pipiens*) repellency in the short-term test type.

Short-term repellency							Standard error of the mean					
Compound identity	**15 minutes**	**30 minutes**	**60 minutes**	**90 minutes**	**120 minutes**	**150 minutes**	**15 minutes**	**30 minutes**	**60 minutes**	**90 minutes**	**120 minutes**	**150 minutes**
N,N-Diethyl-3-methylbenzamide	53	55	75	80	82.5	82.5	9	6	10	7	9	11
Citronellol	32	53	65	80	85	88	9	3	3	4	3	3
Nerol	97.5	100	97.5	97.5	97.5	95	3	0	3	3	3	5
Geraniol	75	90	95	90	100	85	10	6	5	6	0	10
Perillyl alcohol	90	90	95	95	100	100	6	6	5	5	0	0
Citronellyl pivalate	25	7.5	70	90	95	92.5	6	7	9	17	19	14
Citronellyl myristate	45	45	37.5	52.5	50	52.5	5	9	15	16	20	16
Citronellyl cyclopropane carboxylate	22.5	65	87.5	92.5	90	90	8	11	9	7	7	6
Citronellyl cyclobutane carboxylate	37.5	45	70	80	82.5	72.5	9	9	5	3	7	5
Farnesyl pivalate	60	70	67.5	77.5	75	77.5	8	9	13	9	5	6
Farnesyl isovalerate	87.5	85	90	87.5	92.5	92.5	6	3	4	3	5	3
Geranyl isobutyrate	95	92.5	92.5	87.5	82.5	87.5	3	5	5	5	6	5
Neryl isovalerate	72.5	90	92.5	95	95	95	8	4	3	3	3	3
Menthyl isovalerate	37.5	55	50	42.5	47.5	27.5	5	6	9	10	8	3
Ethyl perillyl carbonate	60	75	82.5	92.5	90	90.45	11	9	11	5	0	4
Ethyl citronellyl carbonate	57.5	72.5	72.5	90	90	90	3	5	8	6	4	4

Values represent the mean percentage repellency. The standard error of the mean for each cell is represented on the right in a corresponding cell.

TABLE 3.2 Mosquito (*Cx. pipiens*) repellency in the long-term test type.

Long-term repellency							Standard error of the mean					
Compound identity	15 minutes	30 minutes	60 minutes	90 minutes	120 minutes	150 minutes	15 minutes	30 minutes	60 minutes	90 minutes	120 minutes	150 minutes
N,N-Diethyl-3-methylbenzamide	73	78.5	81.5	81.5	84	81	10	11	12	12	7	7
Citronellol	30	34	22	21	27	27	11	12	11	14	13	11
Citronellyl pivalate	28	20	8	18	8	20	6	7	9	17	19	14
Citronellyl cyclopropane carboxylate	70	80	82.5	80	80	62.5	8	11	9	7	7	6
Citronellyl cyclobutane carboxylate	22.5	22.5	30	37.5	45	35	6	5	4	6	3	3
Menthyl isovalerate	35	27.5	20	15	32.5	30	6	9	11	19	8	9
Ethyl perillyl carbonate	70	93	90	93	93	90	6	3	6	7	3	6
2-Undecanone	13	26	26	10	8	8	18	19	21	15	30	31

Values represent the mean percentage repellency. The standard error of the mean for each cell is represented on the right in a corresponding cell.

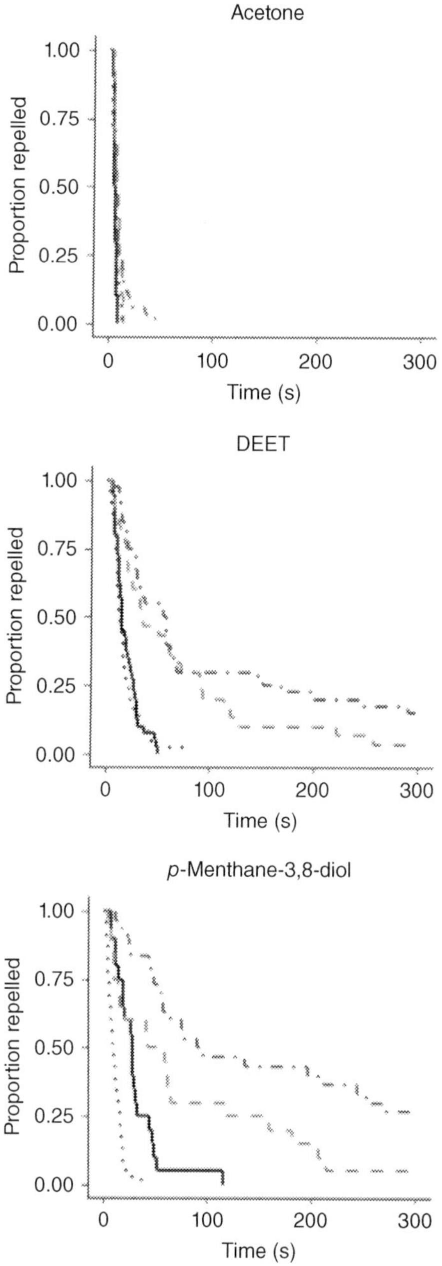

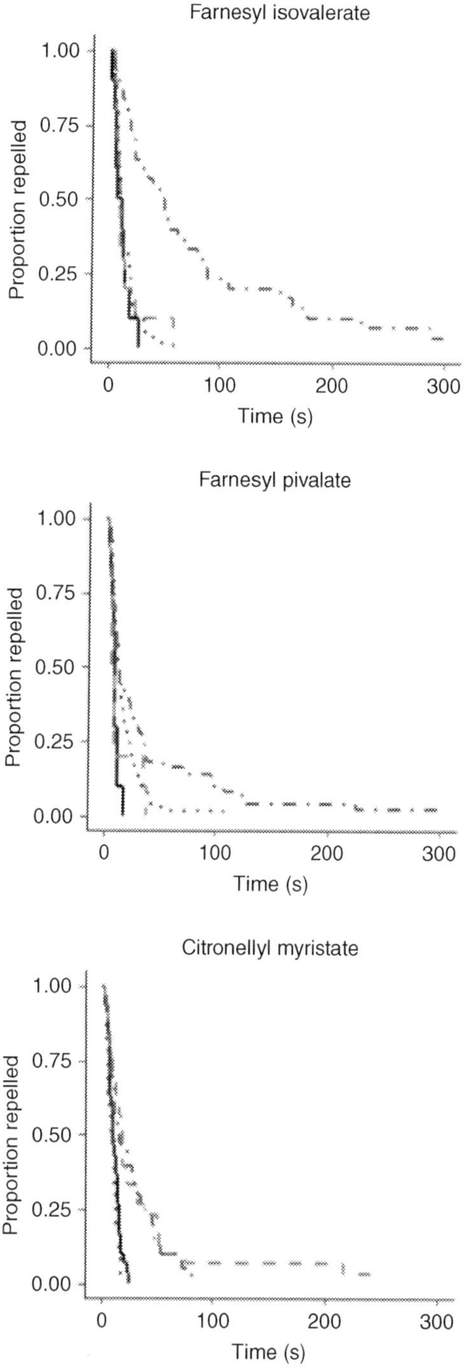

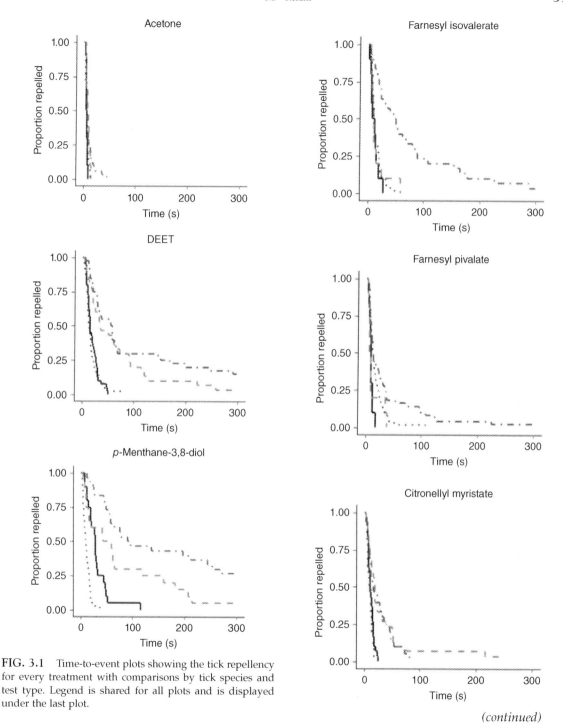

FIG. 3.1 Time-to-event plots showing the tick repellency for every treatment with comparisons by tick species and test type. Legend is shared for all plots and is displayed under the last plot.

(continued)

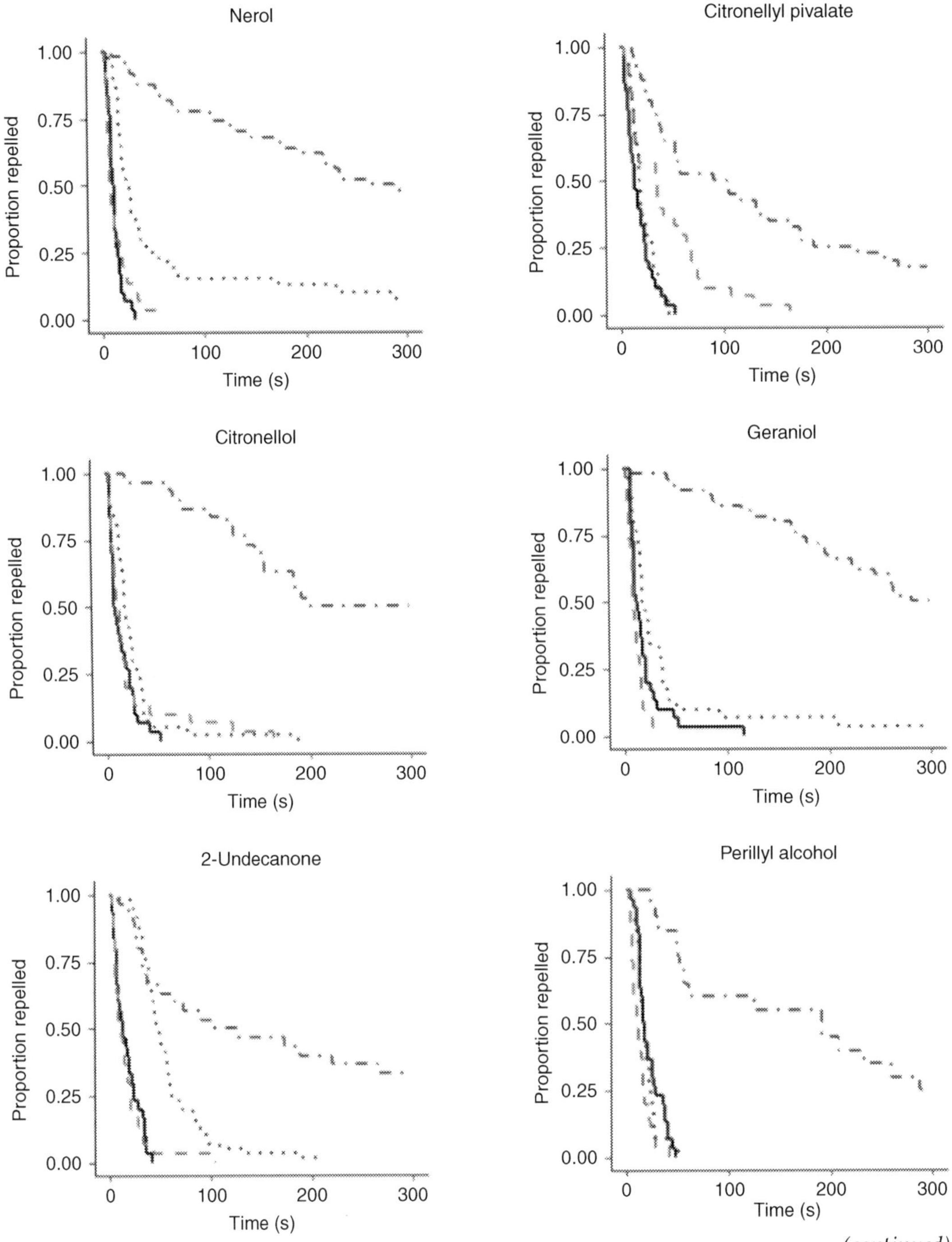
Nerol
Proportion repelled
1.00
0.75
0.50
0.25
0.00
0
100
200
300
Time (s)
Citronellyl pivalate
Citronellol
Geraniol
2-Undecanone
Perillyl alcohol

(continued)

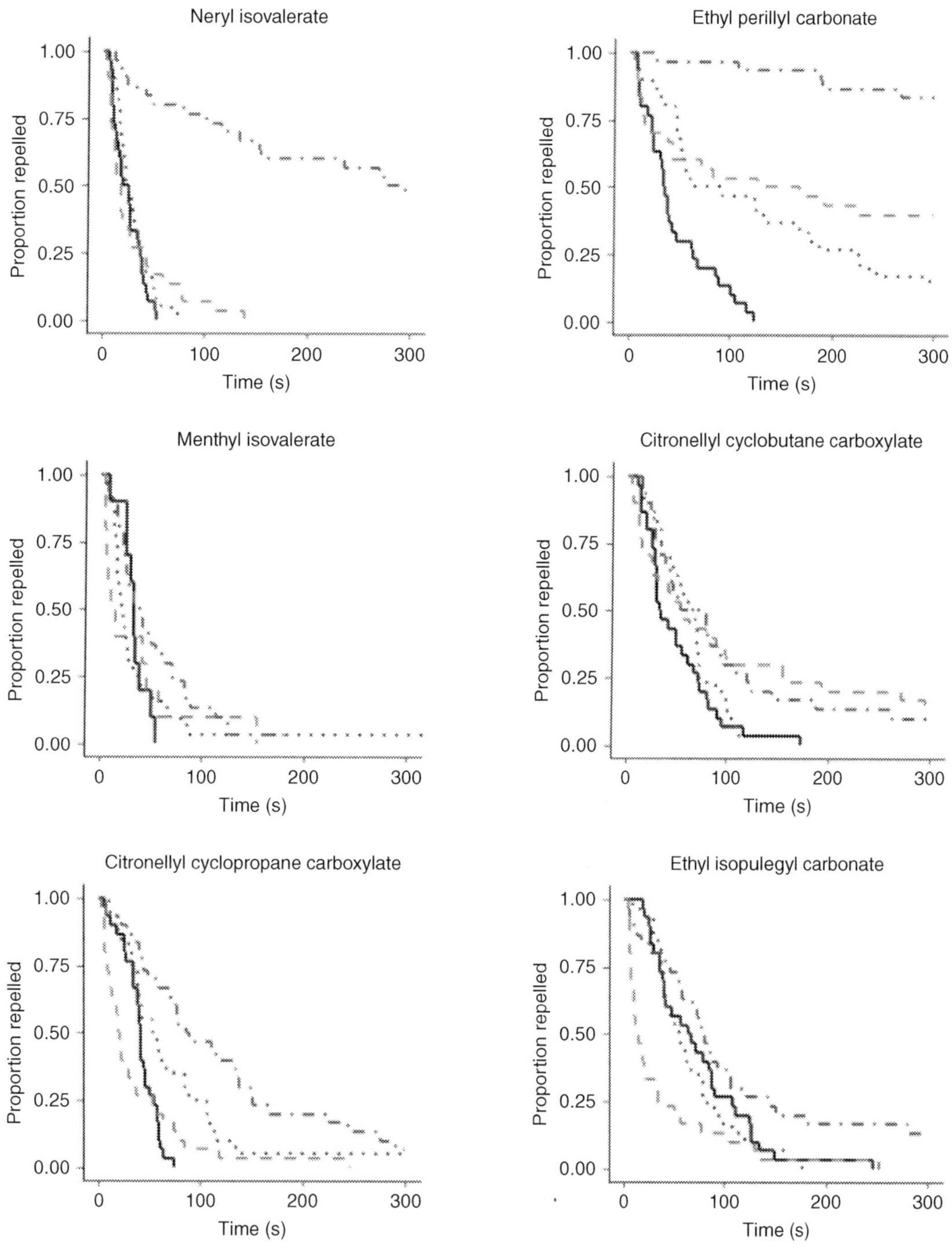
Neryl isovalerate
Ethyl perillyl carbonate
Menthyl isovalerate
Citronellyl cyclobutane carboxylate
Citronellyl cyclopropane carboxylate
Ethyl isopulegyl carbonate
Proportion repelled
1.00
0.75
0.50
0.25
0.00
Time (s)
0
100
200
300

(continued)

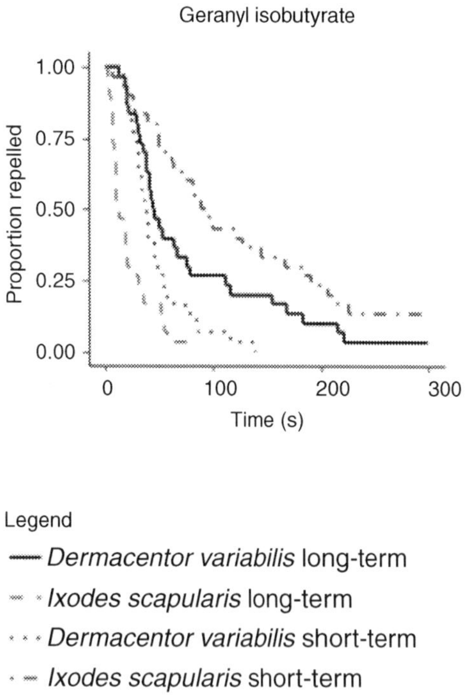

The natural terpene alcohols (geraniol, nerol, citronellol, and perillyl alcohol) exhibited strong repellency against *I. scapularis* in the short-term tests, with four of the top 6 RT_{50} values. This trend was not observed with the *D. variabilis* ticks or into the long-term testing with either species. In the short-term tests with *D. variabilis*, the terpene alcohols had greater RT_{50} values than DEET, but only nerol was significantly greater (Table 3.4).

The natural terpene diol, PMD, did not follow the same trend as the alcohols. PMD performed relatively well in the long-term tests (seventh highest RT_{50} for *D. variabilis* and third highest for *I. scapularis*). Among the *D. variabilis* ticks, PMD showed greater repellency in the long-term tests than in the short-term tests, which was uncommon.

At the low end, the farnesyl derivative biorationals, farnesyl pivalate, and farnesyl isovalerate were weak among all treatment combinations. They were, however, still significantly more effective than their respective control treatments.

3.4 Discussion

3.4.1 Tick repellent treatments

The repellents generally showed greater efficacy during the short-term tests than after they had been aged for 5 hours, which matched our expectations. DEET, which is known for its longevity as a repellent in the field (Bissinger and Roe, 2010), held up in the long-term tests, which agrees with what is noted in other studies. The fact that the acetone control times were different between tick species was unexpected and shows that this bioassay is not strong for making direct comparisons between different species. On the other hand, because the acetone controls were significantly distinct from any compound in their respective treatment groups indicates that the bioassay is sensitive enough to resolve small differences.

The natural terpene alcohols tested were very effective at the chosen concentrations against *I. scapularis* ticks in the short-term, but less against *D. variabilis* ticks. Several of the biorational compounds were of equal or greater effect even in the short-term experiments. In the long-term tests, the terpene alcohols were basically ineffective. This is likely due to the high volatility of the compounds, leaving little compound remaining after 5 hours for the ticks to encounter. Repellents using these small terpene molecules in the field need to address this longevity issue using either greater initial concentrations or slower-release formulations.

The biorational molecules offered some of the greatest repellency. Many were significantly more repellent than DEET at both time points and for both species. DEET, however, is used at very high concentrations in current commercial products (Jensenius et al., 2005) and would likely be as effective if tested at those concentrations. The efficacy of the biorational molecules

TABLE 3.3 Ranking of treatments from most to least repellent.

Ticks				Mosquitoes	
I. scapularis short-term	*I. scapularis* long-term	*D. variabilis* short-term	*D. variabilis* long-term	*Cx. pipiens* short-term	*Cx. pipiens* long-term
Ethyl Perillyl Carbonate	Ethyl Perillyl Carbonate	Ethyl Perillyl Carbonate	Geranyl Isobutyrate	Nerol	Ethyl Perillyl Carbonate
Geraniol	Cit. Cybut. Carboxylate	Cit. Cyprop. Carboxylate	Ethyl Iosopulegyl Carbonate	Perillyl Alcohol	DEET
Citronellol	p-Menthane-3,8-diol	Ethyl Iosopulegyl Carbonate	Cit. Cybut. Carboxylate	Neryl Isovalerate	Cit. Cyprop. Carboxylate
Nerol	DEET	Nerol	Ethyl Perillyl Carbonate	Geraniol	Cit. Cybut. Carboxylate
Neryl Isovalerate	Citronellyl Pivalate	Cit. Cybut. Carboxylate	Cit. Cyprop. Carboxylate	Farnesyl Isovalerate	Citronellol
Perillyl Alcohol	Cit. Cyprop. Carboxylate	2-Undecanone	p-Menthane-3,8-diol	Geranyl Isobutyrate	Menthyl Isovalerate
2-Undecanone	Ethyl Iosopulegyl Carbonate	Geranyl Isobutyrate	Menthyl Isovalerate	Farnesyl Pivalate	Citronellyl Pivalate
p-Menthane-3,8-diol	Citronellyl Myristate	Geraniol	Neryl Isovalerate	Ethyl Citronellyl Carbonate	2-Undecanone
Citronellyl Pivalate	Menthyl Isovalerate	Menthyl Isovalerate	Perillyl Alcohol	Cit. Cyprop. Carboxylate	
Geranyl Isobutyrate	Neryl Isovalerate	Neryl Isovalerate	Geraniol	Citronellyl Pivalate	
Cit. Cyprop. Carboxylate	Citronellol	Citronellol	DEET	DEET	
Ethyl Iosopulegyl Carbonate	Geranyl Isobutyrate	Citronellyl Pivalate	Citronellyl Pivalate	Ethyl Perillyl Carbonate	
DEET	2-Undecanone	Perillyl Alcohol	2-Undecanone	Citronellol	
Cit. Cybut. Carboxylate	Farnesyl Isovalerate	Farnesyl Pivalate	Citronellol	Cit. Cybut. Carboxylate	
Farnesyl Isovalerate	Nerol	DEET	Nerol	Citronellyl Myristate	
Menthyl Isovalerate	Perillyl Alcohol	Farnesyl Isovalerate	Citronellyl Myristate	Menthyl Isovalerate	
Farnesyl Pivalate	Farnesyl Pivalate	p-Menthane-3,8-diol	Farnesyl Isovalerate		
Citronellyl Myristate	Geraniol	Citronellyl Myristate	Farnesyl Pivalate		

Comparison of different treatment compounds separated by test type and by arthropod. Ranking order for ticks is determined by the median repellent time. Ranking order for mosquitoes is determined by the cumulative ranking across all of the time points within a test. Cit. Cyprop. Carboxylate is citronellyl cyclopropane carboxylate. Cit. Cybut. Carboxylate is citronellyl cyclobutane carboxylate.

TABLE 3.4 Median repellent times (RT_{50}) and pairwise significance for tick repellency tests.

	D. variabilis Short-term		*I. scapularis* Short-term		*D. variabilis* Long-term		*I. scapularis* Long-term	
Compound	**Median (s)**	**Significance**	**Median (s)**	**Significance**	**Median (s)**	**Significance**	**Median (s)**	**Significance**
Acetone	6.5	A	10.4	A	5.9	A	9.1	A
Farnesyl Pivalate	18.0	CD	47.4	BC	9.3	B	12.9	ABC
Farnesyl Isovalerate	14.7	BC	75.9	DE	11.6	BC	16.2	ADE
Citronellyl Myristate	10.4	B	26.5	B	12.0	B	36.9	EH
Nerol	63.0	HJK	210.4	HI	12.4	B	14.0	AC
Citronellol	26.2	DEFG	214.6	HI	14.8	BD	23.7	ADEF
2-Undecanone	57.5	JK	154.2	FH	17.1	CD	17.2	AE
Citronellyl Pivalate	21.5	CEF	126.4	F	17.7	CDE	43.7	FGH
DEET	17.9	CD	100.4	DF	19.5	DE	68.2	GI
Geraniol	39.9	EHI	232.0	I	20.3	CDE	11.5	A
Perillyl Alcohol	19.6	CE	167.3	FH	22.1	DF	13.6	AC
Neryl Isovalerate	28.9	GH	206.1	GHI	24.5	EF	29.6	DEF
Menthyl Isovalerate	38.0	FHI	52.3	CD	28.0	FH	35.3	AEGH
Citronellyl Cyclopropane Carboxylate	72.6	JKL	151.0	FG	31.6	FG	80.6	HI
Ethyl Perillyl Carbonate	126.6	L	118.5	EF	40.9	GH	37.9	DGH
Citronellyl Cyclobutane Carboxylate	61.2	JK	275.8	J	45.2	GH	156.3	J
Ethyl Iosopugenyl Carbonate	66.7	K	95.6	EF	49.6	HI	102.8	IJ
Geranyl Isobutyrate	46.6	IJ	108.7	EF	74.8	J	37.7	BEGH
p-Menthane-3,8-diol	11.7	B	124.0	F	79.5	IJ	22.1	CDE

Pairwise significance is constrained within tick species and test duration type.

at concentrations where DEET is less effective supports their further investigation to be developed into future arthropod repellents. The most effective biorationals appeared to be the carbonate esters. Although there were differences between the species, *I. scapularis* ticks were less sensitive, relatively, to the ethyl isopulegyl carbonate than *D. variabilis* ticks. Generally, the biorational compounds showed good repellency at both the long-term and short-term tests, and they performed better than the terpenes or the current market active ingredients.

3.4.2 Mosquito repellent treatments

The spatial repellency (static-air) chamber bioassay used with the mosquitoes has been shown to primarily produce data on spatial repellency rather than contact repellency (Paluch et al., 2009). Notable examples are the results of the current market products, DEET and citronellol. Both active ingredients had low percentage repellency for the first hour of the assay (both short-term and long-term), but reached greater percentages later on (Tables 3.1 and 3.2). This slow action in the spatial repellency chambers suggests that they volatilize slowly and take a long time to create a concentration gradient along with the static-air chamber (Martin et al., 2013). This does agree with the fact that DEET is more often used as a contact arthropod repellent, while citronellol is often used as a spatial arthropod repellent, sometimes with an emission device such as a fan or heat source.

Other monoterpenoids were ranked higher in the short-term assay (Table 3.3). These terpenoids may volatilize more readily which would account for their stronger percentages in the early time points of the assays. Previous explorations of monoterpenoids in this assay suggested that they do not perform as well in the long-term assay, which is corroborated by the results reported here for citronellol (Norris and Coats, 2017).

Examining the biorational compounds in the short-term tests, ethyl citronellyl carbonate, ethyl perillyl carbonate, neryl isovalerate, citronellyl cyclopropane carboxylate, and citronellyl pivalate all had an average percentage repellency of at least 90% for every time point at 90 minutes and after. Some of these, such as citronellyl pivalate and citronellyl cyclopropane carboxylate provided very low percentage repellency at the beginning of the tests, more similar to the monoterpenoid, citronellol. This may indicate that those two biorational esters would best be used as contact arthropod repellents, deployed in emission devices, or paired with faster acting spatial repellents. Geranyl isobutyrate provided very high repellency at the beginning of the short-term assay, but it generated data below an average of 90% repellency for the later time points. Geranyl isobutyrate could be a good candidate to pair with some of the slower biorational molecules to ensure full coverage in a potential mosquito repellent.

The long-term repellency had fewer treatments tested, but of those tested, only ethyl perillyl carbonate and citronellyl cyclopropane carboxylate were good repellents (all time points had average repellent percentages of over 60%). Interestingly, two very similar molecules, citronellyl cyclopropane carboxylate, and citronellyl cyclobutane carboxylate had very different repellent performance, especially in the long-term assay. This difference shows that small changes to an active ingredient's structure can induce significant changes in the repellency of the compound.

3.4.3 Comparison of the results for ticks and mosquitoes

Rankings in Table 3.3 were generated to provide a generalized way to compare results between the open Petri dish tick repellency assay and the closed-chamber spatial repellency mosquito assay. The mosquito repellency assay does not give a single aggregated data point for any one treatment but instead gives several averages over time. While this is important to be able to characterize the active ingredients as spatial arthropod repellents, it does make it difficult to suggest comparisons with different assays. We were unable to perform statistical comparisons between the two experimental bioassays; however, some of the trends seemed clear.

The monoterpenoids generally showed good repellency in the short-term tests for both ticks and mosquitoes. The monoterpenoids were included in the tick long-term repellency testing to determine if their efficacy was diminished over time as had been observed with mosquitoes (Norris and Coats, 2017; Tavares et al., 2018). Citronellol was an example of this loss of efficacy in the long-term tests with the mosquito results. For ticks, RT_{50} values for nerol, geraniol, and citronellol were all lower in the long-term testing than in the short-term (Table 3.4). This trend was particularly pronounced with *I. scapularis* ticks, where the short-term RT_{50} values were quite large. Perillyl alcohol did not exhibit quite as large a difference. Unexpectedly, PMD had a much greater RT_{50} in long-term tests than the short-term with the *D. variabilis* ticks, although, the trend was reversed with the *I. scapularis* ticks. The duration of repellent activity is an important consideration when evaluating natural products that contain monoterpenoids.

There were individual biorational esters and carbonates that had high ranking in both mosquito and tick tests of repellency, but with some that appeared consistently lower on the rankings. However, several of the biorational molecules have inconsistent ranking between species, class and testing type (Table 3.3). Ethyl perillyl carbonate is the most consistently strong repellent across arthropod species and testing duration. Citronellyl myristate had consistently poor rankings in the different repellent tests. The number of citronellyl esters tested allowed for some comparison by the nonterpene moiety of the ester. The two pivalates, citronellyl and farnesyl, had poor repellent rankings. The isovalerates and isobutyrates, of which there were four, were often in the middle of the rankings. The cyclobutane and cyclopropane carboxylates of citronellol were included in the best repellents. For the mosquitoes, the cyclopropane carboxylate was better than cyclobutane carboxylate, but the ticks did not always reflect this trend.

The carbonate esters were the highest-ranked repellents suggesting that this chemical group contains good candidates for further exploration. The carbonate esters, however, took 90 minutes to achieve 90% repellency in the short-term mosquito assays. The short-term tick repellency tests were still strong for the carbonate esters. This may indicate that the carbonate esters tested can exhibit contact repellency in the short-term and spatial repellency in the long-term. Such a repellent profile could be a good candidate for formulation in a lotion for contact repellency or used in an emission device to be used as a spatial arthropod repellent.

3.5 Conclusion

No current commercial active ingredient (DEET, 2-undecanone, PMD, and citronellol) appeared as the best-ranking repellent in any of our experiments (Table 3.3). The commercial repellents had strengths in some of the tests either at certain time points or species. The three natural commercial active ingredients (2-undecanone, PMD, and citronellol) had a better repellent character in the short-term, and citronellol was ranked better in tests with ticks than those with mosquitoes. The monoterpenoids matched expectations to exhibit high levels of repellency in the short-term tests and lower levels in the long-term tests.

Our goal of designing a better repellent consisting of a biorational ester or carbonate was achieved. Ethyl perillyl carbonate was ranked as the best repellent in most of the species-test type combinations. The biorational compounds did not outperform the other mosquito repellents in the short-term tests, but for this test type, many of the compounds tested so high that it is difficult to show improvement. The biorational compounds did show high efficacy with the ticks in both the long and short-term, which is promising for a potential new arthropod repellent active ingredient.

The differences between arthropods and even between the two species of ticks show the need for further understanding the mechanisms behind chemical arthropod repellents. Species-specific targeting of repellents could be of use in particular situations. However, it is more likely that an active ingredient that is repellent to the greatest variety of biting arthropods is the most practical candidate for commercialization. Further investigation of biorational molecules of the type tested here could reveal even more repellent compounds to a greater variety of species and should be considered when bringing new arthropod repellent active ingredients to market.

Testing in the future should include testing on human subjects, once the safety of these biorational compounds is confirmed. An alternate method to test the length of time that a repellent remains effective is to record the time until the first bite of a treated individual. This time-until-first-bite testing should give results that are more relevant to the field as they incorporate the human host cues that could affect the behavior of the biting arthropod when a repellent is deployed.

References

Bissinger, B.W., Roe, R.M., 2010. Tick repellents: past, present, and future. Pestic Biochem. Physiol. 96 (2), 63–79.

Coibion, O., Gorodnichenko, Y., Weber, M., 2020. The cost of the covid-19 crisis: lockdowns, macroeconomic expectations, and consumer spending. No. w27141, NBER Working Paper Series.

Haddi, K., Turchen, L.M., Viteri Jumbo, L.O., Guedes, R.N., Pereira, E.J., Aguiar, R.W., Oliveira, E.E., 2020. Rethinking biorational insecticides for pest management: unintended effects and consequences. Pest Manag. Sci. 76 (7), 2286–2293.

Horowitz, A.R., Ellsworth, P.C., Ishaaya, I., 2009. Biorational pest control—an overview. Biorational Control of Arthropod Pests. Springer, The Netherlands, pp. 1–20.

Jamil, Z., Waheed, Y., Durrani, T.Z., 2016. Zika virus, a pathway to new challenges. Asian Pac. J. Trop. Med. 9 (7), 626–629.

Jensenius, M., Pretorius, A.M., Clarke, F., Myrvang, B., 2005. Repellent efficacy of four commercial DEET lotions against *Amblyomma hebraeum* (Acari: Ixodidae), the principal vector of *Rickettsia africae* in southern Africa. Trans. R. Soc. Trop. Med. Hyg 99 (9), 708–711.

Klimavicz, J.S., Corona, C.L., Norris, E.J., Coats, J.R., 2018. Monoterpenoid isovalerate esters as long-lasting spatial mosquito repellentsAdvances in the Biorational Control of Medical and Veterinary Pests. American Chemical Society, Washington, DC, USA, pp. 205–217.

Maia, M.F., Moore, S.J., 2011. Plant-based insect repellents: a review of their efficacy, development and testing. Malar. J. 10 (1), 1–15.

Martin, N.J., Smith, P.A., Achee, N.L., DeLong, G.T., 2013. Determining airborne concentrations of spatial repellent chemicals in mosquito behavior assay systems. PloS One 8 (8), e71884.

Niesobecki, S., Hansen, A., Rutz, H., Mehta, S., Feldman, K., Meek, J., Niccolai, L., Hook, S., Hinckley, A., 2019. Knowledge, attitudes, and behaviors regarding tick-borne disease prevention in endemic areas. Ticks Tick Borne Dis. 10 (6), 101264.

Norris, E.J., Coats, J.R., 2017. Current and future repellent technologies: the potential of spatial repellents and their place in mosquito-borne disease control. Int. J. Environ. Res. Public Health. 14 (2), 124.

Oliver, J.D., Bennett, S.W., Beati, L., Bartholomay, L.C., 2017. Range expansion and increasing *Borrelia burgdorferi* infection of the tick *Ixodes scapularis* (Acari: Ixodidae) in Iowa, 1990–2013. J. Med. Entomol. 54 (6), 1727–1734.

Paluch, G., Grodnitzky, J., Bartholomay, L., Coats, J., 2009. Quantitative structure—activity relationship of botanical sesquiterpenes: spatial and contact repellency to the yellow fever mosquito, *Aedes aegypti*. J. Agric. Food. Chem. 57 (16), 7618–7625.

Peixoto, H.M., Romero, G.A.S., de Araújo, W.N., de Oliveira, M.R.F., 2019. Guillain–Barré syndrome associated with Zika virus infection in Brazil: a cost-of-illness study. Trans. R. Soc. Trop. Med. Hyg. 113 (5), 252–258.

Rochlin, I., Ninivaggi, D.V., Benach, J.L., 2019. Malaria and Lyme disease-the largest vector-borne US epidemics in the last 100 years: success and failure of public health. BMC Public Health 19 (1), 1–11.

Tavares, M., da Silva, M.R.M., de Siqueira, L.B.D.O., Rodrigues, R.A.S., Bodjolle-d'Almeida, L., Dos Santos, E.P., Ricci-Júnior, E., 2018. Trends in insect repellent formulations: a review. Int. J. Pharm. 539 (1-2), 190–209.

Wong, C., Crystal, K., Coats, J., 2021. Three molecules found in rosemary or nutmeg essential oils repel ticks (*Dermacentor variabilis*) more effectively than DEET in a no-human assay. Pest Manag. Sci. 77 (3), 1348–1354.

Yang, L., Norris, E.J., Jiang, S., Bernier, U.R., Linthicum, K.J., Bloomquist, J.R., 2020. Reduced effectiveness of repellents in a pyrethroid-resistant strain of *Aedes aegypti* (Diptera: culicidae) and its correlation with olfactory sensitivity. Pest Manag. Sci. 76 (1), 118–124.

CHAPTER

4

Evaluating techniques and efficacy of arthropod repellents against ticks

Muhammad Farooq, Rui-De Xue, Steven T. Peper, Whitney A. Qualls

Anastasia Mosquito Control District, St. Augustine, Florida, United States

4.1 Introduction

Globally, humans are at an increased risk of tick-borne diseases and are not adequately prepared to respond to their associated public health threats (Torres and Carey, 2004; De la Fuente and Estrada-Pena, 2012). The United States reported that tick-borne cases have more than doubled from 2004 to 2018 and are currently at an all-time high (Petersen et al., 2019). Tick-borne diseases such as Lyme disease and Rocky Mountain spotted fever can cause serious illness or even death. Lyme disease accounts for 82% of all tick-borne diseases reported in the United States (Rosenberg et al., 2018). In addition to Lyme disease, other tick-borne diseases such as babesiosis and anaplasmosis/ehrlichiosis have become established and are commonly reported in the United States. Based on a report by the US Center for Disease Control and Prevention (CDC), during the last 15 years, the number of vector-borne disease cases has increased dramatically, due to the expanding ranges of vectors and the increasing number of emerging pathogens (Centers for Disease Control and Prevention, 2020a) caused by global warming, climate and environmental changes (Bezirtzoglou et al., 2011), transportation, migration, and globalization. For example, the Asian longhorned tick, *Haemaphysalis longicornis,* causes severe illness and death in humans in other parts of the world as it transmits various pathogens and was identified for the first time in the United States in 2017 and has since spread to 14 states. At least 16 different hosts, including wildlife, livestock, pets, and people have been detected (Centers for Disease Control and Prevention, 2020a). Tick-borne pathogens that have been newly discovered or first reported in the United States since 2004 include *Rickettsia parkeri,* Bourbon virus (likely tick), *Rickettsia*

Advances in Arthropod Repellents.
DOI: https://doi.org/10.1016/B978-0-323-85411-5.00003-0

species 364D, *Ehrlichia muris eauclairensis*, Heartland virus, *Borrelia miyamotoi*, and *B. mayonii*. Personal protection and application of arthropod repellents are a critical measure for the prevention and control of tick bites and tick-borne diseases (Piesman and Eisen, 2008).

This increase in tick-borne diseases identifies a growing public health problem that will require improvements in technology for surveillance, pathogen detection, prevention, and control. Arthropod repellents are an essential strategy for preventing the transmission of vector-borne diseases. Most repellents have been developed against mosquitoes, however, as noted, the dramatic increase in the prevalence and transmission of tick-borne diseases (Madison-Antenucci et al., 2020) has identified a need to evaluate and develop arthropod repellents specific for the prevention of tick bites and tick-borne diseases (Bissinger and Roe, 2010). This chapter describes the evaluation techniques for arthropod repellents against ticks and the effectiveness of available commercial arthropod repellents against ticks.

4.2 A brief history of arthropod repellents used for prevention of tick bites and the transmission of tick-borne diseases

During World War II, thousands of compounds were evaluated for the repellency against mosquitoes and chiggers (Morton et al., 1947; Moore and Debboun, 2007), but little attention was given to tick repellents (Brennan, 1947). However, in the late 1940s and early 1950s, many arthropod repellent compounds that were applied to clothing for use against ticks were identified (Dethier, 1956). Dimethyl phthalate (DMP) is a broad-spectrum repellent that was widely used from the 1940s to 1980s before being phased out. The efficacy of DMP varied by tick species and life stage. Another repellent, Indalone, was used by the military to treat their uniforms. The repellency of Indalone also varied by tick species and life stage. DMP, Indalone, and a compound known as Rutgers 612 were mixed together to form the repellent compound, 6-6-2 which combined the different modes of action of these repellents to extend repellency and increase efficacy. Results varied again by species and life stage. All of these compounds were phased out with the development of N,N-diethyl-3-methylbenzamide (DEET) (Strickman, 2007).

Few advancements in tick repellent development have been made since the identification and registration of DEET for arthropod bite protection. Over 20,000 compounds were screened for repellency against arthropods with none equaling the broad-spectrum range of protection and duration of repellency of DEET (Gupta and Bhattacharjee, 2007). DEET is the most commonly used active ingredient (AI) in commercially available tick repellents used on human skin and is effective against several tick species (Bissinger and Roe, 2010). For decades, the common message for preventing tick exposure has been wearing long sleeves and pants treated with permethrin or picaridin, closing and sealing openings in clothing, and using DEET. Besides DEET, there have been other modern synthetic arthropod repellents for use on human skin that were evaluated against ticks. These included N,N-diethyl-2-phenyl-acetamide (DEPA) (Kumar et al., 1992) and piperidine compounds (McGovern et al., 1978; Solberg et al., 1995; Salafsky et al., 2000; Carroll et al., 2005; Carroll et al., 2007), which includes picaridin (Frances et al., 2007; Carroll et al., 2008). Currently, the Centers for Disease Control and Prevention (2020b) recommends using arthropod repellents containing >20% DEET, 2-(2-hydroxyethyl)-1-piperinecarboxylic acid 1-methylpropyl ester (Picaridin), or ethyl butylacetyl-aminoproprionate (IR3535) on exposed skin for tick bite and disease prevention. It should be noted that an ideal arthropod repellent should provide protection against a broad-spectrum of blood-feeding arthropods for at least 8 h, be nontoxic, nonirritating, odorless, and nongreasy (Fradin, 1998).

Some of the common AIs of topical arthropod repellents used against ticks include DEET, Picaridin, IR3535, DEPA, Oil of Lemon eucalyptus, p-Menthane-3,8-diol (PMD), and 2-Undecanone (Strickman, 2015; Foster et al., 2020). Scott et al. (2016) tested seven commercial arthropod repellent products in the laboratory and also showed that different concentrations of botanical resources, such as soybean, geranium, and lemongrass oils can provide minimal protection against ticks.

Some of the common AIs for spatial arthropod repellents against ticks include Metofluthrin and d-cis-/trans-allethrin which are common compounds found in commercial products, such as OFF! Clip-On and Thermacell mosquito repellents. Both of these repellent products have been tested against ticks (Xue et al., 2016; Bibbs and Xue, 2016; Farooq et al., 2021). These compounds resulted in quick knockdown and mortality of ticks.

4.3 Evaluations of repellency

Evaluation as defined by Cambridge Dictionary is the process of judging or calculating the quality, importance, amount, or value of something. Merriam Webster dictionary describes it as the determination of the value, nature, character, or quality of something or someone. In another way, evaluation can be described as a systematic determination of a subject's merit, worth and significance, using criteria governed by a set of standards. Evaluations of spatial arthropod repellents are an integral part of the developmental process and are required by regulatory agencies. Many times, these evaluations are to be carried out by nonrelated agencies or establishments.

4.3.1 Evaluation and environment

Environment at the time of evaluations, laboratory or field, are an important factor in the evaluation. The effect of the environment on the efficacy of repellent materials or on the effectiveness of delivery systems may not be known; however, it is important for other considerations. The evaluations should be conducted in the environment under which ticks exert their normal behavior, specifically in the absence of any arthropod repellents, such as in a control. Solberg et al. (1995) expressed that field testing of topical repellents on humans is preferable to laboratory testing as it accounts for ambient environmental effects on the repellents, test subjects, and ticks, as well as mechanical effects, such as abrasion of repellent-treated surfaces against vegetation or clothing, washing off of repellents by wet vegetation, etc. The arthropod repellents should be evaluated under conditions specified on the label under which an arthropod repellent is recommended to be used, if any, and under the suitable conditions for the test arthropod and subject, such as atmospheric temperature, relative humidity, and wind speed under which the arthropods and subjects perform normally. For laboratory tests, the suitable conditions should be maintained. For field evaluations, the time and space should be selected to match suitable environmental conditions.

4.3.2 Arthropod repellents

Repellents, as the word indicates, can be described as something which deflects some other object from the third object. In terms of fabrics, these are the substances which when applied to the fabric prevent penetration of any liquids through the fabric and turn the liquids away. It can also be explained by whiter colors deflecting or repelling light while darker colors absorbing lights. In the field of health prevention, arthropod repellents are substances that deflect arthropods from approaching to or attaching on humans or animals.

Spatial arthropod repellents are used to prevent pest arthropods from approaching their targets. The spatial repellent acts as a barrier between the individual, termed as target, and the vector. The extent to which a repellent prevents the vector from coming in contact with the target is termed its efficacy or effectiveness. Topical arthropod repellents, on the other hand, are used to prevent arthropods from attaching to the target once they reach their target. Spatial and topical repellents are evaluated to assess their ability to keep the arthropods away or to make them leave the target, respectively. The methods, procedures, and equipment used for evaluation of the two types of arthropod repellents to assess their abilities will be discussed later in two different sections.

4.3.3 Tick repellency

For ticks, repellency is characterized by an irritant effect, causing the tick to move away from the host or to avoid a bite or feeding and leading it to fall off soon after contact with its host (Halos et al., 2012). According to the European Medicine Agency guidelines, a repellent effect against ticks indicates that no ticks will attach to a host and one already on the host will leave the host soon after treatment (Halos et al., 2012). Stimuli to ticks that induce ambush and hunting behavior include carbon dioxide, butyric and lactic acid, ammonia (from animal wastes), heat, shadows, and vibrations (Sonenshine et al., 2002). Studies have shown that susceptibility to an arthropod repellent varies between tick species (Schreck et al., 1995) and life stages (Evans et al., 1990; Solberg et al., 1995).

4.4 Evaluation methods for spatial arthropod repellents

4.4.1 Objectives of evaluation

The evaluations are generally aimed at assessing the efficacy of the AI used in the arthropod repellent system, the apparatus to deliver the substance or combining both.

4.4.2 Evaluation methods

Ticks are attracted to human or animal hosts through the detection of different stimuli/attractions. Once on the host, ticks find a suitable environment where they can hide and obtain a blood meal at the same time. Spatial arthropod repellents, as defined earlier, block their movement from their habitat to the body of the host by blocking the stimuli or generating vapors around the host which are disliked by the ticks and thus, keeping them away from the host.

The foundation of evaluation of spatial repellent systems is the delivery of the vapors generated by the repellent substance to the ticks and recording their behavioral response to those vapors. Any method to evaluate the efficacy of a repellent substance or delivery device must ensure that vapors generated by the repellent system reach the ticks whose response is to be studied. During the selection of evaluation methods, consideration should be given to the mechanism used in the repellent system to keep the arthropods away from the target. For testing the material only, methodology needs to allow the material to release vapors and direct the vapors to a certain direction or into an area where arthropods may have the option to move into. These are required for an evaluation methodology to be successful. An environment needed for technical materials or formulated products to release vapors should be provided for the evaluation of these types of materials. For example, a repellent substance should not be evaluated at a temperature that is outside the recommended range for use of the substance. For evaluation of delivery systems, the working conditions for the delivery system should be provided for the system to perform accordingly.

Dautel (2004) grouped the methods available for testing putative tick repellents into three broad categories: (1) those that are performed in the absence of hosts or host stimuli, (2) performed in the presence of host stimuli, and (3) performed using a live host.

4.4.3 Laboratory evaluation methods

Jaenson et al. (2006) used a 50-mL Falcon vial (115 by 29 mm centrifugal tube) made of transparent plastic to test the response of single unfed tick nymphs to spatial arthropod repellents. Using this method, one drop (50 μL) of the test substance is applied with pipettes to a cotton cloth. After drying, the cloth is attached with a rubber band to the open end of a Falcon vial. The nymph is placed inside the vial before attaching the cloth. For control, untreated cloths should be used. To simulate host stimuli to attract the nymphs, the observer holds his/her one palm for the blank control and the other for the test substance tight against the outside surface of the cloth for each test period as a stimulus for nymphs. The same palm should not be used for the control or test substance to prevent contamination of the control palm with the test substance. In each replicate, a nymph is first tested with the untreated control for 5 min and then with the test substance for 5 min. To test each substance or dilution, 50–100 replications are used. The position of the nymph, i.e., its location above the tube's bottom is recorded at every minute. Ticks that cling to the cloth 5 min after the start of the test may be counted as attracted, although, Jaenson et al. (2006) considered a tick to be regarded as "attracted" to the cloth, if it has detached all of its legs from the vial's surface. The remaining ticks in the vial may be recorded as "repelled."

In this laboratory experiment, the percentage of repellency was calculated by the following formula:

$$\text{Repellency } (\%) = (\text{Tac} - \text{Tat}) \times 100 / \text{Tac} \quad (4.1)$$

where Tac, ticks attracted in the control test; Tat, ticks attracted in the treatment test.

Jaenson et al. (2006) using this method, evaluated MyggA Natural (Bioglan, Lund, Sweden), a commercially available repellent against blood-feeding arthropods. The repellent contains 30% of lemon-scented eucalyptus oil with a minimum of 50% PMD. The components of MyggA Natural, included the essential oils of lavender, *Lavandula angustifolia* Mill. (Lamiaceae), and geranium, *Pelargonium graveolens* L'Her (Geraniaceae) and were evaluated separately. The results indicated 100% repellency against host-seeking nymphs of the castor bean tick (*Ixodes ricinus* L.) by MyggA Natural. Lavender oil and geranium oil at 30% concentration also provided 100% repellency.

Bibbs and Xue (2016) used 1.2 m long, 3.8 cm interior diameter cylindrical clear plastic tubes as assay tracks. Both ends of the tube are covered with fine mesh to allow airflow. A 118-mL-volume plastic container is affixed to one end as an acclimation chamber with a barrier between the tube and the container. One nymph is transferred into the acclimation chamber per track and allowed 24 h to acclimate to the assay conditions. Ten tracks are aligned in a circle, as shown in Fig. 4.1. A repellent device is suspended 1 m

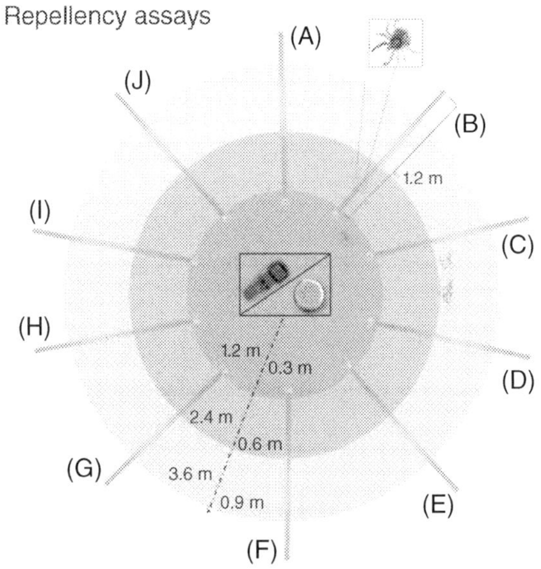

FIG. 4.1 Layout of track tubes in relation to devices to be tested (Bibbs and Xue, 2016).

above the ground or is placed on the ground in the middle. The distance between tube and devise is selected based on the effectiveness range of the device. The distance between track tubes and device can be varied to cover the dispersion distance of the repellent from the device. Bibbs and Xue (2016) used a distance from 0.9 to 4.5 m based on the effective range of the devices tested. To start the assay, the barrier between the acclimation chamber and tube track is removed and a 60 min wait time is allowed for the repellent to dissipate. Then, the distance traveled by the tick in the track is recorded. Controls are conducted in the experimental conditions in absence of the treatment device. The distance traveled by ticks away from the device is an indication of repellency and is proportional to the repellency effect. The delay in start of their movement is also recorded. Longer delay indicates lesser repellency effects. Bibbs and Xue (2016) evaluated the repellency effect of OFF! Clip-On (Environmental Protection Agency Reg: 4822-542, S.C. Johnson & Son, Inc., Racine, Wisconsin) and ThermaCELL appliance (MRGJ06-00, Schawbel Corporation, Bedford, Massachusetts) to lone star tick (*Amblyomma americanum*) nymphs. The results indicated that nymphs were only interested in traveling away from the source of chemical when presented at 0.3 m away during OFF! Clip-on trials and 0.3 m and 1.2 m away during ThermaCELL trials.

Xue et al. (2016) described a method to evaluate the refill cartridge used as a spatial repellent in a laboratory cage bioassay. The cage is constructed from a plastic box (13 × 13 × 21.5 cm) and serves as an open airflow chamber (OAC) for choice evaluations (Fig. 4.2). A clear vinyl tube (3.5 cm outer diameter × 30 cm) can be connected to the OAC through a 3.5 cm diameter hole on the side of the chamber. The other end of the tube can be connected to a 1-L polystyrene disposable storage bottle through a 3.5 cm diameter hole in the center of the bottle lid. The base of the bottle is removed to allow the insertion of a volunteer's hand. For each test, a known number of ticks are placed into the vinyl tube, using feather tip forceps, and secured with fabric mesh held in place by rubber bands to ensure that no ticks escape the tube. Ticks are allowed 2 min to acclimate in the tube before the tube is attached to the OAC and polystyrene bottle. A volunteer then places one hand into the end of the polystyrene bottle and ticks are given 5 min to move to their final position. After 5 min, ticks located between the hand, on the mesh, and up to 15 cm from the hand are counted as "host seeking" or attracted. This first run with no refill cartridge is the control test. Ticks are then given

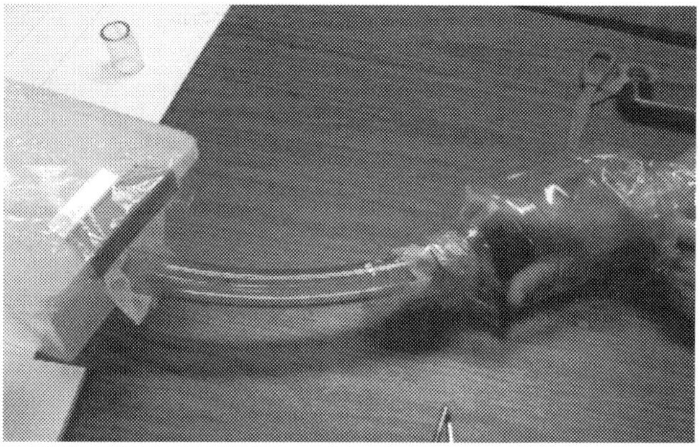

FIG. 4.2 Choice tube laboratory bioassay for evaluating response to a human hand with and without a spatial arthropod repellent (Xue et al., 2016).

a 2-min rest period and the assay is repeated with the volunteer holding a refill cartridge while placing his/her hand in the polystyrene chamber. If evaluating noncartridge type repellents, 1 mL can be put on filter paper and held in volunteer's hand. Ticks are allowed 5 min to locate the volunteer's hand. After 5 min, ticks located on the mesh next to the hand, and within 15 cm of the hand, are counted as attracted to the host. Using this method, Xue et al. (2016) evaluated the repellency of metofluthrin (OFF! Clip-On Mosquito Repellent, AI 31.2%, with fan off) to lone star ticks. Having the fan of the unit turned off allowed the evaluation of the AI as a passive spatial repellent. It was found that 57% of the ticks were significantly repelled from hands that held the OFF! Clip-On refill, compared with 27% of volunteer hands without it (control).

Spatial arthropod repellents can also be evaluated in a wind tunnel as used by Farooq et al. (2021). The tunnel is a suction-type wind tunnel that has a maximum air speed of 0.4 m/s. The testing section of the tunnel is 52 cm wide × 52 cm tall × 156 cm long and all sides are made of clear glass. The upwind side of the section is 47 cm bifurcation module with a screen separating it from the test section. The upwind of the bifurcation section is a filter module that prevents any dust particles from entering the wind tunnel. Downwind of the test section is a 26 cm long arthropod release module followed by a 52 cm long plenum which reduces to 10 cm diameter outlet, connected to an exhaust fan through a duct.

The wind tunnel length is divided into five sections about 30 cm long each. A white paper was placed under the test section with boundaries of each section marked for easy tracking of the ticks' position (Fig. 4.3). The test can be conducted with and without an attractant. Adult ticks are released on the downwind side and a number of ticks in each section are recorded 15 min after release. The test is conducted first with the attractant in the flo (if testing with attractant). Otherwise, the test is run with fresh air through the tunnel. This will also be considered as a control. In the second test, repellent will be placed on the upwind side, with or without attractant. The difference in number of ticks moving toward the upwind section will be an indication of repellence. D-allethrin vapor generated from a personal repellent device (Thermacell MR300) was evaluated for its effectiveness to repel adult ticks in the wind tunnel in the presence of an attractant.

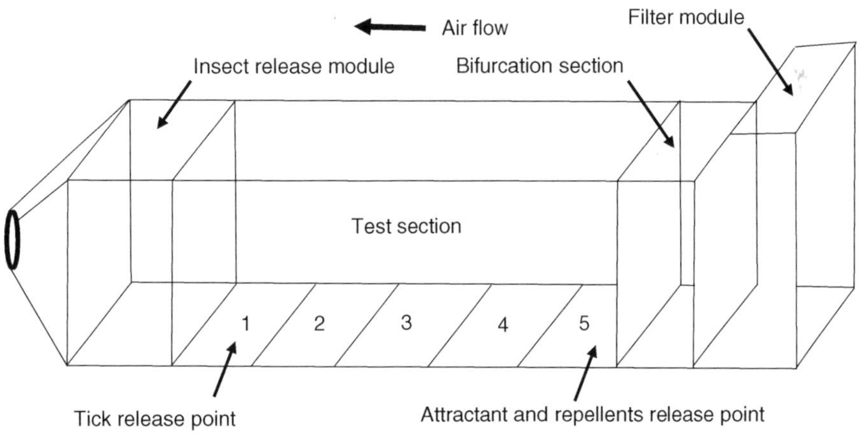

FIG. 4.3 Description of wind tunnel use as an olfactometer (Farooq et al., 2021).

First, the male and female tick's behavior was observed with BG lure cartridge and then by adding Thermacell repellent device after turning it on. The tick count indicated that 48.5% of the ticks moved toward the lure when it was alone compared to 30.8% when the repellent was placed with the lure, indicating the presence of a repellent effect.

Farooq et al. (2021) also evaluated spatial repellents in an olfactometer using a true-choice olfactometer at Anastasia Mosquito Control District, St. Augustine, Florida. The olfactometer consists of flow control valves, odor release chambers, choice chambers, and an acclimation chamber (Fig. 4.4). Flow control valves regulate clean and dry airflow to two odor release chambers (right and left) and to an acclimation chamber. Each of the two odor release chambers feeds to one each of the two choice chambers. The choice chambers are connected to two sides of the acclimation chamber and which, by simple rotation of the choice chambers, can be connected to or disconnected from the acclimation chamber. The odor release chamber may have an attractant on one side and repellent on the other side. Alternately, one chamber may have attractant or repellent and other empty letting fresh air onto the other side. This olfactometer, due to its size, can be used for adults and nymphs. Ticks are released in the acclimation chamber and the attractant or repellent vapors laden air, or fresh air moves through the choice chambers. To start repellency tests, passage between choice chambers and acclimation chamber is opened and ticks are given the choice to move to either side. After 10 min, number of ticks in both choice chambers and acclimation chamber are counted and recorded. All ticks are then removed from olfactometer before the next replication is started. Control tests are conducted in the same way in the absence of repellent in the odor release chamber. D-allethrin vapor generated from a personal arthropod repellent device (Thermacell MR300) was evaluated for its effectiveness to repel ticks in the olfactometer by providing them a choice between the repellent and fresh air. The effect on both adult ticks and nymphs was evaluated in this study. The results indicated strong repellency of d-allethrin vapor to adults but not to nymphs.

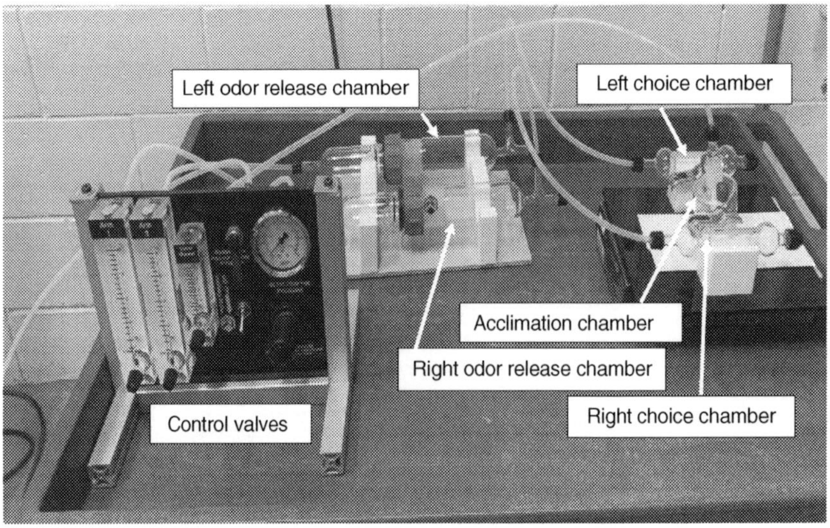

FIG. 4.4 True-choice olfactometer layout (photo taken by Farooq, 2020).

4.4.4 Semi-field and field evaluation methods

4.4.4.1 *Semi-field evaluation*

A test system for the semi-field evaluation of spatial repellents for ticks was developed at Anastasia Mosquito Control District (Fig. 4.5). It used a 135 cm × 89 cm pan covered with white paper sheet as a platform for ticks to move round. The sides of the pan have a strip of glue to keep the ticks from leaving the pan. Bamboo skewers are installed on a 5 cm x 10 cm woodblock to represent grass in the field and are placed in the middle of the pan. A structure built with PVC pipe to stand across the pan at 1.6 m above the pan to hold attractants and repellents at different heights above the skewers. The evaluation can be done in the presence of an attractant. The test starts with hanging the attractant on the frame, releasing ticks in the pan and letting them climb the skewers after smelling the attractant within 15 min. At this point, the number of ticks on skewers and in the pan are recorded. Once the ticks have climbed the skewers, the repellent is hung and the behavior of the ticks is observed. Ticks moving down the skewers will indicate the effectiveness of the repellent. Fifteen min later, the number of ticks on the skewers and in the pan is recorded.

FIG. 4.5 Layout for semifield testing of tick repellency (photo taken by Farooq, 2019).

4.4.4.2 *Field evaluation*

Xue et al. (2016) used a known tick-infested area to evaluate spatial repellents in the field. In their procedure, six human volunteers wore white clothing that fully covered their body from head to toe. Three volunteers served as treatments and three as control. Each treatment person had one repellent device attached to his/her body as recommended by the manufacturer and all sat on a chair 15 m away from the nearest volunteer. All stages of ticks that climbed on the volunteers clothing were collected by scotch tape by the end of the observation period (Fig. 4.6). All volunteers then relocated to a new testing site in the same area and switched treatments, (i.e., previous control volunteers switched to serve as the treatment and the previous treated volunteers now served as the control). In a variation, the volunteers walk at a pace covering 30 m for 15 min (Fig. 4.6). Each volunteer was positioned 15 m from the next volunteer. Ticks were removed from volunteers by the end of each time period. After the 15 min collection period, volunteers relocated to a new test site and switched their roles.

Field trials were conducted with volunteers either sitting or walking while wearing the repellent device OFF! Clip-On Mosquito Repellent, (AI 31.2% of metofluthrin), in areas infested with the lone star tick. Seated volunteers, with the fan running, repelled 89% of host-seeking ticks when compared with volunteers similarly placed without the fan on. In a second field trial, volunteers who walked slowly through the naturally infested area with the repellent fan running provided 28% protection from questing ticks, compared with similar volunteers without the fan on.

FIG. 4.6 Field testing of arthropod repellent against ticks showing the volunteer walking in the tick activity sites, and checking collected ticks on the scotch tape from clothing (photo taken by Xue, 2012).

4.5 Evaluation methods for topical arthropod repellents

4.5.1 Objectives of evaluation

The evaluations are generally aimed at assessing the efficacy of the AI of the arthropod repellent formulation as a product.

4.5.2 Evaluation methods

The topical repellents, unlike spatial repellents, are not developed just to keep ticks away, but to irritate ticks as they come into contact with the human body and prohibit their stay on the body. The evaluation methodology not only should assess their capacity to block ticks, but also their fate if they come in contact with the target.

4.5.3 Laboratory evaluation methods

4.5.3.1 *Fingertip bioassay*

An evaluation procedure using human finger has been described by Carroll et al. (2005). A small amount of repellent is applied to part of the finger in the middle and the boundaries of the treated area are marked with a fine-tipped pen and the solution is allowed to dry for 10 min. A small petri dish (9–10 cm diameter, 1 cm high) is glued to a larger petri dish (15–20 cm diameter, 1.5 cm high) and the intervening space is filled with water to form a moat. Ticks are released in the smaller petri dish. The treated finger is held horizontally and 10 ticks are transferred with forceps to the finger between the nail and the boundary of the treated surface. Once all the ticks were clinging to the finger, the finger was tilted to vertical with the fingertip pointing down. The locations of the ticks were recorded at 10 min after the last nymph was released on the fingertip. Ticks on the untreated fingertip and those that fell or dropped from the finger onto the moated Petri dish 3–4 cm below were considered repelled, whereas ticks on the treated area and those that crossed it were considered not to have been repelled. The same procedure was repeated on the same finger of the other hand without treatment to act as the control. There may be variations to this procedure such as wrapping a piece of cloth around finger and applying the repellent to the cloth instead of applying to the finger as was done by Carroll et al. (2007).

Carroll et al. (2005) studied responses of host-seeking nymphs of the black-legged or deer tick (*Ixodes scapularis* Say) and the lone star tick to the arthropod repellents DEET and (1S, 20S)-2-methylpiperidinyl-3-cyclohexene-1-carboxamide (SS220) using fingertip laboratory bioassays. Both compounds were applied to the skin and strongly repelled both species of ticks at 0.8 and 1.6 μmole of compound/cm^2 skin surface. The ticks were also repelled when two layers of organdie cloth covered the portion of a finger treated with either DEET or SS220. Carroll et al. (2007) evaluated callicarpenal (13, 14, 15, 16-tetranor-3-cleroden-12-al) and intermedeol [(4S,5S,7R,10S)-eudesm-11-en-4-ol], isolated from American beautyberry (*Callicarpa americana* Lamiaceae), in laboratory bioassays for repellent activity against host-seeking nymphs of the black-legged tick and the lone star tick. A strip of organdie cloth treated with test substances was doubly wrapped around the finger. Callicarpenal and intermedeol, at 155 μmole/cm^2 of cloth surface repelled 98% and 96% of black-legged tick nymphs, respectively.

As a small variation to the earlier method, the finger is held vertical in the middle of the petri dish such that the fingertip touches the bottom of petri dish and allows the ticks to climb up the finger as done by Schreck et al. (1995). Twenty-nine repellents were tested on human skin for the duration of activity as protectants against nymphal lone star ticks and against black-legged ticks. Before the actual test, the individual's fingers and ticks can be screened for tenacity and readiness to climb by doing the procedure without repellent. This also acts as control for the test. The subject is considered protected if > 90% ticks stayed away from the treated area. Eleven of these repellents, including DEET, provided >2 h of protection against the lone star tick. One repellent, l-(3-cyclohexenyl-ylcarbonyl) piperidine was effective ≥4 h. Seven repellents that were most effective against lone star ticks, including DEET, were tested against the black-legged tick and none were effective.

Another small variation is when the finger is moved slowly inside the dish (Fig. 4.7) as was done by Pretorius et al. (2003). The efficacy of topically applied 20% lotion of DEET, and KBR 3023, a piperidine compound, was evaluated against laboratory-reared South African bont tick (*Amblyomma hebraeum*) nymphs on treated fingers. Both substances repelled >85% of nymph attacks at 0 and 1 h post-application. At 2, 3, and 4 h, the repellent efficacies of DEET were 84%, 68%, and 71%, whereas those of KBR 3023 were 56%, 55%, and 54%.

FIG. 4.7 Finger bioassay configuration (Photo Taken by Farooq et al., 2021).

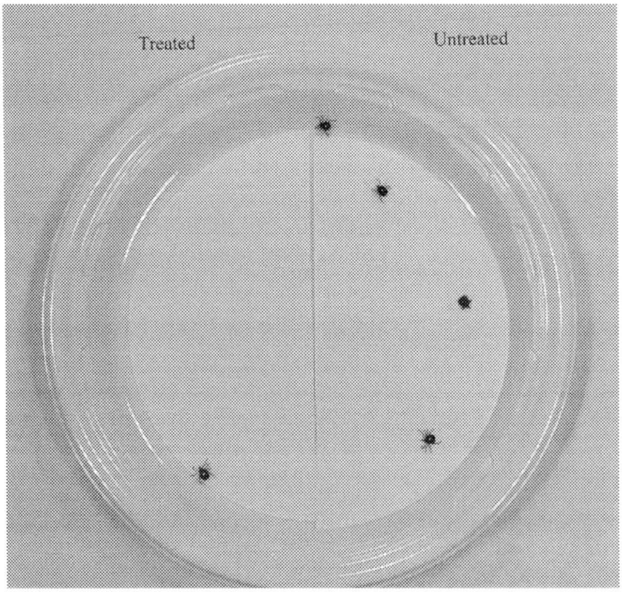

FIG. 4.8 Petri dish bioassay with two filter-paper halves, one untreated (control) and other treated. (Photo taken by Farooq et al., 2021).

4.5.3.2 *Skin tests*

As explained by Witting-Bissinger et al. (2008), ticks are enclosed in a small petri plate lid lined with two layers of cheesecloth with the open-end covered with aluminum screening to prevent the ticks from biting the subject but permitting the ticks to make direct contact with the skin. The lid is held in an incubator at 27 °C, 65% RH, and complete darkness for 30 min before the beginning of each assay. The skin on the inner forearm directly below the elbow of each volunteer is cleansed with 70% ethanol and allowed to dry. An area to place lid on the arm is marked with permanent marker using an empty lid as a template. A lid containing ticks is applied to the untreated marked area on the skin with the aluminum screen against the skin and covered with a piece of foil to exclude light. Distribution of ticks is recorded after 30 min to show that the position of the ticks in the lid as a control. Half of the skin in the test area is treated with repellent and the lid reapplied 2 h after application. The number of ticks on the treated and untreated side is recorded. In trials to determine the repellent activity of BioUD, with the AI 7.75% 2-undecanone, originally derived from wild tomato (*Lycopersicon hirsutum* Dunal f. *glabratum* C. H. Mull) plants, against the American dog tick (*Dermacentor variabilis* Say) on human skin, it was found that BioUD repelled ticks at least 2.5 h after application to human skin.

In a variation to this procedure, used by Erik et al. (2020), the arthropod repellent is applied to the forearm of a volunteer from 3 cm above the wrist up to a point near elbow. Ticks are then placed on the wrist of the volunteer below the treated area. Three min after release, ticks that cross at least 3 cm into the treated area of the forearm are considered not repelled by the tested product. The procedure without repellent was used as the control test. Reluctance of the Asian long-horned tick nymphs to stay in contact with nontreated human skin precluded the use of a human skin bioassay to optimally evaluate repellency.

As reported by Semmler et al. (2011), the upper side of a hand of a test person is sprayed with repellent until the skin appears wet. After drying, one tick is placed onto the treated skin for 10 s. Then the hand is brought into an upright position, so that the ticks had the choice to stay attached or crawl up or down the skin. Their crawling upward or staying on the skin longer than 1 min is noted as nonprotection. These tests were also done with ticks placed onto leather and onto clothes of persons wearing wool or cotton trousers. Results of different tests are summarized in a table by Semmler et al. (2011).

4.5.3.3 *Petri dish test*

In these studies, as described by Erik et al. (2020), the bottom of the petri plate lids is lined with two semicircle pieces of filter paper (Fig. 4.8). For single repellent evaluation, one piece is treated with repellent while the other is left untreated. They evaluated arthropod repellent products containing six AIs recommended by the CDC: DEET, picaridin, IR3535, oil of lemon eucalyptus, PMD, and 2-undecanone for repellency of Asian long-horned tick. All tested arthropod repellent product formulations were highly effective with estimated repellencies ranging from 93% to 97%. For other tests conducted by Witting-Bissinger et al. (2008), using head-to-head comparison, two sides are treated with different repellents. Filter papers are allowed to dry at room temperature before the start of the experiment. Known number of ticks are placed in the middle and are allowed to choose either filter paper surface. Tick distribution was recorded 30 min after the beginning of each test. In one such test, when BioUD and 15% DEET were applied on two sides of filter paper, ticks spent significantly more time on the DEET-treated surface than the BioUD-treated surface. Witting-Bissinger et al. (2008) also used a double layer of cheese cloth instead of filter paper in another test. In one trial, no differences in mean repellency were found through 8 days after application on cloth with BioUD and 7% DEET.

4.5.3.4 *Vertical contact irritancy bioassay*

Eisen et al. (2017) explained a vertical assay where ticks are introduced onto textile and their upward movement is observed. A solid piece of nontreated white cotton textile is sewn onto a playing card covering the entire card. Then, an additional 13-mm horizontal strip of nontreated or treated test textile is sewn on the upper edge on top of the nontreated background textile. The assay is performed with the textile-covered card positioned at a 45° angle. Ticks are introduced just below the treated strip and a finger is held along the top of the card to provide a stimulus for upward movement. Ticks are given 5 min to approach and attempt to cross horizontal test textile strips to reach the finger. Test strips were made from nontreated textile, permethrin-treated textile, or previously nontreated textile sprayed to saturation with 25% DEET from a pressurized spray can and then allowed to air dry for 15–30 min. The ticks not approaching the treated fabric were considered as spatially repelled. The ticks moving toward the fabric and dislodging were considered as showing contact repellency. All nymphal ticks approaching DEET showed spatial repellency compared with 6% and 4% to permethrin-treated textile strip and a nontreated strip. Of the nymphal ticks that made contact with permethrin-treated textile or nontreated textile, 74% dislodged from the permethrin-treated textile as compared with 0% from the nontreated textile. For permethrin-treated and nontreated textile, 0% and 88% remained on the textile.

Erik et al. (2020) evaluated a permethrin-treated T-shirt for contact irritancy against the Asian long-horned tick nymphs. The purpose of the bioassay modified from Eisen et al. (2017) was to evaluate the behavior of ticks following their introduction to a vertically oriented permethrin-treated textile; mimicking a host-seeking tick moving from vegetation to a treated piece of clothing. Groups of five Asian long-horned tick nymphs were placed at the center of a rectangular canvas frames oriented at a 45° vertical covered with textile from either the permethrin-treated 100% cotton T-shirt, or a nontreated 100% cotton T-shirt (control). The number of ticks remaining on the textile was recorded every minute for 5 min. The contact irritancy manifested as agitated movements of ticks attempting to dislodge from the treated textile by flipping and tumbling off the bottom of the textile-covered frame. In contrast, ticks placed on a nontreated textile were expected to display normal movement and remain on the textile. After 1 and 3 min, 72% and 96% ticks, respectively, had dislodged from the treated textile. In contrast, 84% of the ticks remained on the nontreated control textile over the full 5 min observation period.

4.5.3.5 *Other tests*

Scott et al. (2016) used white poster paper, attached to the table as a surface for arthropod repellent application. Multiple papers are attached at least 1 m apart on the table for replications or control. Three concentric circles of 5.5, 14.5, and 15.5 cm diameter are drawn onto each poster paper. Following label, repellent is painted in between largest and middle circles and allowed to dry for 10 min. Then place ticks with forceps onto the center of the smallest circle. Volunteer sits in a chair and places his /her arms to two sides of the outer circle, ensuring no contact with repellent. After 3 min, the ticks which moved to the treated area are considered as locating the volunteers and not repelled. Ticks on the control poster board followed the same bioassay procedure as the treated ones. Seven different repellents (BioUD, Bio Block-Organic Outdoor, Bio Block-Organic Pest Control, Bio Blocker-Organic Insect Repellent, OFF! Botanicals Insect Repellent, OFF! Deep Woods Insect Repellent, and Repel Insect Repellent Sportsman Gear Smart) were tested on four volunteers for efficacy against unfed nymphal lone star ticks. At 10 min postapplication, Bio Block Pest control was the most effective product at repelling ticks (85%) and Repel the least effective (30%). At 120 min, Bite-Blocker-Insect Repellent, OFF! Deepwoods, and Bio Block-Organic Outdoor provided ≥ 55% protection from ticks.

As reported by Farooq (Personal communication), the repellency is evaluated spraying a 7.5 cm-wide outer band of the 30.0 cm diameter plastic plate and inner 15.0 cm circle is left unsprayed. For this, place an inverted 15.0 cm plate in the middle of the 30.0 cm plate and spray with a spray pen at the recommended rate. Treat control plates with water at the same rate. Grains of lure taken out by opening the BG Lure cartridge (Biogents, Regensberg, Germany), are placed around the large plate in group of 10 at 8 locations to create an attraction for the ticks (Fig. 4.9). Five ticks are released in the middle of the plate on untreated surface and movement of the ticks is observed for 15 min. Any tick which crosses the sprayed area and leaves the plate is collected and recorded as not repelled. In addition, the number of ticks, if any, on the sprayed surface or in the middle circle is recorded separately. BIGSHOT Maxim Concentrate Botanical Mosquito and Agricultural Pest Control and clove oil were tested at three different rates to repel adult tick males, females, and nymphs. In all tests, ≤10% ticks stayed off the treated area. The rest, either stayed on the treated surface or crawled out of the test area.

4.5.4 Field evaluation methods

4.5.4.1 *Drag cloth*

Field tests can be conducted in the absence of a host by comparing the number of questing ticks collected on treated and untreated cloths dragged over the ground in tick-infested habitats (Granett and French, 1950). Drag cloths are generally used as surveillance tool to estimate the tick population in an area. By impregnating the cloth with repellent, it is used to evaluate efficacy of tick repellents in a known infested area. Granett and French (1950) applied 2 g of test repellent per square foot on a square yard of heavy felt cloth. The application rate can be varied based on the recommended application

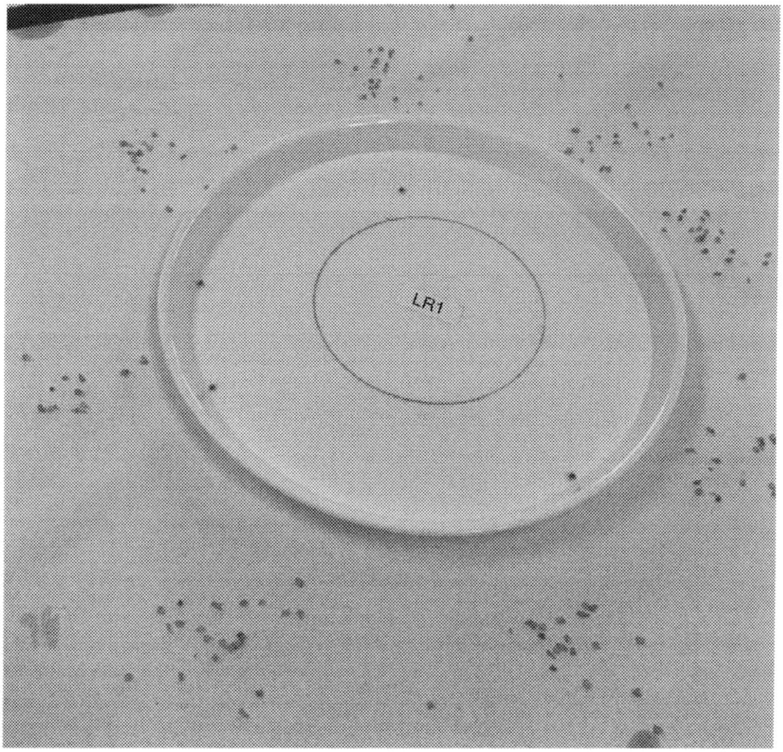

FIG. 4.9 Experimental setup showing ticks on a plastic plate (photo taken by Farooq, 2020).

rate of a repellent. To evenly apply the repellent to the cloth, it could be diluted in a suitable diluent, especially for lower application rates. An untreated similar cloth is dragged as control. The number of ticks attached to both is recorded and the difference will indicate the repellency of the product being tested. As reported by Jaenson et al. (2006), the repellent is applied on one side of the cloth and is dragged on vegetation for 10 min before the ticks on the cloth are counted. Using this procedure, more than one repellent can be evaluated at the same time. Five different repellents were tried in this study and all had 76%–92% repellency over untreated cloths.

Jaenson et al. (2006) evaluated repellency of MyggA Natural and lemon-scented eucalyptus (*Corymbia citriodora*, Hook.) in the tick-infested fields by the blanket-dragging technique for 4 days during a 6-day period. The repellencies were 74% and 85% for the two arthropod repellent products on day 1 which declined significantly from day 1 to day 6 (74%–45% for MyggA Natural; 85%–42% for *C. citriodora* oil).

4.5.5 Trouser test

Jaenson et al. (2006) reported a procedure in which the repellent is applied to the legs of white cotton trousers in two, 2-cm wide rings. One ring is at 30 cm above ground level, just below the knee, and one ring at the lower edge, i.e., at 0 cm above ground level of the test trousers. Untreated trousers are used as control. In each test, the person slowly walks 100 m through the vegetation and ticks attached to the trousers are counted and collected from both legs at stops made every

10 m. In a variation, they used five rings on a leg. The rings were at 0, 20, 40, 60, and 80 cm above ground level. With two rings of MyggA Natural on each leg of the treated trousers, significantly less (29 nymphs) were caught compared to 50 nymphs on the untreated legs resulting in 42.0% repellency. With five rings, 13 nymphs were caught on treated legs compared to 32 on the control legs providing 59.3% repellency.

In a variation, as reported by Solberg et al. (1995), the repellent was applied to the exposed leg of the volunteer and all entries of ticks into clothes were blocked. One leg was treated, while the other was considered as a control and legs were switched on subsequent test days. Volunteers wore light-colored shorts to help in detection of the ticks. The tests were performed at different times after application to study the longevity of the treatment. The volunteer walked slowly through the test site for 30 min. When a tick made contact with the skin, its behavior was observed for 5 min. In 5 min or less, if the tick successfully traversed the skin to the shorts, or it attached to the skin, it was considered as not repelled. Ticks that climbed onto the leg and dropped off or moved down the leg in <5 min were considered repelled. Repellent efficacy of DEET and piperidine, (AI3-37220), was evaluated topically on human volunteers against lone star tick nymphs and adults in the field. AI3-37220, at 0.5 mg/cm^2 application, provided >90% repellency against adult and nymphal ticks over a 6-h test period and showed significantly better repellent efficacy than DEET. On the other hand, DEET, at the same concentration, provided 85% repellency at 0 h and 55% at 6 h.

In another variation, Granett and French (1950), the lower portions of the trouser legs, approximately 75 cm, were cut from cotton trousers, impregnated with the diluted repellent and placed on the volunteer again. The trouser on the other leg was treated as control. Ten different chemicals were evaluated using this procedure. The application rates of 20 and 40 grams/m^2 resulted in repellency in the ranges of 53%–100% and 74%–99%.

As reported by Evans et al. (1990), full uniforms are treated with arthropod repellents to evaluate the effectiveness of uniforms against ticks. The volunteer wearing that uniform is exposed to field populations of ticks during a series of 30-min field trials. Volunteer walks for 5 min to cover 110 m long stretch, sits down for 10 min and then ticks on uniform are counted and removed. Six volunteers clothed in either untreated, DEET-treated, permethrin-impregnated (PI), or permethrin-sprayed (PS) uniforms were exposed to field populations of ticks during a series of 30-min field trials. The rank order of the mean number of ticks for each uniform was untreated > DEET > PI > PS. Overall, the mean number of ticks found on DEET, PI, and PS uniforms were 60, 97, and 98%, respectively, lower than the mean number found on untreated uniforms. This general trend also was observed when individual species were analyzed independently.

4.6 Challenges and recommendations

Ticks and tick-borne diseases have been on the rise and the demand for tick repellents for personal protection has increased. Usually, most arthropod repellents include DEET, picaridin, IR3535, PMD, and 2-Undecanone as they provide effective protection against tick bites. Due to safety issues with synthetic arthropod repellents and environmental concerns, botanical-based arthropod repellents have been in high demand. However, the current and most marketed botanical arthropod repellents, except lemongrass oils, do not show effective protection against tick bites. Additionally, many of these botanical-based products have unpleasant odors. Most of the arthropod repellent products on the market include one or a combination of different concentrations of DEET, Picaridin, IR3535, PMD, 2-Undecanone, Oil of Lemon Eucalyptus, coconut oil, and geranium oil, etc., as AIs. Current topical clothing arthropod repellent products include the permethrin-treated textiles, clothing and uniform, and the spatial arthropod repellents include metofluthrin, transfluthrin, and allethrin as AIs.

There are many testing methods/tools/bioassays for screening and selection of tick repellents in the laboratory, semi-field, and field. Regardless of the type of testing methods/tools used, using identical methods and tools is critical and important to compare the efficacy among different kinds of tick repellents. To save time and reduce the cost for making new arthropod repellents available, the screening and selection of tick repellents should be conducted from the arthropod repellents available on the market. Testing tick repellents should include laboratory, semi-field, and field bioassays. The common and medically important hard (Ixodid) and soft ticks (Argasid) should be used for evaluating arthropod repellents. Care should be taken to use certified disease-free populations of ticks for such testing. Both the nymphal and adult stages of both sexes should be tested based on availability and condition of the ticks. For human protection and prevention from tick bites and tick-borne diseases, the most practical and applicable testing requires the use of human volunteers. However, utilizing human volunteers requires ethical approval and informed consent due to the consideration about safety and protection of the volunteers.

4.7 Conclusion

Personal protection measures and application of arthropod repellents play a critical role in the prevention and control of tick bites and tick-borne diseases. Currently, the CDC recommends using arthropod repellents containing >20% DEET, picaridin, or IR3535 on exposed skin for tick bite and vector-borne disease prevention. In addition to these arthropod repellents, permethrin-treated clothing and uniform are available for personal protection against tick bites. The realization of the use of arthropod repellents as one of the means to control the spread of vector-borne diseases and due to the difficulty of regulatory approval for new AIs, more natural products are being tested as repellents against ticks which have to go through extensive evaluation procedures. Topical and spatial arthropod repellents are two types of arthropod repellents that have different bite prohibition mechanisms; hence, their evaluation schemes are quite different. The laboratory, semifield, and field evaluation methods for two types of arthropod repellents are available. The evaluations can be performed in the absence of hosts or host stimuli, in the presence of host stimuli, or using a live host. The laboratory methods are mainly in the presence of host stimuli while field evaluations are mainly done with live hosts.

References

Bezirtzoglou, C., Dekas, K., Charvalos, E., 2011. Climate changes, environment and infection: facts, scenarios, and growing awareness from the public health community within Europe. Anaerobe 17, 337–340.

Bibbs, C.S., Xue, R.D., 2016. ThermaCELL and OFF! Clip-On devices tested for repellency and mortality against *Amblyomma americamum* (Acari: Ixodida: Amblyommidae). J. Med. Entomol. 53, 861–865.

Bissinger, B.W., Roe, R.M., 2010. Tick repellents: Past, present, and future. Pestic. Biochem. Phys. 96, 63–79.

Brennan, J.M., 1947. Preliminary report on some organic materials as tick repellents and toxic agents. Public Health Rep. 62, 1162–1165.

Centers for Disease Control and Prevention, 2020. A National Public Health Framework for the Prevention and Control of Vector-Borne Diseases in Humans. https://www.cdc.gov/ncezid/dvbd/pdf/Brochure_National_Framework_VBDs-P.pdf [accessed October 2020].

Centers for Disease Control and Prevention, 2020. Preventing Tick Bites. https://www.cdc.gov/ticks/avoid/on_people.html [accessed October 2020].

Carroll, J.F., Klun, J.A., Debboun, M., 2005. Repellency of deet and SS220 applied to skin involves olfactory sensing by two species of ticks. Med. Vet. Entomol. 19, 101–106.

Carroll, J.F., Cantrell, C.L., Klun, J.A., Kramer, M., 2007. Repellency of two compounds isolated from *Callicarpa americana* (Lamiaceae) against *Ixodes scapularis* and *Amblyomma americanum* ticks. Exp. Appl. Acarol. 41, 215–224.

Carroll, J.F., Benante, J.P., Klun, J.A., White, C.E., Debboun, M., Pound, J.M., Dheranetra, W., 2008. Twelve-hour duration testing of cream formulations of three repellents against *Amblyomma americanum*. Med. Vet. Entomol. 22, 144–151.

Dautel, H., 2004. Test systems for tick repellents. Int. J. Med. Microbiol. 293, 182–188.

Dethier, V.G., 1956. Repellents. Annu. Rev. Entomol. 1, 181–202.

De la Fuente, J., Estrada-Pena, A., 2012. Tick and tick-borne pathogens on the rise. Ticks Tick Borne Dis. 3, 115–116.

Eisen, L., Rosea, D., Prosea, R., Breunera, N.E., Dolana, M.C., Thompsonb, K., Connally, N., 2017. Bioassays to evaluate non-contact spatial repellency, contact irritancy, and acute toxicity of permethrin-treated clothing against nymphal *Ixodes scapularis* ticks. Ticks Tick Borne Dis. 8, 837–849.

Erik, F., Amy, C.F., Shelby, L.F., Michael, L.L., Mark, J.D., Rebecca, J.E., Lars, E., 2020. Preliminary evaluation of human personal protective measures against the nymphal stage of the Asian Longhorned tick (Acari: Ixodidae). J. Med. Entomol. 57, 1141–1148.

Evans, S.R., Korch Jr, G.W., Lawson, M.A., 1990. Comparative field evaluation of permethrin and Deet-treated military uniforms for personal protection against ticks (Acari). J. Med. Entomol. 27, 829–834.

Farooq, M., Blore, K., Miah, M., Xue, R.D., 2021. Evaluation of personal d-allethrin in the thermacell mosquito repellent device against the lone star tick under laboratory conditions. J. Fla. Mosq. Control Assoc. 68, 105–109.

Foster, E., Fleshman, A.C., Ford, S.L., Levin, M.L., Delorey, M.J., Eisen, R.J., Eisen, L., 2020. Preliminary evaluation of human personal protective measures against the nymphal stage of the Asian longhorned tick (Acari: Ixodidae). J. Med. Entomol. 57, 1141–1148.

Fradin, M.S., 1998. Mosquitoes and mosquito repellents: a clinician's guide. Ann. Intern. Med. 128, 931–940.

Frances, S.P., 2007. Picaridin. In: Debboun, M., Frances, S., Strickman, D. (Eds.), Insect Repellents: Principles, Methods, and Uses. CRC Press, Boca Raton, FL, pp. 311–326.

Granett, P., French, C.F., 1950. Field tests of clothing treated to repel American Dog Ticks. J. Econ. Entomol. 43, 41–44.

Gupta, R.K., Bhattacharjee, A.K., 2007. Discovery and design of new arthropod/insect repellents by computer-aided molecular modeling. In: Debboun, M., Frances, S., Strickman, D. (Eds.), Insect Repellents: Principles, Methods, and Uses. CRC Press, Boca Raton, FL, pp. 195–228.

Halos, L., Baneth, G., Beugnet, F., Bowman, A.S., Chomel, B., Farkas, R., Franc, M., Guillot, J., Inokuma, H., Kaufman, R., Jongejan, F., Joachim, A., Otranto, D., Pfister, K., Pollmeier, M., Sainz, A., Wall, R., 2012. Defining the concept of 'tick repellency' in veterinary medicine. Parasitology 139, 419–423.

Jaenson, T.G.T., Garboui, S., Palsson, K., 2006. Repellency of oils of lemon eucalyptus, geranium, and lavender and the mosquito repellent MyggA natural to *Ixodes ricinus* (Acari: Ixodidae) in the laboratory and field. J. Med. Entomol. 43, 731–736.

Kumar, S., Prakash, S., Kaushik, M.P., Rao, K.M., 1992. Comparative activity of three repellents against the ticks *Rhipicephalus sanguineus* and *Argas persicus*. Med. Vet. Entomol. 6, 47–50.

Madison-Antenucci, S., Kramer, L.D., Gebhardt, L.L., Kauffman, E., 2020. Emerging tick-borne diseases. Clin. Microbiol. Rev. 33, e00083 -18.

McGovern, T.P., Schreck, C.E., Jackson, J., 1978. Mosquito repellents: alicyclic amides as repellents for *Aedes aegypti* and *Anopheles quadrimaculatus*. Mosq. News 38, 346–349.

Moore, S.J., Debboun, M., 2007. History of insect repellents. In: Debboun, M., Frances, S., Strickman, D. (Eds.), Insect Repellents: Principles, Methods, and Uses. CRC Press, Boca Raton, FL, pp. 3–29.

Morton, F.A., Travis, B.V., Linduska, J.P, 1947. Results of screening tests with materials evaluated as insecticides, miticides, and repellents at Orlando Laboratory, April, 1942 to April, 1947. US Dept. Agr. Bur. Entomol. Plant Quaran E-733, 1–236.

Petersen, L.R., Beard, C.B., Visser, S.N., 2019. Combatting the increasing threat of vector-borne disease in the United States with a national vector-borne disease prevention and control system. Am. J. Trop. Med. Hyg. 100, 242–245.

Piesman, J., Eisen, L., 2008. Prevention of tick-borne diseases. Annu. Rev. Entomol. 53, 323–334.

Pretorius, A.M., Jensenius, M., Clarke, F., Ringertz, S.H., 2003. Repellent activity of DEET and KBR 3023 against *Amblyomma hebraeum* (Acari: Ixodidae). J. Med. Entomol. 40, 245–248.

Rosenberg, R., Lindsey, N.P., Fischer, M., Gregory, C.J., Hinckley, A.F., Mead, P.S., Paz-Bailey, G., Waterman, S.H., Drexler, N.A., Kersh, G.J., Hooks, H., Partridge, S.K., Visser, S.N., Beard, C.B., Petersen, L.R., 2018. Vital signs: trends in reported vector borne disease cases—United States and Territories, 2004-2016. Morb. Mortal. Wkly. Rep. 67, 496–501.

Salafsky, B., He, Y.X., Li, J., Shibuya, T., Ramaswamy, K., 2000. Study on the efficacy of a new long-acting formulation of N,N-diethyl-m-toluamide (DEET) for the prevention of tick attachment. Am. J. Trop. Med. Hyg. 62, 169–172.

Schreck, C.E., Fish, D., McGovern, T.P., 1995. Activity of repellents applied to skin for protection against

Amblyomma americanum and *Ixodes scapularis* ticks (Acari: Ixodidae). J. Am. Mosq. Control Assoc. 11, 136–140.

Scott, J.D., Fulcher, A., Henlzer, J.M., Xue, R.D., 2016. Laboratory evaluation of seven insect repellents against the lone star tick, *Amblyomma americanium*. Tech. Bull. Fla. Mosq. Control Assoc. 10, 81–84.

Semmler, M., Ghaffar, F.A., Al-Rasheid, K.A.S., Mehlhorn, H., 2011. Comparison of the tick repellent efficacy of chemical and biological products originating from Europe and the USA. Parasitol. Res. 108, 899–904.

Solberg, V.B., Klein, T.A., McPherson, K.R., Burge, J.R., Wirtz, R.A., 1995. Field evaluation of deet and a piperidine repellent (AI3-37220) against *Amblyomma americanum* (Acari: Ixodidae). J. Med. Entomol. 32, 870–875.

Sonenshine, D.E., Lane, R.S., Nicholson, W.L., 2002. Ticks (Ixodida). In: Mullen, G., Durden, L. (Eds.), Medical and Veterinary Entomology. Academic Press, San Diego, CA, pp. 517–558.

Strickman, D.A., 2007. Older synthetic active ingredients and current additives. In: Debboun, M., Frances, S.P., Strickman, D. (Eds.), Insect Repellents: Principles, Methods, and Uses. CRC Press, Boca Raton, FL, pp. 361–383.

Strickman, D.A., 2015. Topical repellent active ingredients in common use. In: Debboun, M., Frances, S.P., Strickman, D.A. (Eds.), Insect Repellents Handbook. CRC Press, Boca Raton, FL, pp. 231–238.

Torres, M., Carey, V., 2004. Review of public health advice about ticks. N. S. W. Public Health Bull. 15, 212–215.

Witting-Bissinger, B.E., Stumpf, C.F., Donohue, K.V., Apperson, C.S., Roe, R.M., 2008. Novel arthropod repellent, BioUD, is an efficacious alternative to DEET. J. Med. Entomol. 45, 891–898.

Xue, R.D., Scott, J.M., Fulcher, A., Qualls, W.A., Henlzer, J.M, Gaines, M.K., Weaver, J.H.R., Debboun, M., 2016. Laboratory and field evaluation of OFF! Clip-On mosquito repellent device containing metofluthrin against the lone star tick, *Amblyomma americanium* (Acari: Ixodidae). Tech. Bull. Fla. Mosq. Control Assoc. 10, 85–90.

CHAPTER

5

Evaluation and application of repellent-treated uniform/clothing and textiles against vector mosquitoes

Ulrich R. Bernier[a], *Melynda K. Perry*[b], *Rui-De Xue*[c], *Natasha M. Agramonte*[d], *Amy L. Johnson*[b], *Kenneth J. Linthicum*[a]

[a]United States Department of Agriculture, Center for Medical, Agricultural, and Veterinary Entomology, Gainesville, FL, United States, [b]U.S. Army Combat Capabilities Development Command Soldier Center, Natick, MA, United States, [c]Anastasia Mosquito Control District, St. Augustine, Florida, United States, [d]DeKalb County Board of Health, Decatur, GA, United States

5.1 Introduction: The need for personal protection and arthropod-repellent treated clothing

Historically, nonbattle injuries which include diseases transmitted by mosquitoes and other arthropods have resulted in debilitating illnesses to warfighters, subsequently leading to failed missions with substantial medical costs to treat these individuals. In the American Revolutionary War, the ratio of nonbattle-related injuries and death outnumbered those received from battle by a ratio of 13:1. During the American Civil War and World Wars I and II, the ratio was closer to 2:1. Prior to and continuing through World War II, the US Department of Defense developed a strategy to protect their military personnel when deployed on missions worldwide. This strategy included proper use of personal protectants, i.e., topically applied arthropod repellents and repellent-treated uniforms. This system provided remarkable protection from insect bites (Schreck et al., 1984; Sholdt et al., 1988; Schreck and Kline, 1989). Over the

Advances in Arthropod Repellents.
DOI: https://doi.org/10.1016/B978-0-323-85411-5.00002-9

past 70 years, clothing treatments have been explored for arthropod repellent use, and the most effective treatments have been insecticides. Although pyrethroid insecticides are not recommended for direct treatment to human skin because of dermal toxicity (Brown and Hebert, 1997), their rapid action and low human toxicity make them excellent candidates as contact arthropod repellents in treated textiles, such as clothing/uniforms, hammocks, curtains, and bed nets. Permethrin has been registered for use in military clothing since 1991 and the use of permethrin-treated clothing has recently been shown to reduce the malaria incidence in a local community (Kimani et al., 2006).

As a result of disease incurred by deployed marines and soldiers in 2003 and 2004, the United States Marine Corps and US Army transitioned to factory level production of permethrin-treated combat uniforms in 2007 and 2010, respectively. The decision to move to factory treatment from field treatment was made to alleviate concerns over personnel being noncompliant in the treatment process and to remove situations where unavailability of permethrin prior to and during deployment resulted in service personnel wearing improperly treated or untreated uniforms. The specifications for garment manufacture and treatment of military uniforms are detailed within each uniform's individual purchase description. These documents describe specifically the requirements for permethrin content and efficacy and include the test methods for the chemical and biological analysis to ensure uniforms are properly treated. In 2020, MIL-PRF-32659 – "Textiles, Vector Protection Treatment Of" was published as a standalone document and is currently being incorporated into the relevant uniform purchase descriptions, replacing the original references to permethrin. It included some changes to testing requirements and allows for use of other active ingredients that meet all regulatory and efficacy requirements.

The registered label rate for application of permethrin to military uniforms specifies that uniforms be treated at 0.52% ± 10% ($w_{permethrin}/w_{fabric}$) of the fabric weight making up the garment. This level of treatment translates to different target concentrations/dose rates and tolerance ranges depending on the uniform that is treated. There are a variety of uniform types, fabric constructions, weights, fiber compositions, and print patterns that are used by the major service branches of the US Armed Forces. Specifications for the fabrics discussed within this chapter are shown in Table 5.1. Chemical validation is performed on uniforms to determine if they have been treated at the appropriate label rate. The biological efficacy is evaluated to determine if the permethrin to binder ratio is optimized such that the uniform provides maximum protection throughout the expected lifetime of the uniform. Criteria are set for minimum bite protection levels of the treated garments relative to untreated garments and for various laundering intervals: 0x- (unlaundered), 20x-laundered, and 50x-laundered garments where 50x laundering cycles represents the expected lifetime of these uniforms. All laundering is conducted in accordance with American Association of Textile Chemists and Colorists (AATCC) standardized procedures, specifically American Association of Textile Chemists and Colorists TM135—Test Method for Dimensional Changes of Fabrics after Home Laundering with the following laundering and drying conditions; permanent press cycle, very hot, and tumble dry permanent press.

5.2 Laboratory methods for evaluation of arthropod repellent treated US military uniforms

Complete descriptions of the chemical analysis and bite protection are found in MIL-PRF-32659. Some of the critical methods are described below.

TABLE 5.1 Fabric specifications.

Fabric identification	Construction	Yarn type	Fiber content	Air permeability[a] CFM	Weight[b] oz./yd² (g/m²)
ACU	Plain weave rip stop	Spinning type not specified	Nylon	6.1	6.7
			Cotton		(227.2)
FR Type I (FRACU)	Plain weave rip stop	Spinning type not specified	FR Rayon	49.4	6.5
			Para-aramid		(220.4)
			Nylon		
FR Type III (FRACU)	Plain weave rip stop	Spinning type not specified	FR Rayon	38.0	6.7
			Para-aramid		(227.2)
			Nylon		
MCCUU Blouse	2 × 1 Twill	Spinning type not specified	Nylon	10.4	6.8
			Cotton		(230.6)
MCCUU Trouser	2 × 1 Twill	Spinning type not specified	Nylon	6.8	7.9
			Cotton		(267.9)
FR Type IV	Twill	Spinning type not specified	FR Rayon	44.1	6.8
			Para-aramid		(230.6)
			Nylon		
MA	Plain weave rip stop	Ring spun	Nylon	41.4	5.6
			Cotton		(189.9)
MB	Plain weave	Core spun	Nylon	108.8	4.8
			Cotton		(162.5)
MC	Plain weave rip stop	Vortex spun	Nylon	53.5	5.5
			Cotton		(186.5)
MD	Plain weave	Twisted filament and staple spun	Nylon	29.1	5.5 (186.5)
			Cotton		
			Polytetrafluoro-ethylene		

[a] *Air permeability must meet fabric specific minimum or maximum requirements. Some variation is expected between different fabric samples of the same type.*
[b] *Weight must fall within specific weight requirement ranges specific to each fabric type. Some variation is expected between different fabric samples of the same type.*

5.2.1 Chemical analysis of treated uniforms

Three or more specimens (19.35 cm^2) are cut from single-ply areas of the treated garments that have been subjected to each test condition (unlaundered (0x), 20x, and 50x laundered). A Dionex Accelerated Solvent Extractor system is used to remove permethrin from the specimens. Permethrin concentration is determined by comparison to a series of permethrin external standards. Analysis is accomplished by gas chromatography/mass spectrometry. The gas chromatography column is a 30 m × 0.25 mm i.d., df = 0.25 µm, DB-5 column. The carrier gas is high purity helium, and the column is operated isothermally at 250 °C. Permethrin content results are calculated and recorded as a dose rate.

5.2.2 Biological efficacy of US military uniforms

Bite protection assays involve the comparison of mosquito bites through a treated specimen compared to an untreated control of the same fabric. The test specimen size for bite protection is a trapezoidal pattern, single-ply in thickness, of approximately 29.25 cm in height for each edge and base lengths of approximately 19.5 cm and 30.5 cm. The specimen edges are sewn to produce a sleeve that will fit tightly over the forearm, as shown in Fig. 5.1. Gloves are worn to prevent bites on the hands. There is no specific requirement to test different camouflage print patterns; however, each specimen must be tested against two mosquito species, *Ae. aegypti* (Linnaeus) and *An. albimanus* (Wiedemann). Each cage (20,000 cm^3) test consists of inserting the arm with a sleeve into 200–250 nulliparous (fed only sugar and water) 6–11-day-old female mosquitoes. The test duration is 15 minutes. For each test condition (0x, 20x, and 50x laundered),

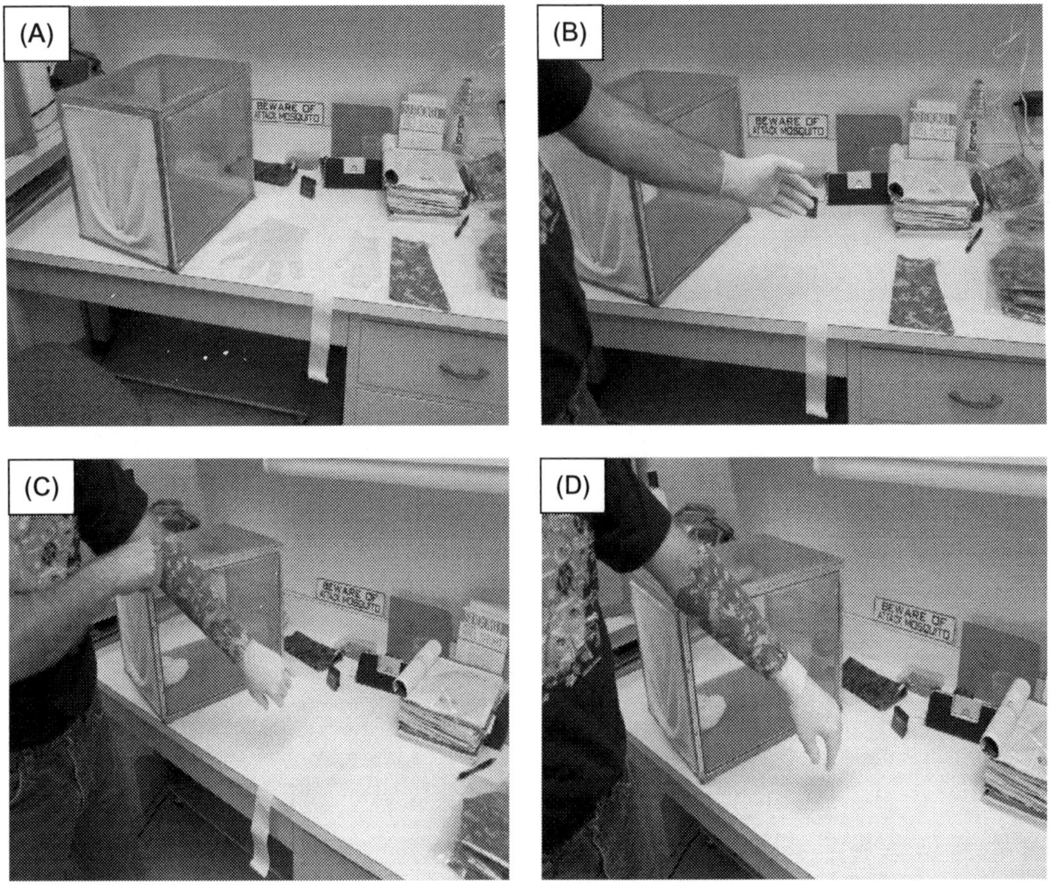

FIG. 5.1 **Bite protection assay.** (A) Equipment used for a bite protection assay consists of cage, gloves, tape, and test sleeve, (B) the hand is double-gloved, (C) the arm is inserted into a sleeve, and (D) tape is used to affix the edge of the sleeve to the glove(s) to prevent mosquito bites to the wrist.

specimens from three different treated samples are tested. The original study conducted with Marine Corps Combat Utility Uniforms (MCCUUs) used four volunteers comprised of at least one female per fabric set. Subsequent studies were changed to use three volunteers per fabric set. Control (untreated) sleeves are tested prior to the treated sleeves, and sleeves are tested in order of decreasing laundering increment. Percent bite protection is calculated by Abbott's formula (Abbott, 1925) as the percentage reduction in bites between the treated specimen and control. The bite protection study had been approved by the University of Florida Institutional Review Board (IRB) from 2006 to 2019. The current studies operate using protocols approved by Western IRB.

5.2.3 Experimental protocol for US Environmental Protection Agency registration of arthropod repellent treated US military fabrics

This section contains the abbreviated protocol for a study conducted in 2015 for registration of etofenprox as a repellent treatment for US military clothing.

5.2.3.1 *Summary of experimental design*

In the US military protocol, each uniform and condition tested requires only three subjects with replicates performed to assess uniformity in the treatment process. In this study, the subject-to-subject variability was expected to be much greater than differences within-subjects; therefore, this study required no within-subject replication but increased the number of subjects from three to a minimum of eight. As presented in Table 5.2, the precision of the overall bite protection value for a treated fabric will depend on the true bite-through rate for the control fabric and the true level of bite protection. The precision of overall bite protection increases as the control bite-through rate and/or the bite protection increase. In each case, the precision also improved with the number of subjects although the benefit per additional subjects is reduced after about 6–8 subjects.

Treated uniform sets were evaluated at specific standardized laundering increments: unlaundered (0x), 20x launderings, 50x launderings, and 75x launderings. Separate fabric specimens for each laundering interval were tested, similar to that described in US military specifications. *Ae. aegypti* and *An. albimanus* were tested separately. The protocol tests only two species because the determinative factor for accuracy is tied to the anthropophilic nature of the mosquito species and their proper response in laboratory assays. As such, prior studies of this nature have shown little difference between these species in their bite protection results, and the addition of a third species (e.g., *Culex* spp.) would not contribute sufficiently distinct data to offset the burden to subjects from participation in this kind of study. Eight subjects (four males and four females) were tested using each fabric and mosquito species combination. Alternates were selected and tested if subjects withdrew prior to completion of data sets. Each subject was tested with only one species on a given day; therefore, it required commitments on two different days to complete all tests. Should a subject withdraw from the study, their data were not used. An alternate would be selected to replace the withdrawn subject. This process continued until eight subjects had completed the study. If it was determined that tests were conducted with mosquitoes that were not behaving properly (very low control biting rates), the subject was asked to repeat the test set with a fresh population of mosquitoes. The experimental design is found in Table 5.3. This study was conducted in accordance with the Environmental Protection Agency Good Laboratory Practice Standards for Federal Insecticide, Fungicide, and Rodenticide Act as defined in 40 Code of Federal Regulations (CFR) Part 160. The Standard Operating Procedures for specific study methods met the requirements of 40 CFR, Part 160, Federal Insecticide, Fungicide, and Rodenticide Act Good

TABLE 5.2 Impact of the number of replications on the number of subjects.

True bite-through rate for control fabric (θC)	**50%**		**20%**	
True bite protection for treated fabric (βT)[a]	*80%*	*95%*	*80%*	*95%*
Number of subjects	Expected half-width of a 95% confidence interval for %bite protection[b]			
3	5.2%	2.7%	8.8%	4.5%
4	4.5%	2.3%	7.5%	3.8%
5	4.0%	2.0%	6.7%	3.4%
6	3.7%	1.9%	6.0%	3.0%
7	3.4%	1.7%	5.6%	2.8%
8	3.2%	1.6%	5.2%	2.6%
9	3.0%	1.5%	4.9%	2.4%
10	2.8%	1.4%	4.7%	2.3%
15	2.3%	1.2%	3.8%	1.9%
20	2.0%	1.0%	3.3%	1.3%

[a]*Bite incidence for treated fabric is calculated from bite protection as θT = θC (1 − βT/100).*
[b]*Average half-width from 1000 simulated datasets. Each dataset consisted of S subjects testing a pair of fabrics (control and treated). For each pair the total number of mosquitoes (M) was a Poisson (200) random variable, and the number of blood-fed mosquitoes was simulated as a binomial (θ, M) random variable. Subject–subject differences were simulated by adding a subject-specific normal (0, 0.3) random variable to the logit of the true incidence for both control and treatment fabrics. For each simulated dataset, a binomial generalized linear model was fit to the data using the GENMOD procedure in SAS. The model specified fixed effects for both subject and test material and used a log link. Bite protection confidence intervals were then obtained by back-transforming the confidence intervals for the contrast log(θT)–log(θC).*

TABLE 5.3 Experimental design.

Fabric and treatment condition[a]	**Number of fabric specimens**	**Number of subjects**	**Number of species**[b]	**Total replicates per fabric type**
Coat untreated unlaundered control[c]	1	8	2	16
Coat treated laundered 75x	1	8	2	16
Coat treated laundered 50x	1	8	2	16
Coat treated laundered 20x	1	8	2	16
Coat treated unlaundered 0x	1	8	2	16
Trouser untreated unlaundered control[c]	1	8	2	16
Trouser treated laundered 75x	1	8	2	16
Trouser treated laundered 50x	1	8	2	16
Trouser treated laundered 20x	1	8	2	16
Trouser treated unlaundered 0x	1	8	2	16

[a]*Fabric treatment conditions are either untreated and unlaundered (control) or treated and unlaundered (0x), treated and laundered 20 times (20x), treated and laundered 50 times (50x), or treated and laundered 75 times (75x).*
[b]*The test species are Ae. aegypti or An. albimanus.*
[c]*Each subject serves as their own control for the bite protection calculation.*

Laboratory Practice Standards. Periodic audits of this study were conducted as required by 40 CFR 160.35.

5.2.3.2 *Subject recruitment*

The study was conducted using participants between the ages of 18 and 62 years, excluding pregnant and lactating women, individuals in poor health or physical condition, those hypersensitive to the test materials, or anyone having open cuts, scrapes or skin conditions on the hands or forearms. Those with a relationship to the study director or sponsor were also excluded. Initial screening was conducted by telephone, followed by in-house interviews. Subjects were asked if they recalled being bitten by one or more mosquitoes in their lifetime and about their physical responses to insect bites and stings. Respondents were asked to disclose any previous known reactions to insect bites and stings. Although rare, instances in which severe allergic reactions to mosquito and fly bites can occur. Individuals who reported allergic reactions beyond discomfort were excluded from the study. Females wishing to participate were required to take a pregnancy test. An initial list of 20 subjects was comprised from respondents, and from that list, four females and four males were selected randomly. The remaining 12 respondents were randomized and selected as alternates. All subjects were required to attend a consent meeting with the study director. Subjects were required to take the Informed Consent Form home prior to signing and instructed that the signing would take place upon their return to the study site.

For data to be valid for use from each subject, there was a minimum requirement that each must run a control sleeve and at least one of the treated sleeves. Ideally, each subject would have completed testing with the control and all four treated specimens for both uniform types (coat and trousers) and for both species (*Ae. aegypti* and *An. albimanus*). A sample study for one person is detailed in Tables 5.4 and 5.5. If subjects did not complete the full set, they were withdrawn from the study and compensated according to their level of participation even though their data were excluded for purposes of the final assessment. The purpose of the untreated control sleeve was to compensate for influences related to the subject's individual attraction level, the general host-seeking response of the test mosquito population, and to correct for the bite-through rate of the fabric. The treatment and control values for a subject were used in Abbott's formula (Abbott, 1925) to calculate the observed bite protection level of the sleeve for that subject.

TABLE 5.4 Testing paradigm using *Aedes aegypti*.

	Subject right arm		Subject left arm	
Test set[a]	Treatment condition	Specimen designation	Treatment condition	Specimen designation
1	Coat untreated unlaundered control[b]	Sleeve 1	Trouser untreated unlaundered control[b]	Sleeve 2
2	Coat treated laundered 75x	Sleeve 3	Trouser treated laundered 75x	Sleeve 4
3	Coat treated laundered 50x	Sleeve 5	Trouser treated laundered 50x	Sleeve 6
4	Coat treated laundered 20x	Sleeve 7	Trouser treated laundered 20x	Sleeve 8
5	Coat treated unlaundered (0x)	Sleeve 9	Trouser treated unlaundered (0x)	Sleeve 10

Each subject's right arm and left arm are tested simultaneously and completes test Set 1–5 for *Ae. aegypti*. Each subject takes a break between test sets when new cages are being filled with mosquitoes. All cages are washed after all test sets for each participant when complete.

[a] *Each test set runs for 15 minutes.*

[b] *Each subject serves as their own control for the bite protection calculation.*

TABLE 5.5 Testing paradigm using *Anopheles albimanus*.

	Subject right arm		Subject left arm	
Test set[a]	Treatment condition	Specimen designation	Treatment condition	Specimen designation
6	Coat untreated unlaundered control[b]	Sleeve 11	Trouser untreated unlaundered control[b]	Sleeve 12
7	Coat treated laundered 75x	Sleeve 13	Trouser treated laundered 75x	Sleeve 14
8	Coat treated laundered 50x	Sleeve 15	Trouser treated laundered 50x	Sleeve 16
9	Coat treated laundered 20x	Sleeve 17	Trouser treated laundered 20x	Sleeve 18
10	Coat treated unlaundered (0x)	Sleeve 19	Trouser treated unlaundered (0x)	Sleeve 20

Each subject's right arm and left arm are tested simultaneously and completes test Set 6–10 for *An. albimanus*. Each subject takes a break between test sets when new cages are being filled with mosquitoes. All cages are washed after all test sets for each participant when complete.

[a] *Each test set runs for 15 minutes.*

[b] *Each subject serves as their own control for the bite protection calculation.*

5.2.3.3 *Preparation of test materials: military sleeves, mosquitoes, and cages*

Fabric cutouts from the treated military garments were removed from single-ply areas at the Combat Capabilities Development Command Soldier Center (CCDC SC) and sewn into "sleeves" at the United States Department of Agriculture-Agricultural Research Service prior to testing. Sleeves were sewn inside out by connecting the left and right ends of the sleeve and sewing the fabric with an approximately 1.25 cm seam so that a "sleeve" was formed. The typical fabric dimensions are illustrated in Fig. 5.2. A range of sleeve sizes was constructed prior to initiation of the study. The fabrics' dimensions were altered by approximately 1.25 cm increments as needed to yield sewn sleeves that fit snug on the forearms of subjects.

Laboratory-reared 6–11-day-old adult mosquitoes were used for the bite protection assay. Because various mosquito species have differing behavior and levels of aggressiveness, females of two of the more aggressive and anthropophilic species were tested. One of these selected species was *Ae. aegypti*, a vector of yellow fever and dengue fever. The second species was *An. albimanus*, a tropical mosquito that is a highly aggressive biter, one of the most tolerant species when tested with topical skin arthropod repellents, and is a competent vector for malaria transmission. Since these mosquitoes were colony reared and not exposed to disease agents, the risk of contracting an infectious disease from bites was negligible. Mosquitoes from the colony were reared through immature stages and delivered to the laboratory testing room just prior to pupal eclosion. Mosquito pupae were received from the colony and allowed to eclose into cages in the laboratory. For those mosquitoes to transmit pathogens such as dengue fever or malaria to the subject, they would need to escape, bite an infected human, and then return back to the cage for selection in the tests. Even if a mosquito escaped, the probability that it could return into a cage and be used in testing was negligible. There is also an incubation period during which the disease pathogen must develop in the mosquito in order to transmit it to another organism. Additionally, all mosquitoes contained in the cages were not allowed to feed on any organism and thus did not have the opportunity to be infected with pathogens prior to testing. The mosquito colonies were reared at the Center for Medical, Agricultural, and Veterinary Entomology (CMAVE), a United States Department of Agriculture facility in Gainesville, Gainesville,

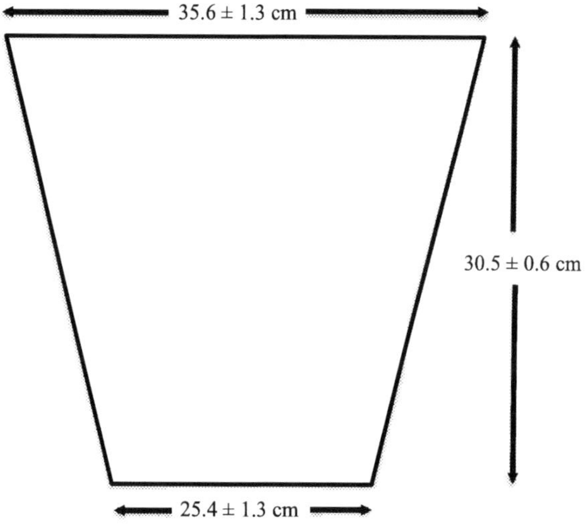

FIG. 5.2 Typical dimensions of military fabrics used for construction of test sleeves.

FL, using a membrane feeding system and bovine blood. The citrated bovine blood was pathogen-free (certified sterile by the supplier). The mosquitoes had not been fed on any humans prior to the study. The colonies used have been maintained in this manner since being established in Orlando in 1952 for *Ae. aegypti*, and in El Salvador in 1974 for *An. albimanus*. There have been periodic introductions of wild types of this species to maintain colony vigor. Mosquitoes from a colony will respond on the whole more aggressively to attractant stimuli than strains reared from freshly collected wild-types. A study conducted in 2012 at CMAVE demonstrated that females reared from eggs collected initially in the field from Puerto Rico required 4–6 generations before they responded as aggressively as the Orlando 1952 strain (Clark et al., 2011).

The test cages were approximately 59,000 cm^3 in volume and each contained 175–225 female mosquitoes (density of ~1 mosquito/300 cm^3). Female mosquitoes were preselected from stock cages by using a specially designed draw box that used odors from the hand of a laboratory staff person to attract mosquitoes upwind into a trap (Posey and Schreck, 1981). The trap containing the mosquitoes was then transferred to the test cage for subsequent testing by subjects.

5.2.3.4 *Detailed stepwise procedures for testing with volunteers*

(1) Specimens with a chain of custody letter were received from CCDC SC as fabrics were precut into the trapezoid shape and labeled both on the fabric and on each individual bag. The label indicated whether the fabric was from a coat or trouser and whether it was treated or the untreated control. The specimens were also labeled with the number of laundering cycles (0x, 20x, 50x, 75x) if treated. A unique identifier code was also added by CCDC SC to allow cross-referencing of data obtained with bite protection with the data collected from their chemical analysis of the fabrics. This chemical information was not provided to CMAVE prior to the bite protection study, CCDC SC did however advise whether or not testing should be aborted (or not conducted) due to over or under-treatment of the fabrics. (This follows the formal procedures used by the US military).

(2) Specimens were sewn into sleeves for testing with an approximately 1.25 cm seam sewn to connect the edges of the trapezoid. Sleeves were sewn with the fabric "inside out" and then are turned "right side out" for testing.

(3) If the participant was a female, she was required to take a pregnancy test. The laboratory technician or study principal investigator requested the subject thoroughly wash their hands with soap and water and then inspected their arms for cuts and scrapes. The principal investigator or technician advised the subject of the scope of the data set, including testing the control specimen first, followed by specimens at the 75x, 50x, 20x, and 0x laundering increments allowing for the lowest concentrations to be tested first. The same order of specimens was supplied to each subject for testing.

(4) The laboratory technician prepared the cages of mosquitoes by removing approximately 175–225 host-seeking female mosquitoes from the draw-box (Posey and Schreck, 1981) (Fig. 5.3A) and loading them into cages (Fig. 5.3B).

(5) The subjects inserted their left hand into a disposable plastic glove, followed by inserting the gloved hand into a latex or nitrile disposable glove (Fig. 5.4A). The subject repeated the process of double gloving for the right hand. The laboratory technician was available to help the subject and ensure the gloves were worn correctly.

(6) Although in Fig. 5.4B, the subject is filling out the data sheet under supervision of the laboratory technician, in the definitive study, the technician recorded all data onto the data sheet to include the test date, subject code, specimen code, test time, species, and age of the mosquitoes. The final two items on the data sheet, mosquito population. and bites (as determined by the number of mosquitoes that were blood-fed) were left blank.

(7) The subject then put their arm(s) into the test sleeves (Fig. 5.4C). The first sleeve tested was the untreated control, and subsequent sleeves were tested in order of laundering from the greatest number of laundering cycles (75x) to the least number of laundering cycles (0x).

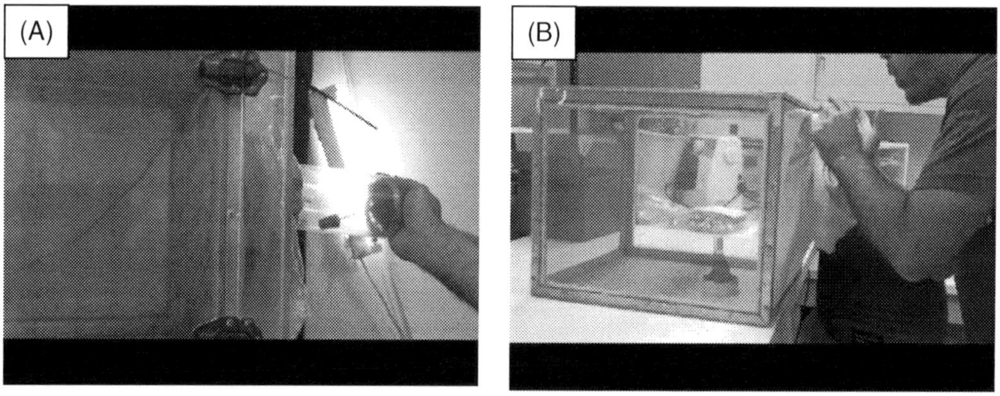

FIG. 5.3 (A) Mosquito collection trap inserted into draw box to collect mosquitoes by having them fly toward odors from the hand and (B) transfer of mosquitoes from trap to testing cage.

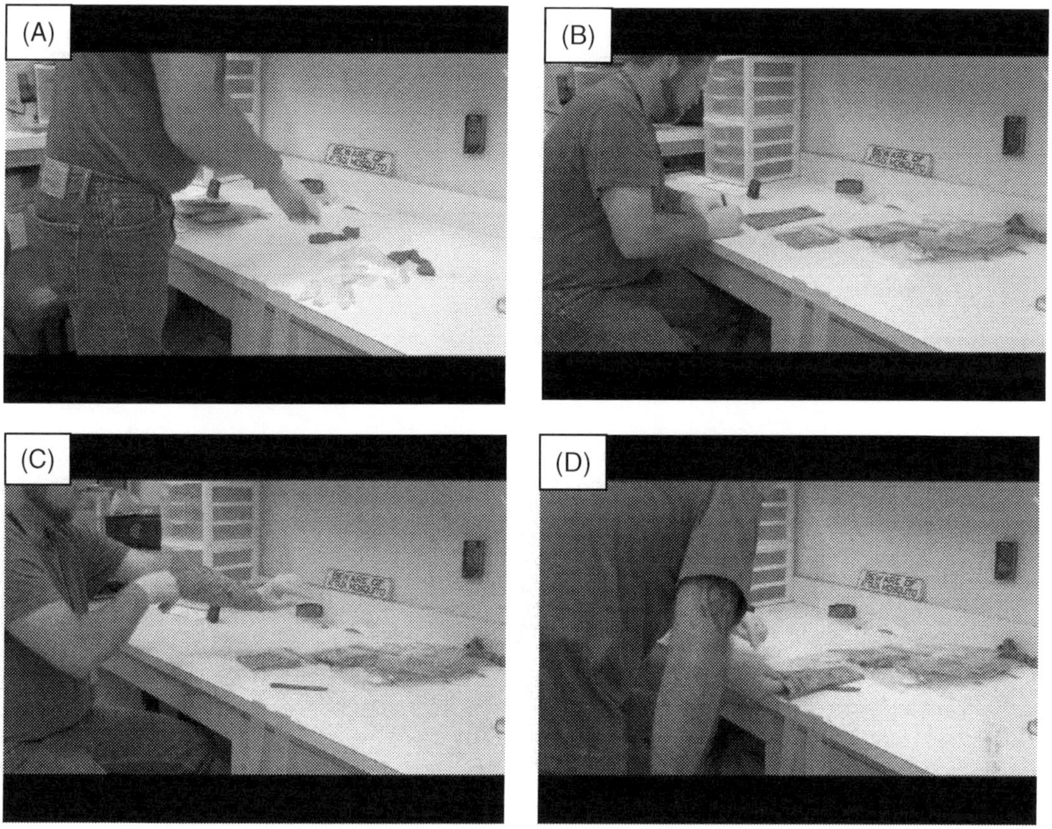

FIG. 5.4 (A) Subject double gloves a hand with a plastic glove, followed by a disposable glove. Laboratory technician oversees the gloving to ensure proper application, (B) laboratory assistant records test date, subject code, time, specimen, species, and age of mosquitoes on the data sheet, (C) subject inserting arm into the sleeve, and (D) laboratory technician records data on data sheet.

(8) The laboratory technician recorded the data on the data sheet and verified that the sleeve code (which specifies the number of laundering cycles) was correct and that the date, time, volunteer code, mosquito species, and age were also correct. (Fig. 5.4D).

(9) The laboratory technician then taped the junction between the sleeves and gloves with a piece of masking tape long enough to cover the entire circumference (Fig. 5.5A) to prevent mosquito access to the skin in the juncture between the glove and wrist. Although most subjects would have been capable of completing this preparative step without assistance from the laboratory technician, the technician performed this step for protocol consistency.

(10) The laboratory technician placed each data test sheet on top of the appropriate cage. The subject inserted the sleeved arms into cages of mosquitoes (Fig. 5.5B). The technician oversaw the insertion of the subject's arms with sleeves into the cages and secured the stocking entrance around the arm. The assistance of a laboratory technician was required to secure the stocking entrance properly around the

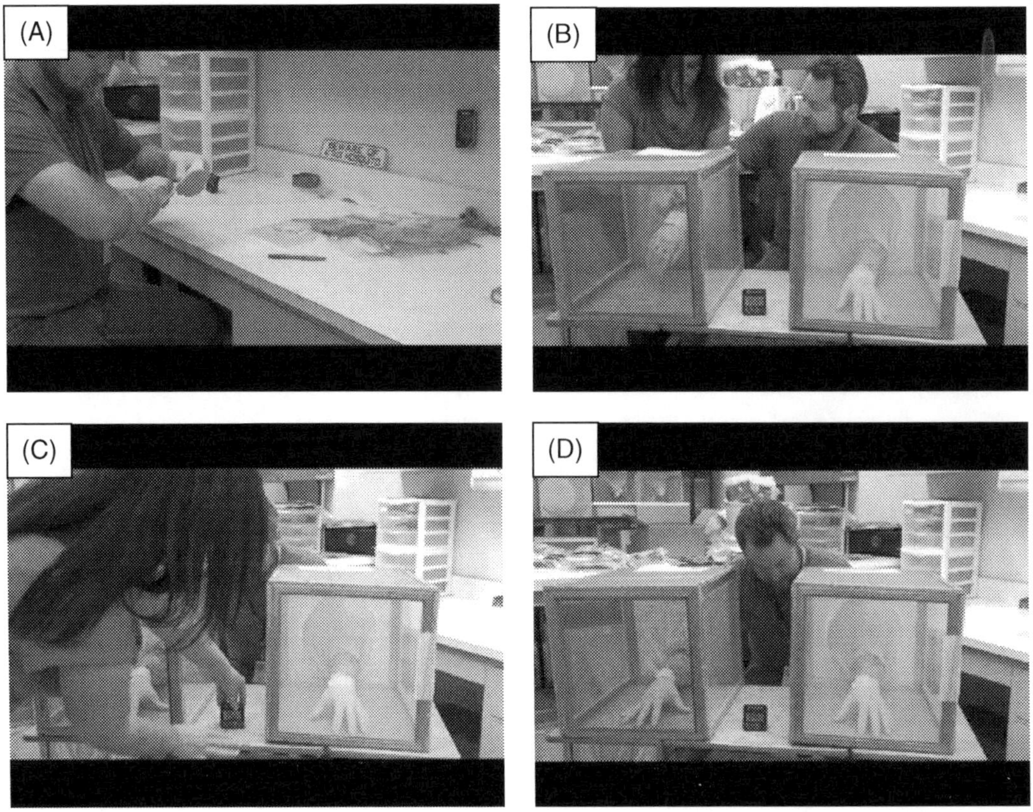

FIG. 5.5 (A) The laboratory technician (instead of the subject) applies the masking tape to juncture in the wrist area where sleeve fabric and glove meet. (B) Subject inserts both sleeved arms into cages of mosquitoes. A laboratory technician is required to secure the entrance stocking properly. (C) The laboratory technician will start the timer for the experiment which is preset for a 15-minute countdown. (D) The subject holds the arms still during the 15-minute test period.

sleeved arm and ensure that mosquitoes did not escape. The laboratory technician would then start the 15-minute timer for the experiment (Fig. 5.5C). The 15-minute test duration was selected because chemicals used for the treatment of clothing require an exposure period before they will produce a repellent effect, knockdown, or mortality. Previous studies have shown that most mosquitoes are intoxicated by permethrin after a 1 to 4-minute exposure. The use of a 15-minute test period ensures that slower responding mosquitoes are provided suitable exposure time once they respond by flying toward and landing on the sleeve surface.

(11) The subject was instructed to hold still during the 15-minute test time. In Fig. 5.5D, the subject has positioned the arms and hands comfortably for the test period. In situations where readjustment of arms was necessary, if done slowly, this could be executed without compromising the validity of the test.

(12) The preselection of host-seeking female mosquitoes from the draw-box (Step 4) provided a test cage population of mosquitoes that should respond with a high affinity toward an attractive source, i.e. the arm of a subject (Fig. 5.6A). For clothing treatments to repel mosquitoes, it is required that the mosquitoes come in contact with the surface of the fabric. The fabric construction impacts the probability that mosquitoes can successfully take a blood-meal by insertion of the stylet through the fabric interstitial space.

(13) At the conclusion of the 15-minute test period, the subject removed their arms from the test cage (Fig. 5.6B). This procedure was performed with the assistance of a laboratory technician to ensure that no test mosquitoes escaped from the cage during this process.

(14) The laboratory technician then removed the cages (with the data sheets on top) from the test table (Fig. 5.6C). Each cage was processed as described below. During this time, another laboratory technician would draw up more test mosquitoes from stock cages and load these mosquitoes into test cages (as previously described in Step 4).

(15) During the transition period between sets of cage tests, the subject would remove the sleeves and return them to the appropriate marked plastic bags (Fig. 5.6D). The gloves

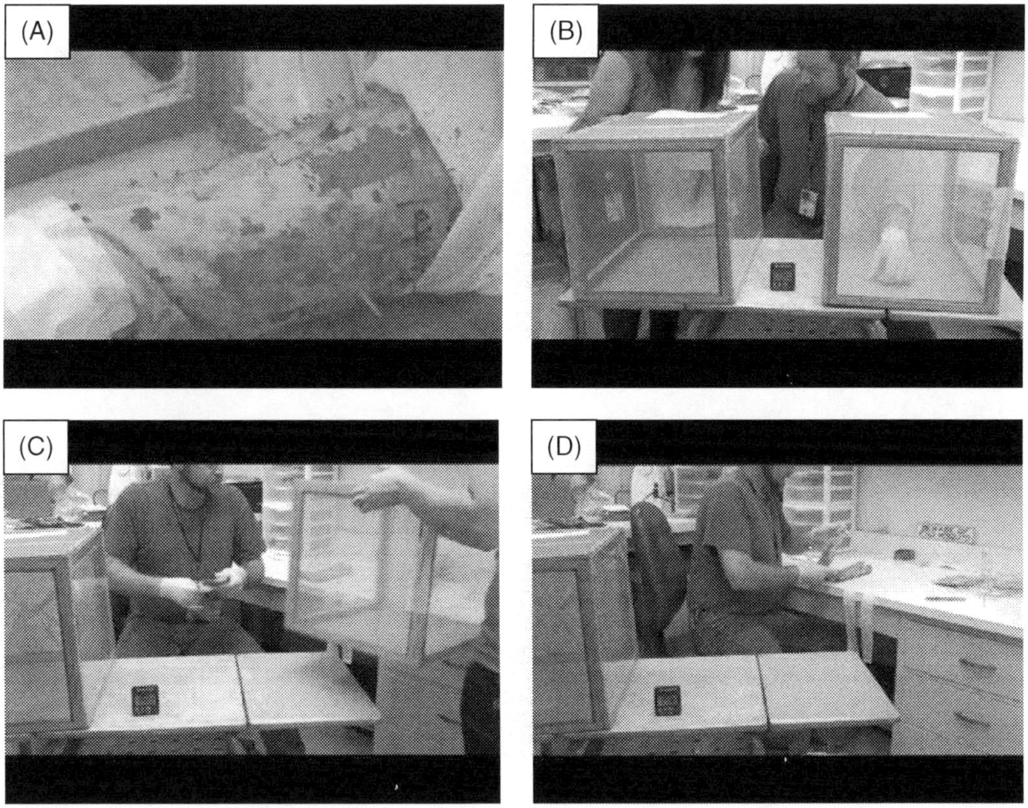

FIG. 5.6 (A) During the test, responding mosquitoes will fly to the surface of the uniform and attempt to take a bloodmeal. (B) A laboratory technician assists the subject with removal of the sleeved arms following the 15-minute test period. (C) The laboratory technician removes the test cages with data sheets for subsequent processing (aspiration and counting). (D) Subject removes the sleeves and returns them to the appropriately marked bags.

remained on the hands throughout this step until the completion of the testing period. Another set of sleeve specimens (in proper sequential order) were then selected, and Steps 5–14 were repeated. The subject was allowed as much time as they needed during the break period, e.g., to get a drink or use restroom facilities.

(16) The laboratory technician moved the test cage with the data sheet to a separate table. The data sheet was removed and placed on a counter by the work space. The laboratory technician removed all of the mosquitoes from the test cage using a mechanical aspirator (Fig. 5.7A).

(17) After aspiration of all mosquitoes from the test cage into a transfer tube, the tube and data sheet were transported into the next room where mosquitoes were counted (Fig. 5.7B).

(18) The counting of mosquitoes by the laboratory technician began initially with "knocking down" of the mosquitoes using house-supplied carbon dioxide (Fig. 5.7C). Once the mosquitoes were knocked down, they were emptied from the transfer tube onto a specially designed "knockdown" table (Fig. 5.7D). The table contains a porous surface where the pores can release carbon

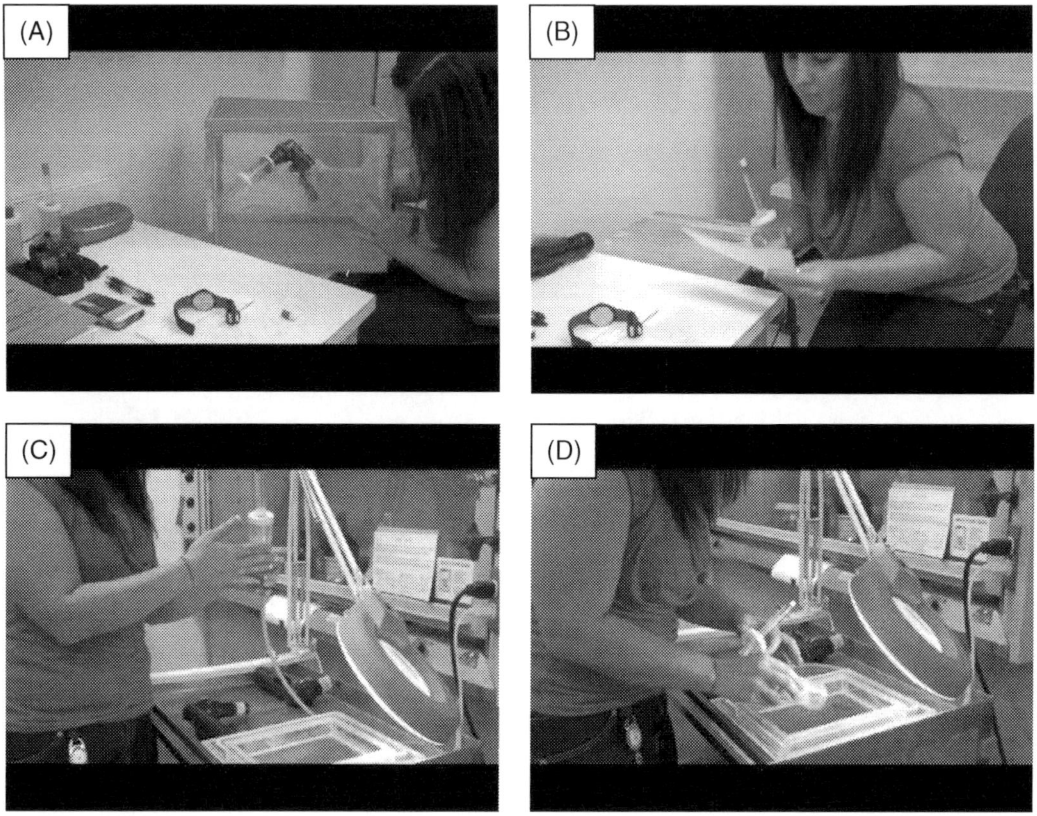

FIG. 5.7 (A) Laboratory technician uses an aspirator to collect all mosquitoes within a test cage. (B) After the mosquitoes are aspirated into the transfer tube, the tube and data sheet are transported to the next room for counting. (C) Mosquitoes are knocked down using carbon dioxide. (D) Mosquitoes remain incapacitated for counting by using a chilled table.

dioxide which is supplied to the table via tubing.

(19) The carbon dioxide is also emitted through the pores of the table surface continuously to keep mosquitoes incapacitated while they were counted by a laboratory technician (Fig. 5.8A). Mosquitoes that were obviously blooded were separated from the rest of the mosquitoes on the cold table. This procedure involved the use of a mounted lighted magnifying glass. All other mosquitoes that appeared to be unblooded were moved to a separate location of the table. The sum of all mosquitoes on the table was recorded as the test population on the data sheet (Fig. 5.8B). The laboratory technician conducted the counting procedure until the same count of mosquitoes was recorded three times for each section: those that were obviously blood-fed, and those that appeared unblooded.

(20) All mosquitoes that were thought to be unblooded were aspirated into the transfer container (Fig. 5.8C) and then emptied onto one side of an 8 1/2″ × 11″ piece of white copy paper that was prefolded lengthwise to produce a crease down the middle of the

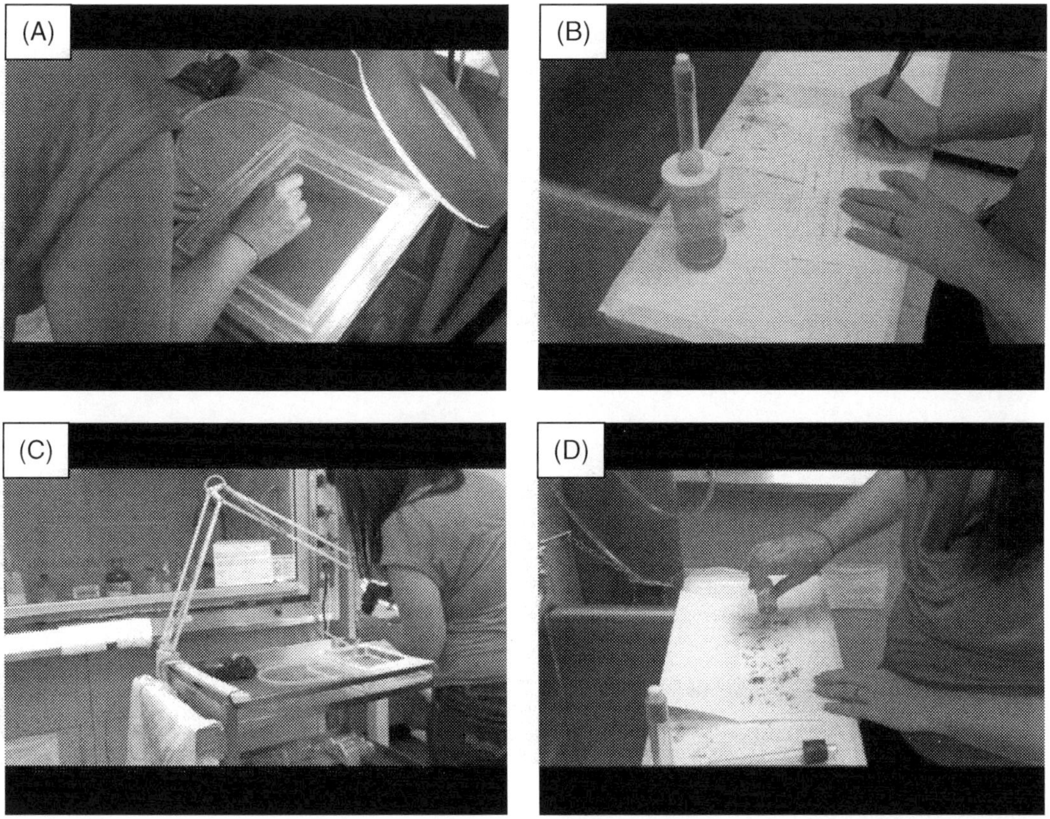

FIG. 5.8 (A) Mosquitoes are separated according to whether they appear to have had a blood-meal or unblooded. At times, a magnifying glass (top right) is used to aid in the determination. (B) The laboratory technician records the total number of mosquitoes aspirated from the cage. (C) Unblooded mosquitoes are aspirated from the chilled table to the transfer tube. (D) Mosquitoes are emptied from the aspirator container to a creased piece of paper.

paper (Fig. 5.8D). Mosquitoes were spread apart on the paper to allow identification of individual mosquitoes in the next phase of the counting process.

(21) After mosquitoes were distributed sufficiently across one side of the paper, it was folded in half along the crease, and a glass rod was used to compress the two sides of the paper together so that mosquitoes contained inside were crushed (Fig. 5.9A).

(22) After crushing the mosquitoes, the paper was reopened and the laboratory technician examined the crushed mosquitoes for evidence of those that had ingested blood (Fig. 5.9B). The number of mosquitoes that had ingested blood from this procedure were summed with those that were separated on the cold table and determined to be obviously blooded by their enlarged, red abdomens. The summed number of blooded mosquitoes were then recorded on the data sheet (Fig. 5.9C). The values from these data sheets were later transcribed into a spreadsheet for calculations and data analysis. Papers containing mosquitoes were disposed of in the container for laboratory waste.

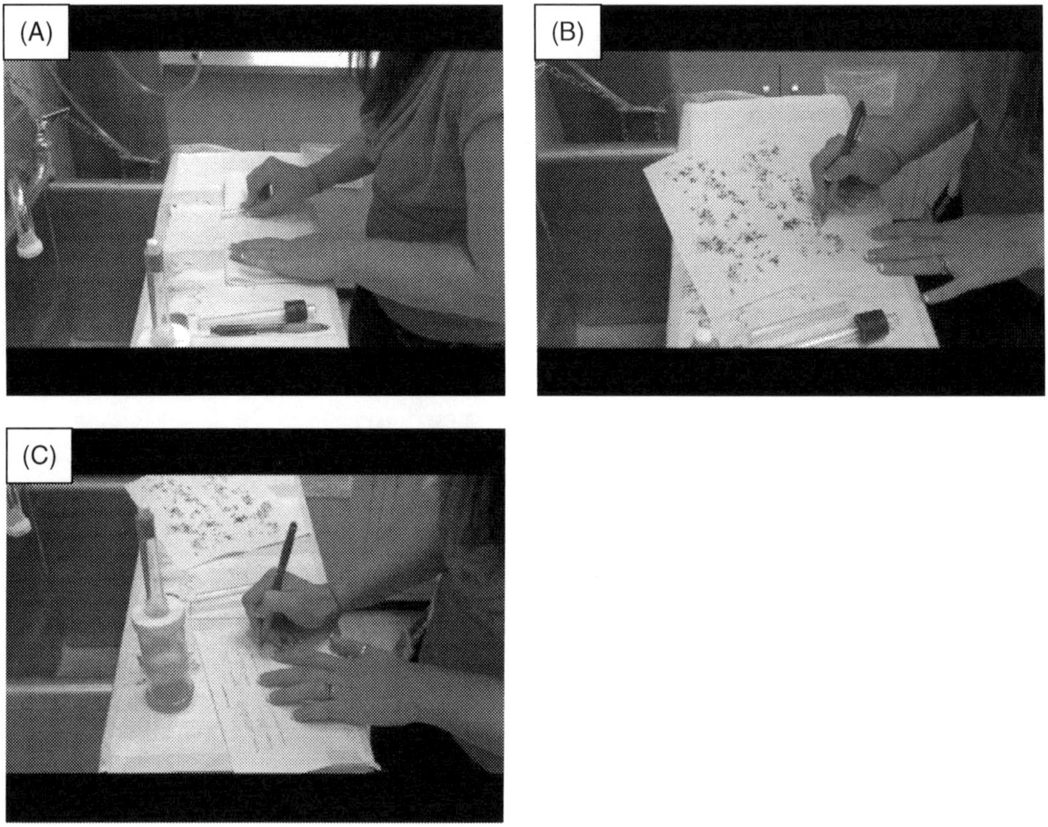

FIG. 5.9 (A) The laboratory technician folds the paper over with mosquitoes inside and used a glass rod to crush the mosquitoes. (B) The paper is reopened and crushed mosquitoes examined for evidence of a blood-meal. (C) The laboratory technician records the final count for the total number of mosquitoes that had a blood-meal.

5.3 Results of efficacy studies with US military uniform fabrics

During these assays, most of the mosquitoes landed on the surface of the fabric regardless of whether or not permethrin was present (Fig. 5.10A). If there was no permethrin or very low amounts of permethrin on the fabric, then mosquitoes attempted to feed for up to the full 15 minutes of the assay. When permethrin was present in sufficient quantity to intoxicate the mosquitoes, most remained in contact with the treated surface for approximately 1–4 minutes before being "repelled" from the treated uniform fabric (Fig. 5.10B).

Up until 2010, field treatment of uniforms with permethrin was more common than factory treatment. One field treatment method that provides excellent surface coverage of a single uniform set is the Individual Dynamic Absorption (IDA) kit. The kit contains plastic gloves, two bottles of permethrin in an emulsifier solution, two bags (one for each garment), and string. The permethrin solution is mixed with water in the bag, the garments are rolled, tied and placed in the respective bag for four hours, and then removed to dry. When the permethrin concentrations were compared at 0x, 20x, and 50x laundering cycles for IDA Kit and factory-treated Army Combat Uniforms (ACU), it was evident that the retention of the factory treatment uniforms with a binder was superior to the IDA kit treatment (Fig. 5.11). The bite protection results were similar at 0 and 20x for all treatments with the IDA kit treated ACU not dropping significantly until reaching 50 launderings. At this time, IDA kit treatment was also being explored for potential use with the Flame Resistant Army Combat Uniform (FRACU), however, due to the low absorption rate of the material and excess solution remaining in the bags when following label instructions, it was determined that factory treatment was the only viable option for this uniform.

The impact of weave tightness (air permeability) on the bite protection levels became evident when comparing data between the IDA Kit treated ACU and factory-treated FRACUs. Despite these materials being similar weight, weave construction (plain weave rip stop), and similar levels of permethrin initially, the bite protection was drastically different between the two, as seen in Fig. 5.12. This is due to the more open construction of the FRACU which leads to larger interstitial spaces between the yarns and a higher air permeability than the ACU (Fig. 5.13). Up until this time, it was assumed that as long as a uniform was treated at similar dose rates, it

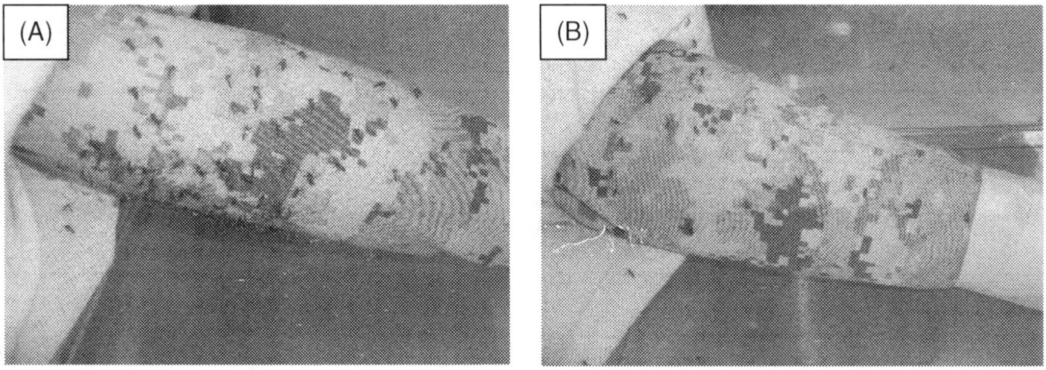

FIG. 5.10 **Bite protection assay (2).** Comparison of mosquito response to (A) untreated control sleeve and (B) treated unlaundered sleeve. The inset photo illustrates that with sufficient permethrin treatment, mosquitoes avoid prolonged contact with the sleeve.

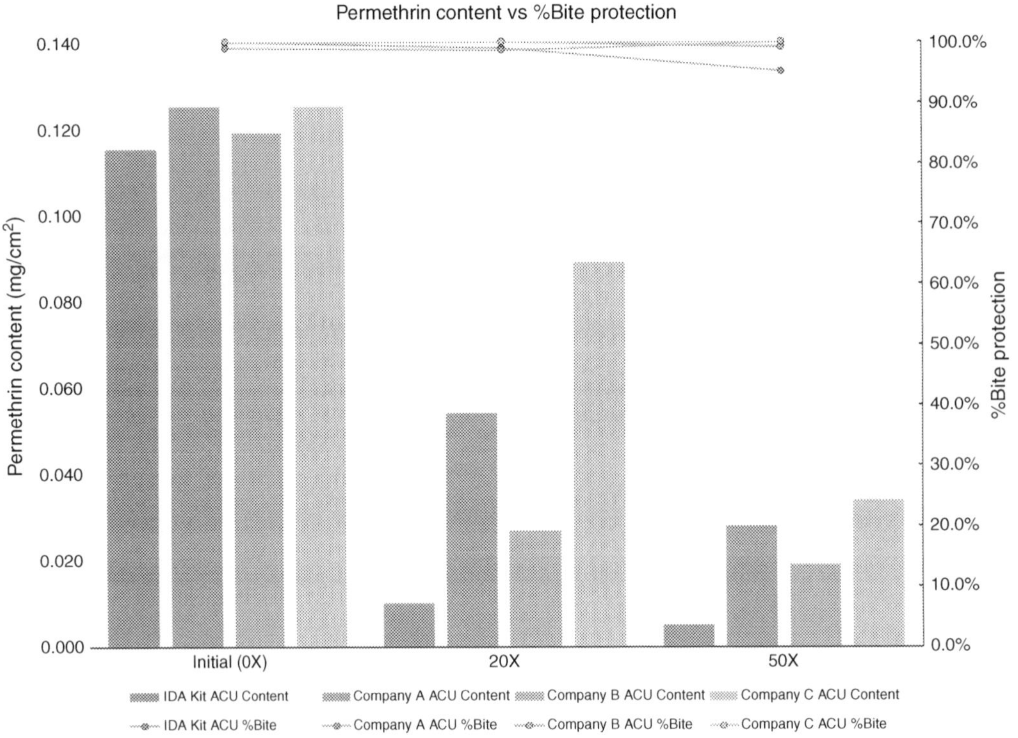

FIG. 5.11 Permethrin content and bite protection of factory-treated Army Combat Uniforms (*ACUs*) for three different permethrin applicators and Individual Dynamic Absorption (*IDA*) Kit treated ACUs.

would result in similar levels of bite protection which is seen in the ACU and MCCUU data shown in Fig. 5.14. Of note, the MCCUU is made up of a twill weave construction with the blouse and trouser being comprised of two different weight materials.

Currently, there are five known fabric variables that can impact bite protection, these are weave construction type, weave tightness (air permeability), thickness, permethrin concentration, and amount of binder. Weave construction becomes a factor because this affects the number of yarn intersections (interstitial space) within the fabric. Generally speaking, the higher the frequency of yarn intersections the easier it is for the mosquito to find and take a blood meal. The effect of weave selection on the number of interstices can be seen when comparing a 2 × 1 twill to a plain weave fabric (Fig. 5.15). Of course the tighter the weave, the smaller the interstitial space, which reduces the ability of a mosquito to insert the stylet. Thus weave construction and tightness are major contributing factors for bite protection of materials. Fig. 5.16 shows a comparison of percent (%) bite through and air permeability of a variety of twill and plain weave fabrics. This figure illustrates a general trend that the more permeable (i.e., open construction of the uniform), the more difficult it is to prevent bites through the uniform fabric, especially as the permethrin surface concentrations decreased, as shown in Fig. 5.17. Not enough data have been collected on materials of different constructions, weights, and thicknesses to indefinitely note that these known variables are the only ones that impact bite protection. As more fabrics are treated with permethrin and tested, it may be determined that other fabric variables may

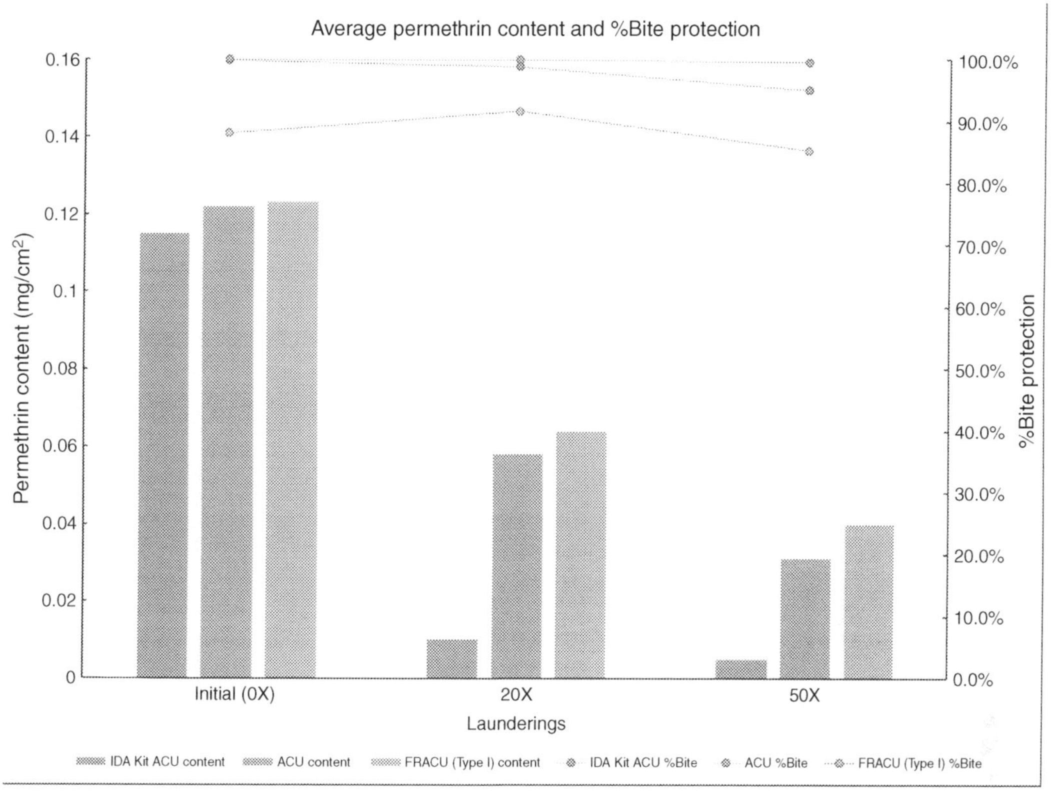

FIG. 5.12 Permethrin Content and Bite Protection of IDA Kit treated ACUs and factory-treated FRACUs. Data shown for factory treatment are the average for multiple permethrin applicators. *ACUs*, Army Combat Uniforms; *FRACU*, Flame Resistant Army Combat Uniform; *IDA*, Individual Dynamic Absorption.

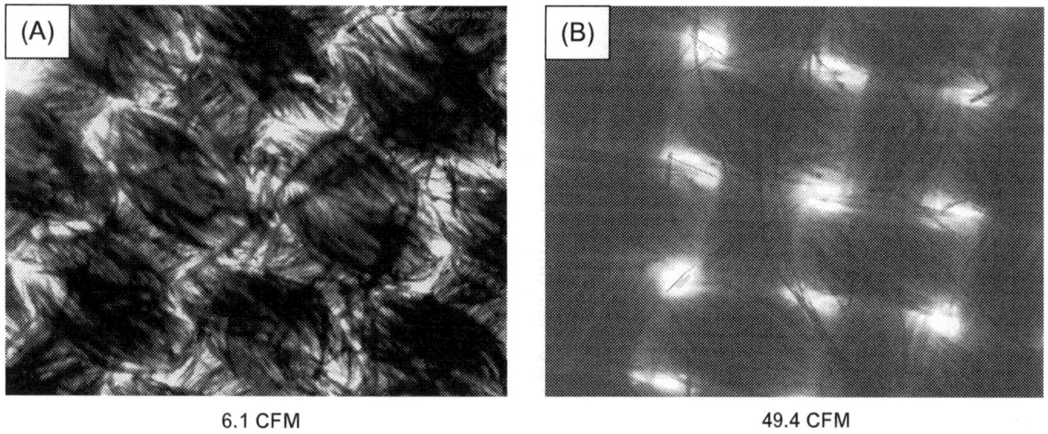

FIG. 5.13 **Comparison of interstitial spaces visible from microscopic examination of uniform fabrics.** The fabrics were (A) ACU fabric with an approximate air permeability of 6.1 CFM and (B) FR Type I (FRACU) fabric with an approximate air permeability of 49.4 CFM. Both fabrics are the same weight and weave construction. *ACU*, Army Combat Uniform; *FRACU*, Flame Resistant Army Combat Uniform.

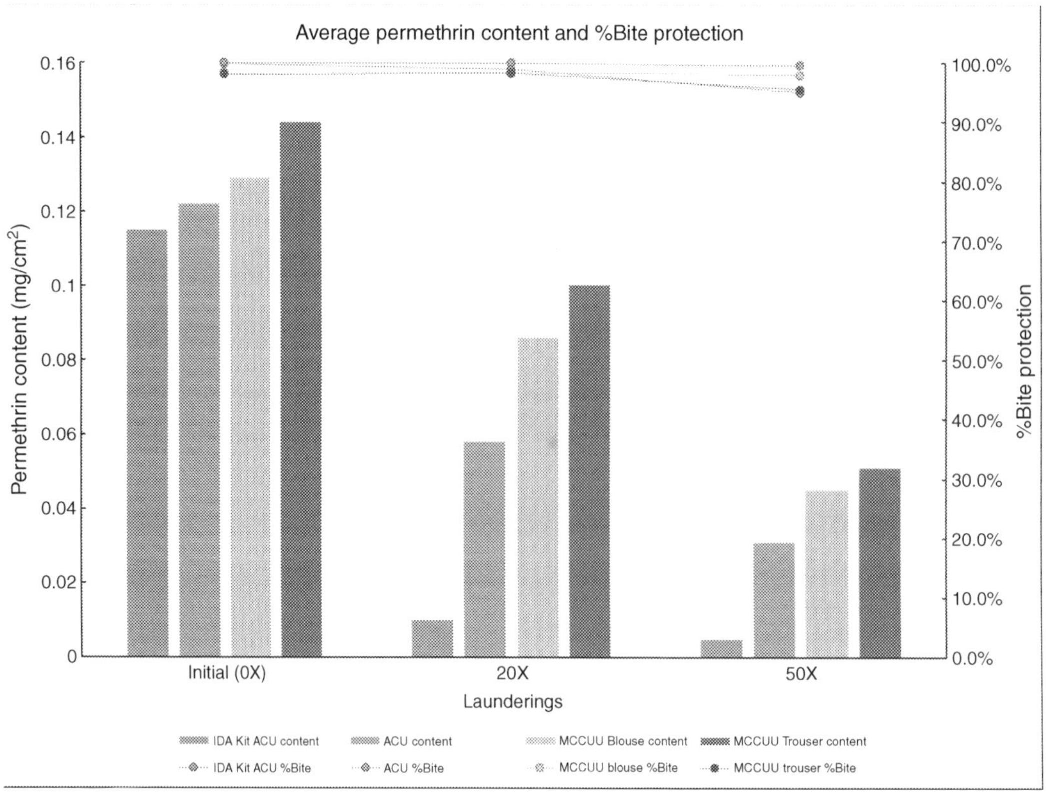

FIG. 5.14 Permethrin content and bite protection of ACUs and MCCUUs. All garments were factory-treated with the exception of the IDA Kit treated ACUs. Data shown for factory treatment are the average for multiple permethrin applicators. *ACUs*, Army Combat Uniforms; *IDA*, Individual Dynamic Absorption; *MCCUUs*, Marine Corps Combat Utility Uniforms.

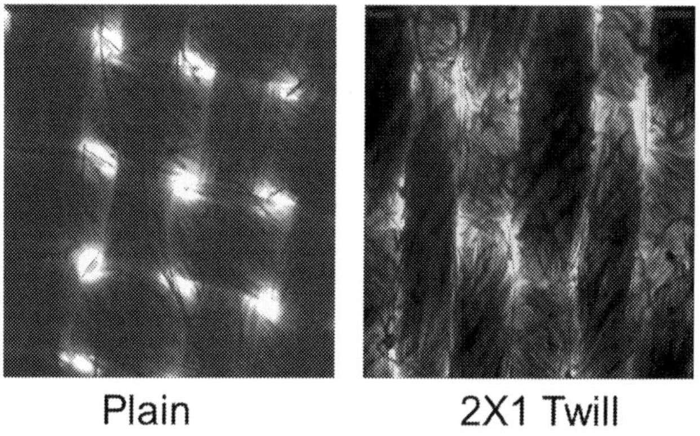

FIG. 5.15 The 1 × 1 weave pattern of the plain weave has the highest frequency of interstitial spaces due to the high number of over and under passes of the weft yarn to the warp yarn. The 2 × 1 twill consists of two weft yarns passing over and under a single warp yarn resulting in the diagonal pattern associated with denim. The looser either of these weaves, the easier it is for the mosquito to take a blood meal.

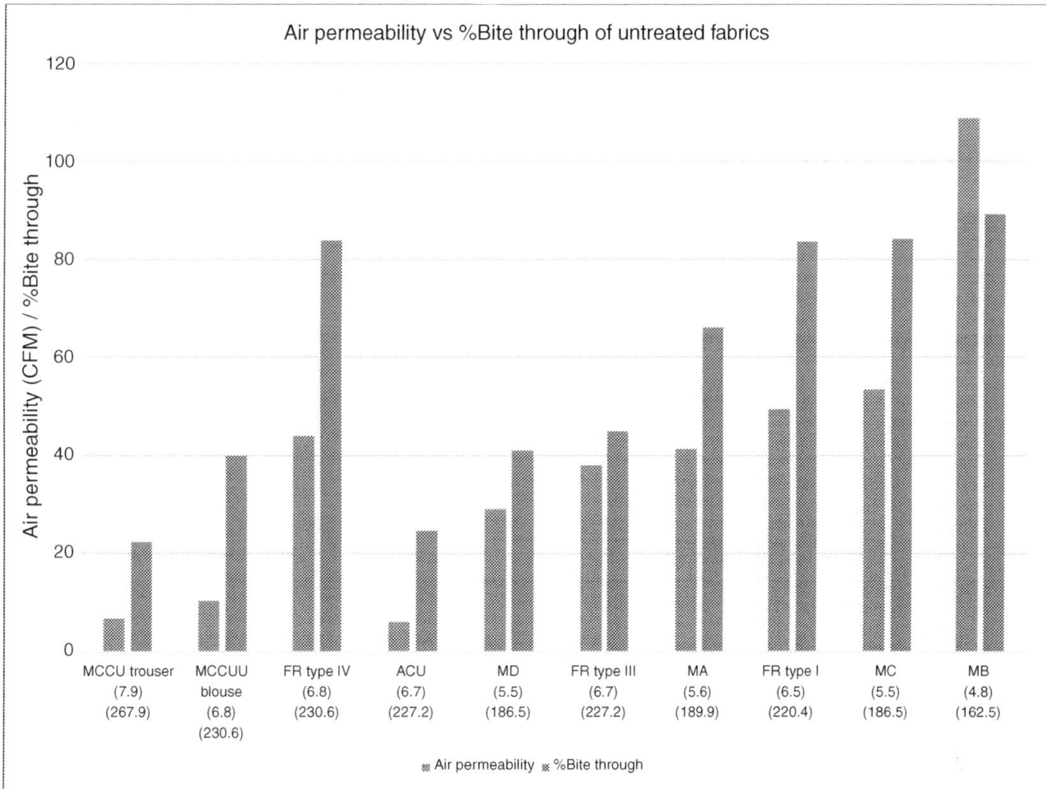

FIG. 5.16 Air permeability and %bite through of various untreated fabrics. Twill weave construction fabrics are on the left and plain weave construction fabrics are on the right in order of increasing air permeability. The weight of the fabric in oz./yd^2 and (g/m^2) is indicated beneath each fabric type.

also impact levels of protection such as weight/thickness, yarn type (filament, staple, wrapped, etc.), yarn hairiness, yarn count, and yarn size. It is not believed that fiber type or blend has an impact on the bite protection, at least directly in terms of fabric construction variables. The fiber type may however impact the permethrin retention thus indirectly resulting in reduction in protection due to reduced permethrin concentrations after laundering. Binder selection and amount could assist in improved retention.

The amount and type of binder added during the permethrin impregnation process is one parameter that can be adjusted during factory treatment. Each permethrin applicator uses their own proprietary treatment formulations, with their binder systems and amounts varying, which enables the comparison of different binder systems within a uniform type. As shown in Fig. 5.18, when comparing data from FRACUs that were treated by three different permethrin applicators, it is evident that treatment formulation will affect the efficacy results. It can be assumed that the lower initial bite protection on the material treated by Company C was due to the lower amount of permethrin applied, however as evidenced by the improved permethrin retention and more stable efficacy results through 50 launderings, it can be theorized that the lower bite protection was also due to the use of a higher level of binder. Companies A and B start off with higher bite protection but have a

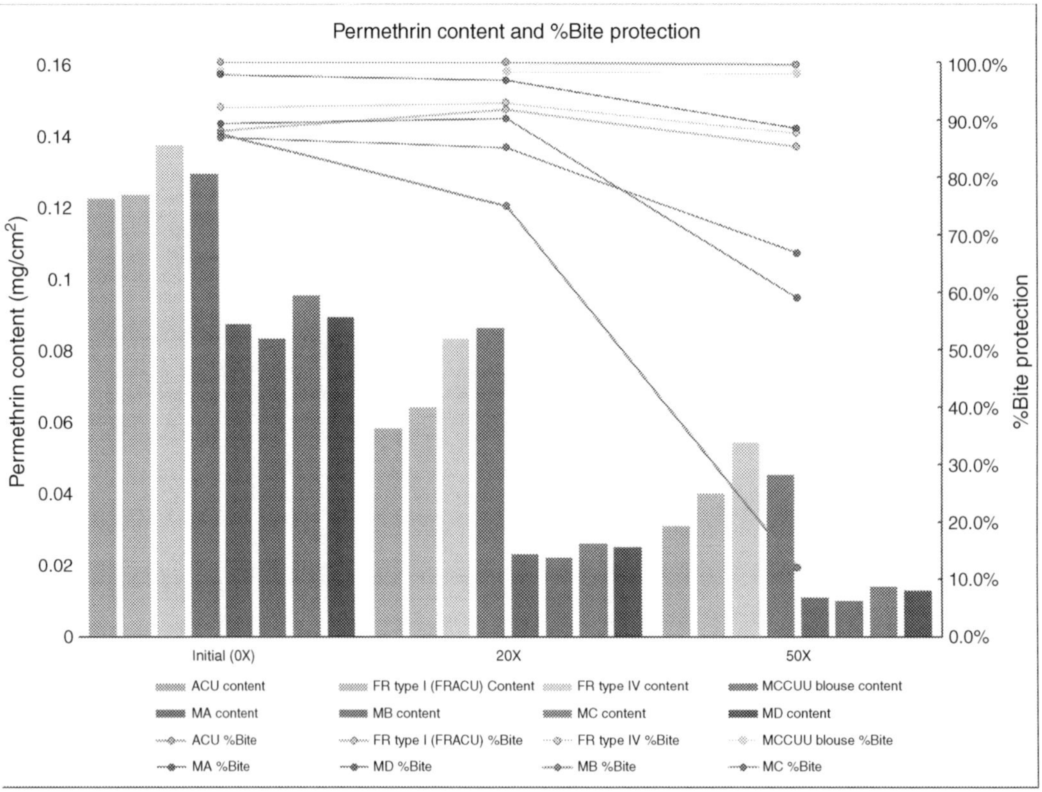

FIG. 5.17 Permethrin content and bite protection of factory-treated fabrics. Data show the average for multiple permethrin applicators.

significant dip in permethrin concentration and bite protection through laundering. A higher amount of binder initially will block any available permethrin on the surface thus limiting contact with the mosquitoes and reducing repellency. As more binder is removed through laundering, more free permethrin becomes available to enable the repellency effect on the mosquitoes coming in contact with the fabric surface. There is a delicate balance between the amount of binder needed during the permethrin application to ensure sufficient efficacy initially and throughout laundering. This can be seen when comparing the company reformulations on the FR Type I material that took place in 2010.

During the qualification phase for FRACUs in 2010, the US Army adopted a bite protection performance level to account for the more open construction of the uniform fabric. The minimum bite protection specifications were set at a minimum of 85% at 0x launderings, 80% at 20x launderings, and 70% at 50x launderings. In 2012, the Army adopted a new material to address durability issues with the FRACU (Type I fabric) in the field. This new fabric construction was identified as Type III in the fabric specification and consisted of a different yarn count and functional finish application to improve tear strength and abrasion resistance. The weave construction (plain weave rip stop) remained the same. The Type III material showed similar bite performance to the Type I material and the minimum bite protection requirements were retained.

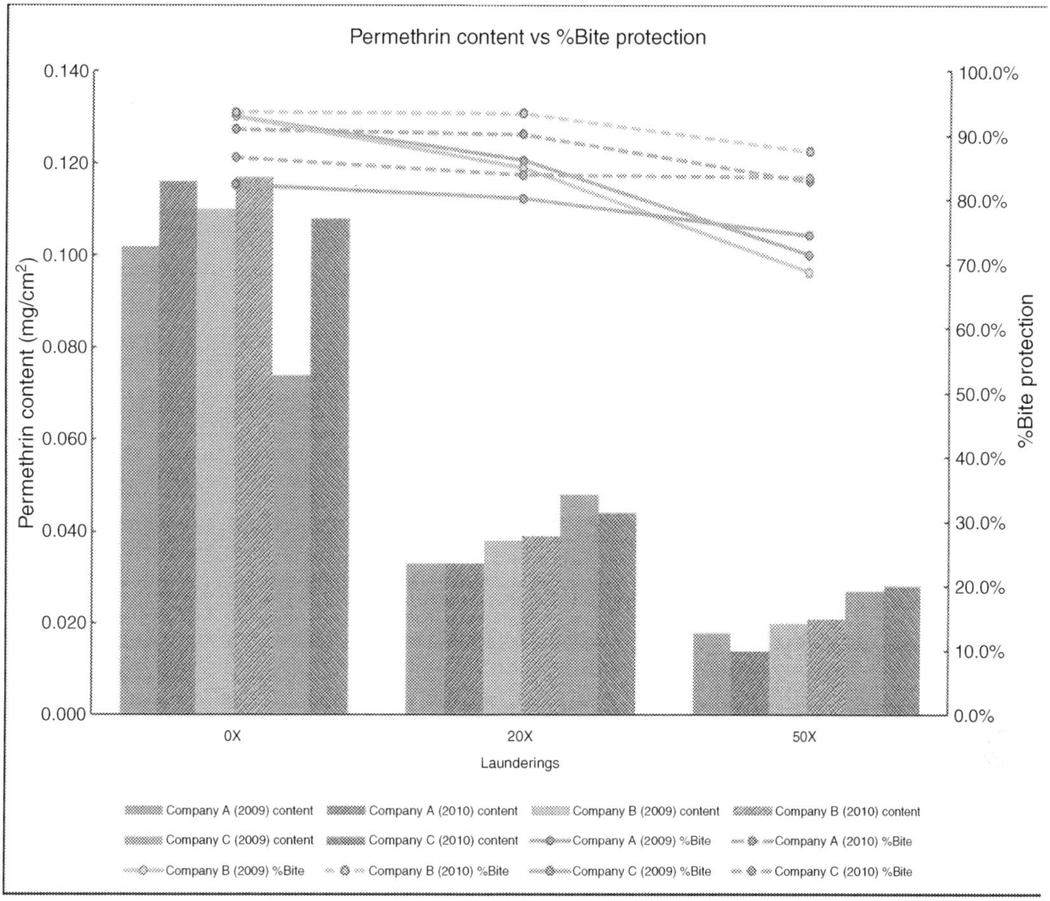

FIG. 5.18 Effect of binder formulation on the bite protection of Factory Treated Flame Resistant Army Combat Uniforms (*FRACUs*). These uniforms were manufactured with the Type I Fabric in 2009 followed by reformulations in 2010.

Regardless of fabric construction, air permeability, fiber type or binder, all treated uniforms have one thing in common; permethrin loss during laundering. Permethrin loss in the wash occurs not only due to free permethrin being removed due to surfactants but also to fiber loss due to surface abrasion. This fiber loss is further compounded during normal wear of the treated garments, resulting in not only lower levels of permethrin, but also fabric degradation and more open interstices leading to additional reductions in efficacy through the life of the uniform.

5.4 Laboratory methods for evaluation of arthropod repellent treated civilian clothing

The laboratory methodology for the evaluation of civilian fabrics is similar to that of military fabrics because these textiles are worn next to skin and the use relies on the combination of fabric construction and activity of the arthropod repellent to confer bite protection. There are a variety of modes of action that contribute to the bite protection and in the case of permethrin and other pyrethroids, this consists of excito-repellency, knockdown, and mortality. The most common test methods for repellent-treated textiles are tests that involve cage, cone, and excito-chamber tests (Anuar and Yusof, 2016; Agramonte et al. 2016; Agramonte et al., 2017). As Anuar and Yusof (2016) reported, test methods using mosquito attractants might imitate the more realistic situation of mosquito biting. Therefore, cage tests using human volunteers are perhaps the most rigorous methodology to assess protection in a standardized manner. The cage test is the most common and direct test method for evaluation of arthropod repellent treated materials. Cage tests are also useful for evaluating the interaction of repellent-treated materials with the emergent issues of insecticide resistance in mosquito control. Agramonte et al. (2017) found that pyrethroid resistance alters the blood-feeding behavior in *Ae. aegypti* mosquitoes exposed to treated fabric, such that higher concentrations of pyrethroids are necessary to deter the pyrethroid-resistant mosquitoes. As noted earlier, use of human volunteers requires ethical approval, IRBs, and documentation of informed consent. Development of a proper study design also requires knowledge of the statistics needed to best design the study, so that replications are sufficient to reduce the data discrepancies caused by individual variations among volunteers (participant race, weight, sex, age, etc.) but also not be so rigorous as to unnecessarily evaluate more individuals than needed to produce a scientifically valid conclusion.

In recent studies of civilian clothing, a screened cage (40 cm^3) containing 200 7–8-day-old starved female mosquitoes (daytime bites) were used for each replicate of the evaluation. Participants wore a long-sleeved rubber glove on one hand/arm with a window cut out (6 cm L × 5 cm W) on the front sleeve of the glove (Fig. 5.19). Next, a piece (7 cm L × 6 cm W) of treated textile was used to cover the window on the glove sleeve. The participant exposed the treated arm into the cage with mosquitoes for 5 minutes. This exposure was repeated at 1-hour intervals until mosquitoes were observed to land and probe through the treated textiles covering the window. Alternatively, similar to US military fabric testing, subjects wore a short-sleeved glove to protect the hand and had their arms wrapped with a large piece of repellent-treated textile such that it covered all parts of the forearm. The covered

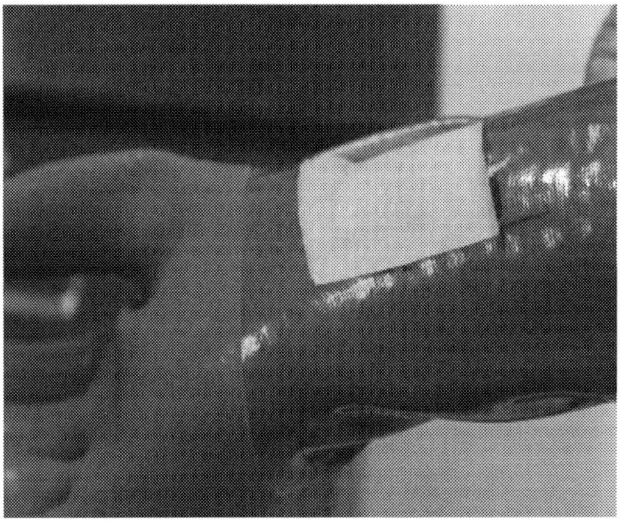

FIG. 5.19 The rubber glove with a cut window (6 cm × 5 cm) on the human volunteer's forearm and the open area was covered with treated textile, then exposed to a testing cage holding 200 female mosquitoes.

forearm was then inserted into the cage of mosquitoes (Fig. 5.20) with a test interval of 5 minutes. Testing was repeated at 1-hour intervals until the landing, probing, and biting occurred.

For civilian clothing evaluations of arthropod repellent treated textiles against mosquitoes, conducting a field test with a group of 6–13 human volunteers is the desired final step. The distance between each volunteer should be ≥10 m. The population density of the target species of mosquitoes, activity peak period, and biting pressure in the testing location should be checked prior to conducting the field testing. The arthropod repellent treated textiles or uniforms are worn by the human volunteers. Based on the textile product and purpose, a specific body part (arm or leg) of the participants will be selected to cover or wrap for testing. The rest of the body is protected by thick, untreated clothing, or by a plastic cover. Usually, repellent-treated head nettings, uniforms, jackets, and t-shirts are tested against mosquito bites in the field. Sometimes a large piece of treated textile is used to cover/wrap the lower part of legs or forearms and a second untreated textile is used to cover or wrap the other leg or forearm as a control. The volunteers sit or stand in the field to expose themselves to mosquitoes for 5 minutes. This exposure is repeated at 1-hour intervals until the mosquitoes land and probe on the treated materials. The number of mosquitoes landing/probing on the treated and untreated body parts (arms or legs) is collected and counted. In addition, the air temperature, humidity, wind speed, and other environmental conditions are detected and recorded. The materials used, arthropod repellent, formulation and amount used, method of treatment, and method of exposure for the field test (Govere and Durrheim, 2006) should be included in the report.

5.5 Conclusion

Past comparisons of the permethrin concentration to the biological responses of knockdown and mortality have not provided adequate correlation of permethrin in uniforms, particularly at lower permethrin concentration levels (Schreck et al., 1982). Therefore, the examination of efficacy of permethrin-treated uniforms is now focused on prevention of bites relative to an untreated control specimen. Because of this, the bite protection protocol was developed for and optimized for use with US military uniforms. It is the standard by which the US Army and United States Marine Corps qualify applicators to treat combat uniforms. Throughout the past decade of research on treated uniforms, it is evident that the fabric construction and

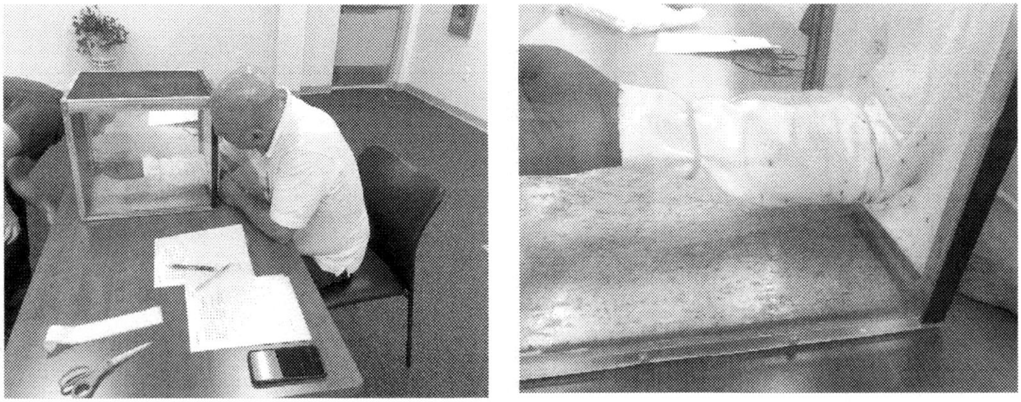

FIG. 5.20 A human volunteer's whole forearm was covered and wrapped by repellent-treated textiles or untreated textiles, and then exposed to the testing cage with 200 female mosquitoes.

openness of the weave plays a significant role in the level at which a permethrin-treated uniform prevents bites. Air permeability can be a good indicator of anticipated level of bite protection when comparing similar weave types. There is a probability that mosquitoes can insert their stylets through the interstitial spaces in fabrics and provided sufficient permethrin is on the surface and the weave is tight enough, mosquitoes will become intoxicated by the permethrin and either be repelled from the surface, or with sufficient exposure, can be knocked down or killed. Other components that can influence the ability of mosquitoes to bite through fabrics are the size and frequency of the interstitial spaces due to yarn size and weave construction and the thickness and weight of the fabric. Mosquitoes may find it more difficult to reach skin capillaries through a thick, heavy fabric compared to relatively thinner materials. Additionally, the fiber composition of the fabrics can affect permethrin absorption and retention; specifically, the flame-resistant fabrics which therefore require factory production to impregnate permethrin using a binder. There is still a lot to be learned about arthropod-treated fabrics and the factors that affect the level of protection. More data is needed to determine which additional factors, if any, will impact the level of protection afforded by the treated fabric.

An important point regarding the evaluation of fabrics in the laboratory is that the methods have been devised to allow for a standard method of assay primarily for comparative purposes between various fabric constructions and treatments. It can be implied that the higher the bite protection, the more protective the fabric. This, however, does not necessarily correlate to an equivalent level of protection in the field. Additional factors are important in the field, such as mosquito species, presence of local mosquito populations that are resistant to the repellent treatment, environmental conditions, and the behavior of the human wearer. Some of these factors have been studied, but large-scale field studies are still needed to correlate the level of protection in the field back to the bite protection (or bite prevention) level that has been determined in the laboratory under standardized conditions using susceptible strains.

References

Abbott, W.S., 1925. A method of computing the effectiveness of an insecticide. J. Econ. Entomol. 18, 265–267.

Agramonte, N.M., Gezan, S.A., Bernier, U.R., 2016. Comparative evaluation of a silicone membrane as an alternative to skin for testing mosquito repellents. J. Med. Entomol. 54, 631–637.

Agramonte, N.M., Bloomquist, J.R., Bernier, U.R., 2017. Pyrethroid resistance alters the blood-feeding behavior in Puerto Rican *Aedes aegypti* mosquitoes exposed to treated fabric. PLoS Negl. Trop. Dis. 11, e0005954.

Anuar, A.A., Yusof, N., 2016. Methods of imparting mosquito repellent agents and the assessing mosquito repellency on textile. Fash. Text. 3, 12.

Brown, M., Hebert, A.A., 1997. Insect repellents: an overview. J. Am. Acad. Dermatol. 36, 243–249.

Clark, G.G., Bernier, U.R., Allan, S.A., Kline, D.L., Golden, F.V., 2011. Changes in host-seeking behavior of Puerto Rican *Aedes aegypti* (L.) following colonization. J. Med. Entomol. 48, 533–537.

Govere, J.M., Durrheim, D.N., 2006. Techniques for evaluating repellents. In: Debboun, M., Frances, S.P., Strickman, D. (Eds.), Insect Repellents Handbook. CRC Press, Boca Raton, FL, pp. 147–159.

Kimani, E.W., Vulule, J.M., Kuria, I.W., Mugisha, F., 2006. Use of insecticide-treated clothes for personal protection against malaria: a community trial. Malar. J. 5, 63.

Posey, KH, Schreck, CE., 1981. An airflow apparatus for selecting female mosquitoes for use in repellent and attraction studies. Mosq. News. 41, 566–568.

Schreck, C.E., Kane, F., Carlson, D.A., 1982. Permethrin impregnations of military fabrics: an evaluation of application rates and industrial methods by bioassay and gas chromatographic analysis. Soap Cosm. Chem. Spec. 58, 36–39.

Schreck, C.E., Haile, D.G., Kline, D.L., 1984. The effectiveness of permethrin and deet, alone or in combination, for protection against *Aedes taeniorhynchus*. Am. J. Trop. Med. Hyg. 33, 725–730.

Schreck, C.E., Kline, D.L., 1989. Personal protection afforded by controlled-release topical repellents and permethrin-treated clothing against natural populations of *Aedes taeniorhynchus*. J. Am. Mosq. Control Assoc. 5, 77–80.

Sholdt, L.L., Schreck, C.E., Qureshi, A., Mammino, S., Aziz, A., Iqbal, M., 1988. Field bioassays of permethrin-treated uniforms and new extended duration repellent against mosquitoes in Pakistan. J. Am. Mosq. Control Assoc. 4, 233–236.

CHAPTER

6

Repelling mosquitoes with electric fields

Ulla Gordon[a], *Farooq Tanveer*[a], *Andreas Rose*[a], *Krijn Paaijmans*[b,c,d]

[a]BioGents AG, Regensburg, Germany, [b]Center for Evolution and Medicine, School of Life Sciences, Arizona State University, Tempe, Arizona, United States, [c]The Biodesign Center for Immunotherapy, Vaccines and Virotherapy, Arizona State University, Tempe, Arizona, United States, [d]ISGlobal, Barcelona, Spain

6.1 Electric fields

Electric fields are physical fields created by electric charges or time-varying magnetic fields. This chapter focuses on electric fields generated by electric charges between two electric conductors and how they can be used to repel mosquitoes.

An electric field is defined as the electric force per unit charge. It emanates from an electric charge and transmits its force on other charges in its vicinity. The electric field around a point charge can be visualized as radial lines of force (orienting away from a positive charge and toward a negative charge, Fig. 6.1) with the resulting field strength decreasing inversely in proportion to the square of the distance, i.e., the electrical field weakens with growing distance to the point charge (Brodie, 2000; Roche, 2016).

Constant, homogenous electric fields can be created between two parallel conductive surfaces with opposite charges, creating a potential difference (voltage) (Fig. 6.2). In this example, the field strength vector is perpendicular to the surfaces and orients from the positive to the negative charge. The electrical field strength is constant; it is proportional to the charge and inversely proportional to the distance between the surfaces (Marinescu, 2009). Thus, the magnitude of the electric field (E) is:

$$E = -\Delta V / d$$

ΔV = potential difference.
d = distance between the surface.

Advances in Arthropod Repellents.
DOI: https://doi.org/10.1016/B978-0-323-85411-5.00015-7

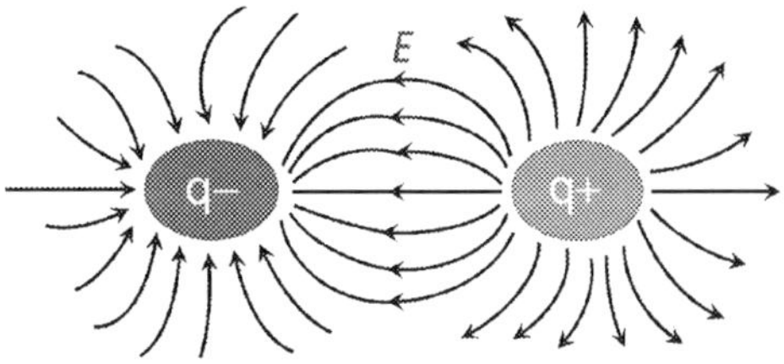

FIG. 6.1 Electric field (E) between two charges. By Farooq Tanveer, Biogents AG.

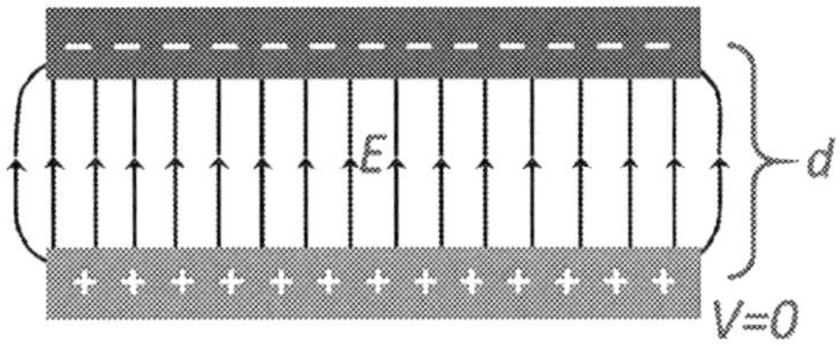

FIG. 6.2 Electric field between two parallel charged metal plates. By Farooq Tanveer, Biogents AG.

6.1.1 Natural occurrence and potential impact

In nature, electric fields are omnipresent. They occur natural or are man-made (anthropogenic). Natural electric fields are found above the surface of the earth, generated by a potential difference between the ground and the ionosphere (König et al., 1981). The earth field has a strength of about 100–300 V/m, depending on time of day and season, local temperature, and humidity, and decreases with height. Common man-made electric fields occur under power transmission lines, are generated by transportation (e.g., electric rail/bus systems), and visual display units such as TV screens, or are caused by charge separation as a result of friction (e.g., walking on nonconductive surfaces). While the short-term exposure to static magnetic fields can induce changes in blood pressure and heartbeat in humans, there is no evidence that the short-term exposure to static electric fields causes acute adverse effects in human health (World Health Organization, 2006). In animals, the exposure to anthropogenic electromagnetic fields may impede their orientation in the magnetic field of the earth, which is especially important for migratory birds and insects (Balmori, 2015).

6.1.2 Electroreception

A variety of animals respond to electric fields, especially those inhabiting the aquatic environment. Electroreception is very common in fish, teleost (bony) fish, amphibians, and dolphins (Von der Emde, 2013). Sharks and rays use weak electric fields that are emitted by other animals in their vicinity to locate their prey; gymnotiform fish, like eels, use lower voltage pulses for navigation, and detection of prey but are also able to generate high voltage electric shocks (Catania, 2015). Two electro sensory structures found in aquatic or semiaquatic organisms have evolved independently. Ampullae are small tubular cavities that detect differences in the electric potential between the inside of the animal and the aquatic environment and trigeminal electroreceptors respond to negative and positive charges (Von der Emde, 2013). Both structures depend on water as an electrically conductive medium. Up to the late 2000s, little was known about electroreception in terrestrial animals which must navigate in an electrically nonconductive environment (air). In humans, body hairs seem to be involved with the perception of electric fields, mainly through hair movement elicited by the electric field (Chapman et al., 2005). Comparable mechanisms involving certain body appendages, like wings or antennae, could also contribute to the electroreception in insects (Newland et al., 2008).

Recent research has shown that both honeybees and bumblebees are able to detect electric fields with their antennae (honeybees) (Clarke et al., 2013) and mechanosensory hairs (bumblebees) (Sutton et al., 2016) providing further fascinating insight into the pollination process and plant-pollinator communication. Entomophilic flowers use a variety of visual and chemical cues to attract pollinators and most flowers also exhibit a negative electric potential (Corbet et al., 1982) mainly on the edges of petal, stigma, and anthers (Clarke et al., 2017). Pollinators are usually positively charged, thus, during their interaction with the flower, the potential difference between the insect´s surface and the stigma promotes efficient pollen transfer and adhesion (Vaknin et al., 2000). As a consequence of the deposition of pollen, the electric potential of the flower changes and so does the pollination status. Bumblebees are able to discriminate between rewarding and non-rewarding electric fields, which greatly contributes to a rapid and dynamic communication between flowers and their pollinators (Sutton et al., 2016).

Antennae and mechanoreceptive hairs are not unique to bees and it is quite possible that other arthropods use the same structures to detect electric fields (Clarke et al., 2017). In a laboratory set-up, cockroaches (*Periplaneta americana*) avoided static electric fields of 8 kV/m and above. They were able to detect the electric field with their antennae and hair plates at the base of the scape. It was found that the random charge on the cockroaches' body changed in close proximity to a positively charged electrode, with negative charges being attracted to the long antennae. As a consequence, the antennae were drawn to the positively charged electrode and bent which caused an avoidance behavior (Newland et al., 2008). A similar mechanism might also apply to other insects making static electric fields an interesting tool for alternative pest control strategies. Research on other arthropods supports the hypothesis that, in general, insects are repelled by static electric fields (Hunt et al., 2005; Maw, 1961, 1962; Newland et al., 2008). A study from Japan investigated the avoidance behavior of a variety of insect species and spiders to an electric field screen that was generated between negatively charged conductor wires and a positively charged earthed metal net (Matsuda et al., 2015). All tested species (including the orders Coleoptera, Hemiptera, Diptera, Hymenoptera, Lepidoptera, and Blattodea) were deterred by the electric field, however, the Voltage (kV) required to cause the avoidance effect differed. The Asian tiger mosquito, *Ae. albopictus* was found to be repelled by 1.7 kV/cm while certain beetle species responded to voltages above 13 kV/cm. The authors also reported that insects contacted the negatively charged screen with their antennae and subsequently turned away; the voltage required to induce this behavior depended on the length of the antennae and body size.

6.2 Challenges in mosquito control

Mosquito-borne diseases remain a major threat to human health all around the globe. Mosquito control programs are constantly challenged by novel arboviruses, growing insecticide resistance, and the spread of invasive mosquito species (Benelli et al., 2017). When there is no (prophylactic) drug and/or vaccine available, control of the mosquito-borne disease relies on the control of the vector. Vector control is primarily carried out through intertwining measures such as (1) source reduction (elimination of breeding sites), (2) killing the vector in the larval or adult stage, and (3) creating physical or chemical barriers between the vector and its host (Norris and Coats, 2017). Killing the adult vector is still widely done using synthetic insecticides, however, the available toolbox is limited to a few classes of chemicals with comparable action and their success in controlling vectors like *Ae. aegypti*, *An. gambiae*, or *Cx. pipiens* is limited due to the emergence and rapid spread of insecticide resistance in these species (Fonseca-González et al., 2011; Marcombe et al., 2011, 2014; Weill et al., 2003). This is why research for alternative vector control strategies continues. Over the last 15 years, a variety of promising new approaches have been investigated to augment existing measures, including lethal traps (Barrera et al., 2014; Degener et al.,

2014; Kröckel et al., 2006), genetically modified mosquitoes (Thomas et al., 2000); release of mosquitoes with Wolbachia (Rasgon et al., 2003; Sinkins and O´Neill, 2000), attractive targeted sugar baits (Fiorenzano et al., 2017; Lea, 1965); spatial repellents (Achee et al., 2012; Bibbs and Kaufmann, 2017; Norris and Coats, 2017; Ogoma et al., 2014); push–pull (Menger et al., 2015; Obermayr et al., 2015; Paz-Soldan et al., 2011) and investigating the potential of "green chemistry," including entomopathogenic fungi (Scholte et al., 2004) and plant terpenoids (Norris et al., 2018).

The use of electric fields to repel mosquitoes is enticing as it would offer a means of controlling mosquitoes physically thereby circumventing a potential adaptation or resistance in the target organism. Here, we introduce and summarize an approach that uses static electric fields to elicit avoidance behavior in *Ae. aegypti* and also investigate whether females´ exposure to electric fields has an impact on their reproductive fitness.

6.3 Assessing the repellency of electric fields in the laboratory

6.3.1 Test mosquitoes *Ae. aegypti*

Five to ten days old *Ae. aegypti* females were used for all laboratory tests. The colony was originally obtained from BAYER AG (Monheim, Germany) in 1998 and has been maintained in the Biogents facilities over the past 22 years. Mosquitoes were reared at a temperature of 27 ± 0.5 °C and 70 ± 5% relative humidity (RH) under a photoperiod of 12:12 (L:D). The light period (full spectrum LEDs, 450 Lux) was set from 8:00 to 20:00. After hatching of the eggs, larvae were kept in water basins (30 × 30 × 10 cm) filled with a 1:1 mixture of deoxygenized tap water and deionized water and fed with Tetramin fish food flakes (Tetra GmbH, Melle, Germany). Pupae were transferred to breeding cages (40 × 30 × 20 cm) for adult emergence. Adult mosquitoes were provided with a 10% sugar solution (dextrose) on filter paper. Nulliparous females were used for all tests, they were selected based on host-seeking behavior, as described by Obermayr et al. (2015). The breeding cage contained a circular opening covered by fine mosquito netting in the left wall, while the right wall was fitted with a port and rotating door, where a transport container could be attached. The transfer container consisted of a Perspex cylinder with a rotating door on one end and a cover made from fine mosquito netting at the other end. A fan running at 7.5 V was connected to the opening in the left wall of the breeding cage, while a human hand was held against the mosquito netting of the transfer container on the opposite side of the cage and rotating doors were opened. Female mosquitoes seeking a blood meal flew upwind into the transfer container, attracted to the skin odors, and were immediately used in the experiments.

6.3.2 Barrier assays in cage tests

The repelling potential of different static electric fields (generated between two parallel plate electrodes, the conductors) was evaluated in a specifically designed laboratory cage test set-up. All experiments were performed under standardized conditions in a climatized room without windows. The temperature and RH of the room air were set to 27.5 ± 0.5 °C and 75 ± 5%. The room was illuminated with full-spectrum LED light tubes (intensity 450 Lux).

Two cubic mosquito rearing cages with a volume of 27 L (BugDorm-1 Insect Rearing Cage, from Watkins & Doncaster, Herefordshire, United Kingdom) were connected by a glass tunnel that allowed mosquitoes to fly from one cage to the other. To produce uniform electric fields, five copper plate electrodes were placed parallel in the center of the tunnel (20 × 10 × 9.6 cm) and connected to an adjustable high-voltage

(HV) device (Spellman, Model: V6 DC 15 KV POS 2MA W/O RS232) in such a way that negatively charged plates and grounded copper plates were alternating. The distance between the plate electrodes was 2 cm and was chosen because it ensures the generation of strong electric fields and still allows mosquitoes to fly through (preliminary experiments, data not shown). The opposing walls of cages I and II contained circular openings covered by fine gauze for air entry/exit. Airflow between the cages was created by a commercial DC fan (12 V) that was placed in front of the gauze opening of cage II to gently suck the air from cages I to II. In this way, attracting volatiles emitted from the palm of the hand of the experimenter (male, 28 years) that was held to the gauze opening in cage I reached cage II and motivated mosquitoes to fly into the tunnel to reach the stimulus source (Fig. 6.3).

In each experiment, a total of 25 *Ae. aegypti* females, preselected for host-seeking behavior, were used to assess the efficacy of different static electric fields generated by voltages of 0.5–4.0 kV (with corresponding field intensities of 0.25–2 kV/cm). In control experiments, no electric field was applied (0.0 kV). Each voltage was tested in four repetitions following the same procedure, i.e., test mosquitoes were allowed to settle and adapt to the test environment for 5 min. In the absence of human odors, mosquitoes were found to remain in cage II and resting on the walls, not attempting to fly into the tunnel. After 5 min, the electric field was generated, the ventilator switched ON and the hand held to the gauze opening. The number of mosquitoes flying through the tunnel and into cage I (with attractive stimulus), hence passing the electrodes, was counted over a time period of 5 min. The electric field was then switched OFF and the number of mosquitoes now entering cage I was documented in the same way for another 5 min. Mosquito flight was also recorded by a GoPro camera (GoPro Hero3+ Black Edition, GoPro Inc., San Mateo, CA, United States), placed above the glass tunnel. Mosquitoes that did not respond to the attractive volatiles and instead remained in cage II were considered inactive. The repelling potential of the generated electrical field was expressed as an attraction reduction to the volatile stimuli emitted by the hand and calculated according to the following formula:

$$\%R = 100 - \left[\left(n_E \div n_0\right) \times 100\right]$$

R = Repellency.

n_E = number of mosquitoes passing the electrical field.

n_0 = number of mosquitoes passing in control tests (no electrical field).

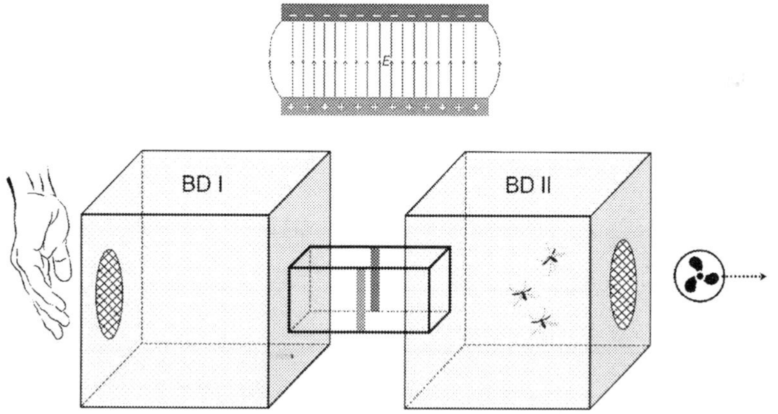

FIG. 6.3 Laboratory cage test set-up. Two BugDorm-1 cages were connected by a glass tunnel that held the plate electrodes. Mosquitoes were released in cage II, the positive stimuli emitted into cage I, by Dr. Ulla Gordon, Biogents AG.

During control experiments, mosquitoes quickly responded to the positive stimuli; an average of 75% entered cage I within the first 5 min. In the second half of the experiment (6–10 min) no further migration was observed in control experiments, thus an average of 25% of the tested mosquitoes was resting in cage II and considered inactive. The weakest electric field-tested had an intensity of 0.25 kV/cm. In these tests, mosquitoes passed the electrodes relatively easily, resulting in an average response rate of 69% during the first 5 min. Once the field was turned off (6–10 min), another 3% oriented toward the attracting stimuli. An incipient repelling effect was observed while testing an electric field with an intensity of 0.5 kV/cm: in these experiments, an average of 43% passed the electrical field (which corresponded to an attraction reduction of 42.7% compared to control experiments). Once the electric field was switched OFF, an additional 34% migrated from cages I to II, leading to an overall response rate of 77%. The repelling potential of electric fields with intensities of 0.75–2.0 kV/cm was noticeably stronger: compared to control tests, the repellency reached 84% (0.75 kV/cm) to 97.3% (2.0 kV/cm). Once the electric field was switched OFF, test mosquitoes were still strongly attracted to the human odors, indicating that the 5 min exposure to the electric field did not induce a behavioral change in *Ae. aegypti´s* response to the host stimuli. Response rates in the second half of each experiment (6–10 min) ranged between 68% and 74%, creating overall response rates (0–10 min) of 70%–80% (Fig. 6.4).

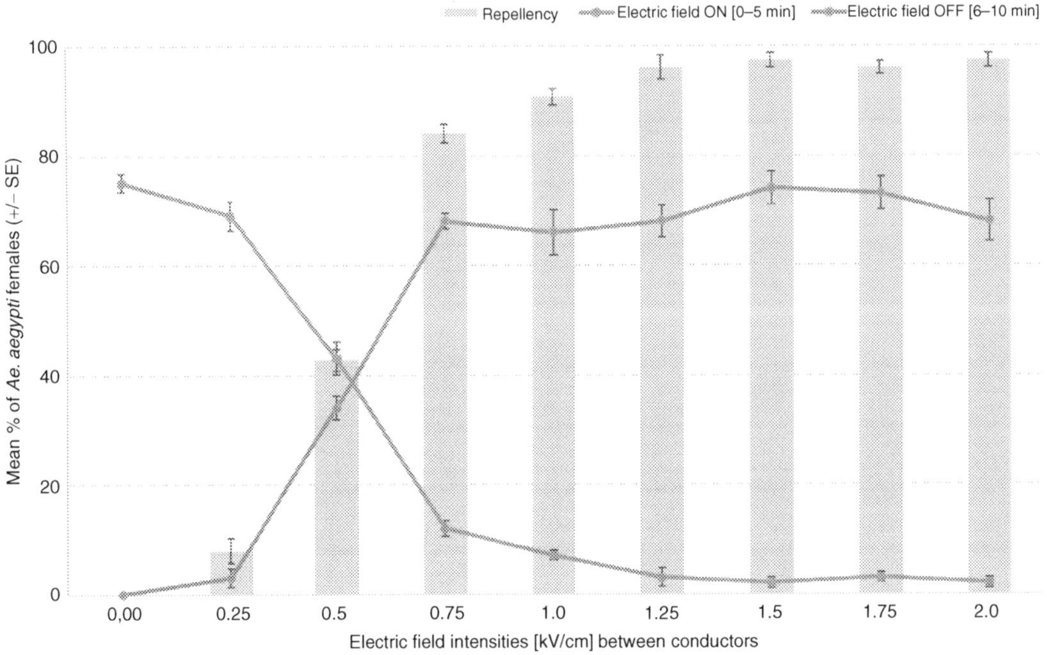

FIG. 6.4 Electric field barrier assays in cage tests. The x-axis shows the tested electric field intensities, the y-axis gives mean percentages of *Ae. aegypti* females passing the plate electrodes (*orange*: electric field ON; *green*: electric field OFF) and mean percentages of repelled individuals (*grey columns*) avoiding the electric field. The standard error (*SE*) is in bars (*n* = 4).

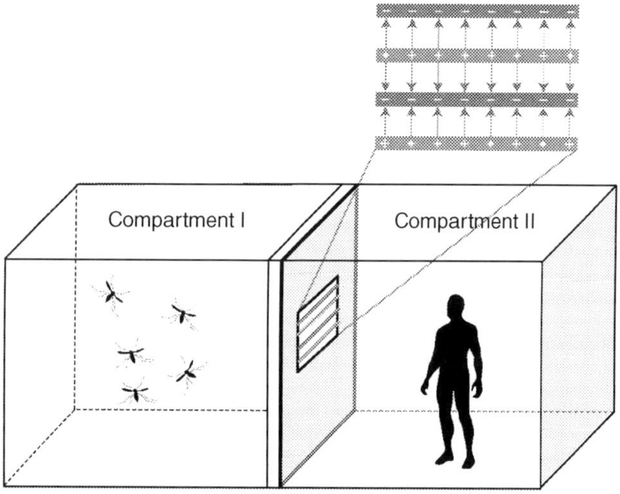

FIG. 6.5 Room test set-up. The two test compartments were separated by a portable wall. An opening in the portable wall was covered by window blinds which were positively and negatively charged. Mosquitoes were released into compartment I while the positive volatile stimuli (human volunteer) were emitted from compartment II, by Dr. Ulla Gordon, Biogents AG.

6.3.3 Barrier assays in room tests

Next, we tested if electric fields could prevent mosquitoes from reaching their host in a more practical set-up, which was tested in Biogents large free-flight rooms. The room is 37 m^3 and was divided into two compartments of 18.5 m^3 by a portable wall. The wall contained a window opening (35 × 35 cm) that allows mosquitoes to move from compartment I to II (Fig. 6.5). Both compartments were set to a temperature of 27.5 ± 0.5 °C and a RH of 75 ± 5%. The light intensity in both compartments was 450 Lux (full spectrum LED light tubes). All experiments were performed on host-seeking *Ae. aegypti* females, as described earlier.

Commercially available aluminum blinds (Jalousie Basic, Bauhaus AG, Regensburg, Germany) were placed in the window opening between the two compartments. Electric fields were created between the slats, which were spaced 2 cm. Slats were connected to an adjustable HV source (Spellman, Model: V6 DC 15 KV POS 2MA W/O RS232) in a way that positively and negatively charged slats were alternating, and voltages of 0.5–3.0 kV (with corresponding field intensities of 0.25–1.5 kV/cm) were tested. In each experiment, 50 host-seeking *Ae. aegypti* females were released into compartment I. In control experiments, no electric field was applied (0.0 kV). Each voltage was tested in five repetitions following the same procedure, i.e., test mosquitoes were allowed to settle and adapt to the test environment for 5 min, afterward the experimenter (male, 28 years) entered compartment II and switched ON the electric field. For 1 h, the number of mosquitoes flying through the window blinds and landing on the experimenter was counted. Mosquitoes that entered compartments II and landed on the experimenter were killed using a commercial electric insect swatter (Basetech eSwatter, from Conrad Electronics, Regensburg, Germany) to avoid counting the same mosquito twice and prevent the mosquitoes from biting the volunteer.

Mosquitoes that were found in compartment I at the end of an experiment were considered inactive. The repelling potential of the generated

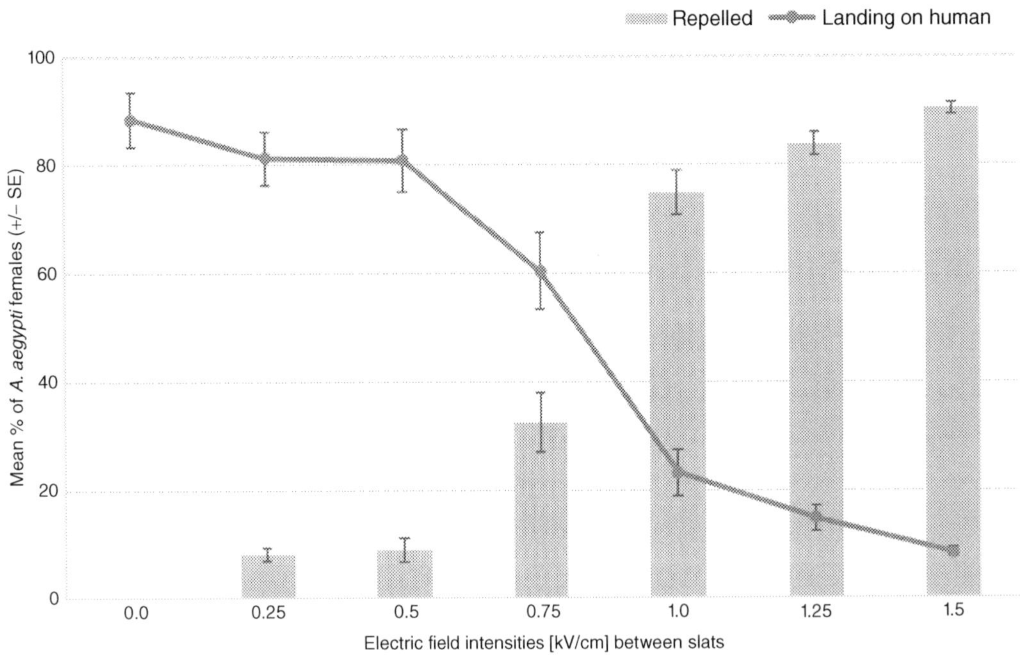

FIG. 6.6 Electric field barrier assays in room tests. The x-axis shows the tested electric field intensities, the y-axis gives mean percentages of *Ae. aegypti* females passing the windows blinds and landing on the volunteer (*green line*) as well as mean percentages of repelled individuals (*grey columns*). The standard error (*SE*) is in bars ($n = 5$).

electric fields was expressed as an attraction reduction to the volatile stimuli emitted by the volunteer and calculated according to the following formula:

$$\%R = 100 - \left[\left(n_E \div n_0\right) \times 100\right]$$

R = Repellency.

n_E = number of mosquitoes passing the electrical field.

n_0 = number of mosquitoes passing in control tests (no electrical field).

The response rate to the human odors in control tests was high, with an average of 88.4% of the test mosquitoes passing the blinds and landing on the volunteer within 1 h. In contrast to the barrier assays conducted in cages, avoidance behavior in room tests was not noticeable until electric fields with an intensity of 0.75 kV/cm and above were tested. Compared to control tests, the number of *Ae. aegypti* passing the window blinds was reduced by 32.4% and an average of 60.3% landed on the volunteer. At field intensities of 1.0 kV/cm and 1.25 kV/cm, the repellency was prominent. In these tests, the average number of mosquitoes passing through the electrically charged window blinds decreased from 23.2% to 14.8% as voltage increased. This led to an increase in repellency from 74.8% to 84.4% compared to control tests. The strongest effect was observed during tests of electric fields intensities of 1.5 kV/cm. In these tests, human landing rates were reduced by an average of 90.3% (Fig. 6.6).

6.3.4 Assessing the effect of electric fields on *Ae. aegypti* female reproductive rates

Results from initial laboratory cage tests indicated, that the short-term exposure to strong electric fields did not elicit changes in the host-seeking behavior of *Ae. aegypti* females, as once the electric field was switched OFF, test mosquitoes were still highly attracted to the human odors and showed regular flight maneuvers. Could the exposure to electric fields, however, have an impact on reproductive rates? In *Drosophila melanogaster*, the exposure to pulsed electromagnetic fields with an intensity of 4 kV/cm caused a slight increase in reproductive rates compared to control groups (Panagopoulos and Margaritis, 2003). When wheat aphids (*Sitbion avenae*) were exposed to static electric fields at intensities of up to 6 kV/cm, long-term adverse effects on the developmental duration and longevity were observed (He et al., 2014).

Potential effects of exposure to electric fields on the reproductive rates of *Ae. aegypti* were assessed by exposing batches of 20 females to static electric fields with an intensity of 1 kV/cm for 5 min. The procedure was based on the barrier assay presented earlier: 20 female mosquitoes preselected for host-seeking behavior were released into a BugDorm-1 cage that was connected to a second one via a glass tunnel that held four copper electrodes. Mosquitoes were allowed to settle for 5 min, after which the electric field was switched ON and a human palm was held next to the second cage to motivate mosquitoes to approach the copper electrodes. After the exposure, mosquitoes were gently collected from the cages with an aspirator and transferred into 0.5 L incubation cups. Control groups were treated in the same way, with the electric field switched OFF (0 kV/cm). A total of six experiments, each consisting of one exposed and one control batch, were conducted under standardized conditions at 27.5 ± 0.5 °C, 75 ± 5%rH, and a light intensity of 450 Lux (full spectrum LED light tubes). Within 1 h after the exposure, mosquitoes were offered an artificial blood meal using sterile bovine blood (Fiebig Nährstofftechnik GbR, Idstein-Niederauroff, Germany) and Hemotek feeding devices (Hemotek Ltd., Blackburn, United Kingdom). Females were allowed to engorge blood for 30 min, afterwards the feeding device was removed, and the number of blood-fed mosquitoes counted. The following 4 days, mosquitoes were incubated at 27 ± 1 °C and 80 ± 5%rH and had access to sugar water (10% dextrose). The incubation cup also provided an oviposition site, a plastic tube filled with tap water and lined with filter paper. At the conclusion of oviposition (day four after blood-meal), the filter paper with eggs was removed, dried, and stored in a sealed plastic container for at least 6 days. To induce larval hatching, a filter paper with eggs was submerged in a 0.5 glass jar filled with 375 mL of deoxygenized water. Larvae were fed with fish food flakes (TetraMin, Tetra GmbH, Melle, Germany), once they transformed into pupae, jars were placed inside BugDorm-1 cages for adult emergence. After adult emergence was completed, both cages were placed in a freezer at −20 °C for 1 h and adult mosquitoes were counted. The emergence rate was calculated according to the following formula:

$$ER = n_M \div n_{BF}$$

ER = Emergence rate.

n_M = number of adult mosquitoes emerged (F1 generation).

n_{BF} = number of blood-fed mosquitoes (F0 generation).

The number of blood-fed individuals in treatment and control groups was comparable after each experiment. In control experiments (0 kV/cm), an average of 16.7 ± 1.3 females had engorged blood while an average of 16.5 ± 1.0 females were found to be blood-fed in treatment groups (1 kV/cm). The number of

TABLE 6.1 Overview of reproductive experiments with *Ae. aegypti* ($n = 6$).

Experiment	Adults emerged from eggs collected from control group	Adults emerged from eggs collected from treatment group
1	228	260
2	47	73
3	481	517
4	185	157
5	363	54
6	129	248
Σ	1424	1309
Emergence rates (±SE)	14.3 ± 3.5	14.2 ± 4.5

Total number of adults emerged in exposed and control groups is given for each experiment as well as mean emergence rate (±standard error).

adult mosquitoes emerging from eggs collected from control and treatment groups varied between experiments, however, resulting emergence rates were comparable (Table 6.1). These results indicated that a short-term exposure to static electric field intensities of 1kV/cm had no adverse effects on *Ae. aegypti* reproductive rate.

6.4 Practical application of electric fields: an approach

Results from initial laboratory assays showed that strong electric fields with intensities of ≥1 kV/cm can cause an avoidance behavior in mosquitoes, hence preventing vector-host contact. How could the basic set-up used in these experiments be transformed into a practical, application-oriented solution, that considers (1) user safety, (2) cost efficiency, and (3) easy usability?

6.4.1 Development of a high-voltage prototype

Subsequent experiments focused on the development of a prototype device that generates an output voltage of at least 4kV but operates on 12 V DC input (Fig. 6.7). This is achieved through the implementation of a flyback converter, an isolated power converter that uses mutually coupled inductors to store energy when current passes through and releases the energy when power is removed. In a typical application, a switching device such as a transistor is turned on and off to control the direction of energy flow. In the on state, the energy is transferred from the input voltage source to the transformer. The

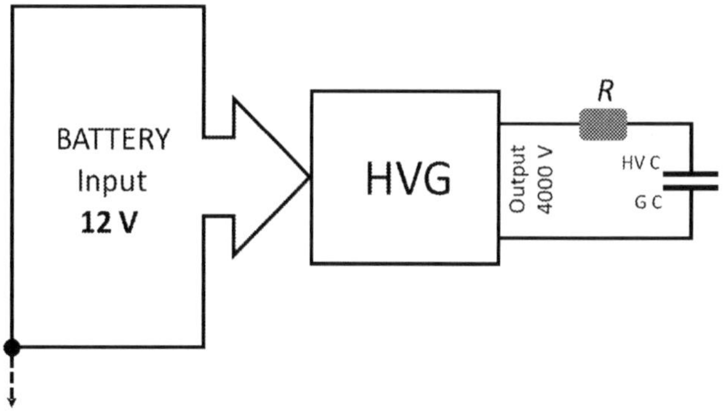

FIG. 6.7 Schematic drawing of the high-voltage (*HV*) prototype. The high-voltage generator (*HVG*) is supplied by 12 V DC power from a battery and generates an output of 4000 V. The voltage is used to create an electric field between the HV and grounded (*G*) conductor (*C*). The resistor (*R*) makes the system safe to the touch.

diode in the second winding is reverse-biased, thus current does not flow and instead the energy is stored in the transformer until a switching device, e.g., a metal-oxide-semiconductor-field-effect transistor, is turned off. Now, the stored energy produces a current that is forward biasing the diode which results in the production of an HV DC output. For safety and short circuit protection, a large resistor of ≥1 mega-ohm was integrated with a series of the output. According to Ohm´s law, the electric current I (A) is the quotient of the voltage (V) across a conductor and the resistance R (Ω) of the conductor. With an output voltage of 4 kV and a resistor of 22 mega-ohm, the electric current through the conductor is 0.1 mA. Currents below 10 mA are considered safe for humans, causing only mild sensations upon touch while 16 mA is the maximum current "an average man can grasp and let go" (Fish and Geddes, 2009).

The developed HV prototype can be supplied with 12 V DC power from a variety of power sources, including primary and secondary batteries (Lead Acid, Lithium-ion), plug-in power supplies using AC to DC converters or solar panels. Important factors that impact the decision on a specific power source are longevity and associated costs for both, acquisition and operating. So far, the technical development was based on the application of continuous electric fields. The implementation of pulsed electric fields could, however, be interesting as pulses offer certain advantages: reduction of operating costs, extension of battery life, and counteracting potential behavioral adaptations in exposed mosquitoes. Follow-up experiments therefore investigated whether pulsed electric fields still caused an avoidance behavior in *Ae. aegpyti*.

6.4.2 Repellency of pulsed electric fields

Laboratory barrier assays in cage tests were performed according to the protocol described earlier including the following modification: the HV device was connected to a function generator through a controller. The function generator produced pulses of desired shape, frequency, and duty cycle, and the controller sent these low voltage pulses to the input of the HV device. The HV device increased their amplitude to convert them into HV pulses (≈3.5 kV) that were monitored on a laptop screen using the software PC LAB 2000LT (Velleman Instruments, Gavere, Belgium). Through output terminals of the HV device, pulse durations of 0.25, 0.5, 1, 2, and 3 s with varying duty cycles (between 0% and 90%) were applied on the copper plate electrodes. A duty cycle of 10% means that the electric field was on for 10% and off for 90% of the pulse duration, i.e., it was on for 0.025 s and off for 0.225 s in trials with a pulse duration of 0.25 s. Each combination of pulse duration and duty cycle was tested in three repetitions with 25 host-seeking *Ae. aegypti* females per experiment.

At duty cycles of 60%–90%, repellency was high in all trials: an average of 84% (pulse duration 3.0 s) to 98% (pulse duration 0.25 s) of the mosquitoes did not pass the plate electrodes. At shorter duty cycles, between 10% and 50%, the repellency was less prominent in longer pulse duration experiments (1.0–3.0 s) compared to shorter pulse durations (0.5 and 0.25 s). When duty cycles were 10%, repellency reached an average of 38% and 29% at pulse durations of 2.0 and 3.0 s, respectively, while it was higher than 94% at pulse durations of 0.25 and 0.5 s. The response rate in control trials (duty cycle 0%, electric field off) was high with an average of 76.9% ± 1.59 of the test mosquitoes responding to the positive stimuli in the absence of electric fields (Fig. 6.8).

Results indicate that pulsed electric fields can repel *Ae. aegypti*, but the efficacy depends on the pulse duration and duty cycle. Best effects were obtained when pulse durations of 0.5 and 0.25 s were used, longer pulse periods (≥1 s) required duty cycles of at least 60 to reach 80% repellency.

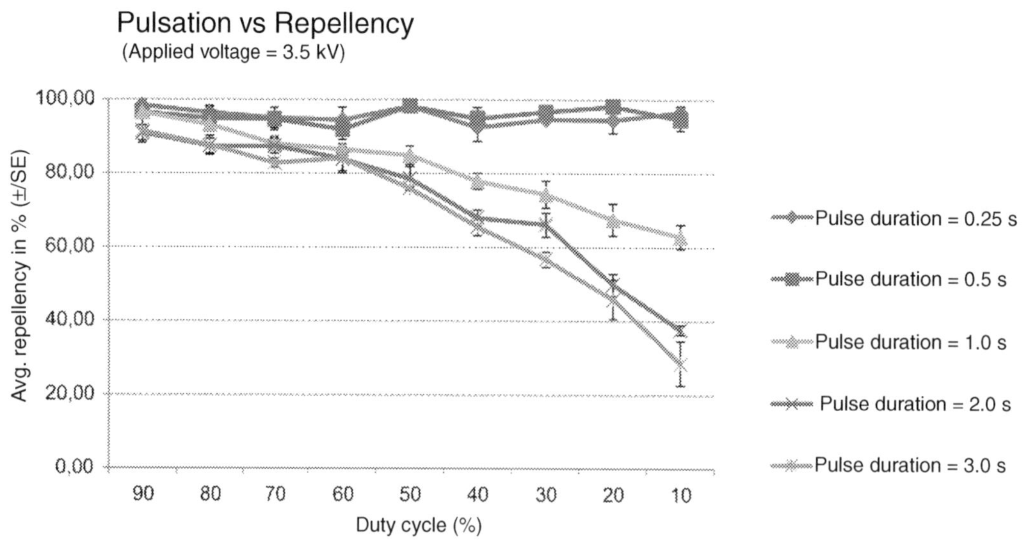

FIG. 6.8 Pulsation experiments in barrier assays with *Ae. aegypti* ($n = 3$). The x-axis shows the tested duty cycles, the y-axis gives mean percentages of test mosquitoes avoiding the electric field. The standard error (*SE*) is in bars.

6.4.3 Cost comparison and envisioned design

How do the previous findings relate to, e.g., battery life? The HV prototype has a maximum current consumption of 65 mA, this means that a battery with a capacity of 12 Ah would be discharged to 50% after approximately 4 days if continuous electric fields are applied.

Battery life is indirectly proportional to the duty cycle; thus, it increases with shorter pulse "on" periods. At duty cycles of 90%, battery life reached approximately 9 days but could be extended to 77 days using duty cycles of 10% (data not shown). In terms of longevity and cost-effectiveness, other power supply sources would also benefit from pulsation. Table 6.2 compares the estimated longevity and associated costs for

TABLE 6.2 Estimated longevity and purchase costs for different 12 V DC power supply sources for the high-voltage prototype.

Type	Purchase costs (US$)[a]	Continuous electric field (24 hours)		50% pulsed electric field (24 hours)	
		Longevity[b]	US$/day	Longevity[b]	US$/day
Batteries 8 × 1.5 V (AA)	7–12	1–2 days	3.5–12	3–4 days	1.75–4
Rechargeable batteries 12 V (Lead-Acid)	20–30	10–15 days	1.33–3	25–30 days	0.67–0.86
Solar power	75–100	15–20 years	0.01–0.02	15–20 years	0.01–0.02
Plug-in power supply (AC to DC transformer)	10–12	1–2 years	0.01–0.03	2–3 years	0.009–0.002

[a]*Estimated costs based on an internet search (https://www.amazon.com) conducted on October 20, 2020.*
[b]*The actual longevity depends on the size of the electrodes (e.g., window blinds).*

eligible power sources when used to generate continuous or pulsed electric fields.

While nonrechargeable batteries and solar panels do not create any additional costs after purchase, running costs for the plug-in solution still need to be considered and depend on the electricity costs at the operation site. When used continuously, the HV prototype device has a 24 h power consumption of 19 W. According to the US Energy Information Administration, the average price for electricity in the United States was 13.26 cents per kWh in July 2020 (https://www.eia.gov/electricity/monthly/epm_table_grapher.php?t=epmt_5_6_a, accessed October 20, 2020) in residential areas; thus, monthly running costs would reach approximately 7.8 cents under continuous use.

Based on the presented findings, an electric field should have an intensity ≥1 kV/cm to elicit avoidance behavior in mosquitoes and it can be applied continuously or pulsed. Potential applications of electric fields as barriers in a home setting could involve charged metal blinds to cover doors and/or windows to prevent mosquitoes from entering (Fig. 6.9). In an outdoor seated space, an electrically charged fence could create a comparable effect (Fig. 6.10). This new method of repelling or preventing mosquitoes from passing through a defined opening, that is largely permeable to ambient air, by the means of an electric field generated by at least two electrodes has been patented in 2017 (Rose et al., 2017) (European Patent Number 17208300.8).

6.5 Discussion

Our idea to use electric fields to control insect pests is not entirely new but had different objectives. Extensive research on the effects of high strength radio-frequency electric fields on stored grain insects started in the 1960s aiming at either killing or sterilizing the target pest (Ponomaryova et al., 2008). More recent research suggests high voltage electric field screens can be used as an air-shielding apparatus to capture airborne

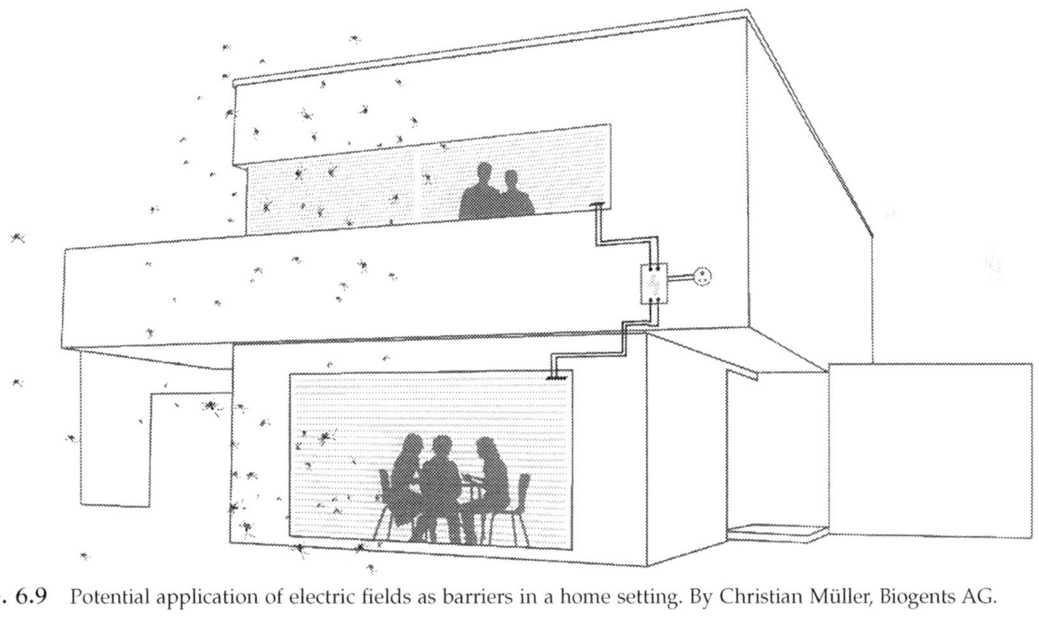

FIG. 6.9 Potential application of electric fields as barriers in a home setting. By Christian Müller, Biogents AG.

FIG. 6.10 Potential application of electric fields as barriers in an outdoor setting. By Christian Müller, Biogents AG.

spores, fungi, and flying greenhouse pests to reduce the use of fungicides and insecticides (Kusakari et al., 2020). A similar application using oppositely charged electric field screens was successful in capturing and trapping *D. melanogaster* and could be implemented in greenhouses or food storage facilities to exclude insect pests (Matsuda et al., 2012). In 2015, the same group presented electric field screens charged by ≥1.2 kV as physical barriers and a means to capture *Ae. albopictus* and *Cx. pipiens* before entering the house (Matsuda et al., 2015).

Our studies investigating the repelling effects of static electric fields on mosquitoes is novel. Our laboratory experiments support the hypothesis that *Ae. aegypti* is able to sense and avoid electric fields at intensities of 1 kV/cm and above. Emphasis was placed on comparing the effects of multiple voltages in laboratory tests. Due to the limited duration of the project, the number of replications in both, cage and room tests is too low for statistical analysis. However, standard errors in all experimental trials were small, indicating low variance within the data sets and allowing us to draw conclusions on the potential of electric fields to repel mosquitoes. The short-term exposure to electric field intensities of 1 kV/cm also did not negatively or positively impact *Ae. aegypti* reproductive rates, the total number of adults emerged from eggs laid by exposed and unexposed females was comparable. Forcing test batches into the electric field, as described by Panagopoulos and Margaritis (2003) in tests with *D. melanogaster*, might have resulted in a different outcome but in our opinion, the bioassay should allow mosquitoes to navigate in close proximity and respond to the electric field just as they would in a realistic setting. The promising outcome of initial laboratory tests needs to be verified, more research needs to be conducted involving different mosquito species, other arthropods of medical importance or food pests and the set-up has to be evaluated under realistic conditions in the field. These studies are currently being prepared to be conducted in Germany and the United States.

A potential benchmark to evaluate the efficacy of our system could be spatial arthropod repellents, chemicals that deter mosquitoes at a distance and inhibit their ability to locate a host (Gouck et al., 1967; Nolen et al., 2002), thereby reducing host- vector contacts. Spatial arthropod repellents are considered effective if they provide a minimum landing inhibition/reduction of 90% in semi-field or field trials (World Health Organization, 2013). In our room tests, such a landing inhibition could be achieved by electric field intensities of 1.5 kV/cm. Mosquitoes that did not pass the charged window blinds but remained in compartment I were still attracted to human odors at the end of a test. Thus, exposure to the electric field did not necessarily alter their host-seeking behavior. This observation is highly interesting and turns electric fields into a potential tool for push–pull strategies. Push–pull combines deterring and attracting stimuli to change the abundance of an insect pest and has been successfully implemented in crop pest management (Cook et al., 2007; Pyke et al., 1987). In mosquito control, suggested push–pull approaches involved the use of spatial repellents such as pyrethroids, transfluthrin, allethrin, or metofluthrin (Kitau et al., 2010; Mmbando et al., 2017; Wagman et al., 2015), catnip, *Nepeta cataria* (Menger et al., 2015; Obermayr et al., 2015; Paz-Soldan et al., 2011), or delta-undecalactone (Menger et al., 2015; Obermayr et al., 2015; Paz-Soldan et al., 2011) as push components in combination with attractive suction traps like the BG Sentinel (Salazar et al., 2012). The use of sublethal doses of volatile pyrethroids usually leads to a significant reduction in human vector contacts (Darbro et al., 2017; Ogoma et al., 2012). However, the neurotoxic action of these compounds might interfere with mosquito host-seeking behavior, therefore rendering the pull-component ineffective (Kitau et al., 2010; Salazar et al., 2013). Non-neurotoxic compounds like catnip on the other hand are less effective in reducing human landing rates when applied in a field setting (Obermayr et al., 2015).

Electric fields represent a physical barrier and provide certain benefits compared to chemical spatial repellents. They also have a high potential in repelling mosquitoes, are odorless, their efficacy does not fade over time, not likely to lead to behavioral adaptation (especially when electric fields are delivered in pulses), and depending on the power source, they are very cost-effective. The developed HV prototype can run on a variety of power sources and the most suitable one will be defined by location/area of application (availability of power supplies, area of coverage), the operating time per day (which will be linked to mosquito biting activity patterns) and personal preferences. While batteries offer a greater portability of the device they have to be recharged or replaced on a regular basis; solar panels represent the most sustainable and environmentally friendly solution but come at greater upfront costs, whereas the plug-in solution is cost-saving but immobile. Regarding safety, the safety resistor in the HV prototype limits the flow of current through the human body to 4 mA in case of accidental touching.

In a potential application, houses would be equipped with charged blinds in windows and even doors to prevent mosquitoes from entering. In this way, human-vector contacts are reduced while indoor spaces remain properly ventilated. In order to reduce the vector population, the system should be combined with attractive traps, e.g., the BG Sentinel, to lure and catch females that are deterred by the electric field. Such a set-up resembles a promising novel push–pull approach for the control of vector mosquitoes.

6.6 Conclusion

In this study, we investigated the repellent potential of strong electric fields on the yellow-fever mosquito, *Ae. aegypti*. In laboratory behavioral assays, host-seeking females had to pass an electric field in order to reach an attractive source (human odors). At field intensities of ≥1.5kV/cm, the response rate to the attractive odors was reduced by at least 90% in both, cage- and room-tests. Once the electric field was switched OFF, females showed regular host-seeking behavior indicating that the exposure to the electric field did not induce any short-term behavioral changes. In contrast to chemical repellents, electric fields are odorless and their efficacy does not fade over time, thus they could be an interesting tool for novel mosquito control approaches, such as Push–Pull. Future research needs to focus on the applicability of such a system in a realistic setting and investigate whether the promising repelling effects observed in the laboratory will persist in the field.

Acknowledgements

This study has been supported by funding from the United States Agency for International Development under the Grant Number AID-OAA-F-16-00092. ISGlobal is supported by the Spanish Ministry of Science and Innovation through the "Centro de Excelencia Severo Ochoa 2019-2023" Program (CEX2018-000806-S), and the Generalitat de Catalunya through the CERCA Program. The funders had no role in study design, data collection and analysis, decision to publish, or preparation of the chapter. We thank Dr. Scott Gordon for reviewing the chapter.

References

Achee, N.L., Bangs, M.J., Farlow, R., Killeen, G.F., Lindsay, S., Logan, J.G., Moore, S.J., Rowland, M., Sweeney, K., Torr, S.J., Zwiebel, L.J., Grieco, J.P., 2012. Spatial repellents: from discovery and development to evidence-based validation. Mal. J. 11. doi:10.1186/1475-2875-11-164.

Balmori, A., 2015. Anthropogenic radiofrequency electromagnetic fields as an emerging threat to wildlife orientation. Sci. Total Environ. 518/519, 58–60. https://doi.org/10.1016/j.scitotenv.2015.02.077.

Barrera, R., Amador, M., Acevedo, V., Hemme, R.R., Félix, G., 2014. Sustained, area-wide control of *Aedes aegypti* using CDC autocidal gravid ovitraps. Am. J. Trop. Med. Hyg. 91 (6), 1269–1276. https://doi.org/10.4269/ajtmh.14-0426.

Benelli, G., Caselli, A., Canale, A., 2017. Nanoparticles for mosquito control: challenges and constraints. J. King Saud. Univ. Sci. 424–435. doi:10.1016/j.jksus.2016.08.006.

Bibbs, C.S., Kaufmann, P.E., 2017. Volatile pyrethroids as a potential mosquito abatement tool: a review of pyrethroid-containing spatial repellents. J. Int. Pest Manag. doi:10.1093/jipm/pmx016.

Brodie, D., 2000. Introduction to Advanced Physics, Hodder Education, London, UK.

Catania, K.C., 2015. Electric eels use high-voltage to track fast-moving prey. Nat. Commun. 6. doi:10.1038/ncomms9638.

Chapman, C.E., Blondin, J.P., Lapierre, A.M., Nguyen, D.H., Forget, R., Plante, M., Goulet, D., 2005. Perception of local DC and AC electric fields in humans. Bioelectromagnetics 26 (5), 357–366. https://doi.org/10.1002/bem.20109.

Clarke, D., Morley, E., Robert, D., 2017. The bee, the flower, and the electric field: electric ecology and aerial electroreception. J. Comp. Physiol. A 203 (9), 737–748. https://doi.org/10.1007/s00359-017-1176-6.

Clarke, D., Whitney, H., Sutton, G., Robert, D., 2013. Detection and learning of floral electric fields by bumblebees. Scienceexpress 340 (6128), 66–69. https://doi.org/10.1126/science.1230883.

Cook, S.M., Khan, Z.R., Pickett, J.A., 2007. The use of push-pull strategies in integrated pest management. Ann. Rev. Entomol. 52, 375–400. https://doi.org/10.1146/annurev.ento.52.110405.091407.

Corbet, S.A., Beament, J., Eisiskowitch, D., 1982. Are electrostatic forces involved in pollen transfer? Plant Cell Environ. 5 (2), 125–129. https://doi.org/10.1111/1365-3040.ep11571488.

Darbro, J.M., Muzari, M.O., Giblin, A., Adamczyk, R.M., Ritchie, S.A., Devine, G.J., 2017. Reducing biting rates of *Aedes aegypti* with metofluthrin: investigations in time and space. Parasites Vectors 10 (1), 1–9. https://doi.org/10.1186/s13071-017-2004-0.

Degener, C.M., Eiras, A.E., Ázara, T.M.F., Roque, R.A., Rösner, S., Codeço, C.T., Nobre, A.A., Rocha, E.S.O., Kroon, E.G., Ohly, J.J., Geier, M., 2014. Evaluation of the effectiveness of mass trapping with BG-sentinel traps for dengue vector control: a cluster randomized controlled trial in Manaus, Brazil. J. Med. Entomol. 51 (2), 408–420. https://doi.org/10.1603/ME13107.

Fiorenzano, J.M., Koehler, P.G., Xue, R.D., 2017. Attractive toxic sugar bait (ATSB) for control of mosquitoes and its impact on non-target organisms: a review. Int. J. Environ. Res. Public Health 14 (4). doi:10.3390/ijerph14040398.

Fish, R.M., Geddes, L.A., 2009. Conduction of electrical current to and through the human body: a review. Eplasty 9, e44.

Fonseca-González, I., Quiñones, M.L., Lenhart, A., Brogdon, W.G., 2011. Insecticide resistance status of *Aedes aegypti* (L.) from Colombia. Pest Manag. Sci. 67 (4), 430–437. https://doi.org/10.1002/ps.2081.

Gouck, H., McGovern, T.P., Beroza, M., 1967. Chemicals tested as space repelents against yellow-fever mosquitoes. I. Esters. J. Econ. Entomol. 60 (6), 1587–1590. https://doi.org/10.1093/jee/60.6.1587.

He, J., Cao, Z., Yang, J., Zhao, H.Y., Pan, W.D., 2014. Effects of static electric fields on growth and development of wheat aphid Sitobion aveanae (Hemiptera: Aphididae) through multiple generations. Electromagn. Biol. Med., Early Online, 1–7.

Hunt, E.P., Jackson, C.W., Newland, P.L., 2005. `Electrorepellancy´ behaviour of *Periplaneta americana* exposed to friction charged dielectric surfaces. J. Electrostat. 63 (6–10), 853–859. https://doi.org/10.1016/j.elstat.2005.03.081.

Kitau, J., Pates, H., Rwegoshora, T.R., Rwegoshora, D., Matowo, J., Kweka, E.J., Mosha, F.W., McKenzie, K., Magesa, S.M., 2010. The effect of Mosquito Magnet® Liberty plus trap on the human mosquito biting rate under semi-field conditions. J. Am. Mosq. Control Assoc. 26 (3), 287–294. https://doi.org/10.2987/09-5979.1.

König, H.L., Krueger, A.P., Lang, S., Sönnig, W., 1981. Biological effects of environmental magnetism, Springer, New York.

Kröckel, U., Rose, A., Eiras, A.E., Geier, M., 2006. New tools for surveillance of adult yellow fever mosquitoes: comparison of trap catches with human landing rates in an urban environment. J. Am. Mosq. Control Assoc. 22 (2), 229–238. https://doi.org/10.2987/8756-971X(2006)22[229:NTFSOA]2.0.CO;2.

Kusakari, S.I., Okada, K., Shibao, M., Toyoda, H., 2020. High voltage electric fields have potential to create new physical pest control systems. Insects 11 (7), 1–14. https://doi.org/10.3390/insects11070447.

Lea, A.O., 1965. Sugar-baited insecticide residues against mosquitoes. Mosq. News 25, 65–66.

Marcombe, S., Darriet, F., Agnew, P., Etienne, M., Yp-Tcha, M.M., Yébakima, A., Corbel, V., 2011. Field efficacy of new larvicide products for control of multi-resistant *Aedes aegypti* populations in Martinique (French West Indies). Am. J. Trop. Med. Hyg. 84 (1), 118–126. https://doi.org/10.4269/ajtmh.2011.10-0335.

Marcombe, S., Farajollahi, A., Healy, S.P., Clark, G.G., Fonseca, D.M., 2014. Insecticide resistance status of United States populations of *Aedes albopictus* and mechanisms involved. PLoS One 9 (7). doi:10.1371/journal.pone.0101992.

Marinescu, M., 2009. Elektrische und magnetische Felder, 3rd ed. Springer, Berlin Heidelberg. https://doi.org/10.1007/978-3-642-25794-0.

Matsuda, Y., Nonomura, T., Kakutani, K., Kimbara, J., Osamura, K., Kusakari, S., Toyoda, H., 2015. Avoidance of an electric field by insects: fundamental biological phenomenon for an electrostatic pest-exclusion strategy. J. Phys., Conf. Series. doi:10.1088/1742-6596/646/1/012003 012003.

Matsuda, Y., Kakutani, K., Nonomura, T., Kimbara, J., Kusakari, S.I., Osamura, K., Toyoda, H., 2012. An oppositely charged insect exclusion screen with gap-free multiple electric fields. J. Phys., Conf. Series 112 (11), 646. https://doi.org/10.1063/1.4767635.

Matsuda, Y., Kakutani, K., Nonomura, T., Kimbara, J., Osamura, K., Kusakar, S., Toyoda, H., 2015. Safe housing ensured by an electric field screen that excludes insect-net permeating haematophagous mosquitoes carrying human pathogens. J. Phys. Conf. Series 646, 012002. https://doi.org/10.1088/1742-6596/646/1/012002.

Maw, M.G., 1961. Behaviour of an Insect on an electrically charged surface. Can. Entomol. 93 (5), 391–393. https://doi.org/10.4039/Ent93391-5.

Maw, M.G., 1962. Behaviour of insects in electrostatic fields. Proc. Entomol. Soc. Manitoba 18, 30–36.

Menger, D.J., Otieno, B., Rijk, M., Loon, W.R., Takken, W., 2015. A push-pull system to reduce house entry of malaria mosquitoes. Mal. J. 13 (1), 119. doi:10.1186/1475-2875-13-119.

Mmbando, A.S., Ngowo, H.S., Kilalangongono, M., Abbas, S., Matowo, N.S., Moore, S.J., Okumu, F.O., 2017. Small-scale field evaluation of push-pull system against early- and outdoor-biting malaria mosquitoes in an area of high pyrethroid resistance in Tanzania. Wellcome Open Res. 112. doi:10.12688/wellcomeopenres.13006.1.

Newland, P.L., Hunt, E., Sharkh, S.M., Hama, N., Takahata, M., Jackson, C.W., 2008. Static electric field detection and behavioural avoidance in cockroaches. J. Exp. Biol. 211, 3682–3690. https://doi.org/10.1242/jeb.019901.

Nolen, J.A., Bedoukian, R.H., Maloney, R.E., Kline, D.L., 2002. Method, apparatus and compositions for inhibiting the human scent tracking ability of mosquitoes in environmentally defined three dimensional spaces. Trademark and Patents Office, U.S. Patent No. 6,362,235.

Norris, E.J., Bartholomay, L., Coats, J., 2018. Present and future outlook: the potential of green chemistry in vector control. In: Advances in the Biorational Control of Medical and Veterinary Pests, 1289, American Chemical Society. ACS Publications, Washington D.C., pp. 43–62. https://doi.org/10.1021/bk-2018-1289.ch004.

Norris, E.J., Coats, J.R., 2017. Current and future repellent technologies: the potential of spatial repellents and their place in mosquito-borne disease control. Int. J. Env. Res. Public Health 14 (2). doi:10.3390/ijerph14020124.

Obermayr, U., Ruther, J., Bernier, U.R., Rose, A., Geier, M., 2015. Evaluation of a push-pull approach for *Aedes aegypti* (L.) using a novel dispensing system for spatial repellents in the laboratory and in a semi-field environment. PLoS One 10 (6). doi:10.1371/journal.pone.0129878.

Ogoma, S.B., Ngonyani, H., Simfukwe, E.T., Mseka, A., Moore, J., Killeen, G. F., 2012. Spatial repellency of transfluthrin-treated hessian strips against laboratory-reared Anopheles arabiensis mosquitoes in a semi-field tunnel cage. Parasites Vectors, 5 (1). doi:10.1186/1756-3305-5-54.

Ogoma, S.B., Ngonyani, H., Simfukwe, E.T., Mseka, A., Moore, J., Maia, M.F., Moore, S.J., Lorenz, L.M., 2014. The mode of action of spatial repellents and their impact on vectorial capacity of *Anopheles gambiae* sensu stricto. PLoS One 9 (12). doi:10.1371/journal.pone.0110433.

Panagopoulos, D.J., Margaritis, L.H., 2003. Effects of Electromagnetic Fields on the Reproductive Capacity of *Drosophila Melanogaster*. Springer, Berlin Heidelberg, pp. 545–578.

Paz-Soldan, V.A., Plasai, V., Morrison, A.C., Rios-Lopez, E.J., Guedez-Gonzales, S., Grieco, J.P., Mundal, K., Chareonviriyaphap, T., Achee, N.L., 2011. Initial assessment of the acceptability of a push-pull *Aedes aegypti* control strategy in Iquitos, Peru and Kanchanaburi, Thailand. Am. J. Trop. Med. Hyg. 84 (2), 208–217. https://doi.org/10.4269/ajtmh.2011.09-0615.

Ponomaryova, I.A., Nino de Rivera, L., Ruiz Sánchez, E., 2008. Interaction of radio-frequency, high-strength electric fields with harmful insects. J. Microw. Power Electromagn. Energy 43 (4), 17–27. https://doi.org/10.1080/08327823.2008.11688621.

Pyke, B., Rice, M., Sabine, B., Zalucki, M.P., 1987. The push-pull strategy: behavioral control of Heliothis. Aust. Cotton Grow 4, 7–9.

Rasgon, J.L., Styer, L.M., Scott, T.W., 2003. Wolbachia-induced mortality as a mechanism to modulate pathogen transmission by vector arthropods. J. Med. Entomol. 40 (2), 125–132. https://doi.org/10.1603/0022-2585-40.2.125.

Roche, J., 2016. Introducing electric fields. Phys. Educ. 51 (5). doi:10.1088/0031-9120/51/5/055005.

Rose, A., Tanveer, F., Paaijmans, K., Garcia, B., Molins, E., 2017. Insect repulsion and/or barrier arrangement and method for repelling insects. European Patent Office (EU Patent Nr. 17208300.8).

Salazar, F.V., Achee, N.L., Grieco, J.P., Prabaripai, A., Eisen, L., Shah, P., Chareonviriyaphap, T., 2012. Evaluation of a peridomestic mosquito trap for integration into an *Aedes aegypti* (Diptera: Culicidae) push-pull control strategy. J. Vec. Ecol. 37 (1), 8–19. https://doi.org/10.1111/j.1948-7134.2012.00195.x.

Salazar, F.V., Achee, N.L., Grieco, J.P., Tuntakon, S., Polsomboon, S., Chareonviriyaphap, T., 2013. Effect of previous exposure of *Aedes aegypti* (Diptera: Culicidae) mosquitoes to spatial repellent chemicals on BG-SentinelTM Trap catches. Parasites Vectors 6, 145.

Scholte, E.J., Knols, B.G.J., Samson, R.A., Takken, W, 2004. Entomopathogenic fungi for mosquito control: a review. J. Insect. Sci. 4. doi:10.1093/jis/4.1.19.

Sinkins, S.P., O´Neill, S.L., 2000. Wolbachia as a vehicle to modify insect populations, Insect Transgenesis: Methods and Applications. CRC Press, London, UK, pp. 271–287.

Sutton, G.P., Clarke, D., Morley, E.L., Robert, D., 2016. Mechanosensory hairs in bumblebees (Bombus terrestris) detect weak electric fields. PNAS 113 (26), 7261–7265. https://doi.org/10.1073/pnas.1601624113.

Thomas, D.D., Donnelly, C.A., Wood, R.J., Alphey, L.S., 2000. Insect population control using a dominant, repressible, lethal genetic system. Science 287 (5462), 2474–2476. https://doi.org/10.1126/science.287.5462.2474.

Vaknin, Y., Gan-Mor, S., Bechar, A., Ronen, B., Eisikowitch, D., 2000. The role of electrostatic forces in pollination. Plant Syst. Evol. 222, 133–142.

Von der Emde, G., 2013. Electroreception, Neurosciences—From Molecule to Behavior: A University Textbook. Springer, Heidelberg.

Wagman, J.M., Grieco, J.P., Bautista, K., Polanco, J., Briceño, I., King, R., Achee, N.L., 2015. The field evaluation of a push-pull system to control malaria vectors in Northern Belize. Central America. Mal. J. 14 (1). doi:10.1186/s12936-015-0692-5.

Weill, M., Luffalla, G., Mogensen, K., Chandre, F., Berthomieu, A., Berticat, C., Pasteur, N., Philips, A., Fort, P., Raymond, M., 2003. Insecticide resistance in mosquito vectors. Nature 423 (6936), 136–137. https://doi.org/10.1038/423136b.

World Health Organization, 2006. Static fields, Environmental Health Criteria, vol. 232, WHO Press, Geneva, Switzerland.

World Health Organization, 2013. Guidelines for efficacy testing of spatial repellents, WHO Press, Geneva, Switzerland.

CHAPTER

7

Multimodal mechanisms of repellency in arthropods

Fredis Mappin and Matthew DeGennaro

Department of Biological Sciences, Biomolecular Sciences Institute, Florida International University, Miami, FL, United States

Despite being developed in the 1940s, N,N-diethyl-3-methylbenzamide (DEET) has remained the most effective repellent for preventing feeding on humans by mosquitoes, flies, ticks, and other arthropods (Staub et al., 2002; Fradin and Day, 2002; Leal, 2014; DeGennaro, 2015). The importance of repellents like DEET in preventing the transmission of vector-borne diseases makes spatial arthropod repellents of considerable interest for those involved in public health and vector-borne disease prevention. Decades of research and developments have yielded a large repertoire of alternative personal protection repellents to DEET (Brown and Hebert, 1997; Katz et al., 2008). These botanically derived and synthetically generated alternatives have demonstrable repellency activity but require additional development to match the efficacy of DEET. The next generation of anthropophilic arthropod repellents will therefore likely need to be rationally designed to target the molecular sensors that enable host attraction and discrimination. A combination of behavioral genetics and electrophysiological approaches suggests that multiple molecular mechanisms of action may disrupt attraction to host cues. In this chapter, we review these possible modes of action providing context for developing the next generation of arthropod repellents focusing on evidence obtained from arthropod studies.

7.1 Toward a more targeted approach

The use of substances such as tars, plant oils, animal dung, and a variety of other natural products to reduce arthropod feeding has ancient origins (Katz et al., 2008; Maia and Moore, 2011). These early topical repellents were most likely discovered through trial and error and by observation of the arthropod's apparent avoidance of its host when used. Over the past 80 years, more systematic approaches have been developed to allow for the discovery and development of significantly more effective personal

Advances in Arthropod Repellents.
DOI: https://doi.org/10.1016/B978-0-323-85411-5.00005-4

protection products (Colucci and Müller, 2018; Kröber et al., 2013). Arthropod repellents in general can be divided into two main categories: botanically derived or synthetic. DEET, the most widely used and effective synthetic repellent, was developed through a laborious screening of over 7000 compounds for insecticide, miticide, and repellent activity by the US Department of Agriculture (USDA) in collaboration with the US military (Knippling et al., 1947; Travis and Morton, 1949). Among the most effective compounds initially screened was N,N-diethylbenzamide, capable of repelling *Ae. aegypti* for more than 3 hours when applied directly to human skin. However, it also was found to cause skin irritation. A safer but equally effective derivative was developed, N,N-diethyl-3-methylbenzamide, known by the acronym DEET for its synonym N,N-Diethyl-*m*-toluamide (McCabe et al., 1954). Since the commercialization of DEET as a personal protection product, it has remained the gold standard of topical repellents (Leal, 2014). Though ultimately effective, the laborious screening approach did little to explain how DEET functions as a disruptor of arthropod host-seeking and feeding behavior.

Essentials oils have long been used as arthropod repellents. Isolation of their components demonstrates that the volatile monoterpenes within them to be primarily responsible for their high bioactivity (Michaelakis et al., 2014; Tisgratog et al., 2016). Of the many monoterpenes screened for repellent properties, geraniol, linalool, citronella, limonene, eucalyptol, and nepetalactone have been identified as compounds that are moderately effective at preventing arthropod feeding behavior (Klocke et al., 1987; Gillij et al., 2008; Müller et al., 2009).

However, the first true rival to DEET was paramenthane 3,8-diol (PMD), which has been endorsed by the Centers for Disease Control and Prevention due to its efficacy and safety (Carroll and Loye, 2006; Goodyer et al., 2020). PMD is a hydro-distillation byproduct of the lemon eucalyptus plant. Besides DEET, the two most commonly used and effective synthetic arthropod repellents are ethyl butylacetyl-aminoproprionate (IR3535) and 2-(2-hydroxyethyl)-1-piperidinecarboxylic acid 1-methylpropyl ester (Picaridin) (Barnard et al., 2002; Naucke et al., 2007). IR3535 was developed in the 1970s by Merck KGaA and is a structure derivative of the natural occurring amino acid β-alanine (Achee et al., 2012). Picaridin was developed almost a decade later by Bayer AG, using a molecular modeling approach, based on the organic compound piperidine which has a pepper-like odor (Boeckh et al., 1996).

The identification of insect olfactory receptors in the late 1990s, and subsequent development of techniques allowing for the study of their odor-gated ion channel function has begun to unravel the mysteries of how repellents work (Vosshall et al., 1999,2000; Melo et al., 2004; Bohbot et al., 2007; Sato et al., 2008; Benton et al., 2009). In addition, recent advancement in genome editing techniques has allowed for the genetic analysis of the sensory pathway in a vector arthropod resulting in a more nuanced parsing of its chemosensory mechanism (Aryan et al., 2013; DeGennaro et al., 2013; Kistler et al., 2015; Raji and DeGennaro, 2017). Advances in genetic methods available for use on arthropods have facilitated the pursuit of three major research objectives: (1) determine the molecular sensors responsible for host detection and discrimination, (2) identify molecular sensors of arthropod repellents, and (3) discriminate the modes of action by which modulation of molecular sensors can result in behavioral disruption.

7.2 The sensory basis for host detection and discrimination

Development of more effective strategies to disrupt host-seeking and feeding behavior requires a better understanding of the neurosensory mechanisms driving these behaviors. Humans emit a wide array of odors that are primarily produced through the metabolic activity of their skin microbiota (Bernier et al., 2000; Verhulst et al., 2010). Anthropophilic insect species such as *Ae. aegypti and An. gambiae* have evolved a strong preference for these human odor cues (Qiu et al., 2006; Logan et al., 2008; McBride et al., 2014) which are detected at distance using olfactory appendages known as the antennae and maxillary palps (Montell and Zwiebel, 2016). These sensory organs house three distinct chemosensory receptor families: odorant receptors (ORs), ionotropic receptors (IRs), and gustatory receptors (GRs) (Suh et al., 2014).

Along with odors humans emit thermal radiation which serves as an important attractive cue at shorter distances (Li et al., 2019). The precise molecular sensor(s) responsible for detecting attractive heat cues is currently unknown in vector species. Transient receptor potential channel, subfamily A, member 1 (TRPA1) has however been shown to be necessary for mosquitoes to avoid objects that exceed human body temperature (Corfas and Vosshall, 2015). TRPA1 can be detected within olfactory and gustatory expressing neurons in insects (Wang et al., 2009; Kang et al., 2012). In Drosophila, the IRs have also been identified with roles in thermosensation and hygrosensation (Enjin et al., 2016; Knecht et al., 2016). If the IRs play a role in thermosensitivity in vector species aiding in host-seeking is an area of ongoing research.

In mosquitoes, carbon dioxide (CO_2) is detected by a group of highly specialized GRs expressed in the capitate peg sensilla of the maxillary palp (Lu et al., 2007). Removal of GR3 protein function is sufficient to eliminate the CO_2 response as determined by single sensillum recordings (McMeniman et al., 2014). The release of CO_2 in the breath of a potential vertebrate host serves as a long-distance chemosensory cue for mosquitoes (McMeniman et al., 2014; Raji and DeGennaro, 2017; van Breugel et al., 2015) Exposure to CO_2 alone has been shown to induce takeoff and sustained flight in female mosquitoes (Dekker and Cardé, 2011) as well as visual searching for hosts (Vinauger et al., 2019). Exposure to CO_2 followed by human odor results in mosquitoes being more behaviorally responsive in flight assays, suggesting CO_2 can sensitize the olfactory system (Dekker et al., 2005). *Gr3* mutants are unable to detect CO_2 and also lose attraction to heat and lactic acid, which suggests that CO_2 gates responses to other sensory cues (McMeniman et al., 2014). These studies clearly demonstrate that carbon dioxide detection is one of the most critically important sensory cues involved in sensory integration, priming, and host attraction.

The ORs are insect-specific, ligand-gated ion channels composed of an odor-tuned receptor and an obligate coreceptor termed Orco (Benton et al., 2006; Larsson et al., 2004). The odor-tuned binding subunits are highly divergent and exhibit varying degrees of sensitivity, with some being narrowly tuned to one particular odor and others showing broad range sensitivity to numerous odorants (Andersson et al., 2015; Carey et al., 2010). The functional characterization of each tuned receptor has become important for understanding the complete olfactory repertoire and the combinatorial code allowing for discrimination of complex volatiles like those involved in host detection. Much of the olfactory repertoire of individual receptors remains uncharacterized in insects, except for *An. gambiae* and *Drosophila melanogaster* (Carey et al., 2010; Hallem and Carlson, 2006). The role of the OR pathway in mosquito host preference was first investigated using zinc-finger nucleases in *Ae. aegypti* to generate targeted null mutations in *orco* (DeGennaro et al., 2013). These *orco* mutants showed no odor-evoked responses in OR-expressing olfactory neurons evaluated with electrophysiological recordings from single sensillia. In two-port choice assays, *orco* mutant females showed little or no response to host odor in the absence of CO_2. However, the presence of CO_2 was capable of rescuing the defect in *orco* mutant attraction, suggesting that CO_2 may activate a redundant mechanism for host detection. In two-port olfactometer choice assays that tested mosquito preference for human subjects versus a live guinea pig, *orco* mutants showed significantly less preference for human hosts than did control mosquitoes. This finding suggests that the OR pathway evolved in mosquitoes to aid in host odor discrimination. A study in *Ae. albopictus*, in which *orco* was knocked down by RNAi, found similar results in the reduction of host-seeking as well as host discrimination (Liu et al., 2016).

Much like ORs, IRs are ligand-gated ion channels that play a major role in chemosensory detection at the periphery. ORs and IRs are believed to have distinct evolutionary origins, with IRs closely related to ionotropic glutamate receptors (iGluR) (Croset et al., 2010). While ORs use a single coreceptor, the IR gene family possesses two known coreceptors, IR8a and IR25a, and a putative third coreceptor, IR76b (Abuin et al., 2011; Benton et al., 2009). In *Ae. aegypti*, IR8a expression is restricted exclusively to the antennae, whereas IR25a and IR76b can be found in the antennae and other chemosensory structures (Matthews et al., 2016). Restriction to the antennae suggests that, unlike the other coreceptors, IR8a functions exclusively in odor-detecting pathways. To determine the function of IR8a in odor detection, null mutants were generated using the Clustered Regularly Interspaced Short Palindromic Repeats (CRISPR)/Cas9 genome editing system (Raji et al., 2019). Electroantennogram recordings of mutants lacking functional IR8a displayed no antennal response to the acidic volatiles found in human odor, including lactic acid. When coexposed with CO_2, lactic acid elicits attraction behavior in *Aedes* and is one of the few known single odor attractants. However, *Ir8a* mutants lose this attraction behavior, in comparison to wild-type. Uniport olfactometer attraction assays with added CO_2 were conducted on human subjects to determine if loss of *Ir8a* impairs host attraction. Wild-type and *orco* mutants showed robust attraction to live hosts, whereas host attraction in *Ir8a* mutants was significantly reduced. In order to isolate the role of olfactory cues, nylon sleeves with trapped human odor were substituted for a live host in the uniport assay. As in the live host assay, wild-type and *orco* mutants showed robust attraction, while *Ir8a* mutants were significantly impaired. Since CO_2 rescues host-seeking behavior in *orco* mutants but not *Ir8a* mutants, we infer that CO_2 modulates the IR8a and Orco pathways differently. Testing *Gr3/Ir8a* double mutants in an olfactometer assay with human odor, determined that the host-seeking defect is enhanced when compared to *Ir8a* mutants alone (Raji et al., 2019). The genetic analysis of the OR, IR, and GR chemosensory pathways demonstrates the fundamental role they have in driving host attraction, this thereby situate the olfactory receptors as attractive targets for informed design of mosquito repellents.

7.3 Proposed mechanisms of olfactory repellency

Researchers have long hypothesized that arthropod repellents modulate sensors in the peripheral nervous system, interrupting normal response to attractive cues (Davis, 1985). The discovery that *orco* mutants are insensitive to volatile DEET in both *D. melanogaster* and *Ae. Aegypti* provided ample evidence to bolster this postulation (DeGennaro et al., 2013; Ditzen et al., 2008). Whether other commercially available synthetic repellents are exclusively dependent on the OR pathway has not been genetically validated in vivo. Since ORs are insect-specific olfactory receptors, it remains to be determined through what receptors DEET is sensed in other arthropods such as ticks, for which DEET also functions as an effective repellent. Recent evidence suggests that GPCRs may also play a role in DEET repellency (Dennis et al., 2018).

Another class of receptors recently identified as having a role in repellent compound perception are the transient receptor potential (TRP) ion channels (Li et al., 2019; Melo et al., 2021). TRP channels have been implicated in a wide variety of sensory perceptions including audition, vision, olfaction, thermosensation, mechanosensation, and nociception in insects (Göpfert et al., 2006; Sakai et al., 2009; Kang et al., 2012; Fowler and Montell, 2013). In mosquitos, thermal and chemical stimuli can evoke response in the TRPA1 ion channels (Corfas and Vosshall, 2015; Li et al., 2019). Among the few known chemicals capable of activating TRPA1 are two essential oil monoterpenes, Citronellal and Nepetalactone which are derive from Citronella oil and Catnip oil, respectfully (Kwon et al., 2010; Li et al., 2019; Melo et al., 2021). In a uniport olfactometer assay Catnip oil acts as a spatial repellent when applied to live host arm resulting in a significant reduction in attraction compared to nontreated control arm. *A. aeypti* TRPA1 mutants run through the same uniprot trials are no longer repelled by the catnip oil on live host arm (Melo et al., 2021). This genetic evidence suggests that some essential oils like catnip act uniquely in term of their molecular target as compared to DEET and do not require activation of ORs, but are sensed by other molecular sensors in the volatile state. Two-choice arm-in-cage assays support this finding, as *orco* mosquitoes and wild-type have similar levels of preference when provided with a choice to approach and probe a catnip treated hand or an untreated hand. Whereas the TRPA1 mutants show a loss of preference for the untreated hand as compared to either the *orco* or wild-type (Melo et al., 2021). How pervasive TRPA1 activation is amongst other botanical repellents as of now is unknown.

While DEET remains the most studied repellent due to its widespread usage and efficacy, other repellents such as picaridin, IR3535, and PMD have also received attention in recent years (Barnard et al., 2002; Carroll and Loye, 2006; Naucke et al., 2007; Colucci and Müller, 2018). Much of the early work seeking to understand the molecular basis of repellency was conducted in the *D. melanogaster* model system (Davis, 1985; Dogan et al., 1999; Ditzen et al., 2008). As the genetic and electrophysiological tools were developed in mosquitoes or by expressing their genes in heterologous systems, work on repellency followed in vector species (Bohbot and Dickens, 2010, 2012; DeGennaro et al., 2013; Afify et al., 2019; Afify and Potter, 2020). These studies in mosquitoes suggested that ORs play a primary role in repellent reception. However, how insect repellents modulate ORs to induce avoidance is not yet fully understood. This has led to an array of competing and highly contested theories for modes of repellent action (Syed and Leal, 2008; Bohbot and Dickens, 2010, 2012; DeGennaro et al., 2013; DeGennaro, 2015; Afify and Potter, 2020).

The discordance amongst these competing theories suggests one universal mode of action does not exist for all currently effective arthropod repellents, and that induction of repellency maybe multimodal. Although a precise mode of repellent action may not be fully understood, three mechanistic repellent classes have been proposed: true repellents, true maskants, and confusants (Fig. 7.1) (Deletre et al., 2016; Pellegrino et al., 2011; Syed and Leal, 2008). True repellents cause-oriented movement away from its source, likely by being perceived as a negative stimulus that should be avoided. True maskants render the host imperceptible by preventing the detection of attractive host cues through inhibition of molecular sensors. Confusants interfere with the ability to host-seek by *changing* the way host cues are being perceived, versus blocking the cues as in maskants. The proposed modes of action of repellents can all theoretically cause a net reduction in pest engagement with their hosts, but the molecular understanding of how these effects are achieved is incomplete.

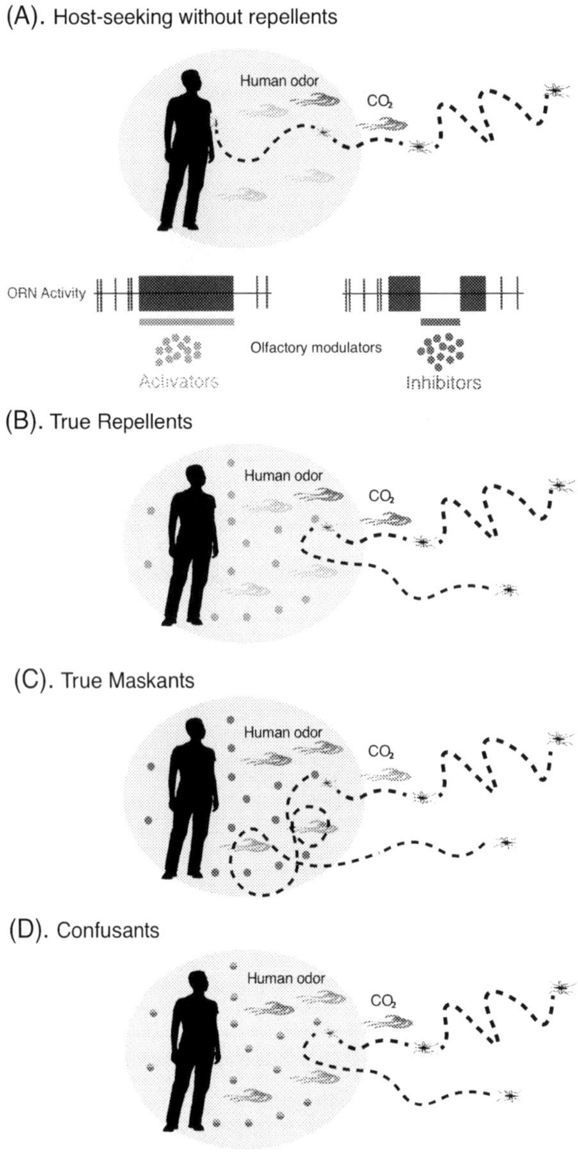

FIG. 7.1 Proposed modes of action of effective mosquito repellents. (A) A host-seeking mosquito detects CO_2 exhaled from human's breath (*blue*). As they move closer, they detect odor emulating from the host (*shades of green*). Integration of these chemosensory cues lead to movement (*dashed lines*) toward the host and the acquisition of a blood meal. (Below) Graphic representation of olfactory receptor neuron (*ORN*) responding to activation or inhibition measured by single sensillum recordings. (B) When mosquitoes detect true repellents, which most likely are ORN activators (*orange*), the repellent evokes movement away from host through direct or combinatorial activation of avoidance response circuits. (C) When mosquitoes detect ORN inhibitors (*purple*) that reduce the sensation of host odor, their behavior is being modified by a true maskant. The loss of chemosensory cues leads to inability to track and locate their host. (D) Mosquitoes detect molecules that activate and/or inhibit ORNs (*orange/purple*). The normally attractive olfactory activation caused by host odor alone is now disrupted by the confusant. This scrambling of olfactory coding changes the way host odor is perceived (*shades of red*). This disorientation could result in the inability to identify the host as well as induce avoidance.

7.3.1 Smell and avoid

Arthropods can detect, discriminate, and behaviorally respond to a wide-range of chemicals found in their environment. Smell-and-avoid mechanisms of repellency suggest that some chemicals are perceived as noxious odors which activate a neural circuit encoding avoidance behavior (Liu et al., 2016). Labeled-line circuits are common coding strategies in olfactory circuits in which salient stimuli are received and processed through dedicated groups of receptors, neurons, and projection fields. In *Drosophila*, one study identified a labeled-line circuit encoding avoidance behavior for the microbial product geosmin (Stensmyr et al., 2012). Geosmin elicited activity in the narrowly tuned receptor OR56a, which projected exclusively into the DA2 glomerulus. Activation of DA2 glomeruli, performed experimentally using the Gal4-UAS binary gene expression system, was shown to be necessary and sufficient to evoke aversion behavior, even in the absence of geosmin. The authors of the study suggested that this highly specialized avoidance circuit in *Drosophila* likely evolved as a warning signal for the presence of toxins often found with the microbes that produce geosmin.

Evidence for labeled-line repellency in mosquitoes has also been documented. A study in *Cx. quinquefasciatus* suggests response to repellents may function via by activating a labeled-line circuit that elicits avoidance behavior (Xu et al., 2014). An OR in the mosquito *Cx. quinquefasciatus*, CquiOR136, responds to methyl jasmonate (Xu et al., 2014), a naturally occurring volatile that has repellent properties across arthropods (Garboui et al., 2007). The study found this receptor was also activated by the commercially available repellents DEET, PMD, picaridin, and IR3535 (Syed and Leal, 2008). Functional knockdown of *CquiOR136* using RNAi resulted in *Cx. quinquefasciatus* attraction to a heated blood feeder surrounded by a paper ring soaked with 0.1% DEET, while mosquitoes without *CquiOR136* knockdown avoided landing on the blood feeder. This result suggests that *CquiOR136* is both necessary and sufficient to detect and avoid DEET in the vapor phase even in the absence of any other olfactory cue. It should be noted that the assay failed to account for the possibility that avoidance behavior was actually the result of contact chemo-repellency with the paper ring. The contact effects of DEET do not depend on ORs in *Ae. aegypti* (DeGennaro et al., 2013; Dennis et al., 2019). One implication of these findings is that *CquiOR136* may have evolved as a generic mosquito repellent detector which volatile DEET, PMD, Picaridin, and IR3535 repellents exploit to drive avoidance behavior. Similarly, this reasoning may serve to explain how extremely diverse stimuli such as high temperature and monoterpenes are able to cause aversion behavior via activation of TRPA1 ion channel in mosquitoes. TRPA1 fundamental role in nociception may have allowed it to evolve into a diverse detector of noxious and irritant environments.

Determining which neurons are activated in the presence of a repellent is a good first step for discovering circuits that drive avoidance behavior. Activation alone, however, is insufficient to determine a neuron's role in evoking a behavioral response. A seemingly paradoxical result of arthropod behavioral assays is that the same odorant can induce attractive or repulsive responses depending the concentration (Afify and Potter, 2020; Xu et al., 2015). At lower concentrations, 1-octen-3-ol elicits attraction behavior in multiple mosquito species, but repellent behavior at higher concentration (Afify and Potter, 2020; Xu et al., 2015). A study using a calcium indicator of neuron activation found that as 1-octen-3-ol concentration increases for the mosquito *A. coluzzii*, so does activation of lower sensitivity olfactory neurons to the odor (Afify and Potter, 2020). One interpretation of this recruitment of new olfactory neurons is that increasing concentration triggers repellent sensing circuits causes a shift in behavior from attraction to repellency. Therefore, discovery of salient olfactory receptor neurons (ORNs) may require a method that allows for activation of isolated populations of neurons without necessarily having knowledge of their odor-ligands. Live imaging of these neurons and the glomeruli they innervate may reveal specific activation patterns necessary for mosquitoes to avoid their hosts (Bui et al., 2019; Riabinina et al., 2016; Zhao et al., 2021).

Olfactogenetics is an emerging strategy for allowing the activation of a defined group of ORNs by a specifically chosen odor (Neff, 2018). When paired with behavioral assay present a powerful tool for correlating neuron activation to behavioral responses. In *Drosophila*, ectopic expression of Or56a neurons confers sensitivity to volatile geosmin in selected neuron groups (Stensmyr et al., 2012; Chin et al., 2018). Ectopic receptor expression allows for odor-guided screening to find behaviorally relevant neurons without knowledge of the ligand affinity of the endogenous receptors. This approach has been used to discover circuits that suppress oviposition behavior in *Drosophila* by ectopically expressing Or56a in selected neurons followed by oviposition assays in the presence of geosmin (Chin et al., 2018). Counter to the idea that repellency could be encoded through a labeled-line mechanism is that the number of potential aversive stimuli an arthropod must respond to outnumbers the total number of olfactory receptors. Therefore, most information is likely to be encoded as a combinational code (Haverkamp et al., 2018). If repellency requires combinational activation of multiple receptors to induce avoidance behavior, the effort to identify the right combination rises with the power of the number of receptors involved. The discovery of behaviorally salient receptors may benefit from targeted generation of multiple olfactory receptor mutations in a single arthropod to allow for compound genetic analysis. Both loss-of-function and gain-of-function approaches can help us identify new molecular targets for repellent design that exploit a potential as-yet undiscovered repellent response circuit.

7.3.2 Global activation

Avoidance behavior can be induced by activation of an innate repellent circuit (Stensmyr et al., 2012) and recruitment of less specifically tuned neurons (Afify and Potter, 2020). It would then follow logically that global activation of ORs might also induce avoidance behavior. During a screen to identify novel arthropod repellents using high-throughput calcium-imaging in HEK293 cells expressing *An. gambiae* ORs, it was discovered that AgOrco could form functional ion channels even in the absence of tuned ORs (Jones et al., 2011). Out of the 118,720 compounds screened, only the synthetic compound N-,2-substituted triazolothioacetamide (VUAA1) elicited macroscopic currents in cells expressing only AgOrco. Further screening of *D. melanogaster, Heliothis virescens, Harpegnathos saltator* Orco orthologs indicated that VUAA1 was a broad-spectrum Orco family agonist capable of activating Orco across insects. In addition to acting as an Orco agonist, when a OR/Orco receptor complex is coexposed to VUAA1 and an odor-ligand, the combination enhances the activation level of the receptor complex by 5-fold to >30-fold over the ligand alone (Tsitoura and Iatrou, 2016). These findings ushered in a new era in repellent design, one focused on a well-defined molecular target. Despite the obvious promise for VUAA1 as a behavioral disruptant, its low volatility and high molecular weight make it less than ideal for usage as an airborne repellent (Chen and Luetje, 2012). In an arm bite protection assay with *An. albimanus, A. quadrimaculatus,* and *Ae. aegypti* mosquito species, VUAA1 was found to offer no protection in preventing mosquitoes from biting (Yang et al., 2020). These authors suggested this lack of effect was most likely due to VUAA1 having low volatility, thus hampering its ability to access Orco binding sites. Evidence that Orco is not involved in contact-mediated repellency may also help explain these results (DeGennaro et al., 2013). Development of other Orco agonists such as OrcoRAM2 (Orco Receptor Activator Molecule 2) has failed to resolve the issue of low volatility due to restrictive structural requirements for Orco agonists (Tsitoura and Iatrou, 2016). Do to this limitation an adequate assessment of a sufficient Orco agonist effect on mosquito behavior is still lacking. In *Drosophila,* however, optogenetics techniques using red-shifted channelrhodopsin to globally activate the OR pathway resulted in enhanced chemotaxis to apple cider vinegar traps rather than repellency (Guo et al., 2020). Current evidence therefore suggests that global activation via Orco agonism may not yield an effective arthropod repellent.

Near-complete global activation could potentially be achieved through another mechanism, activation of ORs by a promiscuous odor ligand. A study of the chemical 2-methyltetrahydro-3-furanone (coffee furanone) found that it acts as a highly promiscuous ligand in the antenna of *D. melanogaster*, detected by approximately 80% of the recorded neurons housing olfactory receptors (Batra et al., 2019). Two-electrode voltage-clamp recordings of Orco-expressing cells confirmed that coffee furanone did not act as an Orco agonist but rather elicited activity only when paired with an odor-tuned receptor (Batra et al., 2019). Additionally, *Drosophila orco* mutants lacked response in specialized neurons expressing IRs and GRs, indicating the response was specific to OR pathway activation. Behavioral assays using Y-tube olfactometer found that coffee furanone was attractive to the cockroach species *Blattella germanica* but repellent in *D. melanogaster* and *Ae. aegypti*. It remains unclear what is structurally unique about coffee furanone to allow it such a high level of receptor promiscuity. Highly promiscuous odors like coffee furanone may provide a novel alternative strategy for hyperactivation of the olfactory system of insects.

7.3.3 Global inhibition

Thus far we have discussed repellency mediated by activation of avoidance circuits. A second possible class of repellency mechanisms is masking host odor through receptor inhibition, thus interfering with host-seeking. In the process of screening VUAA1 analogs, researchers discovered compounds capable of inhibiting homomeric and heteromeric Orco (Jones et al., 2012). These compounds were identified as allosteric modulators of OR signaling capable of broadly inhibiting Orco. In contrast to the low volatility of the known Orco agonist, several Orco antagonists possess higher volatility and fewer structural constraints (Kepchia et al., 2019, 2017). The Orco family antagonist OX1w is a broad-spectrum inhibitor of ORs in *D. melanogaster*, *Cx. quinquefasciatus*, and *An. gambiae* (Kepchia et al., 2017). Vapor phase behavioral inhibition by an Orco antagonist was first demonstrated in a *D. melanogaster* larval chemotaxis assay in which airborne OX1w abolished larval chemotaxis toward ethyl acetate, a known attractant. The larvae retained their normal light avoidance behavior, indicating that OX1w was not a general inhibitor of sensory-mediated behavior (Kepchia et al., 2017). These early successes suggest that further development of Orco antagonists may be a viable strategy in the pursuit of the next generation of arthropod control agents. It may be the case that some commercially available repellents already function as global Orco inhibitors and have yet to be identified as such. In a lepidopteran insect cell-based assay of ethyl cinnamate, carvacrol, cumin alcohol, isopropyl cinnamate, and DEET on Orco homomer-expressing cells, only DEET lacked inhibitory activity (Tsitoura et al., 2015). The remaining naturally derived arthropod repellent compounds showed inhibitory activity in a dose-dependent manner. This study provided the first indication that naturally occurring compounds may function as Orco antagonists in vivo. Prior electrophysiological studies using OrcoRAM2 as an agonist for

Ae. aegypti Orco, coexposed with either DEET or IR3535, yielded no meaningful reduction in activity when compared to OrcoRAM2 alone (Bohbot and Dickens, 2012). The lack of synergy indicates that DEET and IR3535 do not inhibit receptors by acting as Orco agonists. Taken together, these studies suggest that Orco antagonism could potentially be used as a targeted approach to the induction of behavioral avoidance and should warrant further development. Early attempts at finding Orco agonists and antagonists relied on varying certain functional groups from lead compounds (Chen and Luetje, 2012). The computer-aided repellent design offers great potential for overcoming structural restraints (Basak and Bhattacharjee, 2020; Katritzky et al., 2008; Kepchia et al., 2019). Recent work using a simple machine learning approach based on ligand topology has increased the ability to identify candidate Orco antagonists (Kepchia et al., 2019).

7.3.4 Tuning olfactory receptor inhibition

A third possible explanation for the behavioral inhibition of repellents is that they function by masking host odor through selective inhibitions of tuned ORs. The reduction in excitation would presumably diminish host detection and feeding behavior. This mechanism would not require global inhibition of the OR pathway but rather a reduction of odor signal to human odor-tuned glomeruli causing excitation below the threshold necessary for behavioral activation. In the presence of their natural odor-ligand, DEET, IR3535, and Picaridin can inhibit mosquito ORs involved in attraction to human odor (Bohbot and Dickens, 2010). 1-octen-3-ol is a particularly salient odor to mosquitoes, being commonly released with carbon dioxide in vertebrate breath (Xu et al., 2015). OR8 receptor orthologs are highly conserved across numerous mosquito species and narrowly tuned to 1-octen-3-ol (Dekel et al., 2016). A study characterizing response of *A. ageypti* AaOR2 and AaOR8 to DEET, Picardin, and IR3535 determined that, in the absence of an odor-ligand, IR3535 and Picardin inhibited both AaOR2 and AaOR8. DEET, however, responded differently, activating the AaOR2 but not AaOR8. When coexposed with their activating odor-ligand, DEET strongly inhibits AaOR8 but not AaOR2. However, IR3535 and Picaridin were found to strongly inhibit both ORs. Differential response of these two ORs to DEET provides clear evidence that DEET's OR inhibition is a selective rather than global. Another study of ORs from *Ae. aegypti* found that DEET and IR3535 act as insurmountable antagonists to both AaOR8 and another receptor, AaOR10 (Bohbot and Dickens, 2012). These authors suggested that DEET and IR3535 act as silent antagonists that form long-lasting complexes with the orthosteric site of the receptor but alone do not have any intrinsic activity. How ubiquitous is the ability to interact with DEET and IR3535 across the all chemosensory receptors remains unknown.

Lactic acid is another prominent odorant released by humans that serves as a particularly effective mosquito attractant when copresented with carbon dioxide (Smallegange et al., 2005; McMeniman et al., 2014). DEET is capable of reducing the response of lactic acid-excitable neurons as well as increasing inhibition of lactic acid-inhibited neurons, thereby reducing overall neuronal sensitivity to volatile lactic acid (Davis, 1976). Behavioral assays using a repellometer found that DEET is capable of behaviorally inhibiting attraction to lactic acid as well (Dogan et al., 1999). Electrophysiological recordings and behavioral assays using Ir8a mutants have demonstrated IR8a is necessary to detect lactic acid as well as other acidic volatiles and is required to behaviorally respond to lactic acid (Raji et al., 2019). So although DEET has been shown to inhibit lactic acid response it has not been directly implicated in interaction with any IRs (Silbering et al., 2016). Recent preprints have suggested that ORNs can express both ORs and

IRs (Task et al., 2020; Younger et al., 2020). Further analysis of overlapping olfactory receptor activity in ORNs may help explain whether IRs can play a role in the repellency of DEET or other repellents.

An opposing hypothesis to the direct inhibition of receptors is that repellents decrease activation of ORNs indirectly by interacting with the odorant molecules to reduce the quantity of volatile odorants reaching the receptors (Syed and Leal, 2008; Afify et al., 2019). Ca^{2+} imaging revealed that fluorescence response to prestimulation of a neuron by DEET, IR3535, or Picaridin followed by pulses of 1-octen-3-ol was not significantly different than paraffin oil followed by 1-octen-3-ol control. When repellents and 1-octen-3-ol are coexposed however it results in a reduction in fluorescence response. (Afify et al., 2019). This result would suggest that physical mixing of human odor and repellents would be necessary to induce avoidance.

We know that DEET aversion in both *D. melanogaster* and *Ae. aegypti* is Orco-dependent and likely results from interactions of host odors and DEET at the OR-Oorco receptor complexes in the ORNs (Ditzen et al., 2008; Pellegrino et al., 2011; DeGennaro et al., 2013). More evidence is therefore required to determine whether DEET is interacting chemically with host odors to reduce their binding on the receptors.

How global inhibition or partial inhibition of the OR pathway could overcome the reductant cues system driving host attraction is unclear. Mosquitoes with genetic ablation of *orco*, disabling the OR pathway, are still able to host seek (DeGennaro et al., 2013). Thus, a repellent that functions solely by the inhibition of the OR pathway would be only moderately effective at best. To fully disrupt host-seeking, a repellent designed to reduce mosquito attraction to human odor would need to inhibit the production of multiple odorants chemical classes in the host or to inhibit multiple sensory pathways in the mosquito.

7.3.5 Olfactory confusant

DEET can activate, inhibit, or have no effect depending on the OR investigated. Interaction with DEET has been demonstrated on multiple different ORs using physiological recordings from OR-expressing neurons (Ditzen et al., 2008; Pellegrino et al., 2011) and ORs expressed in heterologous systems (Bohbot and Dickens, 2012, 2010). Behavioral assays suggest that DEET alone is not itself a repellent but acts in combination with a secondary host odor (Ditzen et al., 2008). This combinatorial effect has led to the "confusant hypothesis" which posits that volatile DEET itself does not cause changes in arthropod behavior on its own, but instead its repellency is an emergent property of both host odor and DEET (Pellegrino et al., 2011). DEET in the context of the numerous host odor cues changes the normal activation and inhibition of ORNs that would be caused by host odor alone. The DEET-induced change to the activation pattern associated with host attraction thereby disrupts olfactory coding and host recognition.

7.3.6 Discriminating between repellent modes of action

Electrophysiological approaches have played a paramount role in investigating neuron activity to odor presentation but reveals little of how this will translate to behavior responses. Determining how mosquitoes perceive different repellents could be aided by a more nuanced exanimation of their behavioral responses. Repellents can be sorted into the three general classes based on the behavioral response induced through olfaction: true repellents, true maskants, and confusants (Syed and Leal, 2008; Pellegrino et al., 2011; Deletre et al., 2016). True repellents are able to induce oriented movement away from the repellent source. Avoidance circuits are thought to be induced by true repellents. The repellent alone would likely be sufficient to drive avoidance and does not require the presence of a secondary odor source. If attractive odor cues are present, they can be overridden at sufficient concentrations. True maskants on the other hand are not repellent on their own, but rather work by decreasing the attractiveness of the host. In theory, several mechanisms could produce olfactory masking of host odors, however, a true maskant likely would inhibit sensory modalities such as olfactory chemoreception, thereby decreasing attractive host cue signal processing by mosquitoes. Coexposure of a repellent with a normally attractive odor may result in anosmic behavior, which could manifest in a lack of strong preference in choice assay. Confusants fail to fall neatly in either category and would have complex effects on the activation or inhibition at each ORN (Pellegrino et al., 2011). Scrambling of odor coding could potentially result in the inability to identify the host as well as induce avoidance. Under the confusant hypothesis, an arthropod repellent alone could be attractive or have no effect. Avoidance behavior is seen as an emergent property caused by the presence of an arthropod repellent along with multiple host odors that alter the normal perception of host cues by the arthropod. Classifying arthropod repellents by both their efficacy and mode of action will focus the field on how to improve our current tools for personal protection from biting arthropods.

7.3.7 Disrupting skin contact

One of the challenges in comparing behavioral studies of repellents is that many of the studies fail to control for contact between host and arthropod. For example, *orco* mutant *Ae. aegypti* host-seek even in the presence of volatile DEET. Upon contact, however, they are repelled by DEET-treated skin (DeGennaro et al., 2013). Futhermore, *orco* mutants with their tarsal segments occluded by glue are no longer repelled by contact with DEET-treated skin, indicating that the tarsal segments are necessary and sufficient for contact-medicated repellency (Dennis et al., 2019). This study suggests that contact and antifeedant effects of DEET are distinct. The genes that mediate the skin contact effects of DEET remain unknown. To fully characterize a repellent's mode of action requires separating its skin contact effect from its and olfactory effect. Clarifying these modes of action may also provide insight into how repellents like DEET work in arthropods outside of insects such as ticks where contact repellency may be more important than olfactory repellency.

7.4 Acidic volatiles and CO_2 detection pathway modulation

The IR8a coreceptor is crucial for mosquito host-seeking, as it is necessary to detect salient acidic volatiles like lactic acid. The ability of DEET to inhibit attraction to lactic acid may indicate a role of the IRs in detection and response to arthropod repellents (Dogan et al., 1999), however, at this time, minimal research has been conducted on the effects of arthropod repellents on the IR pathway. Multiple characteristics make IR8a an attractive target for repellents. Unlike with *orco* mutants, host-seeking defects in *Ir8a* mutants are not rescued by the presence of CO_2, indicating a different gating relationship and minimal functional overlap with other chemosensory families (Raji et al., 2019). Furthermore, *Ir8a* is highly conserved across arthropods species, making it ideal as a target for a broad-spectrum, general arthropod repellent (Croset et al., 2010). Agonists and antagonists capable of direct IR8a modulation have yet to be identified.

Another attractive target for repellent design is the gustatory receptor Gr3, which is necessary for the detection of CO_2, and which plays a major role in sensory integration and gating other sensory cues (McMeniman et al., 2014). Recent research suggests that *Ae. aegypti* Gr1/Gr2/Gr3 CO_2 receptor complex could be inhibited and activated by compounds other than

CO_2 (Tauxe et al., 2013; Turner et al., 2011). This finding has spurred investigation to develop a new class of arthropod repellents that directly target CO_2 sensation (Potter, 2014). Although Orco, IR8a, and Gr3 present themselves as attractive molecular targets for inhibition, a next-generation arthropod repellent or repellent blend may have to inhibit all of these receptors to fully mask the human odor signature.

7.5 Toward the next generation of targeted arthropod repellents

The conventional starting point for synthetic repellent design has either been a previously known repellent compound or a novel repellent identified by laborious in vivo screens. Derivatives are generated from this original compound and tested for bioactivity against mosquitoes, as well as human safety. In contrast, the next generation of arthropod repellents could be discovered by chemical screens targeting receptors that are either necessary for host attraction or which have been shown to mediate true repellency. The advantage of a targeted approach is that it will allow for high throughput screening of thousands of compounds against a given receptor, significantly increasing screening capability compared to in vivo behavioral approaches. Screens can be devised around the functional expression of ORs in cultured cells, genetically modified to visualize ion channel activity in response to chemical challenge (Tauxe et al., 2013; Turner et al., 2011). Such a screening process has already expanded our current repertoire of Orco modulators (Jones et al., 2012; Kepchia et al., 2019). As active compounds are identified, machine learning and computer-aided chemical design can be applied to identify ligand topology and structural constraints of a potential repellent (Katritzky et al., 2008; Kepchia et al., 2019; Basak and Bhattacharjee, 2020). In addition, this information has tremendous utility for the development of *in silico* chemical prediction models based on structural characteristics (Turner et al., 2011; Devillers et al., 2014).

The proposed multimodal mechanisms of olfactory repellency suggest we have more than one way to repel a mosquito, and therefore more than one screening process is suitable for the isolation of candidates with potential repellent activity. These are some potential approaches: (1) screen for compounds that modulate one or a few receptors which have been identified as being involved in innate avoidance to repellents such as TRPA1, an approach likely to yield a highly specific class of repellents, (2) screen for compounds capable of modulating entire chemosensory pathways by targeting obligate receptor complex components such as Orco, IR8a, GR3, or (3) screen for compounds with highly promiscuous activity that are capable of broad inhibition, broad activation, or both. Taken together these approaches present a potentially significant improvement to conventional methods of identifying behaviorally relevant compounds.

7.6 Conclusion

Topical chemical repellents still remain the most effective personal protection for preventing arthropods diseases transmission. The proposed modes of action of repellents, along with an analysis of the olfactory pathways involved in host-seeking, provides a framework for understanding how the modulation of olfactory receptor activity may impact arthropod behavior. This framework should inform screening strategies for identifying behaviorally salient compounds capable of modulating desired molecular targets to induce avoidance behavior. Design and development of arthropod repellents can benefit greatly from a more targeted approach that takes into account knowledge of important molecular targets involved in host-seeking and mechanisms of how repellents can disrupt attraction.

Acknowledgments

We would like to thank Anthony Bellantuono, Alejandro Acuña, and Philip Stoddard for helpful comments and suggestions on the manuscript. This work was supported by the National Institute of Allergy and Infectious Diseases of the National Institutes of Health under Award Number R21AI142140. M.D. was also supported by The Centers for Disease Control and Prevention (CDC), Southeastern Center of Excellence in Vector-borne Disease (U01CK000510). The content is solely the responsibility of the authors and does not necessarily represent the official views of the National Institutes of Health or the CDC.

References

Abuin, L., Bargeton, B., Ulbrich, M.H., Isacoff, E.Y., Kellenberger, S., Benton, R., 2011. Functional architecture of olfactory ionotropic glutamate receptors. Neuron 69, 44–60.

Achee, N.L., Bangs, M.J., Farlow, R., Killeen, G.F., Lindsay, S., Logan, J.G., Moore, S.J., Rowland, M., Sweeney, K., Torr, S.J., Zwiebel, L.J., Grieco, J.P., 2012. Spatial repellents: from discovery and development to evidence-based validation. Malar. J. 11, 164.

Afify, A., Betz, J.F., Riabinina, O., Lahondère, C., Potter, C.J., 2019. Commonly used insect repellents hide human odors from *Anopheles* mosquitoes. Curr. Biol. 29, 3669–3680.e5.

Afify, A., Potter, C.J., 2020. Insect repellents mediate species-specific olfactory behaviours in mosquitoes. Malar. J. 19, 127.

Andersson, M.N., Löfstedt, C., Newcomb, R.D., 2015. Insect olfaction and the evolution of receptor tuning. Front. Ecol. Evol. 3, 53.

Aryan, A., Anderson, M.A.E., Myles, K.M., Adelman, Z.N., 2013. TALEN-based gene disruption in the dengue vector *Aedes aegypti*. PLoS One 8, e60082.

Barnard, D.R., Bernier, U.R., Posey, K.H., Xue, R.-D., 2002. Repellency of IR3535, KBR3023, para-menthane-3,8-diol, and Deet to black salt marsh mosquitoes (Diptera: Culicidae) in the everglades national park. J. Med. Entomol. 39, 895–899.

Basak, S.C., Bhattacharjee, A.K., 2020. Computational approaches for the design of mosquito repellent chemicals. Curr. Med. Chem. 27, 32–41.

Batra, S., Corcoran, J., Zhang, D.-D., Pal, P.K.P., Kulkarni, R., Löfstedt, C., Sowdhamini, R., Olsson, S.B., 2019. A functional agonist of insect olfactory receptors: behavior, physiology and structure. Front. Cell Neurosci. 13, 134.

Benton, R., Sachse, S., Michnick, S.W., Vosshall, L.B., 2006. Atypical membrane topology and heteromeric function of *Drosophila* odorant receptors *In vivo*. PLoS Biol 4, e20.

Benton, R., Vannice, K.S., Gomez-Diaz, C., Vosshall, L.B., 2009. Variant ionotropic glutamate receptors as chemosensory receptors in *Drosophila*. Cell 136, 149–162.

Bernier, U.R., Kline, D.L., Barnard, D.R., Schreck, C.E., Yost, R.A., 2000. Analysis of human skin emanations by gas chromatography/mass spectrometry. 2. Identification of volatile compounds that are candidate attractants for the yellow fever mosquito (*Aedes aegypti*). Anal. Chem. 72, 747–756.

Boeckh, J., Breer, H., Geier, M., Hoever, F.-P., Krüger, B.-W., Nentwig, G., Sass, H., 1996. Acylated 1,3-Aminopropanols as repellents against bloodsucking arthropods. Pestic. Sci. 48, 359–373.

Bohbot, J., Pitts, R.J., Kwon, H.-W., Rützler, M., Robertson, H.M., Zwiebel, L.J., 2007. Molecular characterization of the *Aedes aegypti* odorant receptor gene family. Insect Mol. Biol. 16, 525–537.

Bohbot, J.D., Dickens, J.C., 2010. Insect repellents: modulators of mosquito Odorant receptor activity. Plos One 5, e12138.

Bohbot, J.D., Dickens, J.C., 2012. Odorant receptor modulation: ternary paradigm for mode of action of insect repellents. Neuropharmacology 62, 2086–2095.

Brown, M., Hebert, A.A., 1997. Insect repellents: an overview. J. Am. Acad. Dermatol. 36, 243–249.

Bui, M., Shyong, J., Lutz, E.K., Yang, T., Li, M., Truong, K., Arvidson, R., Buchman, A., Riffell, J.A., Akbari, O.S., 2019. Live calcium imaging of *Aedes aegypti* neuronal tissues reveals differential importance of chemosensory systems for life-history-specific foraging strategies. BMC Neurosci. 20, 27.

Carey, A., Wang, G., Su, C.-Y., Zwiebel, L.J., Carlson, J.R., 2010. Odourant reception in the malaria mosquito *Anopheles gambiae*. Nature 464, 66–71.

Carroll, S.P., Loye, J., 2006. PMD, a registered botanical mosquito repellent with DEET-like efficacy. J. Am. Mosq. Control Assoc. 22, 507–514.

Chen, S., Luetje, C.W., 2012. Identification of new agonists and antagonists of the insect odorant receptor co-receptor subunit. PloS One 7, e36784.

Chin, S.G., Maguire, S.E., Huoviala, P., Jefferis, G.S.X.E., Potter, C.J., 2018. Olfactory neurons and brain centers directing oviposition decisions in drosophila. Cell. Rep. 24, 1667–1678.

Colucci, B., Müller, P., 2018. Evaluation of standard field and laboratory methods to compare protection times of the topical repellents PMD and DEET. Sci. Rep. 8, 12578.

Corfas, R.A., Vosshall, L.B., 2015. The cation channel TRPA1 tunes mosquito thermotaxis to host temperatures. eLife 4, e11750.

Croset, V., Rytz, R., Cummins, S.F., Budd, A., Brawand, D., Kaessmann, H., Gibson, T.J., Benton, R., 2010. Ancient protostome origin of chemosensory ionotropic glutamate receptors and the evolution of insect taste and olfaction. PLos Genet. 6, e1001064.

Davis, E.E., 1976. A receptor sensitive to oviposition site attractants on the antennae of the mosquito, *Aedes aegypti*. J. Insect. Physiol. 22, 1371–1376.

Davis, E.E., 1985. Insect repellents: concepts of their mode of action relative to potential sensory mechanisms in mosquitoes (Diptera: Culicidae). J. Med. Entomol. 22, 237–243.

DeGennaro, M., 2015. The mysterious multi-modal repellency of DEET. Fly (Austin) 9, 45–51.

DeGennaro, M., McBride, C.S., Seeholzer, L., Nakagawa, T., Dennis, E.J., Goldman, C., Jasinskiene, N., James, A.A., Vosshall, L.B., 2013. Orco mutant mosquitoes lose strong preference for humans and are not repelled by volatile DEET. Nature 498, 487–491.

Dekel, A., Pitts, R.J., Yakir, E., Bohbot, J.D., 2016. Evolutionarily conserved odorant receptor function questions ecological context of octenol role in mosquitoes. Sci. Rep. 6, 37330.

Dekker, T., Cardé, R.T., 2011. Moment-to-moment flight manoeuvres of the female yellow fever mosquito (Aedes aegypti L.) in response to plumes of carbon dioxide and human skin odour. J. Exp. Biol. 214, 3480–3494.

Dekker, T., Geier, M., Cardé, R.T., 2005. Carbon dioxide instantly sensitizes female yellow fever mosquitoes to human skin odours. J. Exp. Bio. 208, 2963–2972.

Deletre, E., Schatz, B., Bourguet, D., Chandre, F., Williams, L., Ratnadass, A., Martin, T., 2016. Prospects for repellent in pest control: current developments and future challenges. Chemoecology 26, 127–142.

Dennis, E.J., Dobosiewicz, M., Jin, X., Duvall, L.B., Hartman, P.S., Bargmann, C.I., Vosshall, L.B., 2018. A natural variant and engineered mutation in a GPCR promote DEET resistance in C. elegans. Nature 562, 119–123.

Dennis, E.J., Goldman, O.V., Vosshall, L.B., 2019. *Aedes aegypti* mosquitoes use their legs to sense DEET on contact. Curr. Biol. 29, 1551–1556. e5.

Devillers, J., Lagneau, C., Lattes, A., Garrigues, J.C., Clémenté, M.M., Yébakima, A., 2014. In silico models for predicting vector control chemicals targeting *Aedes aegypti*. SAR QSAR Environ. Res. 25, 805–835.

Ditzen, M., Pellegrino, M., Vosshall, L.B., 2008. Insect odorant receptors are molecular targets of the insect repellent DEET. Science 319, 1838–1842.

Dogan, E.B., Ayres, J.W., Rossignol, P.A., 1999. Behavioural mode of action of DEET: inhibition of lactic acid attraction. Med. Vet. Entomol. 13, 97–100.

Enjin, A., Zaharieva, E.E., Frank, D.D., Mansourian, S., Suh, G.S.B., Gallio, M., Stensmyr, M.C., 2016. Humidity Sensing in *Drosophila*. Curr. Biol. 26, 1352–1358.

Fradin, M.S., Day, J.F., 2002. Comparative efficacy of insect repellents against mosquito bites. N. Engl. J. Med. 347, 13–18.

Fowler, M.A., Montell, C., 2013. *Drosophila* TRP channels and animal behavior. Life Sci. 92, 394–403.

Garboui, S.S., Jaenson, T.G.T., Borg-Karlson, A.-K., Pålsson, K., 2007. Repellency of methyl jasmonate to *Ixodes ricinus* nymphs (Acari: Ixodidae). Exp. Appl. Acarol. 42, 209–215.

Gillij, Y.G., Gleiser, R.M., Zygadlo, J.A., 2008. Mosquito repellent activity of essential oils of aromatic plants growing in Argentina. Bioresour. Technol. 99, 2507–2515.

Goodyer, L., Grootveld, M., Deobhankar, K., Debboun, M., Philip, M., 2020. Characterisation of actions of p-menthane-3,8-diol repellent formulations against *Aedes aegypti* mosquitoes. Trans. R. Soc. Trop. Med. Hyg. 114, 687–692.

Göpfert, M.C., Albert, J.T., Nadrowski, B., Kamikouchi, A., 2006. Specification of auditory sensitivity by Drosophila TRP channels. Nat. Neurosci. 9, 999–1000.

Guo, H., Kunwar, K., Smith, D., 2020. Multiple channels of DEET repellency in Drosophila. Pest Manag. Sci. 76, 880–887.

Hallem, E.A., Carlson, J.R., 2006. Coding of odors by a receptor repertoire. Cell 125, 143–160.

Haverkamp, A., Hansson, B.S., Knaden, M., 2018. Combinatorial codes and labeled lines: how insects use olfactory cues to find and judge food, mates, and oviposition sites in complex environments. Front. Physiol. 9, 49.

Jones, P.L., Pask, G.M., Rinker, D.C., Zwiebel, L.J., 2011. Functional agonism of insect odorant receptor ion channels. Proc. Natl. Acad. Sci. USA 108, 8821–8825.

Jones, P.L., Pask, G.M., Romaine, I.M., Taylor, R.W., Reid, P.R., Waterson, A.G., Sulikowski, G.A., Zwiebel, L.J., 2012. Allosteric antagonism of insect odorant receptor ion channels. Plos One 7, e30304.

Kang, K., Panzano, V.C., Chang, E.C., Ni, L., Dainis, A.M., Jenkins, A.M., Regna, K., Muskavitch, M.A.T., Garrity, P.A., 2012. Modulation of TRPA1 thermal sensitivity enables sensory discrimination in *Drosophila*. Nature 481, 76–80.

Katritzky, A.R., Wang, Z., Slavov, S., Tsikolia, M., Dobchev, D., Akhmedov, N.G., Hall, C.D., Bernier, U.R., Clark, G.G., Linthicum, K.J., 2008. Synthesis and bioassay of improved mosquito repellents predicted from chemical structure. PNAS 105, 7359–7364.

Katz, T.M., Miller, J.H., Hebert, A.A., 2008. Insect repellents: historical perspectives and new developments. J. Am. Acad. Dermatol. 58, 865–871.

Kepchia, D., Moliver, S., Chohan, K., Phillips, C., Luetje, C.W., 2017. Inhibition of insect olfactory behavior by an airborne antagonist of the insect odorant receptor co-receptor subunit. PLoS One 12, e0177454.

Kepchia, D., Xu, P., Terryn, R., Castro, A., Schürer, S.C., Leal, W.S., Luetje, C.W., 2019. Use of machine learning to identify novel, behaviorally active antagonists of the insect odorant receptor co-receptor (Orco) subunit. Sci. Rep. 9, 4055.

Kistler, K.E., Vosshall, L.B., Matthews, B.J., 2015. Genome engineering with CRISPR-Cas9 in the mosquito *Aedes aegypti*. Cell Rep. 11, 51–60.

Klocke, J.A., Darlington, M.V., Balandrin, M.F., 1987. 1,8-Cineole (Eucalyptol), a mosquito feeding and ovipositional repellent from volatile oil of *Hemizonia fitchii* (Asteraceae). J. Chem. Ecol. 13, 2131–2141.

Knecht, Z.A., Silbering, A.F., Ni, L., Klein, M., Budelli, G., Bell, R., Abuin, L., Ferrer, A.J., Samuel, A.D., Benton, R., Garrity, P.A., 2016. Distinct combinations of variant ionotropic glutamate receptors mediate thermosensation and hygrosensation in *Drosophila*. eLife 5, e17879.

Knippling, E.F., Mcalister, L.C., Jones, H.A., 1947. Results of screening tests with materials evaluated as insecticides, miticides, and repellents at the Orlando. Fla., laboratory. USDA Publication, pp. E–733.

Kröber, T., Bourquin, M., Guerin, P.M., 2013. A standardised in vivo and in vitro test method for evaluating tick repellents. Pestic. Biochem. Physiol. 107, 160–168.

Kwon, Y., Kim, S.H., Ronderos, D.S., Lee, Y., Akitake, B., Woodward, O.M., Guggino, W.B., Smith, D.P., Montell, C., 2010. *Drosophila* TRPA1 channel is required to avoid the naturally occurring insect repellent citronellal. Curr. Biol. 20, 1672–1678.

Larsson, M.C., Domingos, A.I., Jones, W.D., Chiappe, M.E., Amrein, H., Vosshall, L.B., 2004. Or83b encodes a broadly expressed odorant receptor essential for *Drosophila* olfaction. Neuron 43, 703–714.

Leal, W.S., 2014. The enigmatic reception of DEET—the gold standard of insect repellents. Curr. Opin. Insect. Sci. 6, 93–98.

Li, T., Saito, C.T., Hikitsuchi, T., Inoguchi, Y., Mitsuishi, H., Saito, S., Tominaga, M., 2019. Diverse sensitivities of TRPA1 from different mosquito species to thermal and chemical stimuli. Sci. Rep. 9, 20200.

Liu, H., Liu, T., Xie, L., Wang, X., Deng, Y., Chen, C.-H., James, A.A., Chen, X.-G., 2016. Functional analysis of Orco and odorant receptors in odor recognition in *Aedes albopictus*. Parasit.Vectors 9, 363.

Logan, J.G., Birkett, M.A., Clark, S.J., Powers, S., Seal, N.J., Wadhams, L.J., Mordue Luntz, A.J., Pickett, J.A., 2008. Identification of human-derived volatile chemicals that interfere with attraction of *Aedes aegypti* mosquitoes. J. Chem. Ecol. 34, 308–322.

Lu, T., Qiu, Y.T., Wang, G., Kwon, J.Y., Rutzler, M., Kwon, H.-W., Pitts, R.J., van Loon, J.J.A., Takken, W., Carlson, J.R., Zwiebel, L.J., 2007. Odor coding in the maxillary palp of the malaria vector mosquito *Anopheles gambiae*. Curr. Biol. 17, 1533–1544.

Maia, M.F., Moore, S.J., 2011. Plant-based insect repellents: a review of their efficacy, development and testing. Malar. J. 10, S11.

Matthews, B.J., McBride, C.S., DeGennaro, M., Despo, O., Vosshall, L.B., 2016. The neurotranscriptome of the *Aedes aegypti* mosquito. BMC Genomics 17, 32.

McBride, C.S., Baier, F., Omondi, A.B., Spitzer, S.A., Lutomiah, J., Sang, R., Ignell, R., Vosshall, L.B., 2014. Evolution of mosquito preference for humans linked to an odorant receptor. Nature 515, 222–227.

McCabe, E.T., Barthel, W.F., Gertler, S.I., Hall, S.A., 1954. Insect repellents. iii. n,n-diethylamides1. J. Org. Chem. 19, 493–498.

McMeniman, C.J., Corfas, R.A., Matthews, B.J., Ritchie, S.A., Vosshall, L.B., 2014. Multimodal integration of carbon dioxide and other sensory cues drives mosquito attraction to humans. Cell 156, 1060–1071.

Melo, A.C.A., Rützler, M., Pitts, R.J., Zwiebel, L.J., 2004. Identification of a chemosensory receptor from the yellow fever mosquito, *Aedes aegypti*, that is highly conserved and expressed in olfactory and gustatory organs. Chem. Senses. 29, 403–410.

Melo, N., Capek, M., Arenas, O.M., Afify, A., Yilmaz, A., Potter, C.J., Laminette, P.J., Para, A., Gallio, M., Stensmyr, M.C., 2021. The irritant receptor TRPA1 mediates the mosquito repellent effect of catnip. Curr. Biol. 31, 1–7.

Michaelakis, A., Vidali, V.P., Papachristos, D.P., Pitsinos, E.N., Koliopoulos, G., Couladouros, E.A., Polissiou, M.G., Kimbaris, A.C., 2014. Bioefficacy of acyclic monoterpenes and their saturated derivatives against the West Nile vector *Culex pipiens*. Chemosphere 96, 74–80.

Montell, C., Zwiebel, L.J., 2016. Chapter Ten—Mosquito sensory systems. In: Raikhel, A.S. (Ed.), Advances in Insect Physiology, Progress in Mosquito Research, Academic Press, Cambridge, MA, pp. 293–328.

Müller, G.C., Junnila, A., Butler, J., Kravchenko, V.D., Revay, E.E., Weiss, R.W., Schlein, Y., 2009. Efficacy of the botanical repellents geraniol, linalool, and citronella against mosquitoes. J. Vector Ecol. 34, 2–8.

Naucke, T.J., Kröpke, R., Benner, G., Schulz, J., Wittern, K.P., Rose, A., Kröckel, U., Grünewald, H.W., 2007. Field evaluation of the efficacy of proprietary repellent formulations with IR3535 and Picaridin against *Aedes aegypti*. Parasitol. Res. 101, 169.

Neff, E.P., 2018. Stop and smell the geosmin. Lab Animal 47, 270.

Pellegrino, M., Steinbach, N., Stensmyr, M.C., Hansson, B.S., Vosshall, L.B., 2011. A natural polymorphism alters odour and DEET sensitivity in an insect odorant receptor. Nature 478, 511–514.

Potter, C.J., 2014. Stop the biting: targeting a mosquito's sense of smell. Cell 156, 878–881.

Qiu, Y.T., Smallegange, R.C., Van Loon, J.J.A., Ter Braak, C.J.F., Takken, W., 2006. Interindividual variation in the attractiveness of human odours to the malaria mosquito *Anopheles gambiae* s. s. Med. Vet. Entomol. 20, 280–287.

Raji, J.I., DeGennaro, M., 2017. Genetic analysis of mosquito detection of humans. Curr. Opin. Insect. Sci. 20, 34–38.

Raji, J.I., Melo, N., Castillo, J.S., Gonzalez, S., Saldana, V., Stensmyr, M.C., DeGennaro, M., 2019. *Aedes aegypti* mosquitoes detect acidic volatiles found in human odor using the IR8a pathway. Curr. Biol. 29, 1253–1262.e7.

Riabinina, O., Task, D., Marr, E., Lin, C.-C., Alford, R., O'Brochta, D.A., Potter, C.J., 2016. Organization of olfactory centres in the malaria mosquito *Anopheles gambiae*. Nat. Commun. 7, 13010.

Sato, K., Pellegrino, M., Nakagawa, Takao, Nakagawa, Tatsuro, Vosshall, L.B., Touhara, K., 2008. Insect olfactory receptors are heteromeric ligand-gated ion channels. Nature 452, 1002–1006.

Sakai, T., Kasuya, J., Kitamoto, T., Aigaki, T., 2009. The *Drosophila* TRPA channel, painless, regulates sexual receptivity in virgin females. Genes Brain Behav. 8, 546–557.

Silbering, A.F., Bell, R., MÃ¼nch, D., Cruchet, S., Gomez-Diaz, C., Laudes, T., Galizia, C.G., Benton, R., 2016. Ir40a neurons are not DEET detectors. Nature 534, E5–E7.

Smallegange, R.C., Qiu, Y.T., van Loon, J.J.A., Takken, W., 2005. Synergism between ammonia, lactic acid and carboxylic acids as kairomones in the host-seeking behaviour of the malaria mosquito *Anopheles gambiae* sensu stricto (Diptera: Culicidae). Chem. Senses 30, 145–152.

Staub, D., Debrunner, M., Amsler, L., Steffen, R., 2002. Effectiveness of a repellent containing DEET and EBAAP for preventing tick bites. Wilderness Environ. Med. 13, 12–20.

Stensmyr, M.C., Dweck, H.K.M., Farhan, A., Ibba, I., Strutz, A., Mukunda, L., Linz, J., Grabe, V., Steck, K., Lavista-Llanos, S., Wicher, D., Sachse, S., Knaden, M., Becher, P.G., Seki, Y., Hansson, B.S., 2012. A conserved dedicated olfactory circuit for detecting harmful microbes in Drosophila. Cell 151, 1345–1357.

Suh, E., Bohbot, J., Zwiebel, L.J., 2014. Peripheral olfactory signaling in insects. Curr. Opin. Insect. Sci. 6, 86–92.

Syed, Z., Leal, W.S., 2008. Mosquitoes smell and avoid the insect repellent DEET. PNAS 105, 13598–13603.

Task, D., Lin, C.-C., Afify, A., Li, H., Vulpe, A., Menuz, K., Potter, C.J., 2020. Widespread polymodal chemosensory receptor expression in drosophila olfactory neurons. bioRxiv. doi:10.1101/2020.11.07.355651.

Tauxe, G.M., MacWilliam, D., Boyle, S.M., Guda, T., Ray, A., 2013. Targeting a dual detector of skin and CO2 to modify mosquito host seeking. Cell 155, 1365–1379.

Tisgratog, R., Sanguanpong, U., Grieco, J.P., Ngoen-Kluan, R., Chareonviriyaphap, T., 2016. Plants traditionally used as mosquito repellents and the implication for their use in vector control. Acta. Tropica. 157, 136–144.

Travis, B.V., Morton, F.A., 1949. The more effective mosquito repellents tested at the Orlando, Fla., laboratory, 1942–47. J. Econ. Entomol. 42, 686–694.

Tsitoura, P., Iatrou, K., 2016. Positive allosteric modulation of insect olfactory receptor function by Orco agonists. Front. Cell Neurosci. 10, 275.

Tsitoura, P., Koussis, K., Iatrou, K., 2015. Inhibition of *Anopheles gambiae* odorant receptor function by mosquito repellents. J. Biol. Chem. 290, 7961–7972.

Turner, S.L., Li, N., Guda, T., Githure, J., Cardé, R.T., Ray, A., 2011. Ultra-prolonged activation of CO2-sensing neurons disorients mosquitoes. Nature 474, 87–91.

van Breugel, F., Riffell, J., Fairhall, A., Dickinson, M.H., 2015. Mosquitoes use vision to associate odor plumes with thermal targets. Curr. Biol. 25, 2123–2129.

Verhulst, N.O., Andriessen, R., Groenhagen, U., Bukovinszkiné Kiss, G., Schulz, S., Takken, W., van Loon, J.J.A., Schraa, G., Smallegange, R.C., 2010. Differential attraction of malaria mosquitoes to volatile blends produced by human skin bacteria. PLoS One 5, e15829.

Vinauger, C., Van Breugel, F., Locke, L.T., Tobin, K.K.S., Dickinson, M.H., Fairhall, A.L., Akbari, O.S., Riffell, J.A., 2019. Visual-olfactory integration in the human disease vector mosquito *Aedes aegypti*. Curr. Biol. 29, 2509–2516.

Vosshall, L.B., Amrein, H., Morozov, P.S., Rzhetsky, A., Axel, R., 1999. A spatial map of olfactory receptor expression in the *Drosophila* antenna. Cell 96, 725–736.

Vosshall, L.B., Wong, A.M., Axel, R., 2000. An olfactory sensory map in the fly brain. Cell 102, 147–159.

Wang, G., Qiu, Y.T., Lu, T., Kwon, H.-W., Pitts, R.J., Loon, J.J.A.V., Takken, W., Zwiebel, L.J., 2009. *Anopheles gambiae* TRPA1 is a heat-activated channel expressed in thermosensitive sensilla of female antennae. Eur. J. Neurosci. 30, 967–974.

Xu, P., Choo, Y.-M., De La Rosa, A., Leal, W.S., 2014. Mosquito odorant receptor for DEET and methyl jasmonate. Proc. Natl. Acad. Sci. USA 111, 16592–16597.

Xu, P., Zhu, F., Buss, G.K., Leal, W.S., 2015. 1-Octen-3-ol - the attractant that repels. F1000Res 4, 156.

Yang, L., Norris, E.J., Jiang, S., Bernier, U.R., Linthicum, K.J., Bloomquist, J.R., 2020. Reduced effectiveness of repellents in a pyrethroid-resistant strain of *Aedes aegypti* (Diptera: Culicidae) and its correlation with olfactory sensitivity. Pest Manag. Sci. 76, 118–124.

Younger, M.A., Herre, M., Ehrlich, A.R., Gong, Z., Gilbert, Z.N., Rahiel, S., Matthews, B.J., Vosshall, L.B., 2020. Non-canonical odor coding ensures unbreakable mosquito attraction to humans. bioRxiv. doi:10.1101/2020.11.07.368720.

Zhao, Z., Tian, D., McBride, C.S., 2021. Development of a pan-neuronal genetic driver in *Aedes aegypti* mosquitoes. Cell Rep. Meth. 100042.

CHAPTER

8

Finding a repellent against ticks: neurophysiological and behavioral approaches

Zainulabeuddin Syed, Kenneth L. O'Dell Jr.
Department of Entomology, University of Kentucky, Lexington, KY, United States

8.1 Introduction

Annually, vector-borne diseases account for more than 17% of all infectious diseases worldwide, causing more than 700,000 deaths (WHO, 2020). Within the United States, ticks have doubled their distribution range in the past two decades, and at least eight infective microorganisms vectored by ticks have been established during this period (Eisen and Eisen, 2018). Tick control has traditionally relied heavily on pesticides, a successful example being the eradication of cattle tick in the United States by dipping the cattle in pesticide solution (Mullens et al., 2018). However, sustained use of acaricides is resulting in resistant tick populations (Rodriguez-Vivas et al., 2018, Esteve-Gasent et al., 2020, Vilela et al., 2020), and climate change is facilitating the steady expansion of tick distribution. Multipronged management strategies that incorporate innovative control methods are much needed (de la Fuente, 2018). Considering that ticks display robust olfactory-driven behaviors, exploitation of their strong sense of smell will provide effective practices that can be embedded into existing vector control strategies (Allan, 2010, Carr and Roe, 2016, Esteve-Gassent et al., 2016).

Odors, such as pheromones, are increasingly being used to sample and control vector arthropods (Pickett et al., 2010), including ticks (Sonenshine, 2006). On the other hand, odors that induce repulsion/avoidance offer the first line of personal protection against tick bites (Benelli and Pavela, 2018). Therefore, knowledge of how these cues are detected and perceived by ticks will help us elucidate their role in the tick's life cycle, and aid in maximizing the effectiveness of management strategies.

Historically, some of the most successful campaigns against vector-borne diseases have been those that targeted the vectors (Hill et al., 2005; Shaw and Catteruccia, 2019). Ticks detect and respond to a variety of odorants from hosts,

Advances in Arthropod Repellents.
DOI: https://doi.org/10.1016/B978-0-323-85411-5.00001-7

habitats, and mates (Allan, 2010; Syed, 2015; Carr and Roe, 2016). Furthermore, within their environment, ticks parsimoniously use a distinct yet limited range of volatiles which, in various contexts, elicit distinct behaviors such as *attraction* to hosts and mates, and *repulsion/avoidance* from unsuitable sites (Syed, 2015).

We propose that a solid foundation to study these behavioral responses will critically depend on our understanding of the complex chemistries of natural substrates such as essential oils (EOs) that are increasingly being implicated as promising tick repellents. There is a growing appreciation for developing arthropod repellents for personal protection as the first line of defense against hematophagous arthropods. The global repellent market is estimated to be at 3 billion US dollars (Bill and Melinda Gates Foundation, 2007), and the projected global forecasts for mosquito repellents is a *ca.* $4.8 billion by 2022 (Chaturvedi, 2016).

8.2 How arthropod repellents work?

At various points in this book, the term *repellent* will be defined and redefined in different contexts. Based on Dethier et al.'s classical definition of a repellent as "a chemical which causes insects to make oriented movements away from its source (Dethier et al., 1960)," we earlier defined them as "airborne chemicals that induce electrophysiological response from selected olfactory receptor neurons (ORNs), resulting in the release of an innate avoidance response" (Syed, 2014). However, our understanding since then has significantly improved. Extensive physiological and behavioral studies aimed at understanding the mode of action of arthropod repellents, especially those employing the widely used repellent, *N,N*-diethyl-3-methyl benzamide (DEET) in mosquito research, have shown that DEET acts through olfactory and gustatory channels (Klun et al., 2006; Lee et al., 2010; Liu et al., 2010; DeGennaro et al., 2013; Dennis et al., 2019). Though DEET has been proved to be effective against a variety of blood-feeding arthropods—such as mosquitoes (Thavara et al., 2001; Bernier et al., 2005), ticks (Evans et al., 1990), triatomines (Alzogaray et al., 2000; Alzogaray, 2016), and sand flies (Coleman et al., 1993; Klun et al., 2006; Weeks et al., 2019)—a clear understanding of the precise mode of action remains elusive.

A recent study employing genetically modified "smell blind" *Ae. aegypti* mosquitoes and DEET comprehensively evaluated the contribution of smell, ingestion, and contact on mosquitoes' response to DEET, and concluded that DEET-induced repellency is exclusively mediated by the tarsal segments of legs and not the proboscis (Dennis et al., 2019). This work further illustrates the need for dissecting the role of contact/taste modality on influencing the repellence/avoidance function in ticks. Therefore, we will introduce and highlight the two distinct yet overlapping modalities—smell and taste—in understanding the mode of repellence in ticks.

8.3 Chemosensation in ticks

The ability to detect and respond to a chemical is broadly defined as chemosensation. As in other hematophagous arthropods, the chemosensory ability in ticks is defined by two modalities: olfaction and gustation.

8.3.1 Olfaction

Ticks raise and wave their first pair of legs in the air upon odor stimulation and/or in the vicinity of the host. This led the early tick researchers to study the forelegs as potential olfactory organs. A detailed examination of the first pair revealed a capsular structure equipped with multiple hair-like projections (sensilla/setae), and this study implicated the structure in hearing (Haller, 1881). Seminal studies that amputated or masked the forelegs on which the Haller's organ is located demonstrated the role of this organ in odor detection (Hindle and Merriman, 1912, Lees, 1948), and a recent review by Carr et al. further provided the most up-to-date information from across multiple members of tick species (Carr et al., 2017). Ticks' forelegs therefore function similarly to insect antennae. However, unlike insect antennae which have been described and studied in-depth (Leal, 2013; Kohl et al., 2015; Fleischer et al., 2018), the Haller's organ has not been well characterized functionally. Two nonolfactory functions ascribed to the Haller's organ include the detection of *liquid* water (Krober and Guerin, 1999) and infrared detection (Mitchell et al., 2017; Carr and Salgado, 2019).

Detailed ultrastructural analyses of the Haller's organ in hard- (Hess and Vlimant, 1986; Homsher and Sonenshine, 1975) and soft-ticks (Klompen and Oliver Jr., 1993) and, more recently from the field-collected ticks (Josek et al., 2018a) have revealed multiple olfactory sensilla that are present in and around the Haller's organ. In general, *ca.* 20 olfactory sensilla on ticks' forelegs detect and define the chemical landscape, and the Haller's organ contains almost half of those (Hess and Vlimant, 1986). Olfactory sensilla in ticks, as in other hematophagous arthropods, are morphologically diverse. Yet, in ticks, all the olfactory sensilla are wall pore types—either single-walled or double-walled—characterized by multiple pores across the length (Hess and Vlimant, 1986; Tichy and Barth, 1992). The pores allow airborne odorant molecules to pass through toward the dendritic membrane of the ORNs (Leonovich, 1990).

A plethora of chemicals originating from hosts (glandular secretions, breath, etc.) are detected by these ORNs. Ticks such as *Ixodes scapularis* begin their search for hosts and attachment sites by climbing up a grass blade or shrub and waiting for the approaching host animal as they raise up their forelegs (Nicholson et al., 2019). These ticks feed on multiple animal species (e.g., white-tailed deer, white-footed mice, chipmunks, or raccoons) even within one life cycle (LoGiudice et al., 2003). Because *I. scapularis* prefer different hosts during different life stages (Bishopp and Trembley, 1945; Schulze et al., 1986; LoGiudice et al., 2003; Keesing et al., 2009), it is conceivable that they detect a variety of odors (Allan, 2010).

Long-range host detection starts with interactions between odorants and distinct subpopulation(s) of chemoreceptor proteins present in the dendritic membrane of ORNs. The chemoreceptor gene family is comprised of Odorant, Gustatory and Ionotropic receptors (ORs, GRs, and IRs, respectively), which are among the largest gene families in arthropods, and together define the reception of attractive and nonattractive (repellent/deterrent) chemostimuli (Benton, 2015; Robertson, 2019). However, in ticks no ORs have been identified in the genomes (Gulia-Nuss et al., 2016; Jia et al., 2020), and the chemosensation of the odorants therefore appears to be entirely dictated by ~125 IRs identified in the foreleg transcriptome (Josek et al., 2018b). A comparative whole-genome study of six tick species (*I. persulcatus*, *Haemaphysalis longicornis*, *Dermacentor silvarum*, *Hyalomma asiaticum*, *Rhipicephalus sanguineus*, and *R. microplus*)—each with diverse habitats and varying numbers of hosts needed to complete their respective life cycles—annotated only 40–60 IRs (Jia et al., 2020).

8.3.2 Gustation

Taste sensilla, morphologically characterized by a single terminal pore, are present on the distal segment of the palp and chelicerae (Woolley, 1972; Hess and Vlimant, 1986). Foreleg transcriptome analysis in *I. ricinus* revealed the presence of multiple GRs (Josek et al., 2018b). Detailed transcriptome analyses of the palps and chelicerae will reveal potential GRs involved in contact/taste reception. Comparative genomic and transcriptomic analyses will reveal whether or not a tick that attaches to a wide variety of hosts during its three-stage life cycle (e.g., *I. ricinus*) will have different molecular signatures when compared to a one-host tick (*R. microplus*).

8.4 Electrophysiological analyses for repellent discovery in ticks

Besides DEET—a proven repellent against ticks—EOs extracted from medicinal and aromatic plants are increasingly implicated as potential arthropod repellents (Pavela et al., 2016; Benelli and Pavela, 2018). The single sensillum recording (SSR) method is one of the most powerful techniques used to record from individual ORNs inside olfactory sensilla wherein each ORN is challenged with a variety of semiochemicals, and the induced responses are recorded either as excitation or inhibition. This method has widely been used in insect studies (Kaissling, 1995; Olsson and Hansson, 2013). The SSR can also be coupled with gas-chromatography (GC-SSR) to isolate and identify biologically active constituents from the complex odors that stimulate ORNs in blood-feeding arthropods (Reisenman et al., 2016). The GC-SSR protocol has only been used with ticks in limited studies due to the difficulty in securing stable recordings from the strong forelegs that house the olfactory sensilla. The first and only GC-SSR recordings from *Amblyomma variegatum* identified a range of chemostimuli from host odors (Steullet and Guerin, 1992a,b, 1994a,b). A preliminary study in *I. ricinus* identified mostly phenolic compounds as physiologically active (Leonovich, 2004), and in *Boophilus microplus* 2,6-dichlorophenol was isolated and identified as the key electrophysiologically active constituent from the extracts of males, females, and larvae (De Bruyne and Guerin, 1994). We recently developed GC-SSR recordings for *I. scapularis* (Fig. 8.1) that identified a range of electrophysiologically active constituents from host odors that were further evaluated for behavioral responses (Josek et al., 2021). Given the numerical simplicity of olfactory sensilla in ticks, determining the detection properties of each sensillum in ticks offers a robust platform for arthropod repellent identification.

Gustatory responses can be determined by the SSR methods with a slightly modified protocol (Benton et al., 2006; Delventhal et al., 2014). Despite the decent progress in our understanding of the olfaction in ticks, taste remains greatly understudied. Taste is essential to feeding, mating, and biting and is potentially involved in avoiding nonsuitable substrates. The first electrophysiological recordings from the gustatory sensilla on the chelicera in *B. microplus* demonstrated that the sensory neurons respond to blood-constituents (Wallade and Rice, 1977). Recent work increasingly points to the fact that sensilla on palps and chelicerae detect the tastants (Ferreira et al., 2015).

8.5 Behavioral analyses for repellent discovery in ticks

Ideally, the behavioral screening of the electrophysiologically active chemicals identified by SSR (earlier) and subsequent GC-mass spectrometry should be simple, rapid, and cost-effective. A variety of bioassays have been developed over the decades to test repellents in arthropods (Schreck, 1977) and ticks (Dautel, 2004). We would like to urge caution in interpreting the results from different testing paradigms and protocols. For example, a compound tested under different assay systems can yield different results (Dautel et al., 2013); life stages differ in their sensitivity toward a repellent (Kulma et al., 2019), and vertical or horizontal orientation of the same olfactometer yields different results (Zeringóta et al., 2020).

Broadly, the arthropod repellent assays can be divided into two categories: in vivo and in vitro. The in vivo assays involve testing a substance that is applied on a part of the body surface

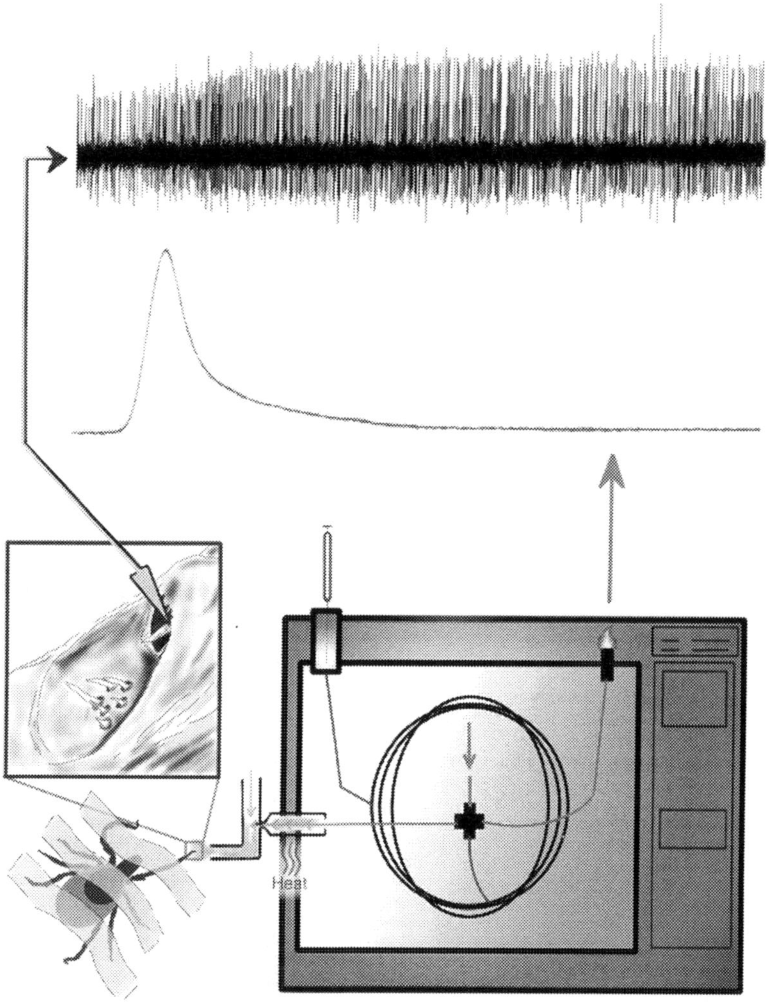

FIG. 8.1 Single unit electrophysiological recordings set-up for *I. scapularis*. Extracelluar single units are obtained from a sensillum in the Haller's organ of a restrained tick. The red trace is the signal from a flame ionization detector.

(such as skin) that is then compared with the untreated part. Besides posing regulatory constraints, these assays do not evaluate whether or not a biologically active chemical is repellent by itself. In vitro assays, on the other hand, offer the advantage of studying the induced effect of a test arthropod repellent alone (Adenubi et al., 2018). While in vivo assays offer more real-life solutions in providing arthropod repellent products that are readily accessible for human use, in vitro assays are critical for extending the neurophysiological analyses of arthropod repellent compounds. Two novel in vitro assays to test DEET-induced behavioral activity in *Cx. quinquefasciatus* (Syed and Leal, 2008) can be extended to test the ticks. Two innovative assays wherein DEET's repellence was established in ticks in absence of any other olfactory stimuli provided evidence that DEET acts as an active arthropod repellent (Dautel et al., 1999; McMahon et al., 2003).

Our laboratory is increasingly using a 4-arm olfactometer—originally developed for studying arthropod parasitoids' response to odors (Vet et al., 1983)—to test olfactory responses. We successfully designed, fabricated, and evaluated a 4-arm olfactometer and have established the attraction of *I. scapularis* to human breath, a well-documented tick attractant (McMahon and Guerin, 2002) (Fig. 8.2). Our olfactometer is equipped with the tracking software EthoVision XT 10 (Noldus Inc.) that measures various associated parameters, such as speed and angular velocity. In a clean environment, the measured speed was 14.8 mm/s, and the speed significantly increased to 18.1 mm/s once human breath was released from one arm. We are in the process of testing the arthropod repellent properties of plant EOs (Pavela et al., 2016; Benelli and Pavela, 2018). Other assays such as "Y" tube and straight tube are also used to test the repulsion in ticks with a demonstrable efficiency and reliability (Carnohan et al., 2017; Carr and Salgado, 2019). However, a mass testing paradigm with automated observation and analyses is few (Dautel et al., 2013; Syed, 2014; Carr and Roe, 2016).

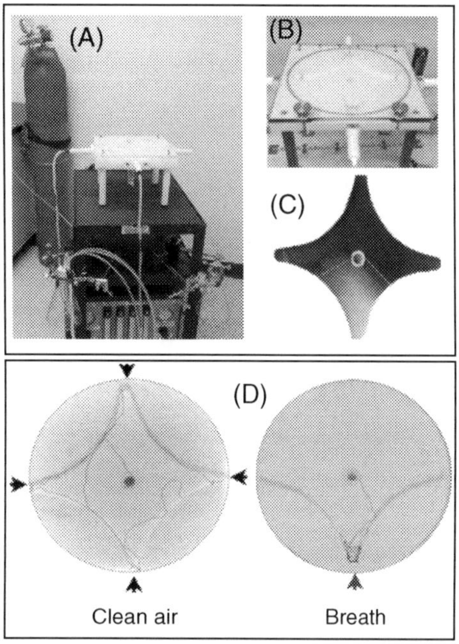

FIG. 8.2 A 4-arm olfactometer set-up for measuring odor induced behaviors in *I. scapularis* (A) an overview, (B) top view of the test arena where a tick is released from the middle aperture, (C) delivery of an odorant visualized by the release of an ammonium acetate smoke that shows the integrity of the plume ("odor") released from one arm confined to that area, and (D) a tick's response to clean air dispensed from all four sides (left) and breath introduced from one (right) tracked with EthoVision XT.

Contact-dependent (gustatory mediated) repellency was initially demonstrated during the development of in vitro feeding assays (Kuhnert et al., 1995), but large-scale screening methods still remain to be realized. Recent work that reports the identification of biologically active substrates by electrophysiological screening followed by in vivo tests (Soares et al., 2013) offers great promise.

8.6 Future directions

Extending our understanding of the odor coding in the chemosensory sensilla at the periphery (Soares et al., 2013; Josek et al., 2021) and how this information is relayed and encoded in the central nervous systems (Borges et al., 2016; Faraone et al., 2019) will provide a fundamental understanding of repellent detection. Comparative analysis of the tick genomes (Jia et al., 2020) is providing insights into the underlying principles of chemosensation in blood-feeding arthropods, including ticks (Allan, 2010; Syed, 2015). Albeit modest progress, there is a tremendous potential to employ pheromones such as those involved in assembly, aggregation, and attachment (Sonenshine, 2006; Allan, 2010) as components of integrated tick management.

8.7 Conclusion

Repellents offer the first line of defense against blood-feeding arthropods, thereby preventing bites and reducing infections. Recent understanding of the ticks' chemical landscapes, and how these volatile signals are detected at the periphery and conveyed to the higher brain regions is providing insights into the rich chemical ecology of ticks. Novel identified chemistries will offer safe and effective repellents and deterrents. Using tick olfactory apparatus as a biological sensor to isolate biologically active molecules from natural substrates has provided us with a robust tool to analyze complex odors from hosts, mates, and oviposition sites.

Acknowledgments

The tick research in our laboratory is supported by funding from the National Institute of Food and Agriculture, US Department of Agriculture (under HATCH Project 2353077000).

References

Adenubi, O.T., McGaw, L.J., Eloff, J.N., Naidoo, V., 2018. In vitro bioassays used in evaluating plant extracts for tick repellent and acaricidal properties: a critical review. Vet. Parasitol. 254, 160–171.

Allan, S.A., 2010. Chemical ecology of tick-host interactions. In: Takken, W., Knols, B.G.J. (Eds.), Olfaction in Vector-Host Interactions, 2, Wageningen Academic Publishers, The Netherlands, 327–348.

Alzogaray, R., Fontan, A., Zerba, E., 2000. Repellency of DEET to nymphs of *Triatoma infestans*. Med. Vet. Entomol. 14, 6–10.

Alzogaray, R.A., 2016. Behavioral and toxicological responses of *Rhodnius prolixus* (Hemiptera: Reduviidae) to the insect repellents DEET and IR3535. J. Med. Entomol. 53, 387–393.

Benelli, G., Pavela, R., 2018. Repellence of essential oils and selected compounds against ticks—a systematic review. Acta Trop. 179, 47–54.

Benton, R., 2015. Multigene family evolution: perspectives from insect chemoreceptors. Trends Ecol. Evol. 30, 590–600.

Benton, R., Sachse, S., Michnick, S.W., Vosshall, L.B., 2006. Atypical membrane topology and heteromeric function of *Drosophila* odorant receptors in vivo. PLoS Biol. 4, e20.

Bernier, U.R., Furman, K.D., Kline, D.L., Allan, S.A., Barnard, D.R., 2005. Comparison of contact and spatial repellency of catnip oil and N,N-diethyl-3-methylbenzamide (DEET) against mosquitoes. J. Med. Entomol. 42, 306–311.

Bill and Melinda Gates Foundation, 2007. Market assessment for public health pesticide products, a report by the Bill and Melinda Gates Foundation and the Boston Consulting Group. Bill and Melinda Gates Foundation, Washington, DC.

Bishopp, F., Trembley, H.L., 1945. Distribution and hosts of certain North American ticks. J. Parasitol. 31, 1–54.

Borges, L.M.F., Li, A.Y., Olafson, P.U., Renthal, R., Bauchan, G.R., Lohmeyer, K.H., d. León, A.A.P., 2016. Neuronal projections from the Haller's organ and palp sensilla to the synganglion of*Amblyomma americanum*. Revista Brasileira de Parasitologia Veterinária 25, 217–224.

Carnohan, L.P., Kaufman, P.E., Allan, S.A., Gezan, S.A., Weeks, E.N., 2017. Laboratory and field evaluation of brown dog tick behavioral responses to potential semiochemicals. Ticks Tick-borne Dis. 8, 226–234.

Carr, A., Mitchell III, R.D., Dhammi, A., Bissinger, B.W., Sonenshine, D.E., Roe, R.M., 2017. Tick Haller's organ, a new paradigm for arthropod olfaction: how ticks differ from insects. Int. J. Mol. Sci. 18, 1563.

Carr, A.L., Roe, M., 2016. Acarine attractants: chemoreception, bioassay, chemistry and control. Pestic. Biochem. Physiol. 131, 60–79.

Carr, A.L., Salgado, V.L., 2019. Ticks home in on body heat: a new understanding of Haller's organ and repellent action. PloS One 14, e0221659.

Chaturvedi,Y., 2016. Mosquito repellent market by product type (spray, coil, cream & oil, mat, vaporizer, other products) and channels of distribution (large retail stores, small retail stores, specialty stores, online)—global opportunity analysis and industry forecast, Allied Market Research. Portland, OR, USA, 2015–2022 p. 120.

Coleman, R.E., Robert, L.L., Roberts, L.W., Glass, J.A., Seeley, D.C., Laughinghouse, A., Perkins, P.V., Wirtz, R.A., 1993. Laboratory evaluation of repellents against 4 anopheline mosquitos (Diptera, Culicidae) and 2 phlebotomine sand flies (Diptera, Psychodidae). J. Med. Entomol. 30, 499–502.

Dautel, H., 2004. Test systems for tick repellents. Int. J. Med. Microbiol. Suppl. 293, 182–188.

Dautel, H., Dippel, C., Werkhausen, A., Diller, R., 2013. Efficacy testing of several *Ixodes ricinus* tick repellents: different results with different assays. Ticks Tick Borne Dis. 4, 256–263.

Dautel, H., Kahl, O., Siems, K., Oppenrieder, M., Muller-Kuhrt, L., Hilker, M., 1999. A novel test system for detection of tick repellents. Entomologia Experimentalis et Applicata 91, 431–441.

De Bruyne, M., Guerin, P.M., 1994. Isolation of 2, 6-dichlorophenol from the cattle tick *Boophilus microplus*: receptor cell responses but no evidence for a behavioural response. J. Insect Physiol. 40, 143–154.

DeGennaro, M., McBride, C.S., Seeholzer, L., Nakagawa, T., Dennis, E.J., Goldman, C., Jasinskiene, N., James, A.A., Vosshall, L.B., 2013. orco mutant mosquitoes lose strong preference for humans and are not repelled by volatile DEET. Nature 498, 487–491.

Delventhal, R., Kiely, A., Carlson, J.R., 2014. Electrophysiological recording from *Drosophila* labellar taste sensilla. J. Visual. Exp., e51355.

Dennis, E.J., Goldman, O.V., Vosshall, L.B., 2019. *Aedes aegypti* mosquitoes use their legs to sense DEET on contact. Curr. Biol. 29, 1551–1556.

Dethier, V., Browne, B.L., Smith, C.N., 1960. The designation of chemicals in terms of the responses they elicit from insects. J. Eco. Entomol. 53, 134–136.

Eisen, R.J., Eisen, L., 2018. The blacklegged tick, *Ixodes scapularis*: an increasing public health concern. Trends Parasitol. 34, 295–309.

Esteve-Gasent, M.D., Rodríguez-Vivas, R.I., Medina, R.F., Ellis, D., Schwartz, A., Cortés Garcia, B., Hunt, C., Tietjen, M., Bonilla, D., Thomas, D., Logan, L.L., Hasel, H., Alvarez Martínez, J.A., Hernández-Escareño, J.J., Mosqueda Gualito, J., Alonso Díaz, M.A., Rosario-Cruz, R., Soberanes Céspedes, N., Merino Charrez, O., Howard, T., Chávez Niño, V.M., Pérez de León, A.A., 2020. Research on integrated management for cattle fever ticks and bovine babesiosis in the United States and Mexico: current status and opportunities for binational coordination. Pathogens 9, 1–23.

Esteve-Gassent, M.D., Castro-Arellano, I., Feria-Arroyo, T.P., Patino, R., Li, A.Y., Medina, R.F., de León, A.A.P., Rodríguez-Vivas, R.I., 2016. Translating ecology, physiology, biochemistry, and population genetics research to meet the challenge of tick and tick-borne diseases in North America. Arch. Insect Biochem. Physiol. 92, 38–64.

Evans, S.R., Korch, G.W., Lawson, M.A., 1990. Comparative field-evaluation of permethrin and DEETt-treated military uniforms for personal protection against ticks (Acari). J. Med. Entomol. 27, 829–834.

Faraone, N., MacPherson, S., Hillier, N.K., 2019. Behavioral responses of *Ixodes scapularis* tick to natural products: development of novel repellents. Exp. Appl. Acarol. 79, 195–207.

Ferreira, L.L., Soares, S.F., de Oliveira Filho, J.G., Oliveira, T.T., de León, A.A.P., Borges, L.M.F., 2015. Role of *Rhipicephalus microplus* cheliceral receptors in gustation and host differentiation. Ticks Tick-borne Dis. 6, 228–233.

Fleischer, J., Pregitzer, P., Breer, H., Krieger, J., 2018. Access to the odor world: olfactory receptors and their role for signal transduction in insects. Cell. Mol. Life Sci. 75, 485–508.

Fuente J, de la, 2018. Controlling ticks and tick-borne diseases…looking forward. Ticks Tick Borne Dis. 9, 1354–1357.

Gulia-Nuss, M., Nuss, A.B., Meyer, J.M., Sonenshine, D.E., Roe, R.M., Waterhouse, R.M., Sattelle, D.B., De La Fuente, J., Ribeiro, J.M., Megy, K., 2016. Genomic insights into the *Ixodes scapularis* tick vector of Lyme disease. Nature Commun. 7, 1–13.

Haller, G., 1881. Vorläufige Bemerkungen über das Gehörorgan der Ixodiden. Zool. Anz 4, 165–167.

Hess, E., Vlimant, M., 1986. Leg sense organs of ticks, Morphology, Physiology, and Behavioral Biology of Ticks. John Wiley & Sons, New York, USA, pp. 361–390.

Hill, C.A., Kafatos, F.C., Stansfield, S.K., Collins, F.H., 2005. Arthropod-borne diseases: vector control in the genomics era. Nature Rev. Microbiol. 3, 262–268.

Hindle, E., Merriman, G., 1912. The sensory perceptions of *Argas persicus* (Oken). Parasitology 5, 203–216.

Homsher, P.J., Sonenshine, D.E., 1975. Scanning electron microscopy of ticks for systematic studies: fine structure of Haller's organ in ten species of *Ixodes*. Trans. Am. Microscopical Soc., 368–374.

Jia, N., Wang, J., Shi, W., Du, L., Sun, Y., Zhan, W., Jiang, J.F., Wang, Q., Zhang, B., Ji, P., Bell-Sakyi, L., Cui, X.M., Yuan, T.T., Jiang, B.G., Yang, W.F., Lam, T.T., Chang, Q.C., Ding, S.J., Wang, X.J., Zhu, J.G., Ruan, X.D., Zhao, L., Wei, J.T., Ye, R.Z., Que, T.C., Du, C.H., Zhou, Y.H., Cheng, J.X., Dai, P.F., Guo, W.B., Han, X.H., Huang, E.J., Li, L.F., Wei, W., Gao, Y.C., Liu, J.Z., Shao, H.Z., Wang, X., Wang, C.C., Yang, T.C., Huo, Q.B., Li, W., Chen, H.Y., Chen, S.E., Zhou, L.G., Ni, X.B., Tian, J.H., Sheng, Y., Liu, T., Pan, Y.S., Xia, L.Y., Li, J., Zhao, F., Cao, W.C., 2020. Large-scale comparative analyses of tick genomes elucidate their genetic diversity and vector capacities. Cell 182, 1328–1340.e1313.

Josek, T., Allan, B.F., Alleyne, M., 2018a. Morphometric analysis of chemoreception organ in male and female ticks (Acari: Ixodidae). J. Med. Entomol. 55, 547–552.

Josek, T., Sperrazza, J., Alleyne, M., Syed, Z., 2021. Neurophysiological and behavioral responses of blacklegged ticks to host odors. J. Insect Physiol. 128, 104175.

Josek, T., Walden, K.K., Allan, B.F., Alleyne, M., Robertson, H.M., 2018b. A foreleg transcriptome for *Ixodes scapularis* ticks: Candidates for chemoreceptors and binding proteins that might be expressed in the sensory Haller's organ. Ticks Tick-Borne Dis. 9, 1317–1327.

Kaissling, K.-E., 1995. Single unit and electroantennogram recordings in insect olfactory organs. In: Spielman, A., Brand., J.G. (Eds.), Experimental Cell Biology of Taste and Olfaction: Current Techniques and Protocols. CRC Press, New York, USA, p. 361.

Keesing, F., Brunner, J., Duerr, S., Killilea, M., LoGiudice, K., Schmidt, K., Vuong, H., Ostfeld, R., 2009. Hosts as ecological traps for the vector of Lyme disease. Proc. R. Soc. B: Biol. Sci. 276, 3911–3919.

Klompen, J., Oliver Jr, J.H., 1993. Haller's organ in the tick family Argasidae (Acari: Parasitiformes: Ixodida). J. Parasitol. 79 (4), 591–603.

Klun, J.A., Khrimian, A., Debboun, M., 2006. Repellent and deterrent effects of SS220, Picaridin, and Deet suppress human blood feeding by *Aedes aegypti*, *Anopheles stephensi*, and *Phlebotomus papatasi*. J. Med. Entomol. 43, 34–39.

Kohl, J., Huoviala, P., Jefferis, G.S., 2015. Pheromone processing in *Drosophila*. Curr. Opinion Neurobiol. 34, 149–157.

Krober, T., Guerin, P.M., 1999. Ixodid ticks avoid contact with liquid water. J. Exp. Biol. 202, 1877–1883.

Kuhnert, F., Diehl, P.A., Guerin, P.M., 1995. The life-cycle of the bont tick *Amblyomma hebraeum* in vitro. Int. J. Parasitol. 25, 887–896.

Kulma, M., Kopecký, O., Bubová, T., 2019. Nymphs of*Ixodes ricinus* are more sensitive to Deet than adult females. J. Am. Mosquito Control Assoc. 35, 279–284.

Leal, W.S., 2013. Odorant reception in insects: roles of receptors, binding proteins, and degrading enzymes. Ann. Rev. Entomol. 58, 373–391.

Lee, Y., Kim, S.H., Montell, C., 2010. Avoiding DEET through insect gustatory receptors. Neuron 67, 555–561.

Lees, A.D., 1948. The sensory physiology of the sheep tick, *Ixodes ricinus* l. J. Exp. Biol. 25, 145–207.

Leonovich, S., 2004. Phenol and lactone receptors in the distal sensilla of the Haller's organ in *Ixodes ricinus* ticks and their possible role in host perception. Exp. Appl. Acarol. 32, 89–102.

Leonovich, S.A., 1990. Fine structural features of sensory systems in ticks and mites: Evolutionary and ethological aspectsSensory Systems and Communication in Arthropods. Springer, Birkhäuser Basel, Switzerland, pp. 44–48.

Liu, C., Pitts, R.J., Bohbot, J.D., Jones, P.L., Wang, G.R., Zwiebel, L.J., 2010. Distinct olfactory signaling mechanisms in the malaria vector mosquito *Anopheles gambiae*. Plos Biol. 8.

LoGiudice, K., Ostfeld, R.S., Schmidt, K.A., Keesing, F., 2003. The ecology of infectious disease: effects of host diversity and community composition on Lyme disease risk. Proc. Nat. Acad. Sci. 100, 567–571.

McMahon, C., Guerin, P., 2002. Attraction of the tropical bont tick, *Amblyomma variegatum*, to human breath and to the breath components acetone, NO and CO_2. Naturwissenschaften 89, 311–315.

McMahon, C., Krober, T., Guerin, P.M., 2003. In vitro assays for repellents and deterrents for ticks: differing effects of products when tested with attractant or arrestment stimuli. Med. Vet. Entomol. 17, 370–378.

Mitchell III, R.D., Zhu, J., Carr, A.L., Dhammi, A., Cave, G., Sonenshine, D.E., Roe, R.M., 2017. Infrared light detection by the hailer's organ of adult american dog ticks, Dermacentor variabilis (Ixodida: Ixodidae). Ticks Tick-Borne Dis. 8, 764–771.

Mullens, B.A., Hinkle, N.C., Fryxell, R.T., Rochon, K., 2018. Past, present, and future contributions and needs for veterinary entomology in the United States and Canada. Am. Entomol. 64, 20–31.

Nicholson, W.L., Sonenshine, D.E., Noden, B.H., Brown, R.N., 2019. Ticks (Ixodida). Medical and Veterinary Entomology. Academic Press (Elsevier), Cambridge, MA, USA, pp. 603–672.

Olsson, S.B., Hansson, B.S., 2013. Electroantennogram and single sensillum recording in insect antennae. Pheromone Signaling, Humana Press (Springer), Totowa, NJ, USA, pp. 157–177.

Pavela, R., Canale, A., Mehlhorn, H., Benelli, G., 2016. Application of ethnobotanical repellents and acaricides in prevention, control and management of livestock ticks: a review. Res. Vet. Sci. 109, 1–9.

Pickett, J.A., Birkett, M.A., Dewhirst, S.Y., Logan, J.G., Omolo, M.O., Torto, B., Pelletier, J., Syed, Z., Leal, W.S., 2010. Chemical ecology of animal and human pathogen vectors in a changing global climate. J. Chem. Ecol. 36, 113–121.

Reisenman, C.E., Lei, H., Guerenstein, P.G., 2016. Neuroethology of olfactory-guided behavior and its potential application in the control of harmful insects. Front. Physiol. 7, 271.

Robertson, H.M., 2019. Molecular evolution of the major arthropod chemoreceptor gene families. Ann. Rev. Entomol. 64, 227–242.

Rodriguez-Vivas, R.I., Jonsson, N.N., Bhushan, C., 2018. Strategies for the control of *Rhipicephalus microplus* ticks in a world of conventional acaricide and macrocyclic lactone resistance. Parasitol. Res. 117, 3–29.

Schreck, C.E., 1977. Techniques for evaluation of insect repellents—Critical-review. Ann. Rev. Entomol. 22, 101–119.

Schulze, T.L., Bowen, G.S., Lakat, M.F., Parkin, W.E., Shisler, J.K., 1986. Seasonal abundance and hosts of *Ixodes dammini* (Acari: Ixodidae) and other ixodid ticks from an endemic Lyme disease focus in New Jersey, USA. J. Med. Entomol. 23, 105–109.

Shaw, W.R., Catteruccia, F., 2019. Vector biology meets disease control: using basic research to fight vector-borne diseases. Nature Microbiol. 4, 20–34.

Soares, S.F., Louly, C.C.B., Marion-Poll, F., Ribeiro, M.F.B., Borges, L.M.F., 2013. Study on Cheliceral sensilla of the brown dog tick *Rhipicephalus sanguineus* (Latreille, 1806) (Acari: Ixodidae) involved in taste perception of phagostimulants. Acta Tropica. 126, 75–83.

Sonenshine, D.E., 2006. Tick pheromones and their use in tick control. Ann. Rev. Entomol. 51, 557–580.

Steullet, P., Guerin, P.M., 1992a. Perception of breath components by the tropical bont tick, *Amblyomma variegatum* Fabricius (Ixodidae).1. CO_2-excited and CO_2-inhobited receptors. J. Comp. Physiol. A 170, 665–676.

Steullet, P., Guerin, P.M., 1992b. Perception of breath components by the tropical bont tick, *Amblyomma variegatum* Fabricius (Ixodidae). 2. Sulfide receptors. J. Comp. Physiol. A 170, 677–685.

Steullet, P., Guerin, P.M., 1994a. Identification of vertebrate volatiles stimulating olfactory receptors on tarsus I of the tick *Amblyomma variegatum* Fabricius (Ixodidae).1. Receptors within the Haller's organ. J. Comp. Physiol. A 174, 27–38.

Steullet, P., Guerin, P.M., 1994b. Identification of vertebrate volatiles stimulating olfactory receptors on tarsus I of the tick *Amblyomma variegatum* Fabricius (Ixodidae). 2. Receptors outside the Haller's organ capsule. J. Comp. Physiol. A 174, 39–47.

Syed, Z., 2014. How repellents work: neurophysiological and behavioral analyses. In: Debboun, M., Frances, S.P., Strickman, D. (Eds.), Insect Repellents Handbook. CRC Press, Boca Raton, FL, USA, p. 43.

Syed, Z., 2015. Chemical ecology and olfaction in arthropod vectors of diseases. Curr. Opin. Insect Sci. 10, 83–89.

Syed, Z., Leal, W.S., 2008. Mosquitoes smell and avoid the insect repellent DEET. Proc. Nat. Acad. Sci. USA 105, 13598–13603.

Thavara, U., Tawatsin, A., Chompoosri, J., Suwonkerd, W., Chansang, U.R., Asavadachanukorn, P., 2001. Laboratory and field evaluations of the insect repellent 3535 (ethyl butylacetylaminopropionate) and DEET against mosquito vectors in Thailand. J. Am. Mos. Control Assoc. 17, 190–195.

Tichy, H., Barth, F.G., 1992. Fine structure of olfactory sensilla in myriapods and arachnids. Micros. Res. Tech. 22, 372–391.

Vet, L.E., Van Lenteren, J., Heymans, M., Meelis, E., 1983. An airflow olfactometer for measuring olfactory responses of hymenopterous parasitoids and other small insects. Physiol. Entomol. 8, 97–106.

Vilela, V.L.R., Feitosa, T.F., Bezerra, R.A., Klafke, G.M., Riet-Correa, F., 2020. Multiple acaricide-resistant *Rhipicephalus microplus* in the semi-arid region of Paraíba State, Brazil. Ticks Tick-borne Dis. 11, 101413.

Wallade, S., Rice, M., 1977. The sensory nervous system of the adult cattle tick *Boophilus microplus* (Canestrini). Part III. Ultrastructure and electrophysiology of the cheliceral receptors. J. Aust. Entomol. Soc. 453, 142–156.

Weeks, E.N.I., Wasserberg, G., Logan, J.L., Agneessens, J., Stewart, S.A., Dewhirst, S., 2019. Efficacy of the insect repellent IR3535 on the sand fly *Phlebotomus papatasi* in human volunteers. J. Vector Ecol. 44, 290–292.

WHO. 2020. Vector-borne diseases, http://www.who.int/news-room/fact-sheets/detail/vector-borne-diseases. (Accessed March 2, 2020).

Woolley, T.A., 1972. Some sense organs of ticks as seen by scanning electron microscopy. Trans. Am. Micros. Soc. 91, 35–47.

Zeringóta, V., de Oliveira Filho, J., Borges, L., 2020. Activation of the ambusher tick *Rhipicephalus microplus* (Acari: Ixodidae) exposed to different stimuli. Med. Vet. Entomol. 34, 236–239.

CHAPTER

9

Arthropod repellents and chemosensory reception

Robert Renthal

University of Texas at San Antonio, Department of Biology, San Antonio, TX, United States

9.1 Arthropod repellents act through chemoreceptor pathways

The main US Environmental Protection Agency-approved arthropod repellents for use on the skin are N,N-diethyl-3-methyl benzamide (DEET), 2-(2-hydroxyethyl)-1-piperidine carboxylic acid 1-methylpropyl ester (Picaridin), ethyl butylacetylaminoproprionate (IR3535), p-menthane-3,8-diol (PMD), methyl nonyl ketone (2-undecanone), catnip oil, and oil of citronella. Biochemical and physiological experiments have shown that these arthropod repellents act through chemoreceptor pathways (Bissinger and Roe, 2010; Bohbot et al., 2011; Bohbot and Dickens, 2010; Du et al., 2015; Grant and Dickens, 2011; Huff and Pitts, 2020; Hull et al., 2020; Kwon et al., 2010; Sparks and Dickens, 2016; Syed et al., 2011; Tsitoura et al., 2015; Xu et al., 2019). An understanding of these pathways at the cellular and molecular level should enhance the development of more effective arthropod repellents and improve the effectiveness of repellents currently in use. The following discussion of arthropod chemoreception highlights some unresolved issues related to the molecular basis of repellency needing further research. The proteins and pathways discussed in this chapter are summarized in Fig. 9.1.

9.2 Chemoreceptor anatomy

Chemosensory tissues are widely distributed throughout arthropod anatomy, including insect antennae, mouth parts, tarsi, wings, and ovipositors (Vosshall and Stocker, 2007), and foretarsi, palps, and cheliceri of Acari (Carr and Roe, 2016; Sonenshine et al., 1986). Odorants and tastants are detected by chemosensory receptor neurons, which send electrical signals to the insect brain or the synganglion of Acari. The dendrites of chemosensory receptor neurons are encased in chitinous projections known as sensilla. Inside the sensilla, an extracellular fluid known as the sensillar lymph surrounds the dendrites.

Advances in Arthropod Repellents.
DOI: https://doi.org/10.1016/B978-0-323-85411-5.00018-2

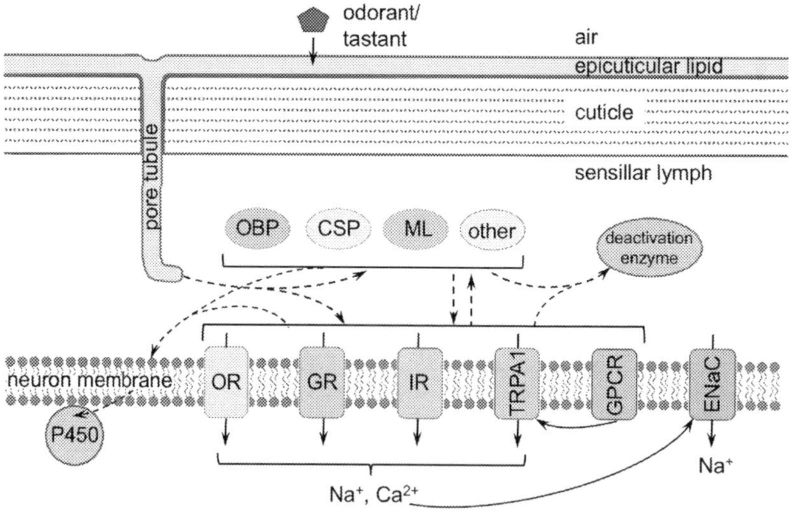

FIG. 9.1 **Chemosensory reception mechanisms.** Odorants and tastants adsorb to epicuticular lipids, diffuse through pore tubules into sensillar lymph, across which they may be carried by binding proteins: OBP, odorant-binding protein; CSP, chemosensory protein; ML, myeloid differentiation-related lipid-recognition proteins, including NPC2 (Neimann-Pick C2) and CheB proteins; and/or other transporters, including TULIPs (tubular lipid-binding proteins) and lipocalins. Odorants and tastants bind to ligand-gated ion channels: OR, odorant receptor; GR, gustatory receptor; IR, ionotropic receptor; TRPA1, transient receptor potential receptor A1, or G-protein coupled receptors (GPCRs). Signal amplification may occur by GPCR activation of TRPA1 channels or Ca2+ activation of epithelial Na+ channels (ENaC) such as pickpocket. Signal termination occurs after odorant/tastant transport, possibly via binding proteins, from receptors to deactivation enzymes in sensillar lymph, or to cytochrome P450s on the neuron membrane. The figure shows all known types of arthropod chemosensory receptor ion channels, but each chemosensory receptor neuron expresses only one or two different types. Dashed arrows indicate proposed steps that need further analysis.

The details of how chemicals reach the chemosensory receptor neurons are not fully known. Many external molecules, either volatile odorants in the air or tastants in oil films on surfaces, are hydrophobic substances that are water-insoluble. Access of this type of external molecule to the receptor neuron dendrites is restricted by two barriers. First, the cuticle covering the sensilla must be crossed, and then the molecules must pass through the aqueous sensillar lymph. Chemosensory sensilla contain cuticular pores. Electron micrographs of thin sections show that the pores, typically about 10 nm in diameter, penetrate the full thickness of the cuticle, providing a path into the sensillar lymph (Steinbrecht, 1997). The pores often appear to be filled with an unknown material that picks up metal stain in electron micrographs. In some single-walled sensilla, this material is contiguous with 20–40 nm diameter tubules that extend into the sensillar lymph and are sometimes observed in contact with the chemosensory receptor neuron dendrites (Keil, 1982). One hypothesis of odorant uptake proposes that the molecules first adsorb onto and dissolve in the hydrophobic epicuticular lipid surface. Then they diffuse to the pores and through the presumably hydrophobic pore tubules into the sensillar lymph (Kanaujia and Kaissling, 1985). The tracer studies that support this also found that sensilla-containing receptors for a moth pheromone adsorb substantially larger amounts of pheromone on their epicuticular surface than did other sensilla (Kanaujia and Kaissling, 1985). This suggested there are differences in epicuticular surface lipid compositions between different sensilla, which was later demonstrated by lipid analyses (Böröczky et al., 2008). For the design of more effective arthropod repellents, it might be valuable to pursue detailed studies of sensillar epicuticular lipid composition. If the compositional differences observed in moth sensilla are general and widespread, then lipid composition differences could be exploited as a design component in developing arthropod repellents that would more efficiently target sensilla-containing repellent receptors. Little work has been done on analysis of sensillar cuticular lipids. Maitani et al. (2010) attempted to detect differences in the lipid composition of pores compared with the sensillar surface using chemical force microscopy. The results were inconclusive, due to confounding effects of surface topology on the chemical force probe response. There are now mass spectrometry tools available for spatially resolved lipid microanalysis at a micrometer scale (Norris and Caprioli, 2013; Yew et al., 2009), but these tools have not yet been applied to analysis of cuticular lipids of sensilla.

How odorants cross the aqueous barrier between the cuticular pores and the chemosensory receptor neuron membrane is still unsettled (discussed in more detail later). The sensillar lymph bathing the chemosensory receptor neuron dendrites is an extracellular compartment that is separate from the hemolymph. This was shown by tracer studies (Keil and Steinbrecht, 1987) and composition analysis (Steinbrecht et al., 1992). Entry of chemicals into the sensillar lymph from the surrounding air is rapid (Kanaujia and Kaissling, 1985), but access to the sensillar lymph from the hemolymph is restricted (Keil and Steinbrecht, 1987). The pore tubules have been considered a potential transport mechanism across the sensillar lymph (Keil, 1982; Larter et al., 2016). However, images of pore tubules in close proximity to the chemosensory receptor neuron membrane do not clearly show a continuous hydrocarbon phase through which nonpolar odorants and tastants could diffuse directly from the epicuticular surface to the neuron membrane. This uncertainty might be resolved by re-examining the pore tubule structures with new microscopy methods (Gwosch et al., 2020; Narayan and Subramaniam, 2015).

For volatile water-insoluble odorants, any transfer mechanism other than direct diffusion through hydrocarbon phases from pore tubule to dendrite membrane probably would require a protein transporter. This premise is based on the known mechanisms by which water-insoluble components of cells, such as lipids and integral membrane proteins, are inserted into and removed from biomembranes. A protein coating appears to form a barrier that surrounds the pore tubule in the sensillar lymph (Keil, 1982), and the pore interior also contains proteins (Steinbrecht, 1992; Steinbrecht et al., 1992). It is unlikely that a nonpolar odorant could spontaneously cross this barrier to reach the dendrite membrane without a transporter.

9.3 Chemosensory receptors

9.3.1 Odorant receptors

Insect ORs were discovered by bioinformatics searches that assumed the ORs would be similar to the vertebrate G-protein coupled odorant receptors (GPCRs) (Clyne et al., 1999; Gao and Chess, 1999), and by searching for low-expression mRNAs in chemosensory appendages (Vosshall et al., 1999). Fortuitously, the insect ORs have seven membrane-crossing segments, the same as GPCRs, but their topology turns out to be the reverse of GPCRs, with the amino terminus in the cytosol and the carboxyl terminus extracellular (Lundin et al., 2007). Unlike GPCRs, insect ORs function as ligand-gated ion channels (Sato et al., 2008; Wicher et al., 2008). The ORs are located in the plasma membranes of OR neuron dendrites, found in antennae, maxillary palps, and labella. Each odor-specific OR is paired with an OR coreceptor subunit known as Orco that is common to all olfactory receptor neurons (Benton et al., 2006). The OR gene family is represented by large expansions of paralogous OR sequences in all insects. However, ORs and Orco are absent in Acari (Eliash et al., 2017; Gulia-Nuss et al., 2016). A detailed phylogenetic analysis indicates that ORs first appeared in Archaeognatha, the earliest apterygotes, and Orco first appeared in Zygentoma (Brand et al., 2018). The quaternary structure of ORs is probably tetrameric, with two Orco subunits and two odor-specific subunits (Butterwick et al., 2018). By itself, Orco forms a tetramer, and its three-dimensional (3D) structure shows the cation channel is lined by amino acids from the C-terminal helix. Hydrophobic amino acids near the extracellular surface form a selectivity filter and gate, and polar amino acids line the rest of the ion pore. On the extracellular surface of each subunit is a pocket of about 1300 $Å^3$ that includes amino acids known to be involved in binding to Orco of the repellent VUAA1 (Vanderbilt University allosteric agonist 1) (Corcoran et al., 2018), and the analogous sequence positions in odor-specific ORs are likely involved in odorant binding. Presumably, odorant-binding to this site is transmitted to the gate to stabilize the open pore, resulting in cation transport through the membrane. The knowledge of this 3D structure offers possibilities for computational methods of insect repellent discovery, as discussed in the section Computational Screening. Many behavioral and electrophysiological studies indicate that ORs in insect antennae respond to arthropod repellents. Mosquito Orco was expressed in frog oocytes paired with ORs, and then tested against a variety of repellents, including DEET. Most were shown to inhibit, or in a few cases activate, the Orco/OR ion channel (Bohbot et al., 2011). In a similar in vitro assay, mosquito ORs that detect carboxylic acids were inhibited by DEET (Huff and Pitts, 2020). The role of OR detection of DEET was demonstrated in behavioral studies of Orco mutant mosquitos (DeGennaro et al., 2013). These mosquitoes had severely inhibited antennal olfactory responses, and they were not repelled by DEET, but were attracted to CO_2, which is not detected by OR receptors. Maxillary palps also have ORs that were shown to respond to DEET, and also to Picaridin and IR3535 (Syed et al., 2011).

9.3.2 Gustatory receptors

The GR proteins have similar amino acid sequences to ORs. The GRs evolved prior to ORs and lack a co-receptor like Orco. The GRs are present in placazoa but absent in chanoflagellates and sponges (Robertson, 2015). They probably first appeared along with the evolution of multicellular organisms and were lost with the evolution of vertebrates. The GRs are expressed in sensilla in insect antennae, mouthparts, tarsi, and also in other tissues involved in chemosensory reception such as the pharynx (Joseph and Carlson, 2015; Joseph and Carlson, 2015). They are also expressed in mite foretarsi (Eliash et al., 2017). Different GRs specifically respond to various sugars (sweet), toxins (bitter), and ions (salt), and also to some pheromones (Chen and Dahanukar, 2020; Ling et al., 2014; Sparks and Dickens, 2017). Activation of neurons that sense bitter taste triggers arthropod repellent behavior, so the bitter-sensing GRs are obvious targets for developing new arthropod repellents. DEET was shown to activate a GR in fly labella (Lee et al., 2010), and anosmic mosquitoes are inhibited from feeding on sugar-containing DEET (Dennis et al., 2019). It also repels anosmic mosquitoes from surface contact, but this activity was shown to be distinct from bitter receptors (Dennis et al., 2019).

9.3.3 Ionotropic receptors

Related to ionotropic glutamate receptors, these IRs form a large family of chemosensory receptors in antennae, mouthparts, and tarsi. The IRs were first identified in olfactory sensilla that did not express ORs (Benton et al., 2009). Sequence alignments with ionotropic glutamate receptors suggest each IR sequence contains three membrane crossing segments, a pore loop, two extracellular domains that form an odorant-binding site, and a large extracellular N-terminal domain. By analogy with ionotropic glutamate receptors, the IRs probably form heterotetramers. Several IR subunits have been identified to have coreceptor function. Ionotropic glutamate receptors are found in basal eukaryotes and prokaryotes, but chemosensory IRs first appear in protostomes (Croset et al., 2010). Some IRs are involved in naturally aversive behavior that might be used as a framework for developing arthropod repellents. For example, IR76b is expressed in *Drosophila melanogaster* labellar GR neurons, and it stimulates aversion to high Na^+ content in food (Lee et al., 2017).

9.3.4 Transient receptor potential channels and other ion channels

Transient receptor potential (TRP) channels constitute a large family of ion channels that transduce force and stretch into electrical signals. A subset of TRP channels, including many TRPA1 channels, are sensitive to extracellular irritants, such as the *D. melanogaster* TRPA1 receptor that responds to wasabi (Al-Anzi et al., 2006). The TRP proteins typically contain six membrane-crossing segments and have a pore loop that constitutes the ion selectivity filter. A four-subunit quaternary structure forms the channels (Liao et al., 2013). Insect TRPA1 channels are important in contact repellency because of their activation by irritants. For example, *D. melanogaster* can detect the repellent citronellal via a TRPA1 channel, triggering avoidance behavior (Kwon et al., 2010). The TRPA1 channels in *D. melanogaster* are expressed in the labellum, brain, antenna, and tarsi (Leung et al., 2020). Amplification is a distinctive feature of the vertebrate chemosensory receptors, compared to arthropods. The vertebrate chemosensory GPCRs link to ion channels via pathways that amplify the chemical signal. By contrast, the direct ligand-gated chemosensory channels of arthropods would seem to be inherently less sensitive. However, signal amplification has been found in at least two different insect ionotropic mechanisms. First, the TRPA1 channels are not only directly gated by ligands, but they can also be activated by GPCRs via the G_q-phospholipase C-protein kinase C pathway (Lapointe and Altier, 2011). In this way, TRPA1 chemoreception in insects is similar to the GPCR-activated vertebrate olfactory receptors and also the GPCR-activated insect visual receptor rhodopsin. Remarkably, Montell and co-workers found that rhodopsin is expressed in *D. melanogaster* labellar GR neurons and binds to the irritant aristolochic acid to activate a G_q pathway that opens TRPA1 channels at low aristolochic acid concentrations (Leung et al., 2020). Higher aristolochic acid concentrations directly gate the TRPA1 channel. Also, a rhodopsin-like GPCR sequence was identified in the foreleg transcriptome of the tick *Dermacentor variabilis* (Carr et al., 2017), although only weak evidence was found for a foreleg TRPA1 receptor. Second, some ORs, GRs, and IRs have been found to activate another type of channel via extracellular Ca^{2+} influx through the chemosensory ligand-gated channels. These secondary Ca^{2+}-activated channels are members of the ENaC family, known in *D. melanogaster* as PPK (Ng et al., 2019). PPKs are also direct taste receptors for Na^+ in some insect mouthparts. It may be useful to test the arthroopod amplification mechanisms in bitter taste receptor pathways for sensitivity to nonbitter substances that might activate the downstream ENaC channels. Such substances could have the valuable property of activating arthropod repellent behavior without being chemical irritants.

9.4 Hydrophobic ligand transport proteins

9.4.1 Odorant-binding proteins

The discovery of odorant- and pheromone-binding proteins (OBPs) in sensillar lymph provided a logical molecular conveyance for odorants between the pore tubules and the ORs on the dendrite membranes (Kaissling, 2009; Ronderos and Smith, 2009; Steinbrecht et al., 1992; Vogt and Riddiford, 1981). In the past 30 years, large numbers of insect OBPs have been identified, and their structures and functions have been examined in detail (Leal, 2005; Pelosi et al., 2018). The OBPs typically have 120–130 amino acids with molecular weights in the 14–15 kDa range, and they usually have six cysteines that form three disulfide bonds. All insect species examined have many paralogous OBP genes, ranging from fewer than 20 to more than 100, usually with highly divergent sequences. Two OBP-like proteins also were identified in tick tarsi and palps (Renthal et al., 2017), and other OBP-like sequences were identified in chelicerate and myriopod genomes (Vizueta et al., 2018).

In many insects, behavioral evidence pointed to the function of OBPs as an essential component of chemoreception. Expression of specific OPBs was knocked down by RNA interference (RNAi), and as a result, specific behavioral changes were observed, including flies' loss of aversion to bitter tastants (Swarup et al., 2014), male moths' loss of attraction toward a pheromone-bated lure (Dong et al., 2017), decreases in flies' attraction or aversion to odorants (Swarup et al., 2011), and diminished feeding by male locusts (Li et al., 2016). The RNAi knockdowns of OBPs also decreased electroantennographic (EAG) responses to specific odorants (Biessmann et al., 2010; Liu et al., 2020; Pelletier et al., 2010; Rebijith et al., 2016; Zhang et al., 2019; Zhang et al., 2017), suggesting that OBPs are involved in the molecular pathway that results in neuronal electrical responses to odorants.

Because of their affinity for odorants and the ease of analyzing their activity, the OBPs have been considered prime targets for manipulating insect behaviors, either by using OBPs to identify new attractants or repellents (Affonso et al., 2013; Kröber et al., 2018; Thireou et al., 2018), or by targeting OBPs using RNAi knockdowns to suppress attractant behavior. Insect OBPs bind to pheromones and odorants with dissociation constants in the 100 nM to 10 mM range (Ban et al., 2003; Campanacci et al., 2001; Du and Prestwich, 1995; Gong et al., 2009; Honson et al., 2003; Iovinella et al., 2011; Katti et al., 2013; Li et al., 2008; Pesenti et al., 2009; Plettner et al., 2000; Riviere et al., 2003; Leal et al., 2005), although some of the reported dissociation constants have been measured without taking into account the phase separation of insoluble pheromones in water as discussed in Katti et al. (2013). The ligand specificities of OBPs are relatively broad (Katti et al., 2013; Lagarde et al., 2011; Lautenschlager et al., 2007; Mao et al., 2010; Oldham et al., 2001; Willett and Harrison, 1999; Wojtasek et al., 1998). Many 3D structures of OBPs have been determined (https://www.rcsb.org/), and all of them show a large interior hydrophobic cavity. 3D structures have also been determined for many OBP-ligand complexes, showing that the odorant or pheromone binding site is in the interior cavity. Some OBPs, when bound to an odorant or pheromone, have a different conformation than the ligand-free OBP (Horst et al., 2001; Laughlin et al., 2008). For one OBP family, the conformational change occurs at low pH, suggesting a pheromone/odorant release mechanism, considering that the surface pH of the receptor neuron dendrite is likely to be acidic compared with the bulk sensillar lymph (Horst et al., 2001).

Arthropod repellents have been tested for binding to OBPs, with the idea that OBPs may be involved in transport of repellents to olfactory and taste receptors, or that repellents may disrupt OBP-attractant interactions. DEET binds to a mosquito OBP, AgamOBP1 (Murphy et al., 2013; Tsitsanou et al., 2012). The binding site is near the surface entrance to the interior hydrophobic binding pocket. DEET has a relatively low affinity for this OBP, raising a question about the relevance of this interaction to DEET repellency. Picaridin binds to the same site, and it also binds to a second site in the interior hydrophobic pocket (Drakou et al., 2017). The affinities for both sites are relatively weak. The repellent 2-undecanone binds, with relatively high affinity, to a mosquito OBP, AfunOBP1 (Xu et al., 2010), and citronellal binds with relatively high affinity to AgamOBP1 (Murphy et al., 2013). Bitter tastants normally repel flies from sucrose feeding, but when OBP49a is knocked out in labellar sensilla, the flies are not repelled, indicating the involvement of this OBP in repellency (Jeong et al., 2013). OBP49a was shown to interact both with bitter tastants and also with sugar-sensitive GRs (Jeong et al., 2013). This indicates that the repellency properties of bitter tastants in the presence of food signals occur at the receptor level in flies and is caused by the direct suppression of attraction signals via OBPs.

The idea that OBPs are odorant transporters gained support from the RNAi knockdown experiments (described earlier), but there is also contrary evidence. Although RNAi knockdown of OBPs decreased the odorant-induced voltage changes in the electroantennogram (EAG), the EAG does not resolve the contributions of individual sensilla to the voltage signal. Until recently, it was not known what the OBP expression pattern was for individual sensilla in an entire insect antenna, so single-sensillum recordings (SSRs) could not be meaningfully measured after OBP knockdowns. In contrast to the EAG, the SSR measures fast trains of action potential spikes that occur in receptor neuron dendrites immediately after exposure to an odorant. Carlson and co-workers (Larter et al., 2016; Xiao et al., 2019) recently used in situ hybridization to locate the sensilla expressing the ten most abundant OBPs in the antenna of *D. melanogaster*. Combinations of five OBP-coding genes were then deleted from basiconic sensilla using CRISPER/Cas9, and SSR measurements were done. The results showed essentially no effect of the absence of specific OPBs on the spike frequency of the SSR response to a wide range of odorants. Similar results were reported by Scheuermann and Smith (2019) for *D. melanogaster* deletion mutants of two other OBPs in eight different trichoid and intermediate sensilla, except that some sensilla showed delayed recovery after the odor pulse ended. These experiments seem to rule out any role for the OBPs in odorant transport from cuticular pores to the dendrite membrane. Although not mentioned by the authors, there is a hint, in the SSR traces, of differences in the SSR spike amplitudes, depending on whether OBPs are present or not. Normally in SSRs, shortly after an odorant pulse, an abrupt decrease occurs in spike amplitudes, and on a second time scale, the spike amplitudes return to their initial levels (Martin and Alcorta, 2016). The size and time course of this effect were shown to parallel the EAG signal, suggesting that they are related. The magnitude of the EAG response also parallels the amount of OBP present, as shown, for example, in studies of circadian rhythm-dependent OBP synthesis (Rund et al., 2013). The recent OBP knockout experiments seem to show, in sensilla lacking OBPs, significantly less of a decrease in the SSR amplitude after the odorant pulse (cf. Fig. 3C and D in Xiao et al. (2019). In other words, the EAG and SSR amplitude effects are consistent in the earlier OBP RNAi knockdowns and the more recent OBP knockouts. The EAG results perhaps had been misinterpreted as showing a loss of odorant response. Assuming that the decrease in EAG with lower amounts of OBP is related to the smaller amplitude decrease in OBP knockouts, it would be important to find out what the origin of this effect is at the molecular level. The outcome of research on the amplitude effect may provide insight into the behavioral phenotypes observed with OBP knockdowns, and this could form the basis for designing arthropod repellents that target OBPs.

9.4.2 Other hydrophobic ligand-binding proteins

If some or all chemosensory sensilla do not use OBPs to transport odorants and tastants to the receptor neurons, there are other hydrophobic ligand transporters known in chemosensory tissues that could be involved.

9.4.2.1 Chemosensory proteins

CSPs contain about 100 amino acids with four cysteines in two disulfide bonds, and they have molecular weights of about 11–12 kDa (Pelosi et al., 2018). Insects contain many paralogous genes coding for CSPs, but the sequence similarities are greater among CSPs than OBPs. The CSPs are expressed in many different tissues, including fat body, secretory glands, wings, mouthparts, tarsi, and antennae. Structure analysis shows that a large conformational change occurs, opening an internal cavity to bind to hydrophobic ligands (Campanacci et al., 2003; Lartigue et al., 2002). A few experiments suggest an involvement of CSPs in chemosensory reception. A CSP expressed in the antenna during the gregarious phase of migratory locusts may be involved in detection of attraction pheromones. After RNAi knockdown of this CSP, gregarious phase locusts displayed less mutual attraction (Guo et al., 2011). A CSP expressed in carpenter ant antennal sensilla was shown to bind cuticular hydrocarbons that constitute nestmate recognition signals (Ozaki et al., 2005). There have not been many studies of the interaction of CSPs with arthropod repellents. An antennal CSP expressed in the antennae of female oriental armyworm moths was found to bind 2-undecanone with relatively high affinity, but the same level of binding was found for many nonpolar substances (Younas et al., 2018). Only weak binding of 2-undecanone to a locust antennal CSP was detected (Ban et al., 2002).

9.4.2.2 Myeloid differentiation-related lipid-recognition proteins

The ML protein family contains both intracellular and extracellular members. An extracellular ML, NPC2 was detected in basiconic sensilla of carpenter ants, and it was found to bind hydrophobic molecules that produce EAG signals (Ishida et al., 2014), suggesting a role in chemosensory reception. The NPC2s are widely distributed in arthropods (Zhu et al., 2018), but detailed studies of their role in chemosensory reception have not been reported. Another ML protein, CheB, is expressed in insect gustatory sensilla (Torres-Oliva et al., 2016), and it may interact with ENaC channels (Ben-Shahar et al., 2010). There is no information about the interaction of ML proteins with arthropod repellents.

9.4.2.3 Lipocalins

This functionally diverse protein family includes vertebrate OBPs and many proteins found in arthropod hemolymph and saliva (Flower, 1996). However, there is only one report of an insect lipocalin involved in chemosensory detection: a female swallowtail butterfly tarsal protein that was shown to be involved in the detection of oviposition sites, and evidence was presented for the ligand being aristolochic acid (Tsuchihara et al., 2000, 2009), which for some insects is an irritant (see section earlier on TRP Channels).

9.4.2.4 Tubular lipid-binding proteins

Many transcriptomes and proteomes of insect chemosensory tissue include numerous sequences that are members of the TULIP family (Dauwalder et al., 2002; Fujikawa et al., 2006; Hagai et al., 2007; Lin et al., 2017a, 2017b; Mohapatra and Menuz, 2019; Schwinghammer et al., 2011; Shah and Renthal, 2020; So et al., 2000; Vanaphan et al., 2012). The TULIPs contain about 230 amino acids and have molecular weights of about 27 kDa, occasionally occurring as tandem sequences of two connected TULIPs.

Proteins in this family have a wide range of functions, including phospholipid transfer between cytoplasmic membranes, cholesterol ester transfer between serum lipoproteins, and antibacterial activity (Alva and Lupas, 2016; Wong and Levine, 2017). The TULIPs are difficult to recognize by sequence comparisons because of extreme sequence diversity. However, their 3D structures contain a distinctive 4-stranded β-sheet "super-roll" that wraps around an α-helix. Most TULIP crystal structures contain an elongated internal cavity where the lipid ligands bind. It is not known whether the vertebrate, arthropod, and plant TULIPs are evolutionarily derived from a common ancestor or have developed by convergent evolution. Two insect TULIPs have been studied in detail: the juvenile hormone-binding protein (JHBP) (Kolodziejczyk et al., 2008), and a circadian rhythm-dependent antennal protein called takeout (TO) (So et al., 2000). Automated annotations of insect genomes often list TULIPs as "juvenile hormone (JH)-binding protein," "takeout-like protein," or "circadian clock-controlled protein," although the actual functions of these proteins are rarely known. In cases where sequence divergence is too great for an identification, it is relatively straight-forward to recognize TULIP sequences by 3D homology modeling (SwissModel, HHPred, Modeller).

JHBP carries the hydrophobic JH molecule through the hemolymph from the tissue where it is synthesized to receptors in JH-sensitive cells. It was discovered (Kramer et al., 1974) and sequenced (Lerro and Prestwich, 1990) many years before the determination of its 3D structure connected it to the larger protein family now known as TULIPs (Kolodziejczyk et al., 2008). JH is the only known functional ligand for any arthropod TULIP. Its binding interaction is strong, with the measured dissociation constant in the submicromolar range (Kramer et al., 1974). TO was identified in *D. melanogaster* as a protein having expression levels tied to the circadian cycle, and its expression affects feeding, survival, and courtship (Sarov-Blat et al., 2000; Saurabh et al., 2018). TO is expressed in the antenna and labella, and it is also expressed in the fat body adjacent to the brain in males. The fat body expression is responsible for a decreased courtship phenotype in a low-expression TO mutant (Dauwalder et al., 2002). The native ligands for TO are not known, but it is likely to have relatively nonspecific binding interactions. A moth antennal TO-like protein was expressed heterologously from two different cell types, and the proteins were found to contain two highly dissimilar hydrophobic ligands, picked up from the host cells (Hamiaux et al., 2013). Also, in the *D. melanogaster* mutant TO courtship phenotype, a wide variety of TOs or different TULIPs from other insects, having very dissimilar sequences, could rescue the mutants (Saurabh et al., 2018).

The role of TULIPs in arthropod chemosensory reception has not yet been investigated. The most complete sequence information available comes from *D. melanogaster* antennal transcriptomes. The data of Shiao et al. (2013) show that 24 TULIP genes are transcribed in the *D. melanogaster* antenna. Of these, three high-expression TULIPs were on the list of antennal TULIPs derived by Mohapatra and Menuz (2019). One of the three was determined to be expressed in coeloconic sensilla. Mohapatra and Menuz limited their analysis to the most highly expressed mRNAs, so many of the other TULIPs were not analyzed, and it is not yet known whether they are expressed in sensilla. In the antennal proteome of the red imported fire ant, *Solenopsis invicta*, we identified eight antenna-specific TULIP proteins (Shah and Renthal, 2020), and in the tarsal proteome of the stable fly, *Stomoxys calcitrans*, we identified 15 TULIPs (Renthal and Olafson, unpublished). Combining these sequences with selected TULIPs from the NCBI database, we generated a phylogenetic tree comparing antennal TULIPs with TULIPs having unknown or nonchemosensory functions (Fig. 9.2).

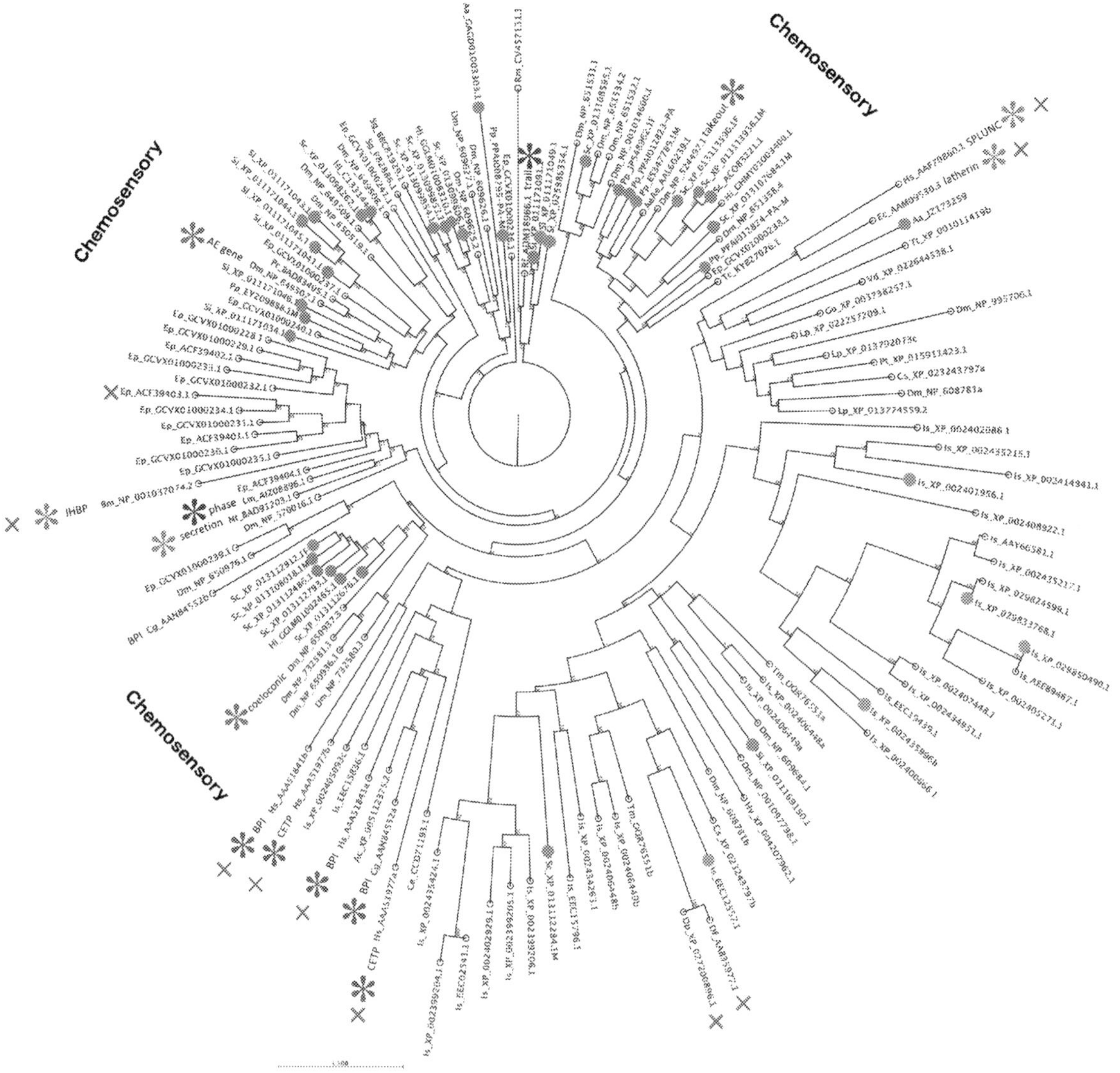

FIG. 9.2 **Phylogenetic tree of TULIPs.** Insect antennal proteins are marked with green nodes; insect and tick tarsal proteins with brown nodes; and tick salivary/midgut proteins with blue nodes. Three *D. melanogaster* TULIPs known to have exclusively antennal expression are marked with red asterisks. Their expansions are likely to contain TULIPs involved in chemosensory reception. TULIPs with altered chemosensory behavior after RNAi knockdown are marked with purple asterisks; secretory function, green asterisks; surfactant function, brown asterisks; innate immunity, blue asterisks. Known 3D structures marked by X. Colors can be viewed in the online version of this chapter.

The tree has many large expansions that generally have good internal boostrap support but poor bootstrap support for the connections between expansions. Eight branches contain predominantly insect TULIPs, including three that each contain one of the *D. melanogaster* TULIPs shown by Mohapatra and Menuz to be expressed exclusively in the antenna. One of these branches includes *D. melanogaster* TO along with many TULIPs identified in fly tarsal proteomes. Two of the insect branches also include TULIPs showing chemosensory-like behavioral phenotypes after RNAi experiments (discussed later). These tree associations support the idea that many insect TULIPs have a role in chemosensory reception. One of the noninsect branches contains many sequences from the black-legged tick, *Ixodes scapularis*, including four sequences that we detected in *I. scapularis* salivary and midgut secretions (Renthal and Seshu, unpublished), suggesting that this group may not have chemosensory functions.

There are reports of a few behavioral experiments that point to a role for TULIPs in chemosensory reception. A low-expression TO mutant of *D. melanogaster* displays a feeding phenotype involving hyperphagy when food is abundant, and a lack of increased food intake after starvation. These effects are paralleled by reduced sensitivity to sugar of gustatory neurons on the proboscis (Meunier et al., 2007), indicating the involvement of TO for signal processing in chemosensory appendages. In a second behavioral study, a termite TULIP that is expressed in the antenna and other tissues was knocked down by RNAi, causing altered trail-following behavior: the workers made large deviations from the trail, suggesting difficulty in detecting the experimentally presented trail pheromone (Schwinghammer et al., 2011). In a third behavioral study, Guo et al. (2011) screened *Locusta migritoria* genes involved in the transition between solitary and gregarious phase. One of the genes most strongly downregulated in the solitary to gregarious transition codes for a TULIP protein named TO1, which is highly expressed only in the antenna and legs of solitary phase locusts. On the TULIP phylogram (Fig. 9.2), TO1 is close to two insect TULIPs that are known to be involved in transporting secretions. When TO1 was knocked down by RNAi in solitary phase locusts, the solitary locusts showed a significantly higher attraction for gregarious locusts. A possible interpretation of this experiment is that TO1 is involved in chemoreception of a repellent substance produced by both the solitarious and gregarious phase locusts, but only detectable by solitarious locusts. Alternatively, TO1 could be involved in the deactivation of an attractant produced by gregarious phase locusts, or in the secretion by solitarious phase locusts of a substance that repels other solitary phase locusts.

TULIPs could be involved in chemosensory reception in several different ways. First, TULIPs might transport nonpolar odorants and tastants through the sensillar lymph to the dendrite membrane. Second, TULIPs might remove odorants and tastants from the chemosensory receptor and transport them to the plasma membrane of dendrites or support cells for metabolic deactivation. A transport activity of this type would be essential for odorant and tastant deactivation by cytochrome P450-dependent monooxygenase enzymes. Cytochrome P450s are membrane-bound and are most likely on the cytoplasmic surface of the plasma membrane, along with the cytochrome P450 reductase (Maibeche-Coisne et al., 2004). Cytochrome P450 substrates have access to the active site from the lipid bilayer, so a simple transport mechanism from the chemosensory receptor binding site to the lipid bilayer would be sufficient for P450 deactivation of water-insoluble odorants and tastants. Third, there is evidence that components of the phospholipase C pathway (G_q, phospholipase C, and protein kinase C, PKC) affect some insect chemosensory responses (Kim et al., 2018; Leung et al., 2020; Sargsyan et al., 2011; Yao and Carlson, 2010). Some TULIPs are known to transport diacylglycerol (DAG), a key component of the phospholipase C pathway (Saheki et al., 2016). Therefore, TULIPs might serve as DAG-binding buffers that deactivate PKC in sensilla. This may at first seem unlikely, considering that the cytoplasmic side of the dendrite plasma membrane is the site where DAG is formed by phospholipase C and where PKC binds to DAG, whereas TULIPs are located in the sensillar lymph that contacts the extracellular side of the membrane (Fujikawa et al., 2006). However, recently it was shown that DAG rapidly and spontaneously crosses lipid bilayers ("flip-flop") (Schuhmacher et al., 2020), unlike phospholipids and glycolipids, which require transporters and energy sources to cross bilayers. Therefore, TULIPs in the sensillar lymph could pull DAG out of the cytoplasmic G_q signaling pathway. In order to test whether any of these possible TULIP functions actually occur in arthropod chemosensory reception, it will be necessary to determine first what is the sensillar expression pattern of the antennal and tarsal TULIPs. Experiments similar to those on OBP function reported by Carlson and co-workers (Larter et al., 2016; Xiao et al., 2019) could then be done on antennal TULIPs to assess their role in chemosensory reception.

When binding to a ligand, TULIPs probably undergo a conformational change to open the interior hydrophobic cavity to admit the ligand, and after binding, another conformational change would be needed to release the bound ligand to a receptor. Three different types of TULIP conformational changes have been described. Conformational changes in JHBP were detected by comparing two structures under different conditions (Suzuki et al., 2011). A latch, consisting of approximately 40 amino acids at the amino terminus, is flexible in the unbound (apo) form of JHBP, and it closes to cover JH when it binds. The amino terminal region, the carboxyl-terminal region, and a loop in the middle of the protein sequence all come together at one end of the TULIP structure to enclose bound JH. These same three sequence regions are highly flexible in the apoprotein. The binding of JH to JHBP was much weaker at pH 4.5 than at pH 7 (Dupas et al., 2020). Spectroscopic evidence for a pH-dependent conformational change of JHBP was observed, and JH binding was inhibited by anionic membranes, suggesting that low surface pH may trigger JH release at membrane surfaces (Dupas et al., 2020). The second type of conformational change was identified by studies of the surfactant properties of a family of vertebrate TULIPs known as bacterial permeability-increasing proteins (BPIs). The BPI fold-containing family member A1 (BPIFA1, also known as SPLUNC1) had weaker surfactant properties after mutations were made in loops at the opposite end of the TULIP structure from the sequences involved in the JHBP latch (Walton et al., 2016). This end of the TULIP structure was first identified as a conserved motif in TO (So et al., 2000), and is near a sequence insertion unique to surfactant TULIPs (such as BPIFA1 and latherin) that was proposed as a trigger for unrolling the entire TULIP 4-stranded β-sheet into a planar structure at the air-water interface (Vance et al., 2013). Large-scale reversible opening may be a general property of TULIPs, which appear to be less stable to thermal- and denaturant-induced unfolding (Bülow et al., 2018; Dobryszycki et al., 2004) compared to other proteins with an α-helix wrapped in a β-sheet (Fukuda et al., 2000). The fact that some members of the TULIP protein family function by interacting with air–water interfaces, combined with the uncertainty of how odorants and tastants reach the receptor neurons, underscores the need for obtaining additional basic information about the conformational states of the chemosensory appendage TULIP proteins.

9.5 High throughput screening methods for repellent discovery

Discovery of new arthropod repellents by screening chemical libraries using behavioral assays cannot be done in a high-throughput mode. However, two alternative approaches have shown promising results, using molecular components of arthropod chemosensory systems.

9.5.1 Microplate assays

Microplate readers can automatically screen tens of thousands of individual samples of cultured cells or purified proteins against large chemical libraries. An impressive example of this approach is the discovery of the repellent VUAA1. Cultured human kidney cells were cotransfected with mosquito Orco and an odor-specific OR and then tested with 100,000 chemicals, using a calcium indicator to assay for ligand-gated ion channel activity. The substance called VUAA1 was found to have allosteric agonist activity, opening cation channels in Orco tetramers, and also stimulating channel opening in Orco/OR tetramers (Jones et al., 2011, 2012; Rinker et al., 2012). Although not volatile, VUAA1 showed strong repellent activity in a mosquito larva mobility assay, comparable to DEET but at a thousand-fold lower concentration (Taylor et al., 2012).

9.5.2 Computational screening

Excellent software is available for prediction of relative binding of small molecules to receptors. Programs such as Autodock (http://vina.scripps.edu/) have user-friendly interfaces and can be run on ordinary small computers. Server-based docking software such as SwissDock (http://www.swissdock.ch/) is also simple to use. These programs can predict the likelihood that a repellent candidate binds strongly to a particular protein target, such as an OR. However, for high throughput screening of large molecular libraries, considerable computer expertise is necessary. A proof of principle for high throughput screening using Autodock was implemented on a supercomputer (Ellingson et al., 2013). In order to carry out docking, the 3D structures of both the candidate repellent and the target receptor must be known. The small molecule structures are easily available. For example, the Cambridge Crystallographic Data Center (https://www.ccdc.cam.ac.uk/) contains more than 1 million molecular structures. For small molecules of unknown 3D structure, docking can be done if the structural formula of the molecule is known. The molecule can be built from the known bonding, using software such as the Molefacture routine in Visual Molecular Dynamics (https://www.ks.uiuc.edu/Research/vmd/), and then Autodock can be run in a mode that allows sampling of all conformations of the rotatable bonds. Selected bond rotations in the receptor protein can also be implemented to account for protein conformational changes when the ligand binds. A major roadblock in implementing computational screening is the lack of known 3D structures for most of the protein targets that need to be assessed. One possible approach is to use 3D homology modeling to obtain the structures. However, the predictive value of the simulation heavily depends on the quality of the model. Homology models are best when the amino acid sequence of the modeled protein is close to the modeling template protein. In protein families such as OBPs, TULIPs, and ORs, the sequence divergence within the family is so great that obtaining high-quality models is very challenging. Quality parameters such as Qmean can be calculated to assess the geometry and energy of a homology model (Benkert et al., 2008). A key to obtaining high-quality homology models is the amino acid sequence alignment between the target protein and the known protein structure (the template). Sequence alignment algorithms such as HHblits (https://toolkit.tuebingen.mpg.de/tools/hhblits) are useful for homology modeling of proteins with divergent sequences. However, even when optimal alignments are used, there is not yet an accurate way to predict the 3D structures of the variable loops that form the ligand-binding sites in proteins such as ORs. Many structures in the Protein Data Bank (https://www.rcsb.org/), including the cryo-EM structure of Orco, have unstructured segments. To use the Orco structure for computational docking would require some de novo modeling because it lacks the atomic coordinates for 17 amino acids in a loop that likely forms part of the VUAA1 binding site.

Loops are often disordered in the unliganded state of receptors, and so the coordinates may be missing in the 3D structure. Progress in computational arthropod repellent screening probably will have to wait for new cryo-EM structures of the odorant-gated ion channels bound to ligands, and for advances in accurate de novo modeling of protein structures.

9.6 Conclusion

Arthropod repellents act through chemosensory reception pathways. Clarification of several steps in these pathways would provide opportunities to develop new or improved arthropod repellents. Among the issues needing further research: (1) Do unique epicuticular lipid compositions in chemosensory sensilla promote specific adsorption of odorants and tastants? (2) Can the high-resolution structure and molecular composition of pore tubules provide insight into the mechanism of odorant/tastant transfer to receptor neurons? (3) What are the functional roles of chemosensory ligand-binding proteins in chemosensory reception? (4) In the signal amplification pathways for bitter tastants, are there steps that can be activated by externally applied nonirritating substances? (5) Would high-resolution molecular structures of chemosensory receptors complexed with ligands provide sufficient information to support computational high-throughput screening for new repellents?

References

Affonso, R.D., Guimaraes, A.P., Oliveira, A.A., Slana, G.B.C., Franca, T.C.C, 2013. Applications of molecular modeling in the design of new insect repellents targeting the odorant binding protein of *Anopheles gambiae*. J. Braz. Chem. Soc. 24, 473–482. https://doi.org/10.5935/0103-5053.20130059.

Al-Anzi, B., Tracey, W.D., Benzer, S., 2006. Response of *Drosophila* to wasabi is mediated by painless, the fly homolog of mammalian TRPA1/ANKTM1. Curr. Biol. 16 (10), 1034–1040. https://doi.org/10.1016/j.cub.2006.04.002.

Alva, V., Lupas, A.N., 2016. The TULIP superfamily of eukaryotic lipid-binding proteins as a mediator of lipid sensing and transport. Biochim. et Biophys. Acta—Mol. Cell Biol. Lipids 1861 (8), 913–923. https://doi.org/10.1016/j.bbalip.2016.01.016.

Ban, L., Scaloni, A., D'Ambrosio, C., Zhang, L., Yan, Y., Pelosi, P., 2003. Biochemical characterization and bacterial expression of an odorant-binding protein from *Locusta migratoria*. Cell. Mol. Life Sci. 60 (2), 390–400. https://doi.org/10.1007/s000180300032.

Ban, L., Zhang, L., Yan, Y., Pelosi, P., 2002. Binding properties of a locust's chemosensory protein. Biochem. Biophys. Res. Commun. 293 (1), 50–54. https://doi.org/10.1016/S0006-291X(02)00185-7.

Benkert, P., Tosatto, S.C.E., Schomburg, D, 2008. QMEAN: a comprehensive scoring function for model quality assessment. Proteins: Structure, Function and Genetics 71 (1), 261–277. https://doi.org/10.1002/prot.21715.

Ben-Shahar, Y., Lu, B., Collier, D.M., Snyder, P.M., Schnizler, M., Welsh, M.J., 2010. The *Drosophila* gene CheB42a is a novel modifier of Deg/ENaC channel function. PLoS One 5 (2). doi:10.1371/journal.pone.0009395.

Benton, R., Sachse, S., Michnick, S.W., Vosshall, L.B., 2006. Atypical membrane topology and heteromeric function of *Drosophila* odorant receptors in vivo. PLoS Biol. 4 (2), 240–257. https://doi.org/10.1371/journal.pbio.0040020.

Benton, R., Vannice, K.S., Gomez-Diaz, C., Vosshall, L.B., 2009. Variant ionotropic glutamate receptors as chemosensory receptors in *Drosophila*. Cell 136 (1), 149–162. https://doi.org/10.1016/j.cell.2008.12.001.

Biessmann, H., Andronopoulou, E., Biessmann, M.R., Douris, V., Dimitratos, S.D., Eliopoulos, E., Guerin, P.M., Iatrou, K., Justice, R.W., Kröber, T., Marinotti, O., Tsitoura, P., Woods, D.F., Walter, M.F., 2010. The *Anopheles gambiae* odorant binding protein 1 (AgamOBP1) mediates indole recognition in the antennae of female mosquitoes. PLoS One 5 (3). doi:10.1371/journal.pone.0009471.

Bissinger, B.W., Roe, R.M., 2010. Tick repellents: past, present, and future. Pestic. Biochem. Physiol. 96 (2), 63–79. https://doi.org/10.1016/j.pestbp.2009.09.010.

Bohbot, J.D., Dickens, J.C., 2010. Insect repellents: modulators of mosquito odorant receptor activity. PLoS One 5 (8). doi:10.1371/journal.pone.0012138.

Bohbot, J.D., Fu, L., Le, T.C., Chauhan, K.R., Cantrell, C.L., Dickens, J.C., 2011. Multiple activities of insect repellents on odorant receptors in mosquitoes. Med. Vet. Entomol. 25 (4), 436–444. https://doi.org/10.1111/j.1365-2915.2011.00949.x.

Böröczky, K., Park, K.C., Minard, R.D., Jones, T.H., Baker, T.C., Tumlinson, J.H., 2008. Differences in cuticular lipid composition of the antennae of *Helicoverpa zea*, *Heliothis virescens*, and *Manduca sexta*. J. Insect Physiol. 54 (10–11), 1385–1391. https://doi.org/10.1016/j.jinsphys.2008.07.010.

Brand, P., Robertson, H.M., Lin, W., Pothula, R., Klingeman, W.E., Jurat-Fuentes, J.L., Johnson, B.R., 2018. The origin of the odorant receptor gene family in insects. ELife 7. doi:10.7554/eLife.38340.

Bülow, S., Zeller, L., Werner, M., Toelge, M., Holzinger, J., Entzian, C., Schubert, T., Waldow, F., Gisch, N., Hammerschmidt, S., Gessner, A., 2018. Bactericidal/permeability-increasing protein is an enhancer of bacterial lipoprotein recognition. Front. Immunol. 9, 2768. https://doi.org/10.3389/fimmu.2018.02768.

Butterwick, J.A., del Mármol, J., Kim, K.H., Kahlson, M.A., Rogow, J.A., Walz, T., Ruta, V., 2018. Cryo-EM structure of the insect olfactory receptor Orco. Nature 560 (7719), 447–452. https://doi.org/10.1038/s41586-018-0420-8.

Campanacci, V., Krieger, J., Bette, S., Sturgis, J.N., Lartigue, A., Cambillau, C., Breer, H., Tegoni, M., 2001. Revisiting the specificity of *Mamestra brassicae* and *Antheraea polyphemus* pheromone-binding proteins with a fluorescence binding assay. J. Biol. Chem. 276 (23), 20078–20084. https://doi.org/10.1074/jbc.M100713200.

Campanacci, V., Lartigue, A., Hallberg, B.M., Jones, T.A., Giudici-Orticoni, M.-T., Tegoni, M., Cambillau, C., 2003. Moth chemosensory protein exhibits drastic conformational changes and cooperativity on ligand binding. Proc. Natl. Acad. Sci., 5069–5074. doi:10.1073/pnas.0836654100.

Carr, A.L., Mitchell, R.D., Dhammi, A., Bissinger, B.W., Sonenshine, D.E., Roe, R.M., 2017. Tick Haller's organ, a new paradigm for arthropod olfaction: how ticks differ from insects. Int. J. Mol. Sci. 18, 1563. doi:10.3390/ijms18071563.

Carr, A.L., Roe, M., 2016. Acarine attractants: chemoreception, bioassay, chemistry and control. Pestic. Biochem. Physiol. 131, 60–79. https://doi.org/10.1016/j.pestbp.2015.12.009.

Chen, Y.C.D., Dahanukar, A., 2020. Recent advances in the genetic basis of taste detection in *Drosophila*. Cell. Mol. Life Sci. 77 (6), 1087–1101. https://doi.org/10.1007/s00018-019-03320-0.

Clyne, P.J., Warr, C.G., Freeman, M.R., Lessing, D., Kim, J., Carlson, J.R., 1999. A novel family of divergent seven-transmembrane proteins: candidate odorant receptors in *Drosophila*. Neuron 22 (2), 327–338. https://doi.org/10.1016/S0896-6273(00)81093-4.

Corcoran, J.A., Sonntag, Y., Andersson, M.N., Johanson, U., Löfstedt, C., 2018. Endogenous insensitivity to the Orco agonist VUAA1 reveals novel olfactory receptor complex properties in the specialist fly *Mayetiola destructor*. Sci. Rep. 8 (1). doi:10.1038/s41598-018-21631-3.

Croset, V., Rytz, R., Cummins, S.F., Budd, A., Brawand, D., Kaessmann, H., Gibson, T.J., Benton, R., 2010. Ancient protostome origin of chemosensory ionotropic glutamate receptors and the evolution of insect taste and olfaction. PLos Genet. 6 (8). doi:10.1371/journal.pgen.1001064.

Dauwalder, B., Tsujimoto, S., Moss, J., Mattox, W., 2002. The *Drosophila* takeout gene is regulated by the somatic sex-determination pathway and affects male courtship behavior. Genes Dev. 16 (22), 2879–2892. https://doi.org/10.1101/gad.1010302.

DeGennaro, M., McBride, C.S., Seeholzer, L., Nakagawa, T., Dennis, E.J., Goldman, C., Jasinskiene, N., James, A.A., Vosshall, L.B., 2013. Orco mutant mosquitoes lose strong preference for humans and are not repelled by volatile DEET. Nature 498 (7455), 487–491. https://doi.org/10.1038/nature12206.

Dennis, E.J., Goldman, O.V., Vosshall, L.B., 2019. *Aedes aegypti* mosquitoes use their legs to sense DEET on contact. Curr. Biol. 29 (9), 1551–1556.e5. https://doi.org/10.1016/j.cub.2019.04.004.

Dobryszycki, P., Kołodziejczyk, R., Krowarsch, D., Gapiński, J., Ozyhar, A., Kochman, M., 2004. Unfolding and refolding of juvenile hormone binding protein. Biophys. J. 86 (2), 1138–1148. https://doi.org/10.1016/S0006-3495(04)74188-0.

Dong, K., Sun, L., Liu, J.T., Gu, S.H., Zhou, J.J., Yang, R.N., Dhiloo, K.H., Gao, X.W., Guo, Y.Y., Zhang, Y.J., 2017. RNAi-induced electrophysiological and behavioral changes reveal two pheromone binding proteins of *Helicoverpa armigera* Involved in the perception of the main sex pheromone component Z11–16:Ald. J. Chem. Ecol. 43 (2), 207–214. https://doi.org/10.1007/s10886-016-0816-6.

Drakou, C.E., Tsitsanou, K.E., Potamitis, C., Fessas, D., Zervou, M., Zographos, S.E., 2017. The crystal structure of the AgamOBP1a: Icaridin complex reveals alternative binding modes and stereo-selective repellent recognition. Cell. Mol. Life Sci 74, 319–33810.

Du, E.J., Ahn, T.J., Choi, M.S., Kwon, I., Kim, H.W., Kwon, J.Y., Kang, K.J., 2015. The mosquito repellent citronellal directly potentiates *Drosophila* TRPA1, facilitating feeding suppression. Mol. Cells 38 (10), 911–917. https://doi.org/10.14348/molcells.2015.0215.

Du, G., Prestwich, G.D., 1995. Protein structure encodes the ligand binding specificity in pheromone binding proteins. Biochemistry 34 (27), 8726–8732. https://doi.org/10.1021/bi00027a023.

Dupas, S., Neiers, F., Granon, E., Rougeux, E., Dupont, S., Beney, L., Bousquet, F., Abdul Shaik, H., Briand, L., Wojtasek, H., Charles, J.P., 2020. Collisional mechanism of ligand release by *Bombyx mori* JHBP, a member of the TULIP/Takeout family of lipid transporters. Insect Biochem. Mol. Biol. 117. doi:10.1016/j.ibmb.2019.103293.

Eliash, N., Singh, N.K., Thangarajan, S., Sela, N., Leshkowitz, D., Kamer, Y., Zaidman, I., Rafaeli, A., Soroker, V., 2017. Chemosensing of honeybee parasite, *Varroa destructor*: Transcriptomic analysis. Sci. Rep. 7 (1). doi:10.1038/s41598-017-13167-9.

Ellingson, S.R., Smith, J.C., Baudry, J., 2013. VinaMPI: facilitating multiple receptor high-throughput virtual docking on high-performance computers. J. Comput. Chem. 34 (25), 2212–2221. https://doi.org/10.1002/jcc.23367.

Flower, D.R., 1996. The lipocalin protein family: structure and function. Biochem. J. 318 (1), 1–14. https://doi.org/10.1042/bj3180001.

Fujikawa, K., Seno, K., Ozaki, M., 2006. A novel takeout-like protein expressed in the taste and olfactory organs of the blowfly, *Phormia regina*. FEBS J. 273 (18), 4311–4321. https://doi.org/10.1111/j.1742-4658.2006.05422.x.

Fukuda, H., Arai, M., Kuwajima, K., 2000. Folding of green fluorescent protein and the Cycle3 mutant. Biochemistry 39 (39), 12025–12032. https://doi.org/10.1021/bi000543l.

Gao, Q., Chess, A., 1999. Identification of candidate *Drosophila* olfactory receptors from genomic DNA sequence. Genomics 60 (1), 31–39. https://doi.org/10.1006/geno.1999.5894.

Gong, Y., Pace, T.C.S., Castillo, C., Bohne, C., O'Neill, M.A., Plettner, E., 2009. Ligand-interaction kinetics of the pheromone-binding protein from the Gypsy Moth, *L. dispar*: insights into the mechanism of binding and release. Chem. Biol. 16 (2), 162–172. https://doi.org/10.1016/j.chembiol.2009.01.005.

Grant, A.J., Dickens, J.C., 2011. Functional characterization of the octenol receptor neuron on the maxillary palps of the yellow fever mosquito, *Aedes aegypti*. PLoS One 6 (6). doi:10.1371/journal.pone.0021785.

Gulia-Nuss, M., Nuss, A.B., Meyer, J.M., Sonenshine, D.E., Roe, R.M., Waterhouse, R.M., Sattelle, D.B., Fuente, Ribeiro, M., J., Megy, K., Thimmapuram, J., Miller, J.R., Walenz, B.P., Koren, S., Hostetler, J.B., Thiagarajan, M., Joardar, V.S., Hannick, L.I., Bidwell, S., Birren, B., 2016. Genomic insights into the *Ixodes scapularis* tick vector of Lyme disease. Nature Comm. 7, 10507. doi:10.1038/ncomms10507.

Guo, W., Wang, X., Ma, Z., Xue, L., Han, J., Yu, D., Kang, L., 2011. CSP and takeout genes modulate the switch between attraction and repulsion during behavioral phase change in the migratory locust. PLos Genet. 7 (2). doi:10.1371/journal.pgen.1001291.

Gwosch, K.C., Pape, J.K., Balzarotti, F., Hoess, P., Ellenberg, J., Ries, J., Hell, S.W., 2020. MINFLUX nanoscopy delivers 3D multicolor nanometer resolution in cells. Nat. Methods 17 (2), 217–224. https://doi.org/10.1038/s41592-019-0688-0.

Hagai, T., Cohen, M., Bloch, G., 2007. Genes encoding putative takeout/juvenile hormone binding proteins in the honeybee (*Apis mellifera*) and modulation by age and juvenile hormone of the takeout-like gene GB19811. Insect Biochem. Mol. Biol. 37 (7), 689–701. https://doi.org/10.1016/j.ibmb.2007.04.002.

Hamiaux, C., Basten, L., Greenwood, D.R., Baker, E.N., Newcomb, R.D., 2013. Ligand promiscuity within the internal cavity of *Epiphyas postvittana* Takeout 1 protein. J. Struct. Biol. 182 (3), 259–263. https://doi.org/10.1016/j.jsb.2013.03.013.

Honson, N., Johnson, M.A., Oliver, J.E., Prestwich, G.D., Plettner, E., 2003. Structure-activity studies with pheromone-binding proteins of the gypsy moth, *Lymantria dispar*. Chem. Senses 28 (6), 479–489. https://doi.org/10.1093/chemse/28.6.479.

Horst, R., Damberger, F., Luginbuhl, P., Guntert, P., Peng, G., Nikonova, L., Leal, W.S., Wuthrich, K., 2001. NMR structure reveals intramolecular regulation mechanism for pheromone binding and release. Proc. Natl. Acad. Sci., 14374–14379. doi:10.1073/pnas.251532998.

Huff, R.M., Pitts, R.J., 2020. Carboxylic acid responses by a conserved odorant receptor in culicine vector mosquitoes. Insect Mol. Biol. 29 (6), 523–530. https://doi.org/10.1111/imb.12661.

Hull, J.J., Yang, Y.W., Miyasaki, K., Brent, C.S., 2020. TRPA1 modulates noxious odor responses in *Lygus hesperus*. J. Insect Physiol. 122. doi:10.1016/j.jinsphys.2020.104038.

Iovinella, I., Dani, F.R., Niccolini, A., Sagona, S., Michelucci, E., Gazzano, A., Turillazzi, S., Felicioli, A., Pelosi, P., 2011. Differential expression of odorant-binding proteins in the mandibular glands of the honey bee according to caste and age. J. Proteome Res. 10 (8), 3439–3449. https://doi.org/10.1021/pr2000754.

Ishida, Y., Tsuchiya, W., Fujii, T., Fujimoto, Z., Miyazawa, M., Ishibashi, J., Matsuyama, S., Ishikawa, Y., & Yamazaki, T. (2014). Niemann-Pick type C2 protein mediating chemical communication in the worker ant. Proceedings of the National Academy of Sciences of the United States of America, 111(10), 3847–3852. https://doi.org/10.1073/pnas.1323928111.

Jeong, Y.T., Shim, J., Oh, S.R., Yoon, H.I., Kim, C.H., Moon, S.J., Montell, C., 2013. An odorant-binding protein required for suppression of sweet taste by bitter chemicals. Neuron 79 (4), 725–737. https://doi.org/10.1016/j.neuron.2013.06.025.

Jones, P. L., Pask, G. M., Rinker, D. C., & Zwiebel, L. J. (2011). Functional agonism of insect odorant receptor ion channels. Proceedings of the National Academy of Sciences of the United States of America, 108(21), 8821–8825. https://doi.org/10.1073/pnas.1102425108.

Jones, P.L., Pask, G.M., Romaine, I.M., Taylor, R.W., Reid, P.R., Waterson, A.G., Sulikowski, G.A., Zwiebel, L.J., 2012. Allosteric antagonism of insect odorant receptor ion channels. PLoS One 7 (1). doi:10.1371/journal.pone.0030304.

Joseph, R.M., Carlson, J.R., 2015. *Drosophila* chemoreceptors: a molecular interface between the chemical world and the brain. Trends Genet. 31 (12), 683–695. https://doi.org/10.1016/j.tig.2015.09.005.

Kaissling, K.E., 2009. Olfactory perireceptor and receptor events in moths: a kinetic model revised. Journal of Comparative Physiology A: Neuroethology, Sensory, Neural, and Behavioral Physiology 195 (10), 895–922. https://doi.org/10.1007/s00359-009-0461-4.

Kanaujia, S., Kaissling, K.E., 1985. Interactions of pheromone with moth antennae: adsorption, desorption and transport. J. Insect Physiol. 31 (1), 71–81. https://doi.org/10.1016/0022-1910(85)90044-7.

Katti, S., Lokhande, N., González, D., Cassill, A., Renthal, R., 2013. Quantitative analysis of pheromone-binding protein specificity. Insect Mol. Biol. 22 (1), 31–40. https://doi.org/10.1111/j.1365-2583.2012.01167.x.

Keil, T.A., 1982. Contacts of pore tubules and sensory dendrites in antennal chemosensilla of a silkmoth: demonstration of a possible pathway for olfactory molecules. Tissue Cell 14 (3), 451–462. https://doi.org/10.1016/0040-8166(82)90039-8.

Keil, T.A., Steinbrecht, R.A., 1987. Diffusion barriers in silkmoth sensory epithelia: application of lanthanum tracer to olfactory sensilla of *Antheraea polyphemus* and *Bombyx mori*. Tissue Cell 19 (1), 119–134. https://doi.org/10.1016/0040-8166(87)90063-2.

Kim, H., Kim, H., Kwon, J.Y., Seo, J.T., Shin, D.M., Moon, S.J., 2018. *Drosophila* Gr64e mediates fatty acid sensing via the phospholipase C pathway. PLos Genet. 14 (2). doi:10.1371/journal.pgen.1007229.

Kolodziejczyk, R., Bujacz, G., Jakób, M., Ozyhar, A., Jaskolski, M., Kochman, M., 2008. Insect juvenile hormone binding protein shows ancestral fold present in human lipid-binding proteins. J. Mol. Biol. 377 (3), 870–881. https://doi.org/10.1016/j.jmb.2008.01.026.

Kramer, K.J., Sanburg, L.L., Kezdy, F.J., Law, J.H., 1974. The juvenile hormone binding protein in the hemolymph of *Manduca sexta* Johannson (Lepidoptera:Sphingidae). Proc. Natl. Acad. Sci. USA 71 (2), 493–497. https://doi.org/10.1073/pnas.71.2.493.

Kröber, T., Koussis, K., Bourquin, M., Tsitoura, P., Konstantopoulou, M., Awolola, T.S., Dani, F.R., Qiao, H., Pelosi, P., Iatrou, K., Guerin, P.M., 2018. Odorant-binding protein-based identification of natural spatial repellents for the African malaria mosquito *Anopheles gambiae*. Insect Biochem. Mol. Biol. 96, 36–50. https://doi.org/10.1016/j.ibmb.2018.03.008.

Kwon, Y., Kim, S.H., Ronderos, D.S., Lee, Y., Akitake, B., Woodward, O.M., Guggino, W.B., Smith, D.P., Montell, C., 2010. *Drosophila* TRPA1 channel is required to avoid the naturally occurring insect repellent citronellal. Curr. Biol. 20 (18), 1672–1678. https://doi.org/10.1016/j.cub.2010.08.016.

Lagarde, A., Spinelli, S., Tegoni, M., He, X., Field, L., Zhou, J.J., Cambillau, C., 2011. The crystal structure of odorant binding protein 7 from *Anopheles gambiae* exhibits an outstanding adaptability of its binding site. J. Mol. Biol. 414 (3), 401–412. https://doi.org/10.1016/j.jmb.2011.10.005.

Lapointe, T.K., Altier, C., 2011. The role of TRPA1 in visceral inflammation and pain. Channels 5 (6), 525–529. https://doi.org/10.4161/chan.5.6.18016.

Larter, N.K., Sun, J.S., Carlson, J.R., 2016. Organization and function of *Drosophila* odorant binding proteins. ELife 5. doi:10.7554/eLife.20242.

Lartigue, A., Campanacci, V., Roussel, A., Larsson, A.M., Alwyn Jones, T., Tegoni, M., Cambillau, C, 2002. X-ray structure and ligand binding study of a moth chemosensory protein. J. Biol. Chem. 277 (35), 32094–32098. https://doi.org/10.1074/jbc.M204371200.

Laughlin, J.D., Ha, T.S., Jones, D.N.M., Smith, D.P, 2008. Activation of pheromone-sensitive neurons is mediated by conformational activation of pheromone-binding protein. Cell 133 (7), 1255–1265. https://doi.org/10.1016/j.cell.2008.04.046.

Lautenschlager, C., Leal, W.S., Clardy, J., 2007. *Bombyx mori* pheromone-binding protein binding nonpheromone ligands: implications for pheromone recognition. Structure 15 (9), 1148–1154. https://doi.org/10.1016/j.str.2007.07.013.

Leal, W.S., 2005. Pheromone reception. The Chemistry of Pheromones and Other Semiochemicals II. In: Schulz, S. (Ed.), Topics in Current Chemistry, 240, Springer, pp. 1–36.

Leal, W. S., Chen, A. M., Ishida, Y., Chiang, V. P., Erickson, M. L., Morgan, T. I., & Tsuruda, J. M. (2005). Kinetics and molecular properties of pheromone binding and release. Proceedings of the National Academy of Sciences of the United States of America, 102(15), 5386–5391. https://doi.org/10.1073/pnas.0501447102.

Lee, M.J., Sung, H.Y., Jo, H., Kim, H.W., Choi, M.S., Kwon, J.Y., Kang, K.J., 2017. Ionotropic receptor 76b is required for gustatory aversion to excessive Na^+ in *Drosophila*. Mol. Cells 40 (10), 787–795. https://doi.org/10.14348/molcells.2017.0160.

Lee, Y., Kim, S.H., Montell, C., 2010. Avoiding DEET through insect gustatory receptors. Neuron 67 (4), 555–561. https://doi.org/10.1016/j.neuron.2010.07.006.

Lerro, K.A., Prestwich, G.D., 1990. Cloning and sequencing of a cDNA for the hemolymph juvenile hormone binding protein of larval *Manduca sexta*. J. Biol. Chem. 265 (32), 19800–19806.

Leung, N.Y., Thakur, D.P., Gurav, A.S., Kim, S.H., Di Pizio, A., Niv, M.Y., Montell, C., 2020. Functions of opsins in *Drosophila* taste. Curr. Biol. 30 (8), 1367–1379. e6. https://doi.org/10.1016/j.cub.2020.01.068.

Li, J., Zhang, L., Wang, X., 2016. An odorant-binding protein involved in perception of host plant odorants in locust *Locusta migratoria*. Arch. Insect. Biochem. Physiol. 91 (4), 221–229. https://doi.org/10.1002/arch.21319.

Li, S., Picimbon, J.F., Ji, S., Kan, Y., Chuanling, Q., Zhou, J.J., Pelosi, P., 2008. Multiple functions of an odorant-binding protein in the mosquito *Aedes aegypti*. Biochem. Biophys. Res. Commun. 372 (3), 464–468. https://doi.org/10.1016/j.bbrc.2008.05.064.

Liao, M., Cao, E., Julius, D., Cheng, Y., 2013. Structure of the TRPV1 ion channel determined by electron cryo-microscopy. Nature 504 (7478), 107–112. https://doi.org/10.1038/nature12822.

Lin, X., Zhang, L., Jiang, Y., 2017a. Characterization of Spodoptera litura (Lepidoptera: Noctuidae) takeout genes and their differential responses to insecticides and sex pheromone. Journal of Insect Science 17 (4). doi:10.1093/jisesa/iex061.

Lin, X., Zhang, L., Jiang, Y., 2017b. Distinct roles of met and interacting proteins on the expressions of takeout family genes in brown planthopper. Frontiers in Physiology 8. doi:10.3389/fphys.2017.00100.

Ling, F., Dahanukar, A., Weiss, L.A., Kwon, J.Y., Carlson, J.R., 2014. The molecular and cellular basis of taste coding in the legs of *Drosophila*. J. Neurosci. 34 (21), 7148–7164. https://doi.org/10.1523/JNEUROSCI.0649-14.2014.

Liu, Z., Liang, X.F., Xu, L., Keesey, I.W., Lei, Z.R., Smagghe, G., Wang, J.J., 2020. An antennae-specific odorant-binding protein is involved in *Bactrocera dorsalis* olfaction. Frontiers in Ecology and Evolution 8. doi:10.3389/fevo.2020.00063.

Lundin, C., Käll, L., Kreher, S.A., Kapp, K., Sonnhammer, E.L., Carlson, J.R., Heijne, G.v., Nilsson, I.M., 2007. Membrane topology of the *Drosophila* OR83b odorant receptor. FEBS Lett. 581 (29), 5601–5604. https://doi.org/10.1016/j.febslet.2007.11.007.

Maibeche-Coisne, M., Nikonov, A.A., Ishida, Y., Jacquin-Joly, E., Leal, W.S., 2004. Pheromone anosmia in a scarab beetle induced by in vivo inhibition of a pheromone-degrading enzyme. Proc. Natl. Acad. Sci., 11459–11464. doi:10.1073/pnas.0403537101.

Maitani, M.M., Allara, D.L., Park, K.C., Lee, S.G., Baker, T.C., 2010. Moth olfactory trichoid sensilla exhibit nanoscale-level heterogeneity in surface lipid properties. Arthropod Structure and Development 39 (1), 1–16. https://doi.org/10.1016/j.asd.2009.08.004.

Mao, Y., Xu, X., Xu, W., Ishida, Y., Leal, W.S., Ames, J.B., Clardy, J., 2010. Crystal and solution structures of an odorant-binding protein from the southern house mosquito complexed with an oviposition pheromone. PNAS 107 (44), 19102–19107. https://doi.org/10.1073/pnas.1012274107.

Martin, F., Alcorta, E., 2016. Measuring activity in olfactory receptor neurons in *Drosophila*: focus on spike amplitude. J. Insect Physiol. 95, 23–41. https://doi.org/10.1016/j.jinsphys.2016.09.003.

Meunier, N., Belgacem, Y.H., Martin, J.R., 2007. Regulation of feeding behaviour and locomotor activity by takeout in *Drosophila*. J. Exp. Biol. 210 (8), 1424–1434. https://doi.org/10.1242/jeb.02755.

Mohapatra, P., Menuz, K., 2019. Molecular profiling of the *Drosophila* antenna reveals conserved genes underlying olfaction in insects. G3. Genes, Genomes, Genetics 9 (11), 3753–3771. https://doi.org/10.1534/g3.119.400669.

Murphy, E.J., Booth, J.C., Davrazou, F., Port, A.M., Jones, D.N.M., 2013. Interactions of *Anopheles gambiae* odorant-binding proteins with a human-derived repellent: implications for the mode of action of N,N-diethyl-3-methylbenzamide (DEET). J. Biol. Chem. 288 (6), 4475–4485. https://doi.org/10.1074/jbc.M112.436386.

Narayan, K., Subramaniam, S., 2015. Focused ion beams in biology. Nat. Methods 12 (11), 1021–1031. https://doi.org/10.1038/nmeth.3623.

Ng, R., Salem, S.S., Wu, S.T., Wu, M., Lin, H.H., Shepherd, A.K., Joiner, W.J., Wang, J.W., Su, C.Y., 2019. Amplification of *Drosophila* olfactory responses by a DEG/ENaC channel. Neuron 104 (5), 947–959.e5. https://doi.org/10.1016/j.neuron.2019.08.041.

Norris, J.L., Caprioli, R.M., 2013. Analysis of tissue specimens by matrix-assisted laser desorption/ionization imaging mass spectrometry in biological and clinical research. Chem. Rev. 113 (4), 2309–2342. https://doi.org/10.1021/cr3004295.

Oldham, N.J., Krieger, J., Breer, H., Svatoš, A., 2001. Detection and removal of an artefact fatty acid from the binding site of recombinant *Bombyx mori* pheromone-binding protein. Chem. Senses 26 (5), 529–531. https://doi.org/10.1093/chemse/26.5.529.

Ozaki, M., Wada-Katsumat, A., Fujikawa, K., Iwasaki, M., Yokohari, F., Satoji, Y., Nisimura, T., Yamaoka, R., 2005. Ant nestmate and non-nestmate discrimination by a chemosensory sensillum. Science, 311–314. doi:10.1126/science.1105244.

Pelletier, J., Guidolin, A., Syed, Z., Cornel, A.J., Leal, W.S., 2010. Knockdown of a mosquito odorant-binding protein involved in the sensitive detection of oviposition attractants. J. Chem. Ecol. 36 (3), 245–248. https://doi.org/10.1007/s10886-010-9762-x.

Pelosi, P., Iovinella, I., Zhu, J., Wang, G., Dani, F.R., 2018. Beyond chemoreception: diverse tasks of soluble olfactory proteins in insects. Biological Reviews 93 (1), 184–200. https://doi.org/10.1111/brv.12339.

Pesenti, M.E., Spinelli, S., Bezirard, V., Briand, L., Pernollet, J.C., Campanacci, V., Tegoni, M., Cambillau, C., 2009. Queen bee pheromone binding protein pH-induced domain swapping favors pheromone release. J. Mol. Biol. 390 (5), 981–990. https://doi.org/10.1016/j.jmb.2009.05.067.

Plettner, E., Lazar, J., Prestwich, E.G., Prestwich, G.D., 2000. Discrimination of pheromone enantiomers by two pheromone binding proteins from the gypsy moth *Lymantria dispar*. Biochemistry 39 (30), 8953–8962. https://doi.org/10.1021/bi000461x.

Rebijith, K.B., Asokan, R., Hande, H.R., Kumar, N.K.K., Krishna, V., Vinutha, J., Bakthavatsalam, N., 2016. RNA interference of odorant-binding Protein 2 (OBP2) of the Cotton Aphid, *Aphis gossypii* (Glover), resulted in altered electrophysiological responses. Appl. Biochem. Biotechnol. 178 (2), 251–266. https://doi.org/10.1007/s12010-015-1869-7.

Renthal, R., Manghnani, L., Bernal, S., Qu, Y., Griffith, W.P., Lohmeyer, K., Guerrero, F.D., Borges, L.M.F., Pérez de León, A., 2017. The chemosensory appendage proteome of *Amblyomma americanum* (Acari: Ixodidae) reveals putative odorant-binding and other chemoreception-related proteins. Insect Sci. 24 (5), 730–742. https://doi.org/10.1111/1744-7917.12368.

Rinker, D.C., Jones, P.L., Pitts, R.J., Rutzler, M., Camp, G., Sun, L., Xu, P., Dorset, D.C., Weaver, D., Zwiebel, L.J., 2012. Novel high-throughput screens of *Anopheles gambiae* odorant receptors reveal candidate behaviour-modifying chemicals for mosquitoes. Physiol. Entomol. 37 (1), 33–41. https://doi.org/10.1111/j.1365-3032.2011.00821.x.

Riviere, S., Lartigue, A., Quennedey, B., Campanacci, V., Farine, J.P., Tegoni, M., Cambillau, C., Brossut, R., 2003. A pheromone-binding protein from the cockroach *Leucophaea maderae*: cloning, expression and pheromone binding. Biochem. J. 371, 573–910.

Robertson, H.M., 2015. The insect chemoreceptor superfamily is ancient in animals. Chem. Senses 40 (9), 609–614. https://doi.org/10.1093/chemse/bjv046.

Ronderos, D.S., Smith, D.P., 2009. Diverse signaling mechanisms mediate volatile odorant detection in *Drosophila*. Fly 3 (4). doi:10.4161/fly.9801.

Rund, S.S.C., Bonar, N.A., Champion, M.M., Ghazi, J.P., Houk, C.M., Leming, M.T., Syed, Z., Duffield, G.E, 2013. Daily rhythms in antennal protein and olfactory sensitivity in the malaria mosquito *Anopheles gambiae*. Sci. Rep. 3. doi:10.1038/srep02494.

Saheki, Y., Bian, X., Schauder, C.M., Sawaki, Y., Surma, M.A., Klose, C., Pincet, F., Reinisch, K.M., De Camilli, P., 2016. Control of plasma membrane lipid homeostasis by the extended synaptotagmins. Nat. Cell Biol. 18 (5), 504–515. https://doi.org/10.1038/ncb3339.

Sargsyan, V., Getahun, M.N., Llanos, S.L., Olsson, S.B., Hansson, B.S., Wicher, D., 2011. Phosphorylation via PKC regulates the function of the *Drosophila* odorant co-receptor. Frontiers in Cellular Neuroscience. doi:10.3389/fncel.2011.00005.

Sarov-Blat, L., So, W.V., Liu, L., Rosbash, M., 2000. The *Drosophila* takeout gene is a novel molecular link between circadian rhythms and feeding behavior. Cell 101 (6), 647–656. https://doi.org/10.1016/S0092-8674(00)80876-4.

Sato, K., Pellegrino, M., Nakagawa, T., Nakagawa, T., Vosshall, L.B., Touhara, K., 2008. Insect olfactory receptors are heteromeric ligand-gated ion channels. Nature 452 (7190), 1002–1006. https://doi.org/10.1038/nature06850.

Saurabh, S., Vanaphan, N., Wen, W., Dauwalder, B., 2018. High functional conservation of takeout family members in a courtship model system. PLoS One 13 (9). doi:10.1371/journal.pone.0204615.

Scheuermann, E.A., Smith, D.P., 2019. Odor-specific deactivation defects in a *Drosophila* odorant-binding protein mutant. Genetics 213 (3), 897–909. https://doi.org/10.1534/genetics.119.302629.

Schuhmacher, M., Grasskamp, A.T., Barahtjan, P., Wagner, N., Lombardot, B., Schuhmacher, J.S., Sala, P., Lohmann, A., Henry, I., Shevchenko, A., Coskun, Ü., Walter, A.M., Nadler, A., 2020. Live-cell lipid biochemistry reveals a role of diacylglycerol side-chain composition for cellular lipid dynamics and protein affinities. PNAS 117 (14), 7729–7738. https://doi.org/10.1073/pnas.1912684117.

Schwinghammer, M.A., Zhou, X., Kambhampati, S., Bennett, G.W., Scharf, M.E., 2011. A novel gene from the takeout family involved in termite trail-following behavior. Gene 474 (1–2), 12–21. https://doi.org/10.1016/j.gene.2010.11.012.

Shah, J.S., Renthal, R., 2020. Antennal proteome of *Solenopsis invicta* (Hymenoptera: Formicidae): caste differences in olfactory receptors and chemosensory support proteins. Journal of Insect Science 20 (5), 1–13. https://doi.org/10.1093/jisesa/ieaa118.

Shiao, M.S., Fan, W.L., Fang, S., Lu, M.Y.J., Kondo, R., Li, W.H., 2013. Transcriptional profiling of adult *Drosophila* antennae by high-throughput sequencing. Zoological Studies 52 (1). doi:10.1186/1810-522X-52-42.

So, W.V., Sarov-Blat, L., Kotarski, C.K., McDonald, M.J., Allada, R., Rosbash, M., 2000. Takeout, a novel *Drosophila* gene under circadian clock transcriptional regulation. Mol. Cell. Biol. 6935–6944. doi:10.1128/MCB.20.18.6935-6944.2000.

Sonenshine, D.E., Taylor, D., Carson, K.A., 1986. Chemically mediated behavior in Acari: adapations for finding hosts and mates. J. Chem. Ecol. 12 (5), 1091–1108. https://doi.org/10.1007/BF01638998.

Sparks, J.T., Dickens, J.C., 2016. Bitter-sensitive gustatory receptor neuron responds to chemically diverse insect repellents in the common malaria mosquito *Anopheles quadrimaculatus*. Science of Nature 103 (5). doi:10.1007/s00114-016-1367-y.

Sparks, J.T., Dickens, J.C., 2017. Mini review: gustatory reception of chemicals affecting host feeding in aedine mosquitoes. Pestic. Biochem. Physiol. 142, 15–20. https://doi.org/10.1016/j.pestbp.2016.12.009.

Steinbrecht, R.A., 1992. Experimental morphology of insect olfaction: tracer studies, Xray microanalysis, autoradiography, and immunocytochemistry with silkmoth antennae. Microsc. Res. Tech. 22 (4), 336–350. https://doi.org/10.1002/jemt.1070220404.

Steinbrecht, R.A., 1997. Pores structures in insect olfactory sensilla: a review of data and concepts. Int. J. Insect

Morphol. Embryol. 26 (3–4), 229–245. https://doi.org/10.1016/S0020-7322(97)00024-X.

Steinbrecht, R.A., Ozaki, M., Ziegelberger, G., 1992. Immunocytochemical localization of pheromone-binding protein in moth antennae. Cell Tissue Res. 270 (2), 287–302. https://doi.org/10.1007/BF00328015.

Suzuki, R., Fujimoto, Z., Shiotsuki, T., Tsuchiya, W., Momma, M., Tase, A., Miyazawa, M., Yamazaki, T., 2011. Structural mechanism of JH delivery in hemolymph by JHBP of silkworm, *Bombyx mori*. Sci. Rep. 1. doi:10.1038/srep00133.

Swarup, S., Morozova, T.V., Sridhar, S., Nokes, M., Anholt, R.R.H, 2014. Modulation of feeding behavior by odorant-binding proteins in *Drosophila melanogaster*. Chem. Senses 39 (2), 125–132. https://doi.org/10.1093/chemse/bjt061.

Swarup, S., Williams, T.I., Anholt, R.R.H, 2011. Functional dissection of Odorant binding protein genes in *Drosophila melanogaster*. Genes, Brain and Behavior 10 (6), 648–657. https://doi.org/10.1111/j.1601-183X.2011.00704.x.

Syed, Z., Pelletier, J., Flounders, E., Chitolina, R.F., Leal, W.S., 2011. Generic insect repellent detector from the fruit fly *Drosophila melanogaster*. PLoS One 6 (3). https://doi.org/10.1371/journal.pone.0017705.

Taylor, R.W., Romaine, I.M., Liu, C., Murthi, P., Jones, P.L., Waterson, A.G., Sulikowski, G.A., Zwiebel, L.J., 2012. Structure-activity relationship of a broad-spectrum insect odorant receptor agonist. ACS Chem. Biol. 7 (10), 1647–1652. https://doi.org/10.1021/cb300331z.

Thireou, T., Kythreoti, G., Tsitsanou, K.E., Koussis, K., Drakou, C.E., Kinnersley, J., Kröber, T., Guerin, P.M., Zhou, J.J., Iatrou, K., Eliopoulos, E., Zographos, S.E., 2018. Identification of novel bioinspired synthetic mosquito repellents by combined ligand-based screening and OBP-structure-based molecular docking. Insect Biochem. Mol. Biol. 98, 48–61. https://doi.org/10.1016/j.ibmb.2018.05.001.

Torres-Oliva, M., Almeida, F.C., Sánchez-Gracia, A., Rozas, J., 2016. Comparative genomics uncovers unique gene turnover and evolutionary rates in a gene family involved in the detection of insect cuticular pheromones. Genome Biology and Evolution 8 (6), 1734–1747. https://doi.org/10.1093/gbe/evw108.

Tsitoura, P., Koussis, K., Iatrou, K., 2015. Inhibition of *Anopheles gambiae* odorant receptor function by mosquito repellents. J. Biol. Chem. 290 (12), 7961–7972. https://doi.org/10.1074/jbc.M114.632299.

Tsitsanou, K.E., Thireou, T., Drakou, C.E., Koussis, K., Keramioti, M.V., Leonidas, D.D., Eliopoulos, E., Iatrou, K., Zographos, S.E., 2012. *Anopheles gambiae* odorant binding protein crystal complex with the synthetic repellent DEET: Implications for structure-based design of novel mosquito repellents. Cell. Mol. Life Sci. 69 (2), 283–297. https://doi.org/10.1007/s00018-011-0745-z.

Tsuchihara, K., Hisatomi, O., Tokunaga, F., Asaoka, K., 2009. An oviposition stimulant binding protein in a butterfly: immunohistochemical localization and electrophysiological responses to plant compounds. Communicative and Integrative Biology 2 (4), 356–358. https://doi.org/10.4161/cib.2.4.8613.

Tsuchihara, K., Ueno, K., Yamanaka, A., Isono, K., Endo, K., Nishida, R., Yoshihara, K., Tokunaga, F., 2000. A putative binding protein for lipophilic substances related to butterfly oviposition. FEBS Lett. 478 (3), 299–303. https://doi.org/10.1016/S0014-5793(00)01838-X.

Vanaphan, N., Dauwalder, B., Zufall, R.A., 2012. Diversification of takeout, a male-biased gene family in *Drosophila*. Gene 491 (2), 142–148. https://doi.org/10.1016/j.gene.2011.10.003.

Vance, S.J., McDonald, R.E., Cooper, A., Smith, B.O., Kennedy, M.W., 2013. The structure of latherin, a surfactant allergen protein from horse sweat and saliva. J. R. Soc., Interface 10 (85). doi:10.1098/rsif.2013.0453.

Vizueta, J., Rozas, J., Sánchez-Gracia, A., 2018. Comparative genomics reveals thousands of novel chemosensory genes and massive changes in chemoreceptor repertories across chelicerates. Genome Biology and Evolution 10 (5), 1221–1236. https://doi.org/10.1093/gbe/evy081.

Vogt, R.G., Riddiford, L.M., 1981. Pheromone binding and inactivation by moth antennae. Nature 293 (5828), 161–163. https://doi.org/10.1038/293161a0.

Vosshall, L.B., Amrein, H., Morozov, P.S., Rzhetsky, A., Axel, R., 1999. A spatial map of olfactory receptor expression in the *Drosophila* antenna. Cell 96 (5), 725–736. https://doi.org/10.1016/S0092-8674(00)80582-6.

Vosshall, L.B., Stocker, R.F., 2007. Molecular architecture of smell and taste in *Drosophila*. Annu. Rev. Neurosci. 30, 505–533. https://doi.org/10.1146/annurev.neuro.30.051606.094306.

Walton, W.G., Ahmad, S., Little, M.S., Kim, C.S.K., Tyrrell, J., Lin, Q., Di, Y.P., Tarran, R., Redinbo, M.R, 2016. Structural features essential to the antimicrobial functions of human SPLUNC1. Biochemistry 55 (21), 2979–2991. https://doi.org/10.1021/acs.biochem.6b00271.

Wicher, D., Schäfer, R., Bauernfeind, R., Stensmyr, M.C., Heller, R., Heinemann, S.H., Hansson, B.S., 2008. *Drosophila* odorant receptors are both ligand-gated and cyclic-nucleotide-activated cation channels. Nature 452 (7190), 1007–1011. https://doi.org/10.1038/nature06861.

Willett, C.S., Harrison, R.G., 1999. Pheromone binding proteins in the European and Asian corn borers: no protein change associated with pheromone differences. Insect Biochem. Mol. Biol. 29 (3), 277–284. doi:10.1016/S0965-1748(99)00003-X.

Wojtasek, H., Hansson, B.S., Leal, W.S., 1998. Attracted or repelled?—a matter of two neurons, one pheromone binding protein, and a chiral center. Biochem. Biophys. Res. Commun. 250 (2), 217–222. https://doi.org/10.1006/bbrc.1998.9278.

Wong, L.H., Levine, T.P., 2017. Tubular lipid binding proteins (TULIPs) growing everywhere. Biochimica et Biophysica Acta—Molecular Cell Research 1864 (9), 1439–1449. https://doi.org/10.1016/j.bbamcr.2017.05.019.

Xiao, S., Sun, J.S., Carlson, J.R., 2019. Robust olfactory responses in the absence of odorant binding proteins. ELife, 8. doi:10.7554/eLife.51040.

Xu, P., Zeng, F., Bedoukian, R.H., Leal, W.S., 2019. DEET and other repellents are inhibitors of mosquito odorant receptors for oviposition attractants. Insect Biochem. Mol. Biol. 113. doi:10.1016/j.ibmb.2019.103224.

Xu, W., Cornel, A.J., Leal, W.S., 2010. Odorant-binding proteins of the malaria mosquito *Anopheles funestus sensu stricto*. PLoS One 5 (10). doi:10.1371/journal.pone.0015403.

Yao, C.A., Carlson, J.R., 2010. Role of G-proteins in odor-sensing and CO2-sensing neurons in *Drosophila*. J. Neurosci., 4562–4572. doi:10.1523/JNEUROSCI.6357-09.2010.

Yew, J.Y., Dreisewerd, K., Luftmann, H., Müthing, J., Pohlentz, G., Kravitz, E.A., 2009. A new male sex pheromone and novel cuticular cues for chemical communication in *Drosophila*. Curr. Biol. 19 (15), 1245–1254. https://doi.org/10.1016/j.cub.2009.06.037.

Younas, A., Waris, M.I., Ul Qamar, M.T., Shaaban, M., Prager, S.M., Wang, M.Q., 2018. Functional analysis of the chemosensory protein MsepCSP8 from the oriental armyworm *Mythimna separata*. Frontiers in Physiology 9. doi:10.3389/fphys.2018.00872.

Zhang, L., Guo, M., Zhuo, F., Xu, H., Zheng, N., Zhang, L., 2019. An odorant-binding protein mediates sexually dimorphic behaviors via binding male-specific 2-heptanone in migratory locust. J. Insect Physiol. 118. doi:10.1016/j.jinsphys.2019.103933.

Zhang, X.Y., Zhu, X.Q., Gu, S.H., Zhou, Y.L., Wang, S.Y., Zhang, Y.J., Guo, Y.Y., 2017. Silencing of odorant binding protein gene AlinOBP4 by RNAi induces declining electrophysiological responses of *Adelphocoris lineolatus* to six semiochemicals. Insect Sci. 24 (5), 789–797. https://doi.org/10.1111/1744-7917.12365.

Zhu, J., Guo, M., Ban, L., Song, L.M., Liu, Y., Pelosi, P., Wang, G., 2018. Niemann-pick C2 proteins: a new function for an old family. Front. Physiol. 9. doi:10.3389/fphys.2018.00052.

CHAPTER

10

Semifield system and experimental huts bioassays for the evaluation of spatial (and topical) repellents for indoor and outdoor use

Mgeni Mohamed Tambwe[a,b,c], *Johnson Kyeba Swai*[a], *Sarah Jane Moore*[a,b,c,d]

[a]Vector Control Product Testing Unit, Ifakara Health Institute, Environmental Health, and Ecological Sciences, Bagamoyo, Tanzania, [b]Vector Biology Unit, Swiss Tropical and Public Health Institute, Basel, Switzerland, [c]University of Basel, Basel, Switzerland, [d]Nelson Mandela African Institute of Science and Technology (NM-AIST), Tengeru, Tanzania

10.1 Introduction

Vector-borne diseases account for more than 17% of all infectious diseases, causing more than 700,000 deaths annually (World Health Organization, 2021). By far the most prevalent vector-borne diseases are malaria with 219 million cases 40,000 deaths and dengue with 3.9 billion cases and 40,000 deaths, annually (World Health Organization, 2021). Both of these diseases are transmitted by mosquitoes: a number of Anopheles species transmit malaria, whereas *Ae. aegypti* and *Ae. albopictus* are the principal vectors of the most common arboviruses including dengue, chikungunya, Zika, and yellow fever.

For public health, insecticides applied on surfaces (Indoor Residual Spray, IRS)/fabrics (Insecticide Treated Nets, ITNs) where vectors regularly rest or feed are used for control of adult vectors while modification of the environment (larval source management) is often deployed to control the aquatic stages (World Health Organization, 2016). IRS and ITNs have been widely used and have proved effective, in controlling malaria transmitted by indoor biting and resting vectors (Alonso et al., 2017). However, these interventions do not protect individuals against mosquito species that bite outside of sleeping hours, that bite outdoors, or that rest outdoors. Larval source management is effective against indoor and outdoor biting mosquitoes and is

Advances in Arthropod Repellents.
DOI: https://doi.org/10.1016/B978-0-323-85411-5.00011-X

recommended for community control of malaria (World Health Organization, 2019) and dengue (World Health Organization, 2012). In addition, the use of repellents and long clothing for personal protection against mosquito bites is recommended (World Health Organization, 2016).

The changing epidemiology of malaria and the global growth of dengue is increasing the importance of bite prevention techniques for the control vector-borne disease. Due to the enormous success in malaria control (World Health Organization, 2020), malaria is now increasingly focal and often clustered in subpopulations with similar social, behavioral, and geographical risk characteristics (Cotter et al., 2013) such as migrants (Kounnavong et al., 2017), forest workers (Sandfort et al., 2020), and people who work outdoors at night (Monroe et al., 2019). In areas where vectors bite in the evening hours and rest outdoors peridomestic malaria transmission often occurs (Lana et al., 2021).

Arboviruses are likely to cause the majority of the vector-borne diseases in the 21st century. These viruses are a growing threat worldwide due to the geographic expansion of vectors and viruses through globalization and urbanization (Brady and Hay, 2020). The mosquito *Ae. aegypti* is the primary tropical vector and has evolved to mate, feed, rest, and lay eggs around urban human habitations: and flourishes in urban environments closely associated with humans (Powell and Tabachnick, 2013). It is a daytime feeder and its peak biting periods early in the morning and before dusk in the evening. Female *Ae. aegypti* frequently bites multiple people during a single feeding period and dengue cases are often clustered and related to the presence of vectors (Liebman et al., 2012). The rapid spread of arboviral infections is a result of demographic and societal changes, importantly rural–urban migration leading to unplanned urban settlements and the introduction of viruses to new areas. Because the global urban population is set to rise to 5 billion by 2030 and land area with urban settlement to 1.2 million km^2 (Seto et al., 2012), it is unlikely that dengue will decline without sustained and effective control measures. This is of great concern, and in the absence of effective vaccines and as yet incomplete roll out of Wolbachia carrying mosquitoes, interventions that reduce contact between humans and vectors are currently the most effective methods of controlling arboviral diseases (Achee et al., 2015a).

The ability of a mosquito to locate a human-host and blood-fed successfully plays an important role in the transmission of disease pathogens. Mosquitoes that feed primarily on humans are the most efficient vectors of human pathogens (Wynne et al., 2020). These mosquitos detect and locate hosts principally through odorant cues released by hosts via their olfactory receptors that are located on their antennae, maxillary palps, and labellum (Takken, 1991; Takken and Knols, 1999). At long range, carbon dioxide signals the presence of the host (Gillies, 1980), and sensitizes mosquito responses to host cues at shorter range (Webster et al., 2015). Mosquitoes use host cues to orient toward hosts using odors that are generated by the decomposition of skin secretions by skin microbiota (Takken and Verhulst, 2017) that are reliable cues for human hosts (Verhulst et al., 2018), in combination with heat and water vapor at close range (Wright and Kellogg, 1962). Visual cues are important particularly in diurnal species (Muir et al., 1992). Advancements in neurobiology and studies of insect olfactory systems have led to the identification and development of numerous behaviorally active compounds that can attract (Smallegange et al., 2011; Okumu et al., 2010; Verhulst et al., 2010) and repel mosquitoes (Rinker et al., 2012; Carey et al., 2010). These compounds can be applied topically (on the skin), on fabric/clothing and or spatially (vapor phased from a point source).

While the effectiveness of topical repellents as public health tools is usually limited (Maia et al., 2018) because people often forget to regularly, or correctly apply them (Gryseels et al., 2015), they remain useful for at-risk populations such as the military (Beiter et al., 2019) and nonimmune travelers (Ahmed et al., 2020), and are recommended for bite prevention by the World Health Organization (World Health Organization, 2019). There is now a growing body of evidence that spatial repellents, in particular volatile pyrethroids, have the potential to provide effective protection against malaria in areas where there is early evening malaria transmission (Syafruddin et al., 2020; Syafruddin et al., 2014; Hill et al., 2014) and arboviruses (Morrison et al., 2021). Due to this proven public health benefit, there is now a renewed research agenda to evaluate new iterations of bite prevention tools. Of particular importance is the development of longer-lasting volatile pyrethroids that can protect multiple users for many weeks, ensuring compliance, because minimal lifestyle modification is needed for an individual to receive protection from these chemicals. Having representative bioassays that allow cost-effective and precise estimates of efficacy are an extremely important component of the product development pathway.

Methods for testing repellents that do not require a human host include (1) use of synthetic human odor baited traps (Salazar et al., 2013; Chauhan et al., 2012); (2) use of animals instead of human volunteers (Vatandoost and Hanafi-Bojd, 2008); (3) laboratory-based artificial blood feeding membrane systems (Debboun and Wagman, 2004); (4) olfactometer experiments (Bibbs et al., 2019); and (5) behavioral response screening systems (Thanispong et al., 2010). Although these methods are highly standardized, can rapidly screen compounds and do not involve human volunteers, the test conditions are not fully representative of what occurs in real-life settings when humans use repellents because these systems do not emit the complete suite of host cues that are important for mosquito landing such as heat and water (Ray, 2015). Additionally, some repellents such as N,N-diethyl-3-methylbenzamide (DEET), ethyl butylacetylaminoproprionate (IR3535), and 2-(2-hydroxyethyl)-1-piperidine carboxylic acid 1-methylpropylester (Picaridin) exert their olfactory mode of action primarily by decreasing the quantity of volatile odorants reaching the odorant receptor neurons and therefore, their repellency is observable only in the presence of host odors (Afify et al., 2019). On the other hand, full-field experiments normally run to verify findings from laboratory tests, use human volunteers who are exposed to potentially infective mosquito bites and should be conducted only once interventions are optimized because they have low throughput and are relatively expensive (Harrington et al., 2020). Therefore, the efficacy of repellents against human host-seeking insects requires a combination of laboratory and field tests (World Health Organization, 2009; WHOPES, 2013).

Using well-characterized bioassays in semifield systems (SFSs) (Ogoma et al., 2014b) and experimental huts (EHs) (Grieco et al., 2000), the efficacies of topical and spatial repellents can be more precisely evaluated against laboratory-reared and field mosquitoes (Fig. 10.1). These bioassays have been proven to have the advantage of helping us to understand the behavioral responses of mosquitoes exposed to the repellents (Smith, 1963), and how to best link data from laboratory tests to that from field tests (Vontas et al., 2014). Here we describe considerations for the design and implementation of experiments to measure the protective efficacy (PE) of repellents in controlled experiments carried out in SFS and EH in the field.

10.2 Semifield system and experimental hut for evaluating repellents

SFS experiments were developed to fill the gap between laboratory and field experiments for ecology studies (Ferguson et al., 2008) and for the evaluation of vector control tools such as indoor residual spraying (Silver and Service, 2008). They have proved extremely useful for the evaluation of behaviorally active odorants in mosquito traps (Schmied et al., 2008; Turner et al., 2011), topical

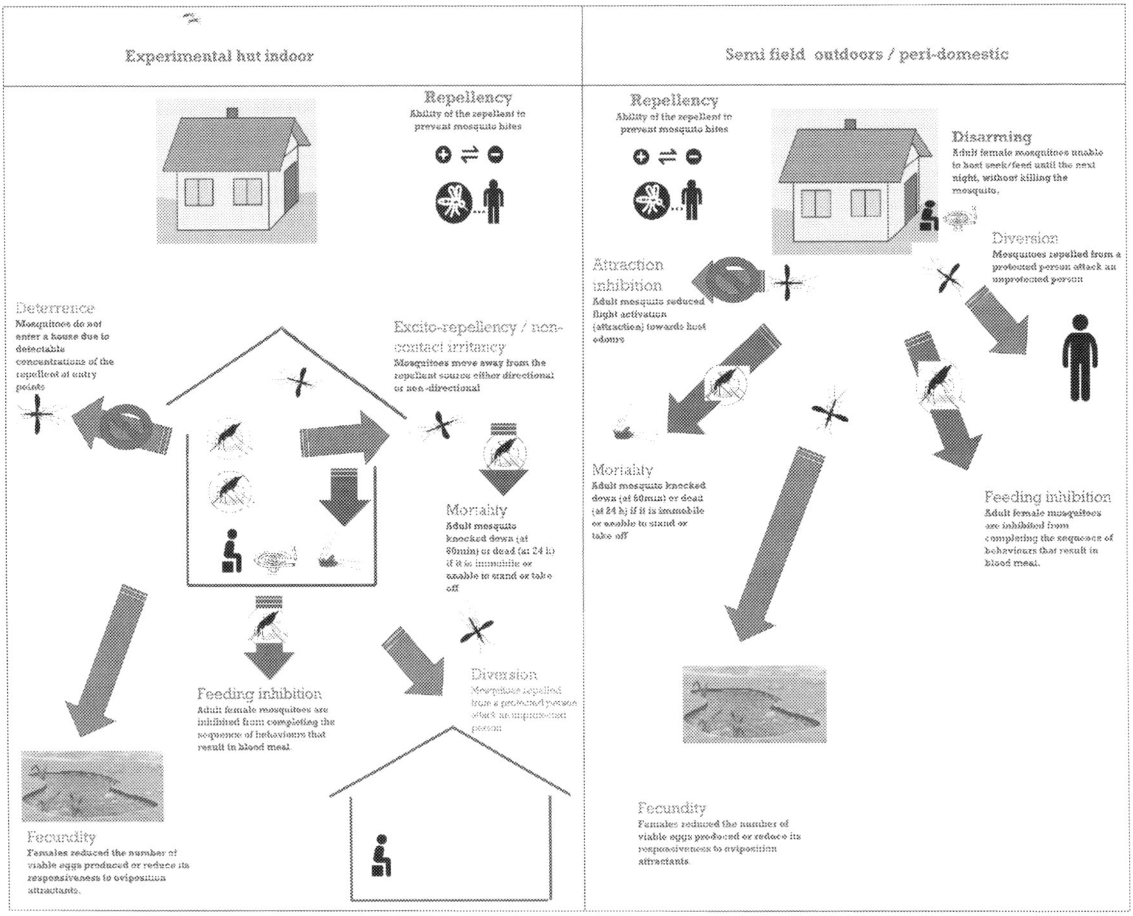

FIG. 10.1 Outcomes measured when testing the mode of action of spatial and topical repellents and how they relate to product efficacy.

repellents (Mbuba et al., 2020), repellent sandals (Sangoro et al., 2020), and spatial repellents (McPhatter et al., 2017).

SFSs and EH evaluations are useful in the developmental pipeline of interventions such as repellents. Such bioassays allow indoor and outdoor evaluations of topical repellents (Sangoro et al., 2014), and spatial repellents (Mmbando et al., 2018; Menger et al., 2014; Menger et al., 2015). They are useful methods to evaluate personal protection because PE estimations have been shown to be similar for semifield and full-field evaluations of spatial repellents (Ogoma et al., 2017; Ogoma et al., 2012b). They allow experiments to be conducted any time using disease-free insectary reared mosquitoes of known physiological and insecticide resistance profiles to give reliable and reproducible data. This provides an opportunity to evaluate the efficacy of repellents against both resistant and susceptible strains of the same species which might not be possible when conducting experiments in the field where most of the disease-vectoring mosquito species are likely to be of the same resistance phenotype (WHOPES, 2013). Findings from SFSs can also satisfactorily used to extrapolate the efficacy of topical repellents when applied in field systems while reducing the risk of acquiring mosquito-borne pathogens from field-testing (Sangoro et al., 2014). Data generated from these studies can be used to improve the performance of repellents before further testing at the community level in randomized control trials as endpoints collected in these studies directly translate to impact on disease (Box with endpoints).

10.2.1 Semifield systems

An SFS is a large screened cage that facilitates controlled experiments with disease-free laboratory-reared mosquitoes under ambient climatic conditions (Ferguson et al., 2008; Ngowo et al., 2017). Generally, an SFS is made of several compartments with walls made of durable netting to approximate ambient microclimatic conditions (Fig. 10.2) (Ferguson et al., 2008). These structures are found in many research institutes on all continents. The SFS structure can be mounted in a concrete base if the area is prone to flooding (Ferguson et al., 2008) and surrounded by a water channel (moat) that restricts entry of ants that would predate on mosquitoes during experiments. Standardized huts of similar construction materials and features to local houses that can be fitted with window and eave exit traps (Okumu et al., 2012) can be constructed within each SFS compartment for evaluations of indoor repellents or to simulate the peridomestic space.

The SFS can also be a long tunnels (semifield tunnel) (100 m × 3.1 m × 2.1 m) made of mosquito netting. The semifield tunnel structure can also be mounted on a concrete base if the area is prone to flooding as well as have a water channel (moat) similar to that of a SFS (Ogoma et al., 2014b).

FIG. 10.2 Picture of semifield systems (*SFS*).

BOX 10.1

Definition of primary outcomes in the semi-field system and experimental huts for evaluation of bite prevention tools for use against mosquitoes

1. Repellency: Adult female mosquitoes move away from an otherwise attractive host. This may be due to contact or noncontact irritancy or an avoidance reaction. Mosquitoes may be able to continue host-seeking and divert to another host is the ability of the repellent active ingredient to prevent mosquito bites.
2. Excitorepellency (noncontact irritancy): Mosquitoes move away from a host due excitation that results in them moving away from the repellent source. Excitorepellents induce three types of movement: (1) taxis, that is, directional movement of mosquitoes away from the treated space, (2) unidirectional movement due to orthokinesis (change in flight speed), and (3) unidirectional movement due to klinokinesis (change in turning during flight) that results in random movement of mosquitoes away from the treated space, often toward light.
3. Feeding inhibition: Adult female mosquitoes are inhibited from completing the sequence of behaviors that result in a blood meal. This is a result of the inhibition or excitation of olfactory receptor neurons.
4. Landing inhibition: Reduction in the number of female mosquitoes that land on a volunteer performing human landing catches in the control relative to the treatment. Landing inhibition is always used in repellent evaluation as a proxy for feeding success.
5. Disarming: Adult female mosquitoes are incapacitated through (1) knockdown (reversable incapacitation due to sublethal exposure to neurotoxic compounds) or (2) prolonged disruption of odor receptor neurons by spatial repellents. Mosquitoes are unable to host-seek until the next night but are not killed, providing both personal and community protection.
6. Diversion: Repulsion of adult female mosquitoes away from a protected person toward an unprotected person. Disarmed mosquitoes are unable to divert.
7. Toxicity: Mosquito knockdown or mortality. A mosquito is classified as knocked down (at 60 min) or dead (at 24 hours) if it is immobile or unable to stand or take off. It is important to consider time of killing: killing before biting (pre-prandial mortality) provides personal protection and community protection, whereas killing after biting (post-prandial mortality) provides only community protection
8. Attraction inhibition: Exhibition by adult female mosquitoes of reduced flight (attraction) to host odors. This occurs when the mosquito's olfactory receptors are inhibited or excited by a repellent.
9. Deterrence: Reduced likelihood of adult female mosquitoes to enter a treated house because of repellent active ingredient applied inside the house. Deterrence is calculated in experimental hut studies by comparing the number of mosquitoes that enter the control hut to those that enter the treated huts.
10. Effect on fecundity: Decrease in the number of viable eggs produced by a blood-fed adult female mosquito or reduction of its responsiveness to oviposition attractants because of sublethal effects of active ingredients.

10.2.2 Experimental huts (full-field testing)

Using local human houses to evaluate the efficacy of interventions for the control of arthropod vectors poses various limitations known to affect mosquito density and response to interventions. These limitations include (1) differences in the total number of individuals residing in a dwelling and the attractiveness of these occupants to mosquitoes (Mukabana, 2002); (2) lack of uniformity in materials used to construct houses and furniture (Kirby et al., 2008); (3) variation in size, number, and location of openings (Okumu et al., 2012); (4) variation in spatial location of homes in relation to larval habitats (Van Der Hoek et al., 2003); and (5) differences in house size. In the home, it is almost impossible to find knocked down or dead mosquitoes because of the presence of scavenger insects such as ants. To standardize data collected during the evaluation of indoor vector control interventions such as indoor residual spray (IRS), researchers have designed and developed a modified hut, i.e., experimental hut which represent local houses (Okumu et al., 2012). The huts are constructed near natural mosquito larval habitats to increase the availability of mosquitoes in them (Van Der Hoek et al., 2003) and can be positioned inside the SFS (Fig. 10.3).

FIG. 10.3 Picture of experimental huts.

EH has been widely used to study mosquito behavioral responses in the presence of interventions including repellents (Massue et al., 2016; Ogoma et al., 2014a). They allow mosquitoes to enter and then retain the mosquitoes that have entered so that mosquito behavior in response to repellent exposure can be evaluated.

10.3 Considerations for conducting semifield system and experimental huts experiments

To ensure reproducible findings between SFS experiments in multiple sites under similar conditions, several considerations for SFS and EH studies need to be harmonized before running an experiment (Fig. 10.4). This will ensure that mosquitoes can exert their natural behavioral responses that would occur under field conditions, while interacting with human hosts in the presence of repellents (Ogoma et al., 2012a). The factors can be organized in to five main categories: (1) environmental conditions, (2) the product itself, (3) the bioassay, (4) test system (mosquito), and (5) host factors (Table 10.1).

10.3.1 Environmental conditions

Climatic conditions such as wind speed, temperature, and humidity in the treated space determine the performance of repellents (Kawada et al., 2005). Hoffman et al. (2002) demonstrated an increased efficacy of DEET when used in artificially windy conditions generated by a fan (Hoffmann and Miller, 2002). Temperature has been found to affect the vaporization of repellents hence, the concentration of the active ingredient available to mosquitoes. Volatile pyrethroids such as transfluthrin evaporate more readily at higher temperatures (Pettebone, 2014) and have been reported to provide higher protection at temperatures between 21 °C and 30 °C, with a reduction in protection specifically

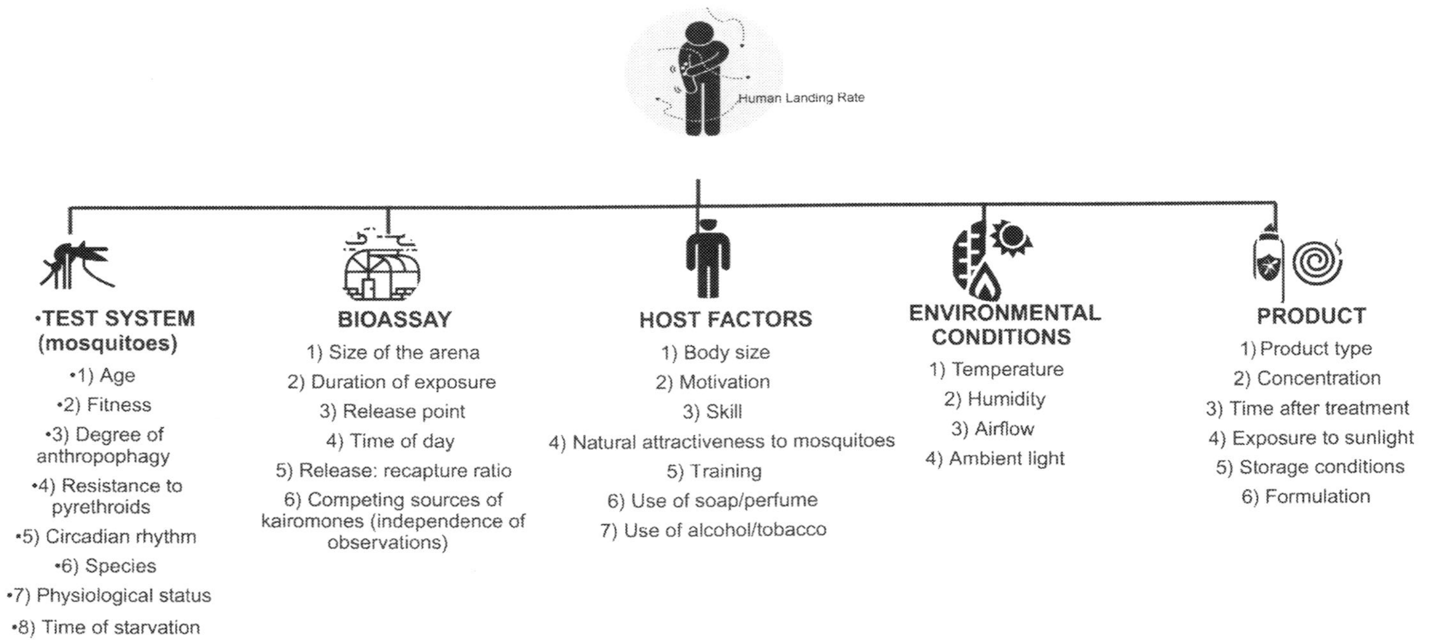

FIG. 10.4 Factors influencing human landing rate that needs to be considered during the designing of an experiment to evaluate the protective efficacy of spatial repellents.

TABLE 10.1 Factors influencing data collected in semifield and experimental hut bioassays.

Factors	Consideration for testing	Recommendation	References
Bioassay	Size of the arena	• Must be suitable for recapturing mosquitoes • Representative of the user case scenario • Baseline experiment needs to be conducted to ensure that a reasonable density of mosquitoes is recaptured	Ogoma et al. (2014b), Njoroge et al. (2021), Mmbando et al. (2018), Masalu et al. (2017)
	Duration of exposure	• Exposure should be representative of the use case - distance from repellent and duration of exposure determine outcome	Bernier et al. (2019), Bibbs et al. (2016)
	Release point	• Remote release mechanism needs to be in place • If several cages used, release mosquitoes simultaneously • Preferred at the four corners of the semifield system (SFS) to mimic mosquitoes approaching a house from a different direction • If multiple hosts are used, must be positioned equidistant from mosquito release points	Njoroge et al. (2021), Tambwe et al. (2020)
	Time of day	• Performed experiment under optimal climatic conditions for mosquitoes, e.g., not at the hottest part of the day • Conducted experiment to coincide with the mosquito circadian rhythm	Xue and Barnard (1996)
	Release/recapture ratio	• Ensure that at least 50% of the mosquitoes are recaptured • For mortality measurement close to 100% recapture is optimal	Tambwe et al. (2020)
	Independence of the observations	• Separate adjacent compartments with polyethylene fabric • Rotate treatments between huts or compartments if possible, to control for locational bias • Use no-choice test setups	Andrés et al. (2015), Ferguson et al. (2008)
Product	Product type	• Select products proven to work against mosquitoes species in lab experiments • Applied according to user instructions • Protected from being accessed by the animal or children • Consider the mode of action in bioassay design	WHOPES (2013)
	Concentration and formulation	• Select dose based on laboratory studies • Higher doses may be equal to low doses for bite prevention • Avoid switching between formulations or types of repellent in a single study • Formulation can also be used to boost volatility or duration of effect	Tambwe et al. (2020), Ogoma et al. (2017)

(continued)

TABLE 10.1 (Cont'd)

Factors	Consideration for testing	Recommendation	References
	Time after treatment	• Label treated emanator with day, month, and year of preparation because of decreased efficacy of spatial and topical repellents over time • Include data for time after deployment in statistical analysis	Ogoma et al. (2017)
	Exposure to sunlight	• Devices should be stored or aged in a shaded location away from direct sun exposure	
	Storage conditions	• Store items according to manufacturer instructions • Between experiments, store the treated devices should be stored at room temperature in shaded under the shade to reflect field age	Tambwe et al. (2020), Ogoma et al. (2017)
Climatic condition	Temperature	• Conduct experiments at the correct temperature for spatial repellent evaporation • Delay experiment until ambient temperature is within the range for repellent evaporation • Place a data logger in the SFS to measure/monitor temperature	Njoroge et al. (2021), Martin et al. (2020)
	Humidity	• Perform experiment when humidity is between 60%–100% because of effect of humidity on mosquito host-searching behavior • Place a data logger in the SFS to measure humidity during the experiment	Andrés et al. (2015)
	Wind	• Perform experiment when wind does not affect host-searching behavior of the mosquitoes • Measure wind speed and direction with a wind anemometer during the experiment because of effect of wind on amount of repellent received by mosquitoes	WHOPES (2013)
	Ambient light	• Conduct experiments at the preferred ambient light for biting	Xue et al. (1996)
Host factors	Body size	• Body size affect mosquitoes attraction to human due to amount of heat released • Allow volunteer rotation to counter individual variation in body size that affects attraction to humans because of the heat release differences • If possible, use individuals with similar body sizes	Takken and Verhulst (2017), Ray (2015)
	Motivation	• Provide meals or drink before the experiment to motivate volunteers to complete data collection • Consider breaks during prolonged experiments	
	Skills	• Select skilled Individuals to perform human landing catches • Train all potential volunteers on HLC technique • Before the experiment, check that each volunteer is able to collect a given number of mosquitoes from a given surrounding; this should be done by the primary investigator	Ogoma et al. (2017)

Factors	Consideration for testing	Recommendation	References
	Natural attractiveness to mosquitoes	• Rotate volunteers between chambers to account for natural differences in attraction to mosquitoes	Lindsay et al. (1993)
	Use of soap, deodorant, alcohol, and tobacco	• Volunteers should avoid using deodorant, soap, alcohol, and tobacco for the test days	de Jong and Knols (1995), Verhulst et al. (2016)
Test system	Age	• The World Health Organization recommends the use of 5 to 8-day-old mosquitoes so that they are optimally avid. • Record mosquito age	Xue et al. (1996), WHOPES (2013), Barnard (1998)
	Fitness	• Regularly asses mosquito body size, i.e., wing size of laboratory-reared mosquitoes • Regularly conduct survival experiment • Conduct baseline experiments to assess mosquitoes' ability to search for the human host in the semifield compartment • Ensure that mosquitoes are health by reducing the possibility of contamination (e.g., fungus, microsporidia) in the insectary to optimize fitness	Xue and Barnard (1996), Njoroge et al. (2021)
	Anthropophagy	• Evaluate repellent using relevant anthropophilic mosquitoes to control for different mosquito host-feeding preferences, which affect human landing rate • If two morphologically identical species, e.g., *A. arabiensis* and *An. gambiae*, are used at the same time, mark them with fluorescent powder	Gillies (1970), WHOPES (2013)
	Resistance to pyrethroids	• Use mosquitoes of known insecticide resistance • Regularly conduct susceptibility tests • To compare landing between resistant and susceptible mosquitoes, use mosquitoes of the same species with different resistance levels	Tambwe et al. (2020)
Test system	Circadian rhythms	• Conduct experiments during natural mosquito biting time (metabolism of insecticides and CYP450 modulators is regulated by circadian rhythms)	Balmert et al. (2014)
	Physiological status	• Nulliparous and starved mosquitoes should be used for experiment	Barnard (1998)
	Time of starvation	• *Anopheles* mosquitoes should be starved for around 5–6 hours • *Ae. aegypti* mosquitoes may be starved for up to 12 hours.	Fernandes and Briegel (2005)

at lower temperatures (Ogoma et al., 2017). The toxicity (knockdown (KD)/mortality) of pyrethroids is temperature dependent (Glunt et al., 2018). Topical repellents evaporate more rapidly when participants are sweating and therefore less efficacious at high temperatures (Khan et al., 1973). When ambient temperatures are low, it is possible that the PE estimates of volatile actives may be low. These factors must be considered when planning experiments under ambient conditions.

Differences in microclimatic conditions may also affect mosquito behavior. The ability of the mosquito to search for human host is reduced at low humidity (<40%) for some species (Takken et al., 1997) which may result in an overestimate of the efficacy of repellents. It is, therefore, important to ensure that the microclimatic conditions in the SFS and EHs are within an acceptable range for precision of estimated efficacy. The impact of climatic conditions underlines the usefulness of conducting controlled experiments under normal use conditions to give a more realistic estimate of efficacy. Microclimatic data loggers can be placed inside the SFS during the evaluation of repellent to constantly monitor the temperature and humidity. During periods of the year when humidity is low, wetting the surface of SFS floor may help to increase humidity.

Generally, insectary mosquitoes are reared at 27± 2 °C and 60%–100% relative humidity with approximately 12:12 light–dark (ambient lighting) (MR4, 2009; Gerberg et al., 1994). Microclimatic differences between inside and outside of houses have been found to influence mosquito abundance and biting behavior (Ngowo et al., 2017). Therefore, the sudden subjection of mosquitoes to a different microclimatic condition when taking them from the insectary to the SFS may alter their biting behavior (Kirby and Lindsay, 2004). To minimize this bias, laboratory-reared mosquitoes used in SFS experiments are transferred from the insectary to the SFS in the released cage or are released 30–60 minutes before the experiment is initiated to allow them to get used to the environmental conditions (acclimatization).

10.3.2 Product type and handling

In this context, product type refers to the device itself, the type of the repellent the device is impregnated with, and/or the repellent formulation. Having this information helps guide the experimental design to evaluate specific endpoints of the repellent (Table 10.2). To ensure a reproducible result, the description of the repellent to be evaluated must be clearly described including the active ingredient (AI), formulation, loading dose or concentration, number of devices used, and method of use (topical application, mats, fabrics, and vaporizers).

Correctly handling repellent-containing bottles and treated devices will help determine the true performance of that device. Storing repellents at 4 °C is often useful during product development to reduce evaporation of volatile components between tests until the product has been formulated to withstand room temperature storage. Bottles and devices should be labeled with the date of receipt and date when it is opened.

It is important to know how long the repellent treated device remained protective. To do this correctly, storage of the treated devices between the experiments needs to be ensured, often in collaboration with the product manufacturer. At a minimum, treated devices should be stored in a shaded environment away from direct sunlight to represent normal "field" storage.

10.3.3 Bioassay

With the development of new tools, bioassays are needed to generate initial efficacy data that enable the prediction of tools' impact on vectorial capacity (Box 10.1). The SFS enables evaluation of the efficacy of vector control tools in a more controlled environment. In the SFS, it is possible to use laboratory-reared mosquitoes to overcome difficulties such as varying mosquito

TABLE 10.2 Outcomes during the evaluation of repellents in the semifield and experimental huts.

End points measured	Formula for measuring	Evaluation in semifield system	Evaluated in experimental hut
Repellency (%)	Protective efficacy = ((C − T)/C) × 100, where C = proportion/number caught in control and T = proportion/number caught in treatment	Yes	Yes
Deterrence (%)	Deterrence = ((C − T)/C) × 100, where C = number of mosquitoes in control and T = number of mosquitoes in treatment	Yes, if huts are inside the SFS	Yes
Excito-repellency (%)	Excito-repellency = ((Te/Tt)-(Ce/Ct))/(1-(Te/Tt)) × 100 where Te = number of mosquitoes in exit trap in treatment, Tt = total number of mosquitoes in treatment hut, Ce = number of mosquitoes in exit trap in control, Ct = total number of mosquitoes in treatment hut	Yes, if huts are inside the SFS	Yes
Landing inhibition (%)	Landing Inhibition = ((C − T)/C) × 100, where C = proportion/number landed mosquitoes in human landing catch in control and T = proportion/number landed mosquitoes in human landing catch in treatment,	Yes	Yes
Feeding inhibition (%)	FI = ((C − T)/C) × 100, where C = proportion/number of fed mosquitoes in control and T = proportion/number of fed mosquitoes in treatment	Yes	Yes
Diversion (%)	Diversion = (T2/(T1+T2))/(C2/(C1+C2)) × 100 Where T1 = number of mosquitoes caught by HLC in position 1 near repellent T2 = HLC in position 2 away from the repellent; and number of mosquitoes caught by HLC in the control C1 = HLC in position 1 near placebo and C2 = HLC in position 2 away from the placebo	Yes	Yes
Disarming (%)	Disarming = ((C-T)/C) × 100 Where C = proportion of mosquitoes that successfully feed when offered blood after exposure in the control, T = proportion of mosquitoes that successfully feed when offered blood after exposure in the treatment. The time in hours of feeding after exposure should be noted.	Yes	Yes, if huts are in SFS
Attraction inhibition	Attraction Inhibition = (T/(T+A)) Where number of mosquitoes move toward = T and away = A	Yes	No
Avoidance reaction	Avoidance reaction = ((A/(T+A)) Where number of mosquitoes move toward = T and away = A	Yes	No
Mortality Corrected for the control (%)	Corrected control mortality (CM24) = ((T − C)/(1 − C)) × 100 where C = proportion of mosquitoes that died in control and T = proportion mosquitoes that died in treatment	Yes	Yes
Fertility (%)	Fecundity = ((C − T)/C) × 100, where T = median eggs per female in treatment and C = median eggs per female in control	Yes if mosquitoes feed	Yes if mosquitoes feed

availability so that tests can be conducted all year round with a known number of mosquitoes. EH are designed to be proxies of local houses and are standardized to minimize heterogeneity in the size, materials, and openings of the EH that affect mosquito density, mosquito behavior, and the amount of AI that they encounter (Massue et al., 2016). EHs have been extensively used to demonstrate the efficacy of vector control tools such as volatile pyrethroids in a variety of formats (Ogoma et al., 2014a; Hudson and Esozed, 1971). Moreover, exposure to insecticide-treated bed nets or IRS with excito-repellent compounds such as DDT has been measured in EH (Grieco et al., 2000).

Results from the SFS and EH are highly dependent on several factors (Fig. 10.1). The size of the arena (where the experiment is conducted) affects results because of the concentration-dependent effects of AI in spatial repellent tests (Achee et al., 2012a). Differences in mortality estimates have been observed between small arena such as taxis boxes where mosquitoes are held close to the source of the repellent (Martin et al., 2020) compared to a larger arena (Tambwe et al., 2021). The duration of exposure may also affect the efficacy of repellents (Bernier et al., 2019). The use of small space or longer exposure for assessment of 24-mortality following transfluthrin exposure may overestimate the result as transfluthrin increases mosquito activity due to excitation and mosquitoes will move away from the source of the transfluthrin under natural outdoor exposure conditions (Sukkanon et al., 2020). Larger chambers enable the use of free-flying mosquitoes and either provide space for the humans to either perform human landing catches (HLCs) or allow mosquitoes to feed in the presence of transfluthrin, which is more representative of what happens in the field (Tambwe et al., 2021). Space and duration of exposure should be considered when designing the experiment to evaluate endpoints of the repellent, in particular, mortality.

Results are also affected by accidental loss of mosquitoes through escape or scavenging ants, which leading to uncertainty of estimates, particularly mortality estimates (Nash et al., 2021). It is therefore important to design the arena to reduce accidental loss of insects (Massue et al., 2019) and to make them ant-proof. Regular monitoring of the proportion recaptured assists in maintaining the quality of bioassays. Releasing a known number of mosquitoes in experimental huts and recapturing them the following morning can be done periodically to check the hut and to assist in training staff to maximise recapture.

Density of the mosquitoes plays an important role in the evaluation of spatial repellents. The rate at which mosquitoes bite human volunteers is density-dependent and mosquitoes can select between hosts at short range (Gillies and Wilkes, 1972; Okumu et al., 2010) and when one of the hosts is protected by a repellent (Moore et al., 2007). In EH, there is often extreme heterogeneity in mosquito densities between huts (Johnson et al., 2014) which can be overcome by rotating treatments and volunteers between huts for many nights. An EH is designed so that all huts allow mosquitoes to enter through set entry points and they are either retained inside or exit into traps. This minimizes interhouse differences in mosquito densities during evaluations (Okumu et al., 2012). The dimensions, structure, construction materials, and location of huts in relation to distance from larval habitats should also be considered. For semifieldwork, the relative positioning of mosquito release points may determine the number of mosquitoes captured by human volunteers conducting HLCs. Mosquitoes should be released at multiple equidistant points around the HLCs or EH (if it is positioned in the SFS) to reflect mosquitoes coming from different directions. If a topical repellents test is conducted by several people, positioning them at least 10 apart and releasing mosquitoes equidistant, and rotating volunteers between collection locations can help to minimize these biases (Mbuba et al., 2020).

To effectively evaluate indoor vector control tools, enough mosquitoes must enter a hut.

During dry seasons, mosquito densities are low and it may not be possible to conduct EH studies. Therefore the huts can be used in the SFS. When the huts are used in the SFS, there is no need to be concerned about the quality and quantity of mosquitoes because a set (high) number of laboratory-reared released mosquitoes are used. However, when experiments are conducted in the field, the density of mosquitoes entering these huts must be optimized. The location of residential houses with respect to larval habitats is a significant factor affecting the density of mosquitoes inside human houses (Van Der Hoek et al., 2003; Okumu et al., 2012; Haddow, 1942). In field experiments, it is optimal for EHs to be equidistant from the larval sites or in areas of high mosquito density. This placement ensures that a sufficient density of mosquitoes enters all the huts because the emerging mosquitoes will travel on average the same distance toward the huts from either the larval habitats or other surrounding locations.

Repellents, particularly spatial repellents work at a distance to exert behavioral changes on exposed mosquitoes (Achee et al., 2012b; Ogoma et al., 2014b). Considering this mode of action, it is important to ensure independence of the SFS compartments when an evaluation is performed (Moore et al., 2007). For example, heavy-duty polyethylene walls may be used to separate compartments, preventing air movement between them and reducing the chance of cross-contamination when working with volatile active ingredients (Ferguson et al., 2008). Also, it is important to consider that there is a huge variation in climatic conditions over 24 hours. Normally, during the day, the temperature is very high accompanied by low humidity, which does not support conducting an experiment. It is therefore important to conduct experiments at the appropriate time of the day (Rund et al., 2016).

10.3.4 Test system

For efficacy evaluations of repellents in the SFS, anthropophilic, zoophillic, susceptible, and resistant strains of the local vectors must be considered (Besansky et al., 2004). This gives a conservative estimate of repellent efficacy before field trials, clinical trials or policy recommendations (World Health Organization, 2013; WHOPES, 2009b).

Mosquitoes use different stimuli such as skin odor, water vapor, heat, and visual cues to locate potential blood sources (Tisgratog et al., 2011; Bibbs et al., 2018a; Takken, 1991; Takken and Knols, 1999). The strength at which mosquitoes are attracted to human cues varies between mosquito species or strains. Differences in attraction to a human are attributed to species' anthrophillic and zoophillic behaviors which tend to be genetically fixed (Mahande et al., 2007). Anthropophillic species such as *An. gambiae* s.s. and *An. funestus* are more likely to blood feed on humans (Costantini et al., 1999) depending on their relative abundance (Asale et al., 2017) or availability (Iwashita et al., 2014) compared to more zoophillic species, such as *A. arabiensis* (Orsborne et al., 2018). The differences in landing rate between these mosquitos species are caused by differences in attraction to human cues (Gillies, 1964). *Ae. aegypti* also feeds almost entirely on humans (Scott et al., 2000). In addition, species vary in their sensitivity to repellents (Van Roey et al., 2014) which result in different doses of repellent active ingredients needed to elicit responses; where among some of the robust species such as *Ae. aegypti* higher concentration of repellents are required (Sukkanon et al., 2020), whereas *Culex* mosquitoes are easier to repel (Lupi et al., 2013). Studies have demonstrated differences in the complete protection time of topical repellents (Schreck, 1977; Curtis et al., 1987). Data suggest that different species of mosquitoes have different behavioral responses to repellents due to differences in repellent-sensing neurons or olfactory neurons (Afify and Potter, 2020).

Previous studies have reported that parity, age, and feeding status can influence host-seeking behavior (Xue and Barnard, 1996). Removing these biases through selection of appropriate mosquitoes is an essential step toward performing a high-quality experiment by ensuring that appropriately aggressive and fit mosquitoes are used (Table 10.1). Avid mosquitoes are selected by placing the palm or warm objects on the side of the cage and aspirating only mosquitoes that are probing (WHOPES, 2013). Various factors may affect avidity or fitness of the adult mosquitoes. Firstly, during the larval stages, environmental variations between bowls such as density and amount of food dispensed potentially influence the fitness and therefore host-seeking behavior of adult mosquitoes (Araújo et al., 2012). Secondly, starved mosquitoes are more likely to be more aggressive than sugar-fed mosquitoes (Fernandes and Briegel, 2005). *Starving* refers to removing sugar solution from the cage containing adult mosquitoes before the evaluation of repellent to optimize the mosquitos' avidity and thereby host-searching behavior. Thirdly, mosquito avidity is related to age. Younger female mosquitoes have lower responses to topical mosquito repellents (Xue and Barnard, 1996), while older mosquitoes are more responsive to repellents (Mulatier et al., 2018). Therefore, younger mosquitoes are preferred because they are most likely to exhibit host-seeking behavior and are less likely to be effected by the repellents (Aldridge et al., 2017) giving the most conservative estimate of repellent activity.

To ensure that mosquitoes used are as heterogeneous as possible, selection is done from different cages using a minimum of three cages. The WHO recommends that mosquitoes need to be nulliparous aged 5–8 days and starved for at least 6–8 hours before the experiment (WHOPES, 2013). The number of mosquitoes to be selected depends on the number estimated by a sample size calculation, which is usually based on parameters measured in previous experiments such as variability between locations, daily variability in mosquito attack rate, and variability in volunteer attractiveness to mosquitoes (Johnson et al., 2014). Sangoro et al. (2014) evaluated the efficacy of topical repellent when 100 mosquitoes were released each hour inside the SFS to match biting pressures experienced in field trials (Sangoro et al., 2014). In another study, 100 *Ae. aegypti* mosquitoes were released inside an SFS during the evaluation of a transfluthrin treated passive emanator with no concerns from the volunteer (Tambwe et al., 2020). However, in a large SFS only a proportion of the total released mosquitoes may be recaptured by HLC and this needs to be accounted for in sample size estimations (Njoroge et al., 2021).

Another factor for consideration is that mosquito species have different circadian rhythms. Although laboratory-reared mosquitoes may be adapted to bite any time throughout the day, it is highly recommended that repellent evaluation should be conducted to coincide with the mosquito's natural biting time. For example, in East Africa, the host-seeking activity of female Anopheles mosquitoes ranges from 18:00~06:00 (Moshi et al., 2017) which mean that evaluation of repellent needs to be done between these times. In addition, circadian rhythms are an important determinant of insecticide detoxification (Balmert et al., 2014) and will therefore impact responses to volatile pyrethroids (Tainchum et al., 2014).

Pyrethroids have been the main class of insecticide used in ITNs and IRS (Zaim et al., 2000). Resistance to these insecticides is now widespread (Mitchell et al., 2012), which poses a threat not only to the efficacy of main vector control tool but potentially to volatile pyrethroids such as transfluthrin and metofluthrin that are used as spatial repellents as they belong to the same chemical class, which could indicate cross-resistance. Wagman et al. (2015) concluded that insensitivity to sublethal doses of transfluthrin against the dengue vector *Ae. aegypti* are heritable and correlate to reduced susceptibility to toxic doses of transfluthrin in CDC bottle (Wagman et al., 2015). Therefore, it is necessary to know the insecticide susceptibility status for the mosquitoes to be used for the evaluation of volatile insecticides.

If multiple strains with different levels of resistance that are not morphologically distinguishable are used in semifield experiments, mosquitoes can be marked with the fluorescent colors to distinguish between strains. Mosquitoes are marked in a cup by dusting the mesh lid of the cup with a brush containing the color pigment thereby creating a cloud of pigment that is transferred to the mosquitoes in small amounts. Preliminary experiments have shown that the fluorescent pigments do not significantly influence mosquito survival or feeding behaviors and can easily identified using an infrared torch (Saddler et al., 2019).

10.3.5 Host factors

Variation in recaptured mosquitoes between human subjects has been demonstrated (Lindsay et al., 1993). Such differences affect the results of repellent evaluations and need to be considered during experimental design (Rutledge and Gupta, 1999).

Mosquitoes use heat emitted by the human host and other vertebrates as a short-range cue to land and bite (Ray, 2015). The amount of heat emitted differs from one individual to another and is largely dependent on body size, which affects individual attractiveness to host-seeking mosquitoes (Carnevale et al., 1978). To account for these variations, study designs should allow the rotation of treatments between volunteers (WHOPES, 2013). Repellents are known to have residual effects thus enough time is needed to allow diffusion of the residual repellent actives before rotation is done to avoid the occurrence of a carry-over effect (Ogoma et al., 2014a). Alternatively, in some circumstances the treatment can remain fixed in one location over the duration of the experiment depending on the experimental design and the insecticides to be evaluated. For example, in a study by Ogoma et al. (2014) where the effect of DDT and airborne pyrethroids were evaluated on entomological parameters of malaria transmission, treatments were not rotated instead, the volunteers rotated between the treatments (Ogoma et al., 2014a). In contrast, both treatment and volunteer were rotated when Andres et al. (2015) evaluated the efficacy of transfluthrin to reduce human landing rate in the peridomestic space (Andrés et al., 2015).

The skill and motivation of mosquito collectors are known to cause variation in the number of mosquitoes collected. To account for this, experienced volunteers can be provided with proper training before the commencement of the experiment. Mosquito landing may also be affected by the use of soap (de Jong and Knols, 1995), deodorant (Verhulst et al., 2016), alcohol (Shirai et al., 2002), and tobacco (Jufri et al., 2016) prior to the experiment which may affect accurate estimation of repellent efficacy. It is recommended that volunteers be educated in advance on how the use of these items may affect repellent evaluation and asked to refrain from the use of soaps and deodorant on the days of testing; and where possible non-smokers and non-drinkers can recruited for tests.

10.4 Study power

Study power refers to the calculation of the representative number of mosquitoes, number of replicates, and number of days required to detect a predefined difference between either the treatment and the control arms or between different treatments.

Sample size calculations, for instance, generalized linear mixed models can be performed before an experiment, using mulation-based power analysis (Johnson et al., 2014) to ensure enough replicates are run to determine the difference between arms with at least 80% power and a significance level of 0.05 for rejecting the null hypothesis. Estimation of the variation between the locations, day, volunteers, mosquito density, and recapture rate as well as a minimum effect difference between arms should be considered in the simulation. These are usually estimated from previous evaluations of similar products. Other factors to consider are whether the study is measuring superiority, noninferiority, or equivalence between the treatment arms.

10.5 Primary outcomes measured in the semifield system/experimental huts and computations

Repellents can induce various behavioral responses when exposed to mosquitoes depending on the dosage, distance from a point source, temperature, and airflow. Outcomes that result in prevention of human-vector contact are defined in Box 10.1 and include repellency, irritancy, deterrence, attraction inhibition (AI), feeding (biting) inhibition, toxicity (KD and mortality), disorientation, or disarming and effects on fertility (Achee, 2012; Sukkanon et al., 2020; Bibbs et al., 2020; Ogoma et al., 2014a). These outcomes can be assessed in SFS and EH experiments (Table 10.2). In generally, these outcome parameters measure the personal protection of the repellent user, while effects on fertility, and toxicity (KD and mortality) can also measure the community protection of the repellents because both users and nonusers benefit from the reduced size and survival of the mosquito population (Magesa et al., 1991; Brady et al., 2016).

Based on these outcomes, it is possible to measure the efficacy of repellents indoors and/or outdoors (in the peridomestic space) in both SFS and EH evaluations. The primary outcomes can be assessed using various study designs such as comparative cross-over designs with wash-out periods, choice and no-choice evaluations, and fully randomized and partially randomized Latin squares are favoured to account for the multiple sources of bias inherent in these tests. The choice of the design depends on the repellent being tested, number of treatment arms, the purpose of the study, and the resources available. For the efficacy evaluations to be robust, it is important that a control (no treatment) arm to be present. The control arm is essential because it allows for the effects of any intervention (s) being tested to be distinguished from natural events that would have occurred even in the absence of intervention and gives an estimation of biting pressure.

Formulae, classical, and inferential statistical analysis can be used to estimate efficacy conferred by repellents. The analysis approach to be used depends on the study design and outcome of interest. Common analyses include parametric and nonparametric t-tests, z-test, and analysis of variance (ANOVA) although most repellent tests generate data that does not follow a normal distribution and tests should be selected based on the distribution of the data obtained. Inferential statistics such as generalized linear mixed model, survival analysis, and binomial regression are usually applied to data generated from these evaluations accounting for sources of variability including temperature, humidity, volunteer attractiveness, and treatment doses. The distribution to be used in the inferential analysis will depend on whether the outcome is proportions (Binomial) e.g. proportion fed or actual numbers (Poisson) e.g. number of landings of mosquitoes. Additionally, it may be necessary to account for any overdispersion (variance > 2 times the mean) or the presence of zeros depending on the distribution of the data collected. Negative binomial regressions or transformation of data may be done to account for this.

Basic formulas for protective efficacy (PE) or other endpoints can also be used in analysis Table 10.2. In regression analysis, the primary outcome effect of the repellent is fitted as the dependent variable, with treatment/volunteer/compartment/location and other factors as fixed independent variables while day or replicate can be included as a fixed or random effect. Inconsistencies in the PE estimates have been observed when using the basic formula and estimates from the model. The reason for this could be that the basic formula is not sensitive enough to capture small differences that may be attributed to other factors that are adjusted for in model estimates of effect. Estimates from statistical models that account for variation are more reliable than basic formula because they account for other variables in the experiment. The simplest way to generate model estimated effect is to use regressions for count data and to describe the PE as 1-the relative rate of the outcome of interest e.g. landing, feeding, exiting.

10.5.1 Contact irritancy

Irritancy is the ability of the repellent to induce directional or nondirectional movements of the mosquitoes away from the treated surfaces resulting from tarsal contact irritancy. This can lead to exiting of mosquitoes that had already entered a hut an occurrence referred as excito-irritancy or excito-repellency (Grieco et al., 2007). This can be measured for treated clothing or irritant insecticides by comparing the relative rate of mosquitoes collected in EH or landing treated clothing with treatment relative to an untreated control (Table 10.2). Mosquitoes can be collected from HLC, indoor resting catches, or exit traps on the huts.

10.5.2 Noncontact irritancy (repellency, excito-repellency and deterrence)

Repellency (spatial repellency) is the ability of the repellent to keep mosquitoes away from the treated space that may occur through a variety of mechanisms (Ogoma et al., 2012a) (Box 10.1) when the mosquitoes come into contact with airborne particles of repellent insecticides. The presence of repellents can reduce (deter) the entry of mosquitoes into the houses—a phenomenon called deterrence (Grieco et al., 2007; Kennedy, 1947; Ogoma et al., 2014b). Both repellency and deterrence can be elicited by spatial repellents. Excito-repellency is noncontact irritancy phenomenon where mosquitoes become overly excited and move nondirectionally from spaces treated with spatial repellents.

Repellency excito-repellency or deterrence is estimated by comparing the relative rate of mosquitoes collected in EH with treatment relative to an untreated hut (Table 10.2). Mosquitoes can be collected from HLC, indoor resting catches, or exit traps on the huts (Fig. 10.3). When risk ratios are used, repellency and its confidence intervals are estimated from the mosquito recapture counts in the control and treatment by replicate for instance for each day (Table 10.2) or using Bayesian methods (Takakura, 2012).

10.5.3 Landing inhibition

Landing inhibition (LI) refers to the reduction in the number of female mosquitoes that come in contact with the host. HLC remains the most commonly used proxy for determining feeding/biting inhibition (WHOPES, 2009a) in both semi-field and full-field. The overall number of mosquitoes landing on the volunteers conducting HLC with the treatment relative to the control is determined. When risk ratios are used, LI and its confidence intervals can be estimated from 1-the relative rate of mosquito recapture counts (Table 10.2).

10.5.4 Feeding (biting) inhibition

Feeding (biting) inhibition refers to the ability of a repellent's AI to inhibit mosquitoes from feeding or biting even after landing on the potential host. HLC is used as a proxy for feeding/biting inhibition (WHOPES, 2009a) in both the semifield and full-field, and is safe in areas where there is no active transmission of vector-borne pathogens provided it is medically supervised (Achee et al., 2015b).

Feeding inhibition (FI) endpoint is determined by comparing the proportion of blood-fed mosquitoes between the control and treated arms. To directly measure this in the SFS, mosquitoes are released in the compartment and the volunteer remains in the chamber or room for the period of interest while allowing mosquitoes to bite. At the end of the experiment, all the mosquitoes are collected from the chamber and proportion of blood-fed mosquitoes in the treatment arm relative to the control is determined (Tambwe et al., 2021).

In EH testing, the overall proportion of blood-fed mosquitoes caught inside the hut including on the floor or wall or in the exit traps of the treatment hut relative to the control is determined. FI is epidemiologically relevant as it is a measure of personal protection (Table 10.2). It should be noted when running regression analysis, either the proportion fed or unfed out of the total recaptured, or the absolute numbers fed can be fitted in the model. When risk ratio of absolute numnbers fed is used, the FI and its confidence intervals can be estimated as 1-the relative rate of mosquitoes fed (Table 10.2).

10.5.5 Diversion

In a scenario of incomplete coverage, that is incomplete application of a topical repellent or members of a group do not use repellent, mosquitoes that are repelled or inhibited from biting can switch to a nearby unprotected area of skin or another host (Moore et al., 2007). Diversion is measured by measuring the ratio of mosquito landings on a protected and unprotected individual in the treatment relative to landings on an unprotected individual in the control (Table 10.2) (Tambwe et al., 2021).

10.5.6 Disarming

Disarming refers to mosquitoes that are unable to complete a feeding cycle that night and cannot divert (feed on other hosts). This can be categorized as immediate or delayed incapacitating effects of repellents, which is assessed in the SFS, where physiological status of released mosquitoes is known. Mosquitoes are incapacitated through; (1) KD, which is a reversible incapacitation due to sublethal exposure to neurotoxic compounds, or (2) prolonged disruption of odor receptor neurons by repellents. This is important because it protects multiple individuals and not just users, i.e., a community effect and also reduces vectorial capacity (Denz et al., 2021). Shorter-term FI occurs when a mosquito can land but does not feed. This is commonly observed with repellents such as DEET that affect odor receptor neurons (DeGennaro, 2015). It is short-lived and mosquitoes are able to divert to an unprotected host in the same feeding cycle. Disarmed mosquitoes return to host-seeking and a start a new feeding cycle within 1 to 3 days.

Disarming is measured by collecting alive unfed mosquitoes, placing them in a holding cage or cups then observing feeding success of the mosquitoes when they are offered a bloodmeal source away from the source of repellent (Ogoma et al., 2014b) as soon as possible after exposure and counting the knocked down mosquitoes that are alive 24 hours later. It can also be determined using mathematical models of mosquito host-seeking and estimating the rates of mosquito feeding, repelling, and disarming from HLC data collected in 15 minutes intervals (Denz et al., 2021).

10.5.7 Toxicity (knockdown and mortality)

Toxicity is the measure of the degree of toxic effect of repellent on exposed mosquitoes. KD and mortality are the two effects of repellent toxicity which depend on initial loading dose on the substrate/surface, environmental factors including the volume of the treated space, distance from the repellent source, release rate, and degradation rates of the repellent. KD is scored if a mosquito is unable to stand or fly in a coordinated manner. Recaptured mosquitoes are placed in a netted cup, supplied with sugar (10% sucrose or glucose) and taken to a climatic controlled room for observation of delayed mortality. At 24 hours postexposure, a mosquito should be re-examined and classified as knocked-down (revived) or dead (knocked-down died) (WHOPES, 2013). Delayed mortality whether before (preprandial) or after (postprandial) a successful blood meal measures the community or personal protection conferred by the repellent (Brady et al., 2016). This can be estimated by comparing the proportion of dead mosquitoes after 24 hours from those captured alive (fed or unfed) in the treatment arm relative to the control. To estimate toxicity in SFS and EH due to exposure to the repellent, the proportion of dead mosquitoes (fed and unfed) in the treatment arm corrected for mortality in the control arm is calculated for each experimental replicate (Table 10.2).

10.5.8 Attraction inhibition

AI refers to the reduced flight activation (attraction) of mosquitoes toward host odors. AI occurs mainly because repellents block or modify responses of the olfactory receptors neurons (ORNs) on mosquitoes' antennae; these receptors are sensitive to specific host attractants (Dickens et al., 2013; Davis and Sokolove, 1976). For example, the application of DEET has been found to reduce orientation of mosquitoes to lactic acid cues produced from human sweat by decreasing the sensitivity of ORN to lactic acid thus, reducing attraction of mosquitoes toward human-host (Dickens et al., 2013). The same effect has also been reported for linalool, dehydrolinalool, catnip oil, and citronella (Ogoma et al., 2014b; Bohbot and Dickens, 2010; Kuthiala et al., 1992). To estimate AI numbers of mosquitoes that moved into the chamber closer or away to the stimulus in a taxis box can be recorded (Lorenz et al., 2013). In addition, the difference between landing and biting may be recorded.

10.5.9 Avoidance reaction

Avoidance reaction (AR) refers to the induced flight activation (negative taxis) of mosquitoes away from host odors. To estimate avoidance, the numbers of mosquitoes moved into the chamber away from the stimulus in a taxis box can be recorded (Lorenz et al., 2013).

10.5.10 Effect on fertility and oviposition

Exposure to sublethal concentrations of repellents has been found to affect subsequent fertility and oviposition behavior in mosquitoes by, (1) decreasing the number of viable eggs produced (Bibbs et al., 2018b), and (2) reducing responsiveness of the ORN sensitive to oviposition attractants (Kuthiala et al., 1992, Bibbs et al., 2018a). This has been observed in *An. gambiae s.l.* mosquitoes when exposed to sublethal concentration of transfluthrin, where the number of eggs produced was reduced (Ogoma et al., 2014a). Similar effects were observed in *Ae. aegypti* and *Ae. albopictus* for which the number of viable eggs laid and their skip oviposition behavior were significantly declined when the mosquitoes were exposed to a repellent (Bibbs et al., 2018a).

Fertility in mosquitoes is measured by the proportion of viable eggs laid while oviposition is the measure of successful laying these viable eggs. Efficacy of repellents in reducing fertility and oviposition success can be assessed by exposing adult females to a repellent during feeding then allowing them to oviposit and counting the number of eggs per female produced in the treatment arm relative to those from the control arm.

10.5.11 Protective distance

Protective distance is the distance between users and the source of the repellent in which repellents can confer repellency, FI, disarm, KD, and kill. This depends on the initial loading dose on the substrate/surface, environmental factors, such as wind speed, release rate, and degradation rates of the repellent. Both SFS and EH experiments can be used to estimate the indoor and outdoor protective distance of a repellent product via HLCs, which can be achieved by having treatment arm/repellent users and control arm/nonusers apart at different distances. To ensure robustness, the volunteers should be rotated and adequate replication conducted to allow precise estimates of protection adjusting for volunteer, location, night, and hour of collection. The distance after which an equal numbers of mosquitoes are captured in the control arms is determined to be the protective distance or radius of the repellent.

10.6 Use of semifield system and experimental hut data for mathematical models

Mathematical models describe the underlying mechanisms that drive a system. They aim to represent a system based on assumptions of its dynamics with a simplified description of the mechanism using assumptions on parameters that often use estimates derived from collected data. The SFS and EHs are often used to parameterize mathematical models because they provide standardized estimates of effects that may not be measured under field conditions, or provide a more cost-effective means to provide model parameters. Models can be used to predict changes in vectorial capacity (Brady et al., 2016) in mosquitoes only as a proxy for transmission or impact on malaria using individual-based stochastic simulations of malaria epidemiology to predict the impacts of interventions on infection, morbidity, mortality, health services use and costs (Smith et al., 2008, Denz et al., 2021, Hellewell et al., 2021).

The two most important factors in determining the effectiveness of a vector control tool are feeding preference and average lifespan. As preference for feeding on humans increases so does the likelihood of parasite transmission. An adult mosquito's lifespan is also critical for malaria transmission as the mosquito must survive long enough for the parasite to complete the period of sporogonic development, which covers the period from the ingestion of gametocytes with the blood meal to the time when infectious sporozoites appear in the salivary glands. When evaluating the effects of repellents there are three main parameters measured in the semifield/EH:

1. Repellency/deterrency: The mosquito is unable to feed or enter a house to feed when it wants to. However, it continues host-seeking through the night, either on the same host/household or it is diverted to a different host/household. At the end of the night, it may have successfully fed (on the same host/household or a different host/household), it may have died later in the night, or it may end the night unfed and alive. The repellency/deterrency is thus treated in the model as a reduction in the availability or the "attractiveness" of the human to the mosquito.
2. Disarmed: The mosquito is removed from the feeding cycle and no longer continues host-seeking that night. This is may be because the (i) mosquito is knocked down; (ii) the mosquito is no longer capable of biting (i.e., it may appear to host-seek and may land on the host but would not bite); or the mosquito enters the resting phase without having fed. The period of disarming would last for at least one night—but may be longer. At the end of the night, disarmed mosquitoes would either be knocked down (and alive 24 hours later) or recaptured unfed alive.
3. Dead: The mosquito is killed by the intervention, either immediately or after a short delay. These mosquitoes could be unfed dead (pre-prandial mortality), fed dead (postprandial mortality), or knocked down (and dead after 24 hours).

Using these relationships described through equations, mathematical models can predict the community-level impact of a repellent by incorporating the outcome parameters measured from the SFS or EH experiments. Outcomes can be used to parameterize models with different species characteristics such as levels of anthropophagy or exophily and how a repellent with a specific mode of action, e.g., LI, KD or repellency may affect malaria or other vector-borne diseases when the tool is applied to an individual (personal protection) or a community (community protection) at different coverage levels.

The sensitivity of effect sizes to proportional changes in the parameters comprising vectorial capacity shows that protecting people from bites through the use of personal protection has a second-order effect because it appears twice in the VC equation (Brady et al., 2016). Therefore reducing 50% of mosquito bites would result in VC that is ½ x ½ = ¼, i.e., only 25% of VC would remain giving a 75% reduction. Importantly, by reducing the human biting rate, especially for a feeding cycle (disarming), fewer mosquitoes will be able to develop eggs and this will have a knock-on effect on VC. As fewer eggs means fewer mosquitoes—this is a first-order effect similar to that seen with larval source management, and combined with the second-order effect of reducing bites an intervention with 50% efficacy would result in ½ x ½ x ½ =12.5% giving a 87.5% reduction in VC. However, the greatest impact of VC is through reducing a mosquito life span. Reducing mosquito daily survival impacts the probability that the mosquito will survive the intrinsic incubation period as well as the number of eggs that the mosquito will lay in its lifetime (lifetime fecundity). It reduces adult density that means that the next generation also contains fewer adults. Therefore if it is assumed that the effect on oviposition and the next generation is also 50% the effect is fourth-order, therefore for an intervention that kills 50% of mosquitoes the effect on VC would be ½ x ½ x ½ x ½ = 6.25% giving a 92.7% reduction in VC. New vector control tools that prevent bites but also stop mosquitoes from feeding as many times in their lifetime through protecting multiple users in a space, disarming mosquitoes from feeding on other hosts postexposure and that cause mosquito mortality are therefore likely to have a far greater impact on disease when utilized for public health than traditional topical repellents that prevent bites for a single user.

10.7 Conclusion

Spatial repellents are an important addition to the vector control toolbox because they protect multiple users within a defined space. Numerous products that are effective for use indoors or outdoors are coming to the market. Repellents, in particular volatile pyrethroid spatial repellents exert many behavioral effects on exposed mosquitoes resulting into multiple outcome parameters to be assessed. These outcome parameters require standardized bioassays so that the data are reliable and reproducible, which helps to generate the robust data sets required to recommend new repellents for use as consumer products or public health tools. Concurrently, robust data analysis is important for data interpretation. Data from SFS and EH experiments can be used to extrapolate the effect of a repellent at the community level through mathematical models, and SFS and EHs provide useful arenas for controlled evaluation of formulated products that gives realistic estimations of field efficacy and that can be used to rapidly and economically measure multiple endpoints relevant to public health, such as mosquito disarming and mortality that are not easily captured in full-field experiments.

References

Achee, N.L., Masuoka, P., Smith, P., Martin, N., Chareonviryiphap, T., Polsomboon, S., Hendarto, J., Grieco, J., 2012a. Identifying the effective concentration for spatial repellency of the dengue vector *Ae. aegypti*. Parasit Vectors 5, 300.

Achee, N.L., Bangs, M.J., Farlow, R., Killeen, G.F., Lindsay, S., Logan, J.G., Moore, S.J., Rowland, M., Sweeney, K., Torr, S.J., Zwiebel, L.J., Grieco, J.P., 2012b. Spatial repellents: from discovery and development to evidence-based validation. Malar. J. 11, 164.

Achee, N.L., Gould, F., Perkins, T.A., Reiner Jr., R.C., Morrison, A.C., Ritchie, S.A., Gubler, D.J., Teyssou, R., Scott, T.W., 2015a. A critical assessment of vector control for dengue prevention. PLoS Negl Trop Dis 9, e0003655.

Achee, N.L., Youngblood, L., Bangs, M.J., Lavery, J.V., James, S., 2015b. Considerations for the use of human participants in vector biology research: a tool for investigators and regulators. Vector Borne Zoonotic Dis. 15, 89–102.

Achee, N.L., Bangs, M.J., Farlow, R., Killeen, G.F., Lindsay, S., Logan, J.G., Moore, S.J., Rowland, M., Sweeney, K., Torr, S.J., Zwiebel, L.J., Grieco, J.P., 2012. Spatial repellents: from discovery and development to evidence-based validation. Malar. J. 11, 164.

Afify, A., Betz, J.F., Riabinina, O., Lahondère, C., Potter, C.J., 2019. Commonly used insect repellents hide human odors from Anopheles mosquitoes. Curr. Biol. 29, 3669–3680 e5.

Afify, A., Potter, C.J., 2020. Insect repellents mediate species-specific olfactory behaviours in mosquitoes. Malar J 19, 127.

Ahmed, S., Reithinger, R., Kaptoge, S.K., Ngondi, J.M., 2020. Travel is a key risk factor for malaria transmission in pre-elimination settings in Sub-Saharan Africa: a review of the literature and meta-analysis. Am. J. Trop. Med. Hyg. 103, 1380–1387.

Aldridge, R.L., Kaufman, P.E., Bloomquist, J.R., Gezan, S.A., Linthicum, K.J., 2017. Application site and mosquito age influences malathion- and permethrin-induced mortality in *Culex quinquefasciatus* (diptera: Culicidae). J. Med. Entomol. 54, 1692–1698.

Alonso, P., Engels, D., Reeder, J., 2017. Renewed push to strengthen vector control globally. Lancet 389, 2270–2271.

Andrés, M., Lorenz, L.M., Mbeleya, E., Moore, S.J., 2015. Modified mosquito landing boxes dispensing transfluthrin provide effective protection against *Anopheles arabiensis* mosquitoes under simulated outdoor conditions in a semi-field system. Malar. J. 14, 255.

Araújo, M., Gil, L.H., E-Silva, A., 2012. Larval food quantity affects development time, survival and adult biological traits that influence the vectorial capacity of *Anopheles darlingi* under laboratory conditions. Malar. J. 11, 261.

Asale, A., Duchateau, L., Devleesschauwer, B., Huisman, G., Yewhalaw, D., 2017. Zooprophylaxis as a control strategy for malaria caused by the vector *Anopheles arabiensis* (diptera: Culicidae): a systematic review. Infect. Dis. Poverty 6, 160.

Balmert, N.J., Rund, S.S., Ghazi, J.P., Zhou, P., Duffield, G.E., 2014. Time-of-day specific changes in metabolic detoxification and insecticide resistance in the malaria mosquito *Anopheles gambiae*. J. Insect. Physiol. 64, 30–39.

Barnard, D.R., 1998. Mediation of deet repellency in mosquitoes (diptera, culicidae) by species, age and parity. J. Med. Entomol. 35, 340–343.

Beiter, K.J., Wentlent, Z.J., Hamouda, A.R., Thomas, B.N., 2019. Nonconventional opponents: a review of malaria and leishmaniasis among united states armed forces. PeerJ 7, e6313.

Bernier, U.R., Kline, D.L., Vazquez-Abad, A., Perry, M., Cohnstaedt, L.W., Gurman, P., D'hers, S., Elman, N.M., 2019. A combined experimental-computational approach for spatial protection efficacy assessment of controlled release devices against mosquitoes (anopheles). PLoS Negl. Trop. Dis. 13, e0007188.

Besansky, N.J., Hill, C.A., Costantini, C., 2004. No accounting for taste: host preference in malaria vectors. Trends Parasitol. 20, 249–251.

Bibbs, C.S., Xue, R.D., 2016. OFF! Clip-on Repellent Device With Metofluthrin Tested on Aedes aegypti (Diptera: Culicidae) for Mortality at Different Time Intervals and Distances. J. Med. Entomol. 53 (2), 480–483.

Bibbs, C.S., Hahn, D.A., Kaufman, P.E., Xue, R.D., 2018a. Sublethal effects of a vapour-active pyrethroid, transfluthrin, on *Aedes aegypti* and *Aedes albopictus* (diptera: Culicidae) fecundity and oviposition behaviour. Parasit Vectors 11, 486.

Bibbs, C.S., Hahn, D.A., Kaufman, P.E., Vectors, J.P., 2018b. Sublethal effects of a vapour-active pyrethroid, transfluthrin, on *Aedes aegypti* and *Aedes albopictus* (diptera: Culicidae) fecundity and oviposition behaviour. Parasit Vectors 11, 1–9.

Bibbs, C.S., Kline, J., Kline, D.L., Estaver, J., Strohschein, R., Allan, S.A., Kaufman, P.E., Xue, R.-D., Batich, C.D., 2019. Olfactometric comparison of the volatile insecticide, metofluthrin, through behavioral responses of *Aedes albopictus* (diptera: Culicidae). J. Med. Entomol. 57, 17–24.

Bibbs, C.S., Kline, J., Kline, D.L., Estaver, J., Strohschein, R., Allan, S.A., Kaufman, P.E., Xue, R.D., Batich, C.D., 2020. Olfactometric comparison of the volatile insecticide, metofluthrin, through behavioral responses of *Aedes albopictus* (diptera: Culicidae). J. Med. Entomol. 57, 17–24.

Bohbot, J.D., Dickens, J.C., 2010. Insect repellents: modulators of mosquito odorant receptor activity. PLoS One 5, e12138.

Brady, O.J., Godfray, H.C., Tatem, A.J., Gething, P.W., Cohen, J.M., McKenzie, F.E., Perkins, T.A., Reiner, R.C. Jr., Tusting, L.S., Sinka, M.E., Moyes, C.L., Eckhoff, P.A., Scott, T.W., Lindsay, S.W., Hay, S.I., Smith, D.L., 2016. Vectorial capacity and vector control: reconsidering sensitivity to parameters for malaria elimination. Trans. R. Soc. Trop. Med. Hyg. 110 (2), 107–117.

Brady, O.J., Hay, S.I., 2020. The global expansion of dengue: how *Aedes aegypti* mosquitoes enabled the first pandemic arbovirus. Annu. Rev. Entomol. 65, 191–208.

Carey, A.F., Wang, G., Su, C.-Y., Zwiebel, L.J., Carlson, J.R., 2010. Odorant reception in the malaria mosquito *Anopheles gambiae*. Nature 464, 66–71.

Carnevale, P., Frézil, J.L., Bosseno, M.F., Le Pont, F., Lancien, J., 1978. [The aggressiveness of Anopheles gambiae A in relation to the age and sex of the human subjects]. Bull. World Health Organ. 56, 147–154.

Chauhan, K.R., Aldrich, J.R., Mccardle, P.W., White, G.B., Webb, R.E., 2012. A field bioassay to evaluate potential spatial repellents against natural mosquito populations. J. Am. Mosq. Control Assoc. 28, 301–306.

Costantini, C., Sagnon, N., Della Torre, A., Coluzzi, M., 1999. Mosquito behavioural aspects of vector-human interactions in the anopheles gambiae complex. Parassitologia 41, 209–217.

Cotter, C., Sturrock, H.J., Hsiang, M.S., Liu, J., Phillips, A.A., Hwang, J., Gueye, C.S., Fullman, N., Gosling, R.D., Feachem, R.G., 2013. The changing epidemiology of malaria elimination: new strategies for new challenges. Lancet 382, 900–911.

Curtis, C.F., Lines, J.D., Ijumba, J., Callaghan, A., Hill, N., Karimzad, M.A., 1987. The relative efficacy of repellents against mosquito vectors of disease. Med. Vet. Entomol. 1, 109–119.

Davis, E.E., Sokolove, P.G., 1976. Lactic acid-sensitive receptors on the antennae of the mosquito, Aedes aegypti. J. Comp. Physiol. 105, 43–54.

De Jong, R., Knols, B.G.J., 1995. Selection of biting sites on man by two malaria mosquito species. Experientia 51, 80–84.

Debboun, M., Wagman, J., 2004. In vitro repellency of N,N-diethyl-3-methylbenzamide and N,N-diethylphenylacetamide analogs against *Aedes aegypti* and *Anopheles stephensi* (Diptera: Culicidae). J. Med. Entomol. 41, 430–434.

Degennaro, M., 2015. The mysterious multi-modal repellency of DEET. Fly 9, 45–51.

Denz, A., Njoroge, M.M., Tambwe, M.M., Champagne, C., Okumu, F., Van Loon, J.J.A., Hiscox, A., Saddler, A., Fillinger, U., Moore, S.J., Chitnis, N., 2021. Predicting the impact of outdoor vector control interventions on malaria transmission intensity from semi-field studies. Parasit Vectors 14, 64.

Dickens, J.C., Bohbot, J.D., 2013. Mini review: mode of action of mosquito repellents. Pestic. Biochem. Physiol. 106, 149–155.

Ferguson, H.M., Ng'habi, K.R., Walder, T., Kadungula, D., Moore, S.J., Lyimo, I., Russell, T.L., Urassa, H., Mshinda, H., Killeen, G.F., Knols, B.G., 2008. Establishment of a large semi-field system for experimental study of African malaria vector ecology and control in Tanzania. Malar J 7, 158.

Fernandes, L., Briegel, H., 2005. Reproductive physiology of *Anopheles gambiae* and *Anopheles atroparvus*. J. Vector Ecol. 30, 11–26.

Gerberg, E.J., Barnard, D.R., Ward, R.A., 1994. Manual for Mosquito Rearing and Experimental Techniques. American Mosquito Control Association, Inc, Lake Charles, LA revised edn.

Gillies, M.T., 1964. Selection for host preference in *Anopheles gambiae*. Nature 203, 852–854.

Gillies, M.T., 1980. The role of carbon dioxide in host-finding by mosquitoes (diptera, culicidae): a review. Bull. Entomol. Res. 70, 525–532.

Gillies, M.T., Wilkes, T.J., 1972. The range of attraction of animal baits and carbon dioxide for mosquitoes. Studies in a freshwater area of West Africa. Bull. Entomol. Res. 61, 389–404.

Gillies, M.T., Wilkes, T.J., 1970. The range of attraction of single baits for some West African mosquitoes. Bull. Entomol. Res. 69, 225–235.

Glunt, K.D., Oliver, S.V., Hunt, R., Paaijmans, K.P., 2018. The impact of temperature on insecticide toxicity against the malaria vectors *Anopheles arabiensis* and *Anopheles funestus*. Mal. J. 17, 131. doi:10.1186/s12936-018-2250-4.

Grieco, J.P., Achee, N.L., Andre, R.G., Roberts, D.R., 2000. A comparison study of house entering and exiting behavior of *Anopheles vestitipennis* (diptera: Culicidae) using experimental huts sprayed with DDT or deltamethrin in the Southern district of Toledo, Belize, CA. J. Vector Ecol. 25, 62–73.

Grieco, J.P., Achee, N.L., Chareonviriyaphap, T., Suwonkerd, W., Chauhan, K., Sardelis, M.R., Roberts, D.R., 2007. A new classification system for the actions of IRS chemicals traditionally used for malaria control. PLoS One. doi:10.1371/journal.pone.0000716.

Gryseels, C., Uk, S., Sluydts, V., Durnez, L., Phoeuk, P., Suon, S., Set, S., Heng, S., Siv, S., Gerrets, R., Tho, S., Coosemans, M., Peeters Grietens, K., 2015. Factors influencing the use of topical repellents: implications for the effectiveness of malaria elimination strategies. Sci. Rep. 5, 16847.

Haddow, A.J., 1942. The mosquito fauna and climate of native huts at Kisumu, Kenya. Bull. Entomol. Res. 33 (2), 91–142.

Harrington, L.C., Foy, B.D., Bangs, M.J., 2020. Considerations for human blood-feeding and arthropod exposure in vector biology research: an essential tool for investigations and disease control. Vector Borne Zoonotic Dis. 20, 807–816.

Hellewell, J., Sherrard-Smith, E., Ogoma, S., Churcher, T.S., 2021. Assessing the impact of low-technology emanators alongside long-lasting insecticidal nets to control malaria. Philos. Trans. R. Soc. Lond. B Biol. Sci. 376, 20190817.

Hill, N., Zhou, H.N., Wang, P., Guo, X., Carneiro, I., Moore, S.J., 2014. A household randomized, controlled trial of the efficacy of 0.03% transfluthrin coils alone and in combination with long-lasting insecticidal nets on the incidence of *Plasmodium falciparum* and *Plasmodium vivax* malaria in Western Yunnan province, China. Mal J. 13, 208.

Hoffmann, E.J., Miller, J.R., 2002. Reduction of mosquito (Diptera: Culicidae) attacks on a human subject by combination of wind and vapor-phase DEET repellent. J. Med. Entomol. 39, 935–938.

Hudson, J.E., Esozed S., 1971. The effects of smoke from mosquito coils on *Anopheles gambiae* Giles and *Mansonia uniformis* (Theo.) in verandah-trap huts at Magugu, Tanzania. Bull Entomol. Res. 61, 247–265.

Iwashita, H., Dida, G.O., Sonye, G.O., Sunahara, T., Futami, K., Njenga, S.M., Chaves, L.F., Minakawa, N., 2014. Push by a net, pull by a cow: can zooprophylaxis enhance the impact of insecticide treated bed nets on malaria control? Parasit. Vectors 7, 52.

Johnson, P.C.D., Barry, S.J.E., Ferguson, H.M., Muller, P., 2014. Power analysis for generalized linear mixed models in ecology and evolution. Methods Ecol Evol. doi:10.1111/2041-210X.12306.

Jufri, M., Irmayanti, E., Gozan, M., 2016. Formulation of tobacco based mosquito repellent to avoid dengue fever. Int. J. Pharmtech. Res. 9, 140–145.

Kawada, H., Yen, N.T., Hoa, N.T., Sang, T.M., VAN Dan, N., Takagi, M., 2005. Field evaluation of spatial repellency of metofluthrin impregnated plastic strips against mosquitoes in Hai Phong City, Vietnam. Am. J. Trop. Med. Hyg. 73, 350 –35.

Kennedy, J.S., 1947. The excitant and repellent effects on mosquitos of sub-lethal contacts with DDT. Bull. Entomol. Res. 37, 593–607.

Khan, A.A., Maibach, H.I., Skidmore, D.L., 1973. A study of insect repellents 2. Effect of temperature on protection time. J. Econ. Entomol. 66, 437–439.

Kirby, M.J., Green, C., Milligan, P.M., Sismanidis, C., Jasseh, M., Conway, D.J., Lindsay, S.W., 2008. Risk factors for house-entry by malaria vectors in a rural town and satellite villages in the gambia. Malar. J. 7, 2 -2.

Kirby, M.J., Lindsay, S.W., 2004. Responses of adult mosquitoes of two sibling species, *Anopheles arabiensis* and *Anopheles gambiae* s.s. (Diptera: Culicidae), to high temperatures. Bull. Entomol. Res. 94, 441–448.

Kounnavong, S., Gopinath, D., Hongvanthong, B., Khamkong, C., Sichanthongthip, O., 2017. Malaria elimination in Lao PDR: the challenges associated with population mobility. Infect. Dis. Poverty. 6, 81.

Kuthiala, A., Gupta, R.K., Davis, E.J., 1992. Effect of the repellent DEET on the antennal chemoreceptors for oviposition in *Aedes aegypti* (Diptera: Culicidae). J. Med. Entomol. 29, 639–643.

Lana, R., Nekkab, N., Siqueira, A.M., Peterka, C., Marchesini, P., Lacerda, M., Mueller, I., White, M., Villela, D., 2021. The top 1%: quantifying the unequal distribution of malaria in Brazil. Malar. J. 20, 87.

Liebman, K.A., Stoddard, S.T., Morrison, A.C., Rocha, C., Minnick, S., Sihuincha, M., Russell, K.L., Olson, J.G., Blair, P.J., Watts, D.M., Kochel, T., Scott, T.W., 2012. Spatial dimensions of dengue virus transmission across interepidemic and epidemic periods in Iquitos, Peru (1999-2003). PLoS Negl. Trop. Dis. 6, e1472.

Lindsay, S.W., Adiamah, J.H., Miller, J.E., Pleass, R.J., Armstrong, J.R.M., 1993. Variation in attractiveness of human subjects to malaria mosquitoes (Diptera: Culicidae) in the Gambia. J. Med. Entomol. 30 (2), 368–373.

Lorenz, L.M., Keane, A., Moore, J.D., Munk, C.J., Seeholzer, L., Mseka, A., Simfukwe, E., Ligamba, J., Turner, E.L., Biswaro, L.R., Okumu, F.O., Killeen, G.F., Mukabana, W.R., Moore, S.J., 2013. Taxis assays measure directional movement of mosquitoes to olfactory cues. Parasit. Vectors 6, 131.

Lupi, E., Hatz, C., Schlagenhauf, P., 2013. The efficacy of repellents against Aedes, Anopheles, Culex and Ixodes spp.—a literature review. Travel Med. Infect. Dis. 11, 374–411.

Magesa, S.M., Wilkes, T.J., Mnzava, A.E., Njunwa, K.J., Myamba, J., Kivuyo, M.D., Hill, N., Lines, J.D., Curtis, C.F., 1991. Trial of pyrethroid impregnated bednets in an area of Tanzania holoendemic for malaria. Part 2. Effects on the malaria vector population. Acta Trop. 49, 97–108.

Mahande, A., Mosha, F., Mahande, J., Kweka, E., 2007. Feeding and resting behaviour of malaria vector, *Anopheles arabiensis* with reference to zooprophylaxis. Malar. J. 6, 100.

Maia, M.F., Kliner, M., Richardson, M., Lengeler, C., Moore, S.J., 2018. Mosquito repellents for malaria prevention. Cochrane Database Syst. Rev. 2018 (2), CD011595. doi:10.1002/14651858.CD011595.pub2.

Martin, N.J., Nam, V.S., Lover, A.A., Phong, T.V., Tu, T.C., Mendenhall, I.H., 2020. The impact of transfluthrin on the spatial repellency of the primary malaria mosquito vectors in Vietnam: Anopheles dirus and anopheles minimus. Malar. J. 19, 9.

Masalu, J.P., Finda, M., Okumu, F.O., Minja, E.G., Mmbando, A.S., Sikulu-Lord, M.T., Ogoma, S.B., 2017. Efficacy and user acceptability of transfluthrin-treated sisal and hessian decorations for protecting against mosquito bites in outdoor bars. Parasit Vectors 10, 197.

Massue, D.J., Kisinza, W.N., Malongo, B.B., Mgaya, C.S., Bradley, J., Moore, J.D., Tenu, F.F., Moore, S.J., 2016. Comparative performance of three experimental hut designs for measuring malaria vector responses to insecticides in Tanzania. Mal J. 15, 165.

Massue, D.J., Lorenz, L.M., Moore, J.D., Ntabaliba, W.S., Ackerman, S., Mboma, Z.M., Kisinza, W.N., Mbuba, E., Mmbaga, S., Bradley, J., Overgaard, H.J., Moore, S.J., 2019. Comparing the new Ifakara Ambient Chamber Test with WHO cone and tunnel tests for bioefficacy and non-inferiority testing of insecticide-treated nets. Malar. J. 18, 153.

Mbuba, E., Odufuwa, O.G., Tenywa, F.C., et al., 2021. Single blinded semi-field evaluation of MAÏA® topical repellent ointment compared to unformulated 20% DEET against *Anopheles gambiae*, *Anopheles arabiensis* and *Aedes aegypti* in Tanzania. Malar. J. 20, 12. https://doi.org/10.1186/s12936-020-03461-9.

Mcphatter, L.P., Mischler, P.D., Webb, M.Z., Chauhan, K., Lindroth, E.J., Richardson, A.G., Debboun, M., 2017. Laboratory and semi-field evaluations of two (transfluthrin) spatial repellent devices against *Aedes aegypti* (l.) (Diptera: Culicidae). US Army Med. Dep. J. (1-17), 13–22.

Menger, D.J., Omusula, P., Holdinga, M., Homan, T., Carreira, A.S., Vandendaele, P., Derycke, J.-L., Mweresa, C.K., Mukabana, W.R., Van Loon, J.J.A., Takken, W., 2015. Field evaluation of a push-pull system to reduce malaria transmission. PLoS One 10, e0123415.

Menger, D.J., Otieno, B., Marjolein De Rijk, W., Van Loon, J.J., Takken, W., 2014. A push-pull system to reduce house entry of malaria mosquitoes. Malar J 13, 119.

Mitchell, S.N., Stevenson, B.J., Muller, P., Wilding, C.S., Egyir-Yawson, A., Field, S.G., Hemingway, J., Paine, M.J., Ranson, H., Donnelly, M.J., 2012. Identification and validation of a gene causing cross-resistance between insecticide classes in Anopheles gambiae from Ghana. Proc. Natl. Acad. Sci. USA 109, 6147–6152.

Mmbando, A.S., Ngowo, H., Limwagu, A., Kilalangongono, M., Kifungo, K., Okumu, F.O., 2018. Eave ribbons treated with the spatial repellent, transfluthrin, can effectively protect against indoor-biting and outdoor-biting malaria mosquitoes. Malar J. 17, 368.

Monroe, A., Moore, S., Koenker, H., Lynch, M., Ricotta, E., 2019. Measuring and characterizing night time human behaviour as it relates to residual malaria transmission in sub-Saharan Africa: a review of the published literature. Malar J. 18, 6.

Moore, S.J., Davies, C.R., Cameron, M.M., 2007. Are mosquitoes diverted from repellent-using individuals to non-users? Results of a field study in bolivia. Trop. Med. Int. Health 12, 1–8.

Morrison, A.C., Reiner, R.C., Elson, W.H., Astete, H., Guevara, C., Del Aguila, C., Bazan, I., Siles, C., Barrera, P., Kawiecki, A.B., et al. Efficacy of a spatial repellent for control of *Aedes*-borne virus transmission: a cluster randomized trial in Iquitos, Peru. medRxiv 2021. DOI: 10.1101/2021.03.03.21252148.

Moshi, I.R., Ngowo, H., Dillip, A., Msellemu, D., Madumla, E.P., Okumu, F.O., Coetzee, M., Mnyone, L.L., Manderson, L., 2017. Community perceptions on outdoor malaria transmission in kilombero valley, Southern Tanzania. Malar. J. 16, 274.

Mr4, 2009. Methods in Anopheles research manual. Malaria Research and Reference Reagent Resource Centre. http://www.Mr4.Org/portals/3/pdfs/protocolbook/methods-anophelesresearchv4c.Pdf. (Accessed May 18th 2011).

Muir, L.E., Kay, B.H., Thorne, M.J., 1992. *Aedes aegypti* (Diptera: Culicidae) vision: response to stimuli from the optical environment. J. Med. Entomol. 29, 445–450.

Mukabana, R.W., Takken, W., Coe, R., Knols, B.G.J., 2002. Host-specific cues cause differential attractiveness of Kenyan men to the African malaria vector *Anopheles gambiae*. Malaria J. 1, 19.

Mulatier, M., Porciani, A., Nadalin, L., Ahoua Alou, L.P., Chandre, F., Pennetier, C., Dormont, L., Cohuet, A., 2018. Deet efficacy increases with age in the vector mosquitoes *Anopheles gambiae s.s.* and *Aedes albopictus* (Diptera: Culicidae). J. Med. Entomol. 55, 1542–1548.

Nash, R., Lambert, B., N'guessan, R., N'gufor, C., Rowland, M., Oxborough, R., Moore, S.J., Tungu, P., Sherrard-Smith, E., Churcher, T.S., 2021. Systematic review of the entomological impact of insecticide-treated nets evaluated using experimental hut trials in Africa. Biorxiv. https://www.medrxiv.org/content/10.1101/2021.1104.1107.21254306v21254301.full.pdf. (Accessed May 25th 2021).

Ngowo, H.S., Kaindoa, E.W., Matthiopoulos, J., Ferguson, H.M., Okumu, F.O., 2017. Variations in household microclimate affect outdoor-biting behaviour of malaria vectors. Wellcome Open Res. 2, 102.

Njoroge, M.M., Fillinger, U., Saddler, A., Moore, S., Takken, W., Van Loon, J.J.A., Hiscox, A., 2021. Evaluating putative repellent 'push' and attractive 'pull' components for manipulating the odour orientation of host-seeking malaria vectors in the peri-domestic space. Parasit Vectors 14, 42.

Ogoma, S.B., Lorenz, L.M., Ngonyani, H., Sangusangu, R., Kitumbukile, M., Kilalangongono, M., Simfukwe, E.T., Mseka, A., Mbeyela, E., Roman, D., Moore, J., Kreppel, K., Maia, M.F., Moore, S.J., 2014a. An experimental hut study to quantify the effect of DDT and airborne pyrethroids on entomological parameters of malaria transmission. Malar J. 13, 131.

Ogoma, S.B., Mmando, A.S., Swai, J.K., Horstmann, S., Malone, D., Killeen, G.F., 2017. A low technology emanator treated with the volatile pyrethroid transfluthrin confers long term protection against outdoor biting vectors of lymphatic filariasis, arboviruses and malaria. PLoS Negl Trop Dis 11, e0005455.

Ogoma, S.B., Moore, S.J., Maia, M.F., 2012a. A systematic review of mosquito coils and passive emanators: defining recommendations for spatial repellency testing methodologies. Parasit Vectors 5, 287.

Ogoma, S.B., Ngonyani, H., Simfukwe, E.T., Mseka, A., Moore, J., Maia, M.F., Moore, S.J., Lorenz, L.M., 2014b. The mode of action of spatial repellents and their impact on vectorial capacity of*Anopheles gambiae sensu stricto*. PLoS One 9 (12), e110433.

Ogoma, S.B., Ngonyani, H., Simfukwe, E.T., Mseka, A., Moore, J, Killeen, G.F., 2012. Spatial repellency of transfluthrin-treated hessian strips against laboratory-reared *Anopheles arabiensis* mosquitoes in a semi-field tunnel cage. Parasit Vectors 5, 54.

Okumu, F.O., Killeen, G.F., Ogoma, S., Biswaro, L., Smallegange, R.C., Mbeyela, E., Titus, E., Munk, C., Ngonyani, H., Takken, W., Mshinda, H., Mukabana, W.R., Moore, S.J., 2010. Development and field evaluation of a synthetic mosquito lure that is more attractive than humans. PLoS One 5, e8951.

Okumu, F.O., Moore, J., Mbeyela, E., Sherlock, M., Sangusangu, R., Ligamba, G., Russell, T., Moore, S.J., 2012. A modified experimental hut design for studying responses of disease-transmitting mosquitoes to indoor interventions: the Ifakara experimental huts. PLoS One 7, e30967.

Orsborne, J., Furuya-Kanamori, L., Jeffries, C.L., Kristan, M., Mohammed, A.R., Afrane, Y., O'reilly, K., Massad, E., Drakeley, C., Walker, T., Yakob, L., 2018. Using the human blood index to investigate host biting plasticity: a systematic review and meta-regression of the three major african malaria vectors. Mal. J. 17, 479.

Pettebone, M.S., 2014. Characterisation of transfluthirn emissions over time in an enclosed space over a range of discreet temperatures. Uniformed Services University of the Health Sciences, MSc Thesis. https://apps.dtic.mil/sti/pdfs/AD1012851.pdf. (Accessed May 25th 2021).

Powell, J.R., Tabachnick, W.J., 2013. History of domestication and spread of*Aedes aegypti*--a review. Mem. Inst. Oswaldo Cruz 108 (Suppl 1), 11–17.

Ray, A., 2015. Reception of odors and repellents in mosquitoes. Curr. Opin. Neurobiol. 34, 158–164.

Rinker, D.C., Jones, P.I., Pitts, R.J., Rützler, M., Camp, G.J., Sun, L., Xu, P., Dorset, D., Weaver, D., Zwiebel, L.J., 2012. Novel high-throughput screens of anopheles gambiae odorant receptors reveal candidate behaviour-modifying chemicals for mosquitoes. Physiol Entomol 37, 33–41.

Rund, S.S., O'donnell, A.J., Gentile, J.E., Reece, S.E., 2016. Daily rhythms in mosquitoes and their consequences for malaria transmission. Insects. 7, 14.

Rutledge, L.C., Gupta, R.K., 1999. Variation in the protection periods of repellents on individual human subjects: an analytical review. J. Am. Mosq. Control Assoc. 15, 348–355.

Saddler, A., Kreppel, K.S., Chitnis, N., Smith, T.A., Denz, A., Moore, J.D., Tambwe, M.M., Moore, S.J., 2019. The development and evaluation of a self-marking unit to estimate malaria vector survival and dispersal distance. Malar. J. 18, 441.

Salazar, F.V., Achee, N.L., Grieco, J.P., Prabaripai, A., Ojo, T.A., Eisen, L., Dureza, C., Polsomboon, S., Chareonviriyaphap, T., 2013. Effect of *Aedes aegypti* exposure to spatial repellent chemicals on BG-sentinel trap catches. Parasit Vectors 6, 145.

Sandfort, M., Vantaux, A., Kim, S., Obadia, T., Pepey, A., Gardais, S., Khim, N., Lek, D., White, M., Robinson, L.J., Witkowski, B., Mueller, I., 2020. Forest malaria in Cambodia: the occupational and spatial clustering of *Plasmodium vivax* and *Plasmodium falciparum* infection risk in a cross-sectional survey in Mondulkiri province, Cambodia. Malar. J. 19, 413.

Sangoro, O., Lweitojera, D., Simfukwe, E., Ngonyani, H., Mbeyela, E., Lugiko, D., Kihonda, J., Maia, M., Moore, S., 2014. Use of a semi-field system to evaluate the efficacy of topical repellents under user conditions provides a disease exposure free technique comparable with field data. Malar. J. 13, 159.

Sangoro, O.P., Gavana, T., Finda, M., Mponzi, W., Hape, E., Limwagu, A., Govella, N.J., Chaki, P., Okumu, F.O., 2020. Evaluation of personal protection afforded by repellent-treated sandals against mosquito bites in South-Eastern Tanzania. Malar. J. 19, 148.

Schmied, W.H., Takken, W., Killeen, G.F., Knols, B.G., Smallegange, R.C., 2008. Evaluation of two counterflow traps for testing behaviour-mediating compounds for the malaria vector *Anopheles gambiae s.s.* Under semi-field conditions in Tanzania. Malar. J. 7, 230.

Schreck, C.E., 1977. Techniques for the evaluation of insect repellents: a critical review. Annu. Rev. Entomol. 22, 101–119.

Scott, T.W., Amerasinghe, P.H., Morrison, A.C., Lorenz, L.H., Clark, G.G., Strickman, D., Kittayapong, P., Edman, J.D., 2000. Longitudinal studies of *Aedes aegypti* (Diptera: Culicidae) in Thailand and Puerto Rico: blood feeding frequency. J. Med. Entomol. 37, 89–101.

Seto, K.C., Guneralp, B., Hutyra, L.R., 2012. Global forecasts of urban expansion to 2030 and direct impacts on biodiversity and carbon pools. Proc. Natl. Acad. Sci. USA. 109, 16083–16088.

Shirai, O., Tsuda, T., Kitagawa, S., Naitoh, K., Seki, T., Kamimura, K., Morohashi, M., 2002. Alcohol ingestion stimulates mosquito attraction. J. Am. Mosq. Control Assoc. 18, 91–96.

Silver, J.B., Service, M.W., 2008. Chapter 16: experimental hut techniquesMosquito Ecology: Field Sampling Methods. Springer, Netherlands.

Smallegange, R.C., Verhulst, N.O., Takken, W., 2011. Sweaty skin: an invitation to bite? Trends Parasitol. 27, 143–148.

Smith, A., 1963. Principles in assessment of insecticides by experimental huts. Nature 198, 171.

Smith, T., Maire, N., Ross, A., Penny, M., Chitnis, N., Schapira, A., Studer, A., Genton, B., Lengeler, C., Tediosi, F., De Savigny, D., Tanner, M., 2008. Towards a comprehensive simulation model of malaria epidemiology and control. Parasitology 135, 1507–1516.

Sukkanon, C., Nararak, J., Bangs, M.J., Hii, J., Chareonviriyaphap, T., 2020. Behavioral responses to transfluthrin by *Aedes aegypti*, *Anopheles minimus*, *Anopheles harrisoni*, and *Anopheles dirus* (diptera: Culicidae). PLoS One 15, e0237353.

Syafruddin, D., Asih, P.B.S., Rozi, I.E., et al., 2020. Efficacy of a Spatial Repellent for Control of Malaria in Indonesia: A Cluster-Randomized Controlled Trial [published correction appears in Am J Trop Med Hyg. 2020 Nov;103(5):2151. Am. J. Trop. Med. Hyg. 103(1), 344–358. doi:10.4269/ajtmh.19-0554.

Syafruddin, D., Bangs, M.J., Sidik, D., Elyazar, I., Asih, P.B., Chan, K., Nurleila, S., Nixon, C., Hendarto, J., Wahid, I., Ishak, H., Bogh, C., Grieco, J.P., Achee, N.L., Baird, J.K., 2014. Impact of a spatial repellent on malaria incidence in two villages in Sumba, Indonesia. Am. J. Trop. Med. Hyg. 91, 1079–1087.

Tainchum, K., Ritthison, W., Sathantriphop, S., Tanasilchayakul, S., Manguin, S., Bangs, M.J., Chareonviriyaphap, T., 2014. Influence of time of assay on behavioral responses of laboratory and field populations *Aedes aegypti* and *Culex quinquefasciatus* (Diptera: Culicidae) to DEET. J. Med. Entomol. 51, 1227–1236.

Takakura, K.I., 2012. Bayesian Estimation for the Effectiveness of Pesticides and Repellents. Journal of Economic Entomology 105(5), 1856–1862.

Takken, W., 1991. The role of olfaction in host-seeking of mosquitoes: a review. Int. J. Trop. Insect Sci. 12, 287–295.

Takken, W., Knols, B., Otten, H.J., 1997. Interactions between physical and olfactory cues in the host-seeking behaviour of mosquitoes: the role of relative humidity. Ann. Trop. Med. Parasitol. 91, S119–S120.

Takken, W., Knols, B.G., 1999. Odor-mediated behavior of Afrotropical malaria mosquitoes. Annu. Rev. Entomol. 44, 131–157.

Takken, W., Verhulst, N.O., 2017. Chemical signaling in mosquito-host interactions: the role of human skin microbiota. Curr. Opin. Insect Sci. 20, 68–74.

Tambwe, M.M., Moore, S.J., Chilumba, H., Swai, J.K., Moore, J.D., Stica, C., Saddler, A., 2020. Semi-field evaluation of freestanding transfluthrin passive emanators and the bg sentinel trap as a "push-pull control strategy" against *Aedes aegypti* mosquitoes. Parasit. Vectors 13, 392.

Tambwe, M.M., Saddler, A., Moore, J.D., Kibondo, U.A., Odufuwa, O.G., Stica, C., Moore, S.J., 2021. The use of Ifakara large ambient chamber test (i-lact) as a modified peri-domestic space bioassay for measuring feeding success and 24-hour mortality of volatile insecticides in the semi field system. Biorxiv. In preparation for Parasites and Vectors.

Tambwe, M.M., Saddler, A., Kibondo, U.A., Mashauri, R., Kreppel, K.S., Govella N.J., Moore S.J., 2021. Semi-field evaluation of the exposure-free mosquito electrocuting trap and BG-Sentinel trap as an alternative to the human landing catch for measuring the efficacy of transfluthrin emanators against Aedes aegypti. Parasit Vectors 14 (1), 265.

Thanispong, K., Achee, N.L., Grieco, J.P., Bangs, M.J., Suwonkerd, W., Prabaripai, A., Chauhan, K.R., Chareonviriyaphap, T., 2010. A high throughput screening system for determining the three actions of insecticides against *Aedes aegypti* (Diptera: Culicidae) populations in thailand. J. Med. Entomol. 47, 833–841.

Tisgratog, R., Tananchai, C., Bangs, M.J., Tainchum, K., Juntarajumnong, W., Prabaripai, A., Chauhan, K.R., Pothikasikorn, J., Chareonviriyaphap, T., 2011. Chemically induced behavioral responses in *Anopheles minimus* and *Anopheles harrisoni* in Thailand. J. Vector Ecol 36, 321–331.

Turner, S.L., Li, N., Guda, T., Githure, J., Carde, R.T., Ray, A., 2011. Ultra-prolonged activation of CO_2-sensing neurons disorients mosquitoes. Nature 474, 87–91.

Van Der Hoek, W., Konradsen, F., Amerasinghe, P.H., Perera, D., Piyaratne, M.K., Amerasinghe, F.P., 2003. Towards a risk map of malaria for Sri Lanka: the importance of house location relative to vector breeding sites. Int. J. Epidemiol. 32, 280–285.

Van Roey, K., Sokny, M., Denis, L., Van Den Broeck, N., Heng, S., Siv, S., Sluydts, V., Sochantha, T., Coosemans, M., Durnez, L., 2014. Field evaluation of picaridin repellents reveals differences in repellent sensitivity between Southeast Asian vectors of malaria and arboviruses. PLos Negl. Trop. Dis. 8, e3326.

Vatandoost, H., Hanafi-Bojd, A.A., 2008. Laboratory evaluation of 3 repellents against anopheles stephensi in the Islamic Republic of Iran. East Mediterr. Health J. 14, 260–267.

Verhulst, N.O., Andriessen, R., Groenhagen, U., Bukovinszkine Kiss, G., Schulz, S., Takken, W., Van Loon, J.J., Schraa, G., Smallegange, R.C., 2010. Differential attraction of malaria mosquitoes to volatile blends produced by human skin bacteria. PLoS One 5, e15829.

Verhulst, N.O., Umanets, A., Weldegergis, B.T., Maas, J.P.A., Visser, T.M., Dicke, M., Smidt, H., Takken, W., 2018. Do apes smell like humans? The role of skin bacteria and volatiles of primates in mosquito host selection. J. Exp. Biol., 221.

Verhulst, N.O., Weldegergis, B.T., Menger, D., Takken, W., 2016. Attractiveness of volatiles from different body parts to the malaria mosquito Anopheles coluzzii is affected by deodorant compounds. Sci. Rep. 6, 27141.

Vontas, J., Moore, S., Kleinschmidt, I., Ranson, H., Lindsay, S., Lengeler, C., Hamon, N., Mclean, T., Hemingway, J., 2014. Framework for rapid assessment and adoption of new vector control tools. Trends Parasitol. 30, 191–204.

Wagman, J.M., Achee, N.L., Grieco, J.P., 2015. Insensitivity to the spatial repellent action of transfluthrin in *Aedes aegypti*: a heritable trait associated with decreased insecticide susceptibility. PLoS Negl. Trop. Dis. 9, e0003726.

Webster, B., Lacey, E.S., Carde, R.T., 2015. Waiting with bated breath: opportunistic orientation to human odor in the malaria mosquito, *anopheles gambiae*, is modulated by minute changes in carbon dioxide concentration. J. Chem. Ecol. 41, 59–66.

World Health Organization, 2009. Guidelines for efficacy testing of mosquito repellents for human skin. WHO/HTM/NTD/WHOPES/2009.4, World Health Organisation, Geneva.

World Health Organization, 2012. Global strategy for dengue prevention and control 2012-2020. who/htm/ntd/vem/2012.5. World Health Organisation, Geneva.

World Health Organization, 2013. Guidelines for efficacy testing of spatial repellents. World Health Organisation, Geneva.

World Health Organization, 2016. Global vector control response 2017–2030. World Health Organisation, Geneva.

World Health Organization, 2019. Guideline for malaria vector control. World Health Organisation, Geneva.

World Health Organization, 2020. The world malaria report 2020. World Health Organisation, Geneva.

World Health Organization. 2021. Vector borne disease fact sheet. https://www.who.int/news-room/fact-sheets/detail/vector-borne-diseases. (Accessed 11/03/2021).

WHOPES, 2009. Guidelines for efficacy testing of mosquito repellents for human skin. World Health Organisation, Geneva.

WHOPES, 2009. Guidelines for efficacy testing of mosquito repellents for human skin who/htm/ntd/whopes/2009.4. World Health Organisation, Geneva http://whqlibdoc.Who.Int/hq/2009/who_htm_ntd_whopes_2009.4_eng.Pdf.

WHOPES, 2013. Guidelines for efficacy testing of spatial repellents. World Health Organisation, Geneva.

Wright, R.H., Kellogg, F.E., 1962. Response of aedes aegypti to moist convection currents. Nature 194, 402–403.

Wynne, N.E., Lorenzo, M.G., Vinauger, C., 2020. Mechanism and plasticity of vectors' host-seeking behavior. Curr. Opin. Insect Sci. 40, 1–5.

Xue, R.D., Barnard, D.R., 1996. Human host avidity in Aedes albopictus: influence of mosquito body size, age, parity, and time of day. J. Am. Mosq. Control Assoc. 12(1), 58–63.

Zaim, M., Aitio, A., Nakashima, N., 2000. Safety of pyrethroid-treated mosquito nets. Med. Vet. Entomol. 14, 1–5.

CHAPTER

11

Semi-field evaluation of arthropod repellents: emphasis on spatial repellents

Daniel L. Kline[a], *Karen McKenzie*[b], *Adam Bowman*[a]

[a]USDA-ARS, Center for Medical and Veterinary Entomology, Gainesville, FL, United States, [b]Woodstream, Melbourne, FL, United States

11.1 Introduction

Arthropod-borne pathogens are a public health concern worldwide owing to their association with several diseases of humans and animals. Arthropods that vector these pathogens include mosquitoes, biting midges, flies, sand flies, ticks, and fleas. Unfortunately, for humans, there are not many effective vaccines against these pathogens and few specific drugs for inhibiting their propagation. As a consequence, controlling the spread of these diseases requires direct targeting of the biting arthropods. Therefore, enormous efforts have been made to develop effective personal protection methods such as arthropod repellents, insecticide-treated clothing, and bed nets for use against these biting arthropods. The main objective of using these repellents is to prevent or discourage arthropod vector-mammalian host contact and prevent bites (Diaz, 2016). Arthropod repellents are thus often used as the first line of defense against biting hematophagous arthropods. There are many reports which document the important role that arthropod repellents have played in reducing the chance of getting vector-borne diseases (Barnard, 2000; Barnard et al., 1998; Barnard and Xue, 2004; Debboun and Strickman, 2013; Hill et al., 2007; Onyango and Moore, 2015; Rowland et al., 2004). Two broadly defined classes of arthropod repellents will be discussed in this chapter: topical and spatial repellents (SRs) with emphasis on the use of semi-field environments to evaluate their efficacy parameters.

11.1.1 Arthropod repellents and intended function in biting arthropod management

Vector-borne diseases from mosquitoes are a concern worldwide, and the ability to control and repel mosquito species that carry diseases

Advances in Arthropod Repellents.
DOI: https://doi.org/10.1016/B978-0-323-85411-5.00007-8

is a necessity for public health. The WHO reports that global malaria cases range from 149 to 303 million annually with 88% of cases occurring in the African region, 10% in South-East Asia, and 2% in Eastern Mediterranean (2015). In addition to malaria, outbreaks of locally acquired dengue fever have occurred in southern France attributed to the invasion of vector species, *Ae. aegypti* populations (Schaffner et al., 2014). Periodic epidemics of West Nile virus, Chikungunya, and Zika virus (all mosquito-transmitted diseases) have also been confirmed in the United States (Rosenberg et al., 2018).

Types of control measures and arthropod repellents can include stationary passive dispersal devices such as bed nets impregnated with pyrethroid insecticides (Cui et al., 2012), passive insecticide dispersal devices like impregnated ignitable coils (Hill et al., 2014) insecticide-soaked cloth decorations, ribbons, and chairs (Masalu et al., 2017, 2020), and topically applied repellents that are sprayed onto the skin, or applied via lotion formula (Barnard and Xue, 2004). When focusing on arthropod repellents, the ability to test the efficacy and keep participants safe from possible vector disease is paramount. While pure laboratory arm-in-cage studies done with disease-free mosquitoes of interest accomplish this, the need for more realistic environmental conditions is necessary to make generalized deductions about repellent protection from target species of mosquitoes; however, pure field studies do open the possibility of exposing participants to mosquito-borne diseases. The semi-field environment (SFE) may fill the need of performing safer, more environmentally realistic assays, as enclosures can be built in situ for areas where mosquito diseases are prevalent but still use laboratory-reared mosquitoes. Semi-field environments have been shown to have comparable results to field testing topical repellents when using laboratory-reared mosquito species (Sangoro et al., 2014). Specifics of topical and SRs will be discussed in detail within the following chapter sections.

11.1.1.1 Topical arthropod repellents defined

In this section, we will highlight the efficacy of topical repellents for protection against vector disease, current testing methodology limitations, and how semi-field (SF) testing may offer some solutions. Typically, topical arthropod repellents are applied as a spray, lotion, or cream directly on exposed skin or, in some cases on clothing, requiring either contact by the arthropods or very close proximity to stop them from biting at the very last stages of their feeding processes. If the application is uneven or otherwise applied incorrectly, an exploitable gap in protection may be present. The inherent efficacy of each product depends on its active ingredients and its formulation. The usual standard is for complete protection from the target arthropod so that relative effectiveness is judged by the duration of protection (Bernier et al., 2007). These compounds interfere with the mosquito's olfactory system preventing them from identifying their hosts (Davis and Sokolove, 1976). However, repellents do not kill the host-seeking mosquitoes, they simply reduce man-vector contact.

Topical arthropod repellents, such as N,N-diethyl-3-methylbenzamide (DEET), ethyl butylacetylaminoproprionate (IR3535), and 2-(2-hydroxyethyl0-1-piperidine carboxylic acid 1-methylpropyl ester (Picaridin), can be used to fill a protection gap from the point a person leaves a physical structure, and residual sprays or stationary protection devices are no longer available. It has been proposed that these topical repellents act as olfactory agonists or antagonists, via modulating olfactory receptors activity, in the absence and presence of indole and octanol, which are specific to these olfactory receptors (Bohbot and Dickens, 2012). Topical repellents are noted as having one of two mutually exclusive outcomes when a user is in a mosquito dense area. Direct physical contact with the mosquito can result in the mosquito avoiding the person without engaging a blood meal, or the mosquito may be repelled by the topical application and not make physical contact (Killeen, 2014). This presents the scenario where the mosquito either is repelled from contact or makes contact and is potentially killed. The difference in these two outcomes often characterize the repellent as a contact irritant, or an SR, respectively (Achee et al., 2012; Grieco et al., 2007; WHO, 2013). Understanding the mode of action of arthropod repellents and how repellents modulate odor-sensing will allow us to design and develop better repellent formulations. It is possible that a topical repellent works both as an SR, and as contact irritant. Regardless of the lack of understanding the mode of action of these products, there are a host of commercially available products that can be used as topical arthropod repellents.

11.1.1.2 *Available products*

Synthetic compounds, such as DEET, Picaridin, and IR3535 are widely used topical repellents that can be applied directly to the skin and have been proven to be excellent tools to protect individuals against the bites of hematophagous arthropods. Permethrin is another synthetic compound that is sometimes classified as a topical repellent and a synthetic Type I pyrethroidal neurotoxic insecticide, obtained from the dried flowers of *Chrysanthemum cinerariifolium*. Its main mechanism of action is axonal sodium channel depolarization causing repetitive nerve impulses as a neurotoxin (van den Bercken and Vijverberg, 1988). Permethrin is recommended to be used as an arthropod repellent on clothing only and direct skin contact should be avoided. Permethrin-treated clothes have been proven to be effective against insects. It can also be applied to bed nets and bedding with high-level protection against mosquitoes, ticks, and other arthropods (Banks et al., 2014).

There are many different forms of chemical topical repellents. Picardin and DEET have been shown to be effective at repelling bites from anopheline and aedine vectors (Frances, 1987; Van Roey et al., 2014), however, a study showed that perceptions (odor, perceived risk) of the repellent lotion may lead to lowered efficacy in areas of high vector disease risk in Cambodia due to nonuse or improper reapplication (Gryseels et al., 2015). One limitation of topical repellents may be willingness to use it and reapply the repellent correctly.

Increased efficacy may be achieved in some cases by the combination of active ingredients in topical arthropod repellents. Lindsay et al. (1998) have shown that a mixture of DEET and thanaka (a paste made from the pulp of the wood apple tree *Limonia acidissima*) extended the protective capability of the topical repellent for *An. minimus* and *An. maculatus* in rural Thailand. Laboratory bioassays also showed that thanaka alone conferred some protection against *Ae. aegypti*. The pregnant women in this study also preferred the topical repellent when mixed with thanaka, perhaps by taking into consideration cultural differences and motivations in regard to chemical application, this may lead to better behavior in using and reapplying topical repellents, thereby making them more effective (Lindsay et al., 1998).

DEET is the most common topical arthropod repellent currently available and provides hours of protection after a single application (Kweka et al., 2012); however, there are some nonsynthetic alternatives. Of 23 essential oils evaluated by Uniyal et al. (2016), Litsea oil showed a 77.26% repellency against *Ae. aegypti*. When DEET was tested in the same study, it was found to have only a slightly higher repellency at 85.48% (Uniyal et al., 2016). Conversely, of 13 commercial repellent products evaluated by Barnard and Rui-De Xue (2004), those with DEET showed the longest protection and efficacy against *Ae. albopictus* a known vector of chikungunya, dengue, and dirofilariasis (Barnard and Xue, 2004).

Barnard and Xue (2004) also noted that of the nine natural (non-DEET) repellents that were tested in their study, they were protective for less than 3 hours on average (Barnard and Xue,

2004). While some alternatives for synthetic topical repellents are available, a limitation of widespread protection could be reapplication of the repellent formulation, especially those with lower protection times. While the methodologies for these field tests do replicate real-world repellent use, the field testing also introduces a number of confounding variables when testing efficacy of topical repellents. Stochastic events like changing temperature, humidity, wind speed, and evaporation rates can be difficult to quantify and standardize across treatment levels in a pure field setting, and these factors do affect the protection time of repellents (Khan et al., 1975). Environmental factors are important to consider in efficacy testing and using a more controlled SF environment could help to augment data in pure field studies and create more generalized experimental designs for topical repellent efficacy that can attempt to standardize environmental variables across treatments. The limitations for laboratory assays to evaluate repellent efficacy are that testing cannot be done including environmental factors mentioned earlier, and while that may be advantageous in some cases (Barnard and Xue, 2004), certain environmental factors may be desired for more robust testing. SF testing methodologies could bridge the gap between pure laboratory studies and pure field studies, as SF testing has been shown to yield similar results to filed work. Sangoro et al. (2014) found that when testing efficacy of topical DEET against *An. arabiensis* in a SF assay, the results were similar to testing DEET against *An. gambiae* in a field study (Sangoro et al., 2014). The enclosures for these semi-field environments consist of large frame structures covered with screen permeable to air, but which traps mosquitoes inside (Sangoro et al., 2014) These structures have the added benefit of being enclosed and modifiable to have more fine control over experimental design factors than a pure field study, such as you would have in a laboratory study, but are able to be erected in the geographical location where mosquito research would be needed for testing. This allows for testing target species with realistic field conditions and adds some flexibility not seen in pure field trials and bioassays. Knols et al. (2002) were able to construct an SF assay in western Kenya that allowed the research team to study the behavioral ecology of *An. gambiae*, a known malaria vector. They were able to collect a wealth of data on the species behavioral and reproductive ecology in a realistic ecosystem that would have been nearly impossible to accomplish under a set of pure field experiments (Knols et al., 2002).

11.1.1.3 Major issues with topical arthropod repellent use

Applications and limitations for using topical repellents: Gryseels et al. (2015) reported that while topical arthropod repellents were made widely available in a Cambodian study site and were effective at repelling mosquito bites, 97%, the repeated daily use was low, i.e., 8%. When participants in the study were surveyed, some of the reasons for nonuse were an aversion to the smell (similar to the study done by Lindsay et al., 1998, where 50% of participants disliked the topical repellent odor), and fear of possible side effects from the chemical substrates (Lindsay et al., 1998). These deterrent concerns were especially notable factors for nonuse in women and for their children. Another delivery device for topical repellents was a commercially available formulation of soap, Mosbar (Frances, 1987; Yap, 1986). In a study done in Afghanistan by Rowland et al. (2004), this soap was shown to be an effective protectant against *falciparum* malaria in age groups from 5 to 20 years old when compared to placebo lotion. The Mosbar treatment also reduced the density of anopheline mosquitoes in the area of sleeping participants outdoors. The majority of the participants in the study also reported the smell of the soap to be acceptable. By considering the motivations to use topical arthropod repellents and adjusting formulations and delivery devices, the nonuse variable that lowers the efficacy of vector disease control could potentially be reduced (Rowland et al., 2004). In more natural-based topical repellents, there is a tradeoff between protection time and number of reapplications needed over time. DEET-based topical repellents provide the longest protection time (Fradin and Day, 2002; Frances et al., 1996), and some natural-based repellents need reapplication after only one to three hours (Barnard and Xue, 2004; Fradin and Day, 2002) so compliance with reapplication of the topical repellent becomes crucial for effective results. In a study done in Beni, Bolivia, the use of plant-based *Eucalyptus maculata citriodon* at a 30% concentration lotion reduced the incidence of *Plasmodium falciparum* malaria parasite by 84% when applied properly in the early evening; however, this treatment was combined with insecticide-treated nets (Hill et al., 2007). In areas of endemic mosquito-borne diseases, topical repellents can fill a protection gap when stationary protection such as residual sprays and treated bed nets are not being used. Also, as noted by Hill et al. (2007), when topical repellents are used in conjunction with some of the stationary repellent devices like bed nets, the incidence of vector-borne diseases can be reduced, thereby increasing protection when sleeping, or stationary. This does require proper instruction in the reapplication of the topical repellent. In DEET-based topical repellents, this is noteworthy as the concentrations and volume of the active substances are directly corelated with the amount of protection time (Frances et al., 1996, 2009; Thavara et al., 2001). The ability of the repellent to remain on the skin during perspiration, and the concentration of the active ingredient are also limitations to effective protection. The effects of human odor combined with topical repellents are an important consideration when efficacy testing. Most studies include this by using human volunteers, in a true field setting, this presents the complication of potentially exposing human subjects to vector-borne diseases. Okumu et al. (2009) have studied the potential for using synthetic human odors, but they also showed that there were considerable inconsistencies in using synthetics, and this may not be a suitable replacement for human volunteers. In using semi-field experimental sites and laboratory-raised mosquitos, it is possible to test human volunteers without the threat of exposure to vector-borne diseases (Sangoro et al., 2014). Semi-field experimental sites can also be constructed in place within geographic areas of concern for mosquito-borne diseases. This provides the ability to design robust experimentation while using environmental conditions in the surrounding area. Ferguson et al. (2008) have constructed a large SF system in Tanzania, and they showed that by using this type of methodology they were able to design experiments that provided benefits of a laboratory study (i.e., controls, volunteer safety), but still replicated environmental field conditions in the surrounding geographic area (Ferguson et al., 2008). The benefits of the SF system also include the ability to alter the terrain, vegetation, and structures within, while still providing environmental realism of the surrounding geography. Chemical safety studies are also an important factor when testing topical repellents either in SF systems, laboratory, or pure field experiments.

11.1.1.4 Safety of topical arthropod repellents

Evidence does support that DEET-based repellent products are the most efficient as topical repellents, widely available in many types of substrates, and have a long time to reapplication rate. This confers protection against vector-borne diseases when the user is mobile. Some limited evidence has been put forth showing possible dangers of DEET use. Briassoulis et al. (2001) have highlighted case analysis of DEET toxicity in children who developed encephalopathy (damage that affects brain function) following exposure to DEET; however, the authors noted that the sample size for this case report analysis was rather low (18 children), the route of exposure was mixed (oral ingestion vs skin exposure), and that there was a circumstantial link to the children's illness and DEET exposure given the mixed case reports. While toxicity testing is warranted for DEET, current case studies of negative patient outcomes cannot be attributed to DEET exposure alone (Briassoulis et al., 2001; Goodyer and Behrens, 1998)

Toxicity testing of the active ingredient DEET has been performed by the United States Environmental Protection Agency (EPA) using animal models, and has found no detrimental effects attributed to dermal exposure of up to 1000 mg/kg body weight (United States Environmental Protection Agency, 1998). Similarly, McGready et al. (2001) studied a group of 897 pregnant women in the third trimester who applied DEET daily in second and third trimesters of pregnancy, and no differences in growth or mortality rates were noted from birth to the first year of life when comparing the control group to the mothers receiving DEET topical repellent (McGready et al., 2001; United States Environmental Protection Agency, 1998).

11.1.1.5 *Summary*

The use of topical arthropod repellents serves as gap protection when the user is traveling or working in field conditions but is also effective when used in conjunction with other repellent strategies such as area dispersal or treated bed nets. By using a SF testing environment, the possibility is present of creating efficacy assays that provide a finer control over variables, as in a laboratory study, but still providing a semblance of environmentally realistic conditions, and providing disease protection for human participants when laboratory-reared mosquitoes are used with the semi-field environment. A study by Sangoro et al. (2014) used lotion-based 15% DEET to test the efficacy against laboratory-reared *An. arabiensis* in a 200 m^2 SFE located in Tanzania. Two experiments were done using human landing catches both in the SFE and in a pure field environment. When comparing the relative risk of being bitten by a mosquito, they found the results were comparable to the SFE experiment, and the pure field experiment (Sangoro et al., 2014). The use of SFEs for topical repellent efficacy is warranted. One consideration in using SFEs to determine topical repellent efficacy is slight changes in wind, temperature, light levels, and other environmental variables that may differ in a control and treatment cage. While the benefits of the SFE experiments are to test in a more natural environment, it is important to standardize these climatic variables across treatment and control as much as possible.

Most examples of SFE experiments done, involve using a single enclosure and rotating the control and treatment experiments at random. Perhaps it should be considered that multiple enclosures run concurrently with control and treatment, and then rotating the treatment and control between enclosures cages would better average the climatic variables. This would be important for topical assays using traps for mosquitoes, as wind speed, temperature, and sunlight could affect the mosquito behavior and subsequent recapture counts. Semi-field enclosures are useful in replicating true field climatic conditions for study of topical repellent efficacy, and if we can test efficacy while retaining the most possible similarities in these climatic conditions between the treatment experiments and control experiments, this could give more statistical power in evaluating the true percentage efficacy of potential topical repellents and insecticides. The mathematical model created by Kiszewskia and Darling (2010) includes repellent efficacy to predict the community-level protection against vector-borne diseases. The model uses the repellent efficacy, product acceptance rate, biting pressure, human infection rate, and sporozoite infection rate (Kiszewski and Darling, 2010). This model could potentially be improved with the addition of accurate topical repellent efficacies. There is potential for the testing of topical efficacies done in SFEs to be incorporated into the mathematical modeling. By testing these efficacy models using data from in situ environmental SFEs, the predictive limitations of the current model could be reduced and provide more accurate estimates and model adjustments to predict the probability of disease vector protection with specific topical repellents. Further, models could be tested and adjusted to fit the geographic area where the semi-field enclosure is located, thus providing multifaceted tools to combat the spread of vector-borne disease by providing general and specific predictive estimates of topical arthropod repellent and insecticide protection.

11.1.2 Spatial arthropod repellents defined

Spatial arthropod repellents (SR) have recently received increased interest as a potential component of novel biting arthropod management systems, but the concept of utilizing SRs is not new. Regardless of how long they have been in use, there is still no consensus on a clear definition of SRs. In their review of the literature, Kline and Strickman (2014) found that Gouck et al. (1967) defined SRs as compounds that could produce repellency at a distance. Nolen et al. (2002) defined SRs as volatile chemical compounds that inhibit the host-seeking behavior of insects in an environmentally defined three-dimensional space. Bernier et al. (2007) used the term "attraction-inhibitors" for human-produced masking chemicals, and other volatile compounds, which interfere with the attraction of mosquitoes to human-produced odors and describe the behavioral effect (inhibition) observed in bioassays. Ogoma et al. (2014) proposed that the term "spatial repellent" be used as a general term for mosquito behaviors that result in a reduction in human–mosquito contact, such as movement away from an airborne chemical stimulus, interfering with host detection by attraction-inhibition, deterrents, and/or feeding inhibition, by creating a space where hosts are safe from bites and potential disease transmission. One key difference between a topical and an SR is that the latter are volatile compounds that become airborne and prevent/discourage blood-seeking arthropods from making physical contact or even entering a defined space occupied by a potential host thus protecting that host from bites. By creating this 'vector-free space' the risk of disease transmission is reduced. Chemicals that have been shown to have SR effects include volatile pyrethroids such as metofluthrin and transfluthrin (TF), botanical compounds such as terpenoids, or volatiles found from human skin and skin bacteria such as 1-methylpiperazine. Historically, DDT was known to have an "excito-repellent" effect in addition to lethality when applied for indoor residual spraying. A unique benefit of spatial repellency is that the safe zone can include areas both indoors and outdoors. The volume of space that is protected, or minimum protection range, will be dependent on the properties of the active ingredient, application platform, and environmental conditions (airflow, temperature, and humidity). Other potential advantages of SRs include protection of several people with a single product, ease of use without the need to make repeated applications to the skin, and a continuous level of protection. Also, some claim, when looking at synthetic pyrethroids as well as other classes of chemicals, that an additional benefit of this type of behavioral modification is a delayed or diminished development in the emergence of insecticide resistance (Amelia-Yap et al., 2018; Corbel and N'Guessan, 2013), but this claim needs further investigation.

While the major emphasis of most research has been on host-seeking and blood-feeding behaviors affected by SRs, recent studies by Bibbs et al. (2018) demonstrated that exposure to sublethal concentrations of TF vapors reduced female mosquito reproduction, including fecundity, fertility (reduction in viable eggs), and the dispersion of eggs (skip oviposition behavior) across potential oviposition sites in both *Ae. aegypti* and *Ae. albopictus*. A complementary effect was observed by Choi et al. (2016) where *Ae. aegypti* displayed increased attraction to oviposition sites after sublethal exposure to TF. When paired with the reduced reproductive performance observed by Bibbs et al. (2018), SRs appear to stimulate container-inhabiting mosquitoes to oviposit in nearby containers urgently (Bibbs et al., 2018; Choi et al., 2016).

11.1.2.1 *Spatial arthropod repellent products*

SR (Strickman, 2007) products disperse a volatile chemical into the air either actively or passively, repelling, or killing biting arthropods from an area designated for protection (Strickman, 2007). Spatial repellent actives have been incorporated into a wide range of devices, which release the active ingredient either actively by using heat, mechanical aerosolization, or volatilization enhanced by airflow (fan) or passively, through unaltered volatilization and diffusion. McPhatter et al. (2017) reviewed the commercial SR products that have been reported in the literature (McPhatter et al., 2017): impregnated plastic or paper strips, (Yayo et al., 2016) coils, (Avicor et al., 2015; Hill et al., 2014; Msangi et al., 2010) candles, (Lindsay et al., 1996; Muller et al., 2008, 2009) fan emanators, (Dame et al., 2014; Lloyd et al., 2013; Revay et al., 2013; Xue et al., 2012a) heat-generating devices, (Collier et al., 2006; Dame et al., 2014; Lloyd et al., 2013), liquid vaporizers, heated mats, and microdispensers (Bernier et al., 2015). Emanators impregnated with either volatile pyrethroids or essential oils have been reviewed by Ogoma et al. (2012) and also reported, that despite differences in evaluation methodologies, coils, and emanators clearly reduce human mosquito contact (Ogoma et al., 2012). They induce mortality, deterrence, repellency, and reduce feeding by mosquitoes on humans. Coils, considered to be the most popular and widely used SR product, use a variety of active ingredients impregnated in a slow-burning flammable matrix to release the gaseous and fine particulate active ingredient into the air. Part of their popularity is based on the low cost to manufacture and the low cost for the consumer. Other emanator products utilize passive-emission that contain an active ingredient sufficiently volatile under ambient conditions that do not require heat, forced air, or other energy inputs to create sufficient concentrations of the repellent chemical. These products incorporate repellent chemicals with a low vapor phase into a substrate such as paper, plastic agar-gel, or vermiculite through impregnation with the active ingredient that enables a passive dispersion of repellent volatiles at ambient temperature. They are intended to release an effective amount of the active chemical or chemicals into the air at as steady a rate as possible over an extended period of time depending on the formulation. Spatial repellents allow the dissemination of active ingredients on small scales, often serving as personal protection devices (Bibbs et al., 2015; Bibbs and Xue, 2016; Ritchie and Devine, 2013). Recently, an alternative emanator delivery format for volatilizing TF at ambient temperatures has proven to be efficacious against mosquito bites, their current design requires that the host is confined to a protected air space, therefore, limiting mobility. In order to impact outdoor biting, these delivery formats must be optimized to protect users wherever they are outdoors (Ogoma, et al., 2014). The large number of products available serves as indicators of public desire for an effective SR product. Current products bear labels that suggest protection from vectors and a broad consensus exists that they have a role in vector control, but at present, most products are targeted to the consumer market and there is limited epidemiological evidence of impact on disease (Logan et al., 2020).

11.1.2.2 Issues associated with use of spatial arthropod repellents

Similar to topical repellents, two of the main concerns are compliance and diversion to users in unprotected spaces. For example, the use of coils requires nightly compliance as well as regular purchasing and the perception is that mosquitoes will move from protected to unprotected individuals. However, in published studies where TF was applied to hessian strips and used outdoors, at least 80% bite prevention was observed consistently over six months without any sign of mosquito diversion to non-users within an 80 m radius (Ogoma et al., 2017).

A critical knowledge gap does exist in our understanding of the extent to which sublethal exposure effects subsequent activities such as host-finding and blood feeding. It needs to be determined how quickly these activities take place once the target species is out of the effective range of an emanator. How quickly, if at all, will they resume blood-feeding, mating, and ovipositing with the same success as an unexposed individual? What happens when an emanator is removed from a treated space or air movement removes the active ingredient below the effective concentration?

It is unknown what happens at the margins of the defined effective zone. Host-seeking biting arthropod species will be less affected and may fully recover if they are able to orientate away from the chemical source or if the treatment is removed. The immediate effects on flight and host orientation need to be determined. The effects may vary with active ingredient used and species of biting arthropod present. Effects may be different for indoor and outdoor use of the arthropod repellent product.

The usage pattern may be affected by availability and affordability. Fortunately, new formats are now available for dispensing SRs without application on human skin or burning coils, thereby minimizing challenges associated with compliance. More recent efforts have been directed toward using locally available materials. These area-wide mosquito repellent formats offer protection in the form of passively dispensed vapors, without any external energy for vaporization, and may be highly applicable in low-income or remote communities (Masalu et al., 2017; Ogoma, Moore, et al., 2012). With regard to TF, which is one of the most widely used SR compounds, hessian-based fabrics have particularly demonstrated a high level of retention for the insecticides and can maintain efficacy for up to half a year (Ogoma, et al., 2012; Ogoma et al., 2017). Generally, SRs create a more uniform usage pattern due in part that only one individual needs to use the product to protect multiple hosts in the vicinity (Cook et al., 2007; Paz-Soldan et al., 2011). This also circumvents failures to apply the repellent to all potential bite sites or to reapply when needed.

Future developments of the SR technology should focus on finding novel active ingredients and delivery systems that are both effective and safe.

Successful laboratory results may not transfer to SF or field settings which may be due to the delivery device not producing/delivering an effective concentration of active ingredient.

There is a need to determine what the efficacy of SRs is against insecticide-resistant populations. Conflicting data have been presented in the published literature.

11.2 Testing guidelines for spatial arthropod repellents

Various methods have been used to evaluate volatile compounds such as essential oils and synthetic pyrethroids as SRs. These methods have been reviewed by Bibbs and Kaufman (2017). They categorized the methods as either "point-source" or "high-throughput screening." The point source approach initially replicated human structures and was often dubbed the "hut design." The principle involves having a point of attraction, such as a host or harborages, in an area of defined boundaries in which mosquitoes are allowed flight (Bibbs and Kaufman, 2017). The environment is freely navigable by the mosquitoes but involves high rates of exposure to a treatment present inside the boundaries. These could take place within actual structures (Achee et al., 2012; Katsuda et al., 2009; Lee, 2007; Ogoma, Lorenz, et al., 2014; Pates et al., 2002; Rapley et al., 2009; Ritchie and Devine, 2013; Wagman et al., 2015) in which the space is partitioned into multiple rooms. Variations are seen where defined borders still exist but are not a dwelling or analogue of living space (Cohnstaedt and Allan, 2011; Ogoma, et al., 2012, 2014b), with examples being tunnels or screened enclosures. All variations are consistent in that a number of mosquitoes are within the defined boundaries alongside a measure of the success of insects in reaching the point of attraction, often a host. Behavioral effects are not severely limited, so factors such as avoidance of the treatment may be observed. This makes an informative design for objectives concerned with spatial components, like repellency, but poor where spatial components interrupt data collection, such as toxicity. In contrast, the high-throughput screening method is essentially a containerized testing environment in which the mosquito passes through or is contained within a device to expose the target to specific conditions. This began with containers composed of simple materials, such as paper, cloth, metal, and glass, where the arthropod is contained for maximum exposure (Abdel-Mohdy et al., 2008; Adanan et al., 2005; Liu et al., 1986; Roberts et al., 1984; Sathantriphop et al., 2014; Stanczyk et al., 2013). Contemporary work has birthed a true high-throughput design with a rather specific construction, and allows optimal control of exposure time, dosage, and contact or noncontact variants (Achee et al., 2009; Thanispong et al., 2009; Wagman et al., 2015). Regardless of type, these are defined by the arthropod's inability to avoid treatment, and often involve close observation of the signs of exposure, making them more informative for sublethal effects, but less informative when requiring environments with competing stimuli. The majority of the discussed effects from repellency, acute symptoms, and sublethal effects have been generated by hut studies and high-throughput studies. Semi-field and field methods are employed as a response to the shortfalls of the prior two methods. These types of test allow competing stimuli and multiple stressors, such as environmental constraints, to factor into treatment outcomes. Semi-field methods are more controlled than field methods, and still allow for a known test group, approximated dosage, and are not impaired by the spatial components of testing (Bibbs et al., 2015; Bibbs and Xue, 2016; Buhagiar et al., 2017a, 2017b; Obermayr et al., 2015). Field studies are the most realistic method, but the most difficult to define. Studies often take place where treatment is employed against known pressures, such as bite contact, and it is then assessed for its ability to impede or remediate those pressures (Xue et al., 2012a, 2012b; Revay et al., 2013b; Dame et al. 2014).

Bibbs and Kaufman (2017) present a diagram of hypothetical exposure outcomes of mosquitoes versus volatile pyrethroids, such as repellency, sublethal oviposition effects, sublethal blood-feeding effects, contact irritancy, disorientation, knockdown, and mortality that can be evaluated in a semi-field environment (Bibbs and Kaufman, 2017).

WHO (2013) published a document that provides guidance and describes steps for laboratory testing and for semi-field and field evaluations of SR products (technical materials and formulated products) designed to provide protection in a specific space (indoor and/or outdoor) against mosquitoes. These guidelines state that the objective of semi-field trials is to extend the results of laboratory efficacy studies of candidate active ingredients and to test formulated products against free-flying populations of one or more target species under simulated indoor or outdoor conditions. The WHO guidelines recommend that these semi-field trials be conducted in screened enclosures (with or without experimental huts) using the release of well-characterized mosquitoes, ideally in the natural ecosystem of a target disease vector. The guidelines also recommend that the dimensions of screened enclosures should be reported in m^3, with a minimum size of $10 \times 10 \times 2\ m^2$ per compartment and, ideally, three identical compartments to evaluate simultaneously: the SR, a negative control and a positive control. It is important to evaluate each treatment independently of the others and to avoid interaction between treatments, especially as SRs may exert an effect over several meters (Achee et al., 2012; Grieco et al., 2007; WHO, 2013). The advantages of using screened enclosures for semi-field evaluation ensures that the mosquitoes are pathogen-free, that a known number of mosquitoes of fixed physiological status (e.g., parity) are used and there is a known distance between the point at which the mosquito populations originate and the source of the chemical stimulus, allowing estimation of the protective area (especially important in outdoor evaluation). These guidelines further recommend that appropriate arthropod containment guidelines be followed. The use of netting around the enclosure allows tests to be conducted in local conditions at ambient temperature, light, humidity, and air movement. The enclosure should be sufficiently large to reflect the area over which the product is intended for use.

These guidelines also recommend that evaluations be conducted in accordance with applicable national ethical regulations. This will affect the type of evaluations that can be conducted. In the United States, if researchers want to publish their data in a peer reviewed journal or use the data for submission to any regulatory agency, e.g., EPA, an Institutional Review Board (IRB) is mandatory for any human participation. This can be an expensive proposition to have a novel active ingredient registered. Therefore, alternative methods may be used such as the use of baited traps as surrogate hosts. Some preliminary screening may take place by a researcher to narrow potential candidate compounds, then have the most promising tested out of country by respected proven research laboratories where only informed consent from multiple human subjects is required. This narrows down the number of candidate compounds for testing using IRBs and good laboratory practices (GLP) at approved testing facilities within the United States. This will reduce costs considerably required to pay for IRBs (and possibly Human Subjects Review Boards) to register a product.

11.3 Semi-field environment defined

Semi-field environment (SFE) containment facilities are an intermediate step between laboratory and field trials that offer a safe, controlled environment that replicates field conditions. Ferguson et al. (2008) defined the semi-field system as:

an enclosed environment, ideally situated within the natural ecosystem of the target disease vector and exposed to ambient environmental conditions, within which all features necessary for its lifecycle completion are present (Knols et al. 2002). In the case of mosquito vectors of human disease, this typically involves a large outdoor cage in which the movement of the disease vector of interest either in or out of the unit is restricted by netting, and within which features such as aquatic larval habitats, blood hosts for adult females, sugar sources (plants) for adults, appropriate resting sites (houses, cattle sheds, etc.) and environmental features (e.g swarm markers to stimulate mating), are present. There are no general guidelines for the appropriate size of such a unit, but ideally it should be large enough to sustain a population of similar density to that encountered in the target environment for numerous generations.

As such, these semi-field systems have been recognized by entomologists as environmentally realistic experimental systems where mosquito vector behavior, ecology, and population dynamics can be studied in a natural context over multiple generations (Ferguson et al., 2008; Knols et al., 2002; Ng'habi, 2010; Ng'habi et al., 2015).

Semi-field environments provide more fine control over experimental design factors than a pure field study, but not as much control over environmental factors as with laboratory studies. An advantage is that these enclosures can be erected in the geographical location where mosquito research is needed for testing to allow for testing target species with realistic field conditions, and adds some flexibility not seen in pure field trials and bioassays. As per WHO (2013) Guidelines, SF studies use mosquitoes reared under standardized conditions. Thus, mosquitoes used in efficacy testing are of known age and physiological status., which is essential to ensure the reliability and reproducibility of data. Outdoor effective dosage, effective range, and duration of protective efficacy can be determined in a semi-field system by using volunteers, baited traps, and/or bioassay cages positioned singly at a collection station at a specified distance from the SR product release point. The duration of the evaluation and sampling periods depends on the product label specifications (Achee et al., 2012; Grieco et al., 2007; WHO, 2013).

Recognized as a suitable evaluation system to bridge the conceptual and methodological gaps between laboratory and field experiments (Ferguson et al., 2008; Knols et al., 2002). The SFEs have been developed and implemented successfully in many regions in the world, particularly in Africa (Facchinelli et al., 2011, 2013; Ferguson et al., 2008; Knols et al., 2002; Ng'habi, 2010; Ng'habi et al., 2015; Ritchie et al., 2011). In the next section, we briefly describe some of these sites and the efficacy testing conducted at each site.

11.3.1 Semi-field environment utilized to evaluate efficacy parameters

Due to the risk involved in testing arthropod repellents, i.e., a topical, spatial, or area repellent, both the efficacy of protection within a given area and the duration of that efficacy are important end points of research that are better suited for SF evaluations. There are other aspects of mosquito control evaluations that do not necessarily require SF environments but are possible and may even be more plausible in this type of environment. The inhibition of feeding behavior can be observed in a laboratory setting after exposing arthropods, in typical laboratory fashion, by topical application or direct contact in cups or small petri dish type arenas, however, in a semi-field environment, arthropods can be exposed to a candidate treatment, in a way that is very similar to how it is meant to be used. This allows the researcher to control the number and species of mosquitoes released as well as have more knowledge of the mosquito behavior in response to treatment (Sangoro et al., 2014). Other aspects of testing that may be more suited for SFEs, include, optimum dosage determination, efficacy over time and distance from a treatment, whether that be a passive emanator or an active SR device. It is also much easier in SFEs that contain simulated homes, to evaluate both exit rate after exposure to an arthropod repellent compound as well as the decrease in the entrance to an area that has been treated (Logan et al., 2020).

11.3.2 Worldwide utilization of various types of semi-field enclosures

Our search of the literature has revealed that many variations of the SFE are being used worldwide. Most of these variations are based on the WHO (2013) Guidelines for Efficacy Testing of Spatial Repellents. For outdoor efficacy evaluations, the WHO recommends that paired SF compartments be used simultaneously to observe landing inhibition for both control and repellent treatments, one volunteer per compartment to either act as resting bait for feeding inhibition tests or to actively participate in the human landing counts for inhibition of landing measurements. The length of time a volunteer participates, and the intervals of collection depend on the claims made by the manufacturer of the product being evaluated. Replicates should have a minimum of 100 mosquitoes in each treatment group that are released from a set distance from human volunteers. It is recommended that treatments are separated by at least a plastic sheet to prevent odors from crossing into another treatment site and rotate treatments between the two compartments, as well as to measure the environmental conditions, each day of testing. Depending on claims of coverage, defined distances should be tested and repeated, over time, to not only determine how far a repellent protects but for how long it protects (Achee et al., 2012; Grieco et al., 2007; WHO, 2013).

The structure and types of experiments that have been conducted in these SFEs are briefly summarized. As this is a review of the literature, as well as discussion of current research that is being conducted, the original terminology for the semi-field environments will be preserved.

11.3.2.1 Albania

Even as early as 1939, there are publications that discuss the use of SF enclosures to study mosquitoes that transmit disease, in the environments that they were prevalent, and Malaria was present. One of these first "cages" was about 10 m long × 5 m wide and 6 m high, it was supported by telegraph poles. This cage was large enough to study the behavior of these vector species with structures built in that could be easily removed or restructured. Different types of settings were designed inside the enclosures, depending on what type of behavior the researchers were studying. Animal stables could be added in for host-seeking behavior studies, as well as small pools with aquatic vegetation to determine if the plants affected oviposition. They found that there were limitations to this new system because they were not able to maintain populations of the mosquito species in the large cages, for various reasons and therefore had to start colonies in small cages (Hackett and Bates, 1939).

11.3.2.2 India

Russell and Rao (1942) also established what they called a large outdoor insectary that was screened and measured less than 75 m^2 in Madras, India. Their SFE was used to study the mating behavior of *Anopheles culicifacies*—to visualize the swarming behavior that was required for successful mating (Russell and Ramachandra-Rao, 1942). Thirty-four years later, Curtis (1976) conducted some of the first studies examining gene transport capability by breeding different strains of *Cx. fatigans* in an outdoor field cage measuring 5 × 3.5 × 2 m that had its own little brick hut with a thatched roof (Curtis, 1976).

These early studies in SFEs, were the basis for what has been developed into newer sophisticated test systems. They were useful for understanding the biology and behavior of mosquitoes and for developing methods of control and vector-borne disease mitigation.

11.3.2.3 East

11.3.2.3.1 Africa

Kenya: In western Kenya, Knols et al. (2002) transformed an existing greenhouse into, what they called, a "MalariaSphere" as part of a project with the Mbita Point Research and Training Center, a division of the International Centre of Insect Physiology and ecology, Fig. 11.1. It was an enclosed environment with all the features believed to be necessary to create a natural malaria vector ecosystem. It measured 11.4 m × 7.1 m and had all glass replaced with "shade netting" or screen material

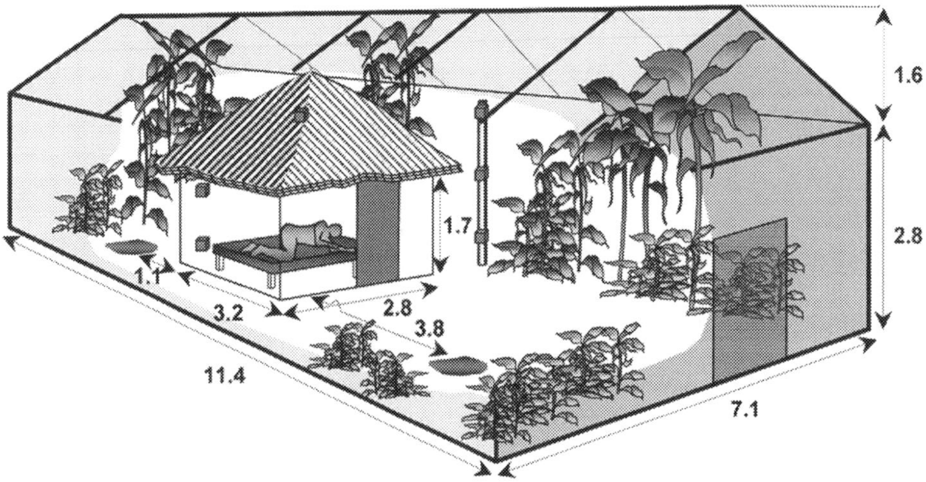

FIG. 11.1 MalariaSphere. A schematic drawing of the MalariaSphere. Courtesy of Menger et al. (2014).

to facilitate air flow, as well as precipitation within the enclosure. The goal was to maintain ambient conditions that were similar to the conditions outside of the sphere. They were successful in rearing *An. gambiae* in the Malariasphere and concluding that this type of enclosure could be used for behavioral studies using both wild and colony-reared malaria vectors (Knols et al., 2002). While this particular study in the Malariasphere centered around both the ecology and biology of the vector *An. gambiae*, subsequent studies were conducted to evaluate both attractants and repellents, sometimes simultaneously (Menger et al., 2014). Their study evaluated the reduction of the entry of mosquitoes into the house, in the presence of attracted baited traps (one inside the home and four on the outside) and/or repellent treatments, as compared to the control, which had only an attractive baited trap inside the house. It was shown that all efforts to discourage mosquitoes from entering the house were successful, with the push-pull method being the most successful, achieving a greater than 95% reduction in house entry.

11.3.2.4 *Tanzania*

11.3.2.4.1 Ifakara Health Institute, Tanzania

Ferguson et al. (2008) established one of the largest SFS for research on both the ecology and control of anophelines, at the Ifakara Health Institute (IHI) Experimental station in Kilombero district, Southeastern Tanzania. This massive 625 m^2 enclosure was established behind the fenced in experimental station to have full access to electricity and control access to the SFS, with the thought of allowing for better control of ongoing experiments, after the SFS was complete (Ferguson et al., 2008). The structure of the SFS was constructed using steel-reinforced concrete, within both the 22 m × 30 m platform and the posts that supported it to allow for water to flow under it during heavy rains that could cause flooding, Fig. 11.2.

The shell of the SFS was created from a greenhouse frame kit made in France. The screen used for the SFS was a finer mesh than standard bed nets and was chosen for its breathability, UV stability, shade factor, and resistance to stretching. Most SFEs are fully enclosed in screen to allow for natural precipitation as well as air to

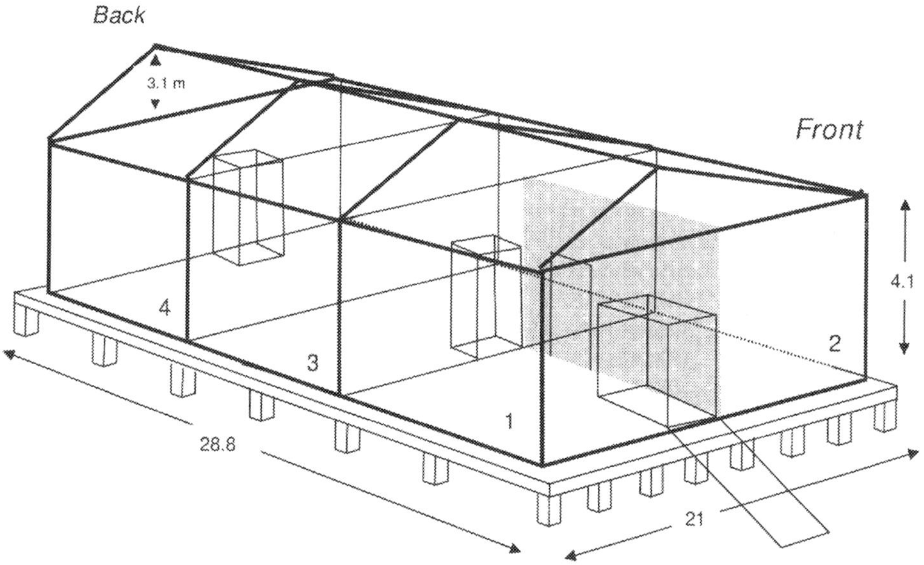

FIG. 11.2 IHI semi-field system diagram. Schematic diagram of the IHI semi-field system (*SFS*) for research on African Anopheles ecology and control. Courtesy of Ferguson et al. (2008).

pass, as it would outside of the enclosure, however, this particular SFS was designed with a polyethylene top to protect from the intense rains that occur in Southeastern Tanzania. It would also allow for studies in which "rain" could be added into a study scenario at varying rates.

The SFS took a little under 2 years to build and allocated specific activities to each of the four sections of the enclosure, based on minimizing the risk of mosquito entry or escape, as well as logistical efficiency for testing activities, Fig. 11.3.

One section was isolated and set up to study chemical ecology and olfaction and was physically separated with plastic barriers to prevent as much air flow as possible from this section to or from the others, Fig. 11.4.

The effort put into creating this state-of-the-art facility has not gone unnoticed. Many studies, a few of which will be discussed in this section, were conducted in this SFS (Ferguson et al., 2008).

Another name for the SF enclosure at the Ifakara Health Institute is the "VectorSphere." The large (625 m^2) screened cage that, at the time of this testing, had three separate areas or chambers, approximately 202 m^2 each, for studies to be conducted. As shown in Fig. 11.5, the chamber used was kept natural with small animals and vegetation to more closely resemble the environment around the sphere. Two huts were constructed inside the study area to simulate local dwellings, with the typical eave openings. The two huts also contained a bed with an untreated bed net.

In this study, strips of hessian (burlap) fabric were treated with an SR and tested as an eave treatment to prevent both indoor and outdoor biting of mosquitoes. Five hundred *An. arabiensis* mosquitoes were used, each night, in the study as they are more variable in their feeding behavior (Fornadel et al., 2010). Human subjects were used in this study to determine the reduction of outdoor biting, by collecting mosquitoes that came to bite an uncovered area of skin

FIG. 11.3 IHI semi-field system photo. Outer structure of the IHI SFS on completion. Courtesy of Ferguson et al. (2008).

(lower leg) in the evening hours and then by entering the hut and sleeping under a bed net with a light trap beside it to collect the mosquitoes that came to feed, Fig. 11.5. The study showed that in the SFE, the treated eave ribbons reduced both indoor and outdoor biting by more than 99% .

The Ifakara Health Institute's Semi-field enclosures were utilized to test the efficacy of repellent treated hessian fabric affixed to men's sandals. Two types of SFEs were utilized for the testing; one was a 553 m^2 area that was 4.5 m tall that had two areas measuring 36 m^2 partitioned off for use. The second SF enclosure was a long tunnel-like screened system that was 110 m long, 2.5 m high and 2 m wide, Fig. 11.6.

Male subjects that were trained to do human landing catches were utilized in this study. Four different trials were carried out in the SF systems with both *An. arabiensis* and *Ae. aegypti* added for one of the trials. There were also two field studies conducted based on preliminary results

FIG. 11.4 Inside of IHI semi-field system. Section for olfaction and chemical ecology research within the IHI SFS. Courtesy of Ferguson et al. (2008).

FIG. 11.5 VectorSphere. Semi-field chamber inside the VectorSphere. Courtesy of Mmbando et al. (2018).

from within the SF studies. Dose-response and surface area of the hessian fabric were evaluated in early tests. One other question that was posed in one of the trials, that could only be done, ethically, in a SFE was whether the wearing of the repellent treated sandals on one person would push the mosquitoes to another person not wearing sandals.

FIG. 11.6 IHI semi-field cages. Pictorial illustration of (IHI) semi-field cages, (A) large multicompartment system; (B) inside sections of the long, tunnel-shaped screened semi-field system; (C) outside view of the long tunnel-shaped screened semi-field system. Courtesy of Sangoro et al. (2020).

This test was successful in elucidating the question of diversion. There was no significant difference in the number of mosquitoes attracted to the control subject next to the one wearing the treated sandals versus the two control subjects without treated sandals. It was also discovered that this particular repellent may have a plateau value for protection, as there was no significant increase in protection with the increasing dose of repellent. It was shown that the use of the treated sandals does offer significant protection in the SF tests as well as the field tests that were conducted. The field tests did have exposure to different mosquitoes than those which were used in the SF tests, but the results translated to a similar level of protection (Sangoro et al., 2020).

11.3.2.4.2 Muheza, Tanzania

At the Amani Research station in Muheza, Tanzania, Kitau et al. (2010) utilized a set of three "Mosquito Spheres" to test the efficacy of a commercial mosquito trap alone and in conjunction with various personal protection options. Each sphere had a 100 m^2 footprint and was covered in screen, to allow the passage of both wind and rain, to closely resemble the ambient conditions from the local environment. The spheres were equipped with a double door system with a corridor to help prevent escape of test mosquitoes or entry of wild ones, Fig. 11.7A. Each sphere is also planted with local vegetation, contains a mud hut with one door and two open windows. The hut is equipped with a Zanzibari-style rope bed and has a thatched roof, for volunteer use during testing, Fig. 11.7B (Kitau et al., 2010).

In this 2010 testing, a propane-based trap was evaluated with and without a lure (a commercially available mosquito attractant containing lactic acid and ammonium bicarbonate) and in conjunction with various commercially available repellent strategies: an allethrin based mosquito coil, a DEET-based roll-on skin repellent, as well as DEET based wearable repellent bands for both wrist and ankle. Both *An. gambiae* and *Cx. quinquefasciatus* were used to determine efficacy, i.e. reduction in human landing rates, of the treatments. Data from trap collections as well as human landing rates were recorded to determine reductions. Both treatments and volunteers were rotated nightly in a Latin Square design to limit the effects of site, weather, and volunteer bias over time. One of the findings of this study was that, in the SFE, the use of a mosquito trap significantly reduced the human landing count, and the use of skin repellents increased the catch rate of the mosquito traps, while the use of the mosquito coil and repellent bands did not (Kitau et al., 2010).

FIG. 11.7 **MosquitoSphere.** (A) MosquitoSpheres in Muheza Tanzania. (B) Close up of Hut in MosquitoSphere. Photos courtesy of Hans Jamet.

FIG. 11.8 Ifakara semi-field system. Semi-field system and experimental hut at the Ifakara health Institute, Bagamoyo, Tanzania. Courtesy of Tambwe et al. (2014).

11.3.2.4.3 Bagamoyo, Tanzania

The SFS in Bagamoyo at the Ifakara health Institute is 22 m × 29 m and is set upon a concrete base with a moat to prevent predator entry, as per WHO guidelines. As designed, the SFS maintains environmental conditions that are very similar to the surrounding landscape and provides a controlled environment in which disease-free mosquitoes can be released and evaluated for human landing catches, much more safely and quickly than a standard field test. For this test, two compartments or test arenas were set up with matching huts and mattresses for volunteer placement. Exit traps were placed on all the openings in each hut and mosquito exit was measured, as shown in Fig. 11.8.

Two SRs, one natural and one not, were tested versus a negative control to determine percent feeding inhibition, mortality, both immediate and 24 hours postexposure. Volunteers were asked not to drink alcohol, smoke, or wear deodorant. They were asked to wear protective clothing, as a bed net was not provided during the study. After each test was set up as either a control or repellent treated experimental hut, volunteers entered and lay on their mats. Fifty starved female *An. gambiae* were released in each experimental hut, with the sleeping volunteers.

Number of blood fed mosquitoes exiting the hut was measured. Ultimately, it was determined that the natural repellent did not offer control against malaria vectors, although some interesting findings were noted (Tambwe et al., 2014).

11.3.2.5 Zambia

In Zambia in 2018, a large SFS similar to ones created at Ifakara were used to evaluate the efficacy of a controlled release device to emit a SR. The SFS used was set on a concrete slab and measured 28.8 m × 21 m with three separate, 9 × 9.5 m, chambers on each side, separated by a central corridor. The chambers on one side of the SFS were used with a moat fitted around them to prevent ants or other scavengers from entering the test areas and removing any knocked down or dead mosquitoes. Each chamber was fitted with a small hut resembling local structures including the open eave design, plastic was used to cover the doors of the huts and untreated mesh was used as curtains for the windows, to more closely resemble what would occur in an occupied hut. This study evaluated the ability of a controlled release device (passively releasing a SR) to reduce the indoor counts of mosquitoes, with or without an occupant indoors. The floor of each chamber was lined with a white sheet to facilitate the recovery and counting of both dead and knocked down mosquitoes, Fig. 11.9. The central chamber was used as a buffer zone to prevent cross-contamination between the control chamber and the treatment chamber. Treatment and control were rotated between the two chambers and repeated five times. Environmental conditions, including wind speed, lunar phase, and illumination level was recorded for each night of testing. There were three different scenarios tested: low-level, high-level, and low-level SR plus human occupant in a hut. For each scenario, there was a baited light trap inside the hut next to a bed net covered mattress, for indoor feeding mosquitoes, and four baited light traps outside the hut at the corners to evaluate outdoor feeding mosquitoes. Each night of testing, between 100 and 300 susceptible *An. gambiae* were released (same number in both the control and treated chamber). The results of this study showed that even the low level of SR release was able to reduce indoor mosquito numbers, as expected when the hut was occupied, both the control and treated indoor numbers were higher. The excito-repellency activity of the SR that was used may explain the increased indoor host-seeking and mortality within the huts. Without a SF environment for this test, this same behavior may have moved mosquitoes out and about without the ability to measure the real numbers or relative distribution after exposure to the SR (Stevenson et al., 2018).

FIG. 11.9 **Macha semi-field system.** The semi-field system (*SFS*) at Macha, Southern Zambia and the artificial huts constructed within the SFS. (A) Exterior view of the SFS. (B) Experimental set up depicting artificial huts and trap arrangement. Courtesy of Stevenson et al. (2018).

FIG. 11.10 **Mali Artificial Hut.** Artificial hut set up in Mali: (A) screened windows, (B) indoors showing bed net, (C) atrium with plants. Photos courtesy of G. Muller.

11.3.2.6 West Africa

11.3.2.6.1 Mali—University of Bamako

While the majority of SFEs that are discussed are permanent structures that are set up with a large footprint and a greenhouse style frame that allows the entire external area to be screened with various types of mesh/net, in areas where they do not have the funding or space to take on such endeavors, alternate facilities are used. At the University of Bamako in Mali, "artificial huts" are used to do indoor studies by screening all windows and covering doors with plastic or screen, Fig. 11.10. Indoor sugar feeding on different plant types can be studied, as well as

FIG. 11.11 Alternate size mali hut. Nonpermanent semi-field enclosure—party tent with screened walls and canvas floor. Photo courtesy of G. Muller.

examining resting behavior. Mosquitoes have a choice to go indoors or to stay in an atrium area with plants (Muller personal communication).

Another strategy used for SF testing in Mali and in other areas of the world, is the use of party tents as the SF enclosures which come in various sizes and colors, as well as with varying degrees of screen coverage, Fig. 11.11.

They need to be set up on-site but are sturdy and can be moved to different sites as needed. Wild mosquitoes can be collected and used for testing in their natural environment or colony mosquitoes can be added to evaluate local plants or other test parameters, Fig. 11.12.

A final strategy utilized by the University of Bamako is the use of "pop-up" screened cages (Fig. 11.13) which are 2 × 2 × 2 m and can be used as small test enclosed sites. Here, researchers are determining the percent of sugar feeding that is occurring immediately after emergence in the presence of various plant species (Muller 2020, personal communication).

11.3.2.7 *Australia*

For intermediary Dengue vector research bridging the gap from laboratory to field, Ritchie et al. (2011) utilized a SF enclosure with biocontainment levels to prevent ingress or egress of insects. The construction of the facility was performed 15 km Northwest of Cairns, Queensland Australia. The category two cages were built on a soil pad with a concrete perimeter with an inner mesh screen and outer mesh screen enclosure for a total interior volume of 465 m^2. The V-shaped SFE was built with a solid concrete wall front, 11 panels for the ceiling, and 5 panels at the rear portion. The individual panels were screened with 0.009 mm mesh wire on inner and outer layers, which were designed to be removable in case of breach. The concrete perimeter also consists of 60 × 10 cm walls at the base. The entire exterior of the cage is covered with 7 cm heavy aluminum security screening, as well as the sides up to 120 cm off the ground. A portion of the sides and the entire roof is also covered with shade cloth to cut down on radiant heat from the sun. A 4.0 × 5.0 × 2.2 m high "Queenslander" structure was erected inside the SFE to simulate a house based on homes in a style common to North Queensland Australia. A yard was replicated using 10–15 cm deep mulch. Common

FIG. 11.12 Second alternate size mali hut. Inside VIEW of nonpermanent semi-field enclosure—with plants for sugar feeding and resting site trials. Photo courtesy of G. Muller.

flowering ornamental plants (*Calyptrocalyx* sp., *Whitfielda longiflora*, *Chlorophytum* sp., *Spathphyllum* sp., (Peace lily), *Jasminium officinale* (Climbing Jasmin), *Calathea warscewiczii*, *Euphorbia* sp. (Diamond frost), *Cuphea* sp., and *Dypsis lutescens* (Golden cane) were planted in plastic pots throughout the yard and watered using a ground-based sprinkler system. The entire simulated

FIG. 11.13 Pop-up semi-field enclosures. Pop-up nonpermanent semi-field enclosures. Photo Courtesy G. Muller.

Queenslander house was under water protective covering. An air conditioning unit was utilized to reduce any differentials in ambient temperature and humidity, and outside conditions. Two identical SFEs were constructed. It is important to note that this facility is unique not only for the climate-controlled SFEs but also for the inside laboratory that is adjacent to the two cages and surrounded by a solid outer wall. This facility is ideal for experimentation on species that are of containment concern.

F1 *Ae. aegypti* were collected from ovitraps in suburban areas in Cairns Australia and were insectary reared on site for a total of 2500 males and 2500 females in each cage for survival trials, and three cohorts of 120 female and 60 male pupae were allowed to emerge in the cages at two-day intervals for the distribution trials.

For distribution trials, *Ae. aegypti* were surveyed, using an aspirator, three to seven days after emergence from pupae. The cage was divided into five sections incorporating areas inside and outside the Queenslander structure. This was done in mid-morning and at dusk to compare the number of mosquitoes captured inside the Queensland structure vs the ones captured in other areas of the cage. For the survival trials, the mosquitoes released were trapped with BGS traps set inside the Queensland structure for 30 min a day and removed from the population, and after 22 days, the final mosquitoes were trapped with BGS traps. Oviposition was allowed during the duration of the trial using buckets set in the simulated garden area of the Queenslander. Daily survival rates were estimated using the mean daily and final trapping counts.

Solar gain, ambient temperatures, and humidity were monitored closely in the SFEs to ensure accurate ambient conditions matching the outside conditions. For the *Ae. aegypti* experiment, to determine preferred resting sites, the mosquitoes were captured in the five sections, at 7, 8, and 9 days after the first release of pupae, and later at 9, 10, and 11 days. The aspirated surveys were performed by one person following a set route through the five predesignated sections of the SFEs. After tallying the mosquitoes, they were released back into the respective SFE. A Fisher's Exact Test was used to compare the counts of mosquitoes within the Queenslander structure to the mosquitoes caught outside the Queenslander. For the survivability survey, the mosquitoes were given a human volunteer blood meal for 10 m at 10 am every day, and BGS traps were run for 30 minutes. The mean number of males and females were enumerated and removed from the population. After 22 days, the remainder of the population was trapped and enumerated. The average male and female trap collections were $\log_e$ transformed and fitted to a linear regression by day. The slopes of the linear regression analysis were used to calculate the daily survival rate. The daily survival rate for the remaining mosquitoes' captures was estimated using the exponential decay equation.

These experiments showed evidence that the constructed SFEs accurately represented a typical urban environment of the area in regard to mosquito behavior, and that the survivability for populations within the cage during testing was high. They also showed strong evidence that the structure adequately contains quarantine insects as none of the *Wolbachia*-infected mosquitoes were detected in traps set in the surrounding area (Ritchie et al., 2011).

11.3.2.8 Asia (South Korea and Thailand)

The structure used for the SF portion of the experiment in Yoon et al. (2014) in South Korea was 12 × 3 × 2.5 m metal structure similar to a shipping container with open windows. The inside of the structure was lined with undisclosed mosquito netting material, or the temperature within the semi-field enclosure during testing. There are what appears to be six total windows on the outer metal container in the figures. It is important to note that while the authors define this design as a "semi-field enclosure" it does not fit the classic definition of a SF enclosure because of the confined nature of the

FIG. 11.14 In Thailand, Similar to Harrington's village set up of a non-permanent SFE, an example of a permanent test site utilizing an experimental tent inside a semi-field enclosure. Photo of AFRIMS Large Mosquito Enclosure (LME) in Thailand illustrates a semi-field enclosure built around an existing structure. The AFRIMS LMEs are covered with a polyester netting which allows both wind and precipitation to enter, creating climatic conditions inside the enclosure which are similar to the surrounding conditions. Photo by personal correspondence courtesy of Alongkot Ponlawat, PhD.

enclosure's design in the study. In Harrington et al. (2008), Thailand, the structure for the SF enclosure was built on-site in a rural village on the Thailand-Myanmar border and consisted of a 10 × 10 × 4 m high bamboo frame surrounded by nylon mesh netting. The entire enclosure was built around an existing rudimentary vacant hut and yard, Fig. 11.14 (Harrington et al., 2008; Yoon et al., 2014).

In a SF experimental design by Ponlawat et al. (2016) in Chanthanburi province, Thailand two 5 × 1 × 1.5 m SF tunnels were constructed with entrance tents at each end measuring 2.4 × 2.4 × 1.9 m. The walls were made of 100 denier polyester mesh netting with wood supports placed every 2 m, Fig. 11.15.

In Yoon et al. (2014), the species tested was the dengue and encephalitis vector, *Ae. albopictus* that was mentioned in the study as being "brought to" the Korean CDC for testing, but it is not clear where they were actually collected or reared.

In the experiment by Harrington et al. (2008), *Ae. aegypti* pupae were collected from Thai villages of Pai Lom and Lao Bao and reared under laboratory conditions. Mated adult females were blooded from volunteers in preparation to be used in the experiment. Ponlawat et al. (2016) used laboratory-reared *Ae. aegypti* and *An. dirus*. Adult mosquitoes were fed a 10% sucrose solution and starved for six hours prior to experimentation.

The study by Yoon et al. (2014) aimed to compare laboratory tests and field tests of 24% DEET with semi-field tests to compare outcomes and judge if the semi-field study performed comparatively with the field and laboratory efficacy studies of DEET.

The experiments by Harrington et al. (2008) were focused on investigating the variables of size, light-dark coloration, and reflectance as physical attributes of oviposition containers that may affect the number of eggs laid by *Ae. aegypti*.

The field studies by Ponlawat et al. (2016) evaluated the long-term efficacy of metofluthrin impregnated nets that are common for use against vector-borne disease vectors.

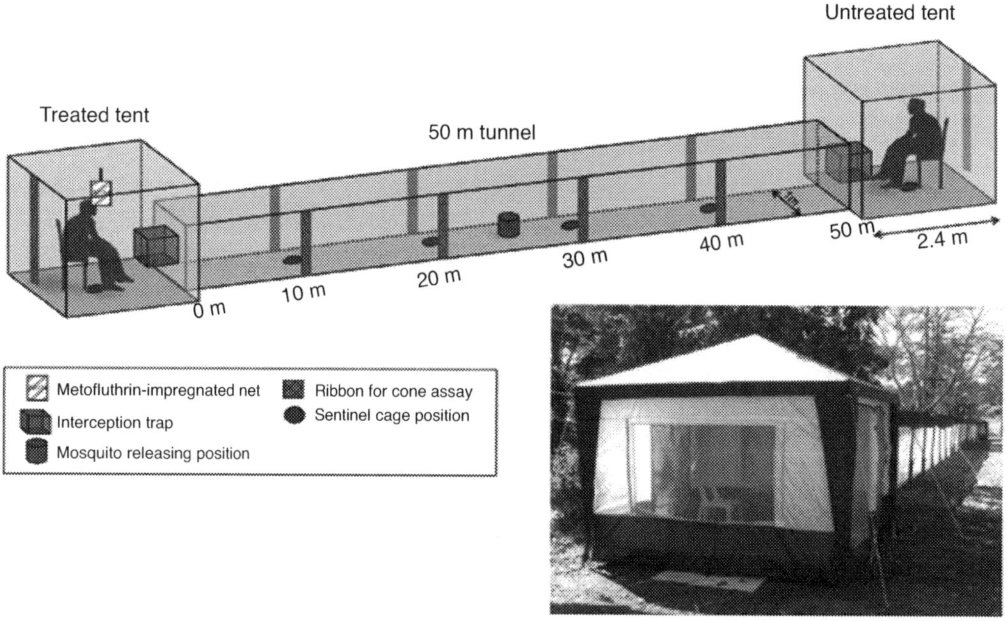

FIG. 11.15 Semi-field enclosure with tunnel. Diagram of semi-field enclosure from Ponlawat et al. (2016).

In Yoon et al. (2014), the pure laboratory study used human volunteers to enumerate the number of mosquitoes landing on an untreated forearm vs on a DEET-treated forearm. The number of mosquitoes landing was counted each 4 for 6 total hours. The complete protection time was calculated by determining the number of minutes from application of DEET to the first mosquito landing or biting the forearm. Repellency was measured each hour by taking the control landing for each volunteer and dividing the control landed minus the total number landed on the DEET treated forearm for each hour per volunteer. Field tests were conducted using CDC traps with a restricted mouse in the attracting area. Treatments traps had 5 × 5 cm cotton balls moistened with 24% DEET. Two DEET treated and two untreated traps were used. Counts of mosquitoes were taken at 2, 4, and 6 hours. The entire experiment was replicated three times. The semi-field testing used the same procedure. The study found evidence to support no difference in efficacy results of the 24% DEET across the methodologies of field, semi-field, and laboratory testing (Yoon et al., 2014).

The experiments by Harrington et al. (2008) used clay and glass storage jars, obtained locally, ranging in diameter from 5 to 45 cm placed in the SFE. Colors of the containers were black, brown, and grey. Containers were filled to 75% capacity. The total water surface area and volume were calculated based on the dimensions of each individual container. Seed germination paper was placed around the rim of each container extending into the water and secured via clips. The containers were placed, at random, around and under the house inside of the SFE. Approximately 200–350 gravid females were released into the enclosure every evening from 1800 to 2000 hours. The oviposition papers were removed, inspected, and replaced three times per day (0800–0930, 1130–1300, 1730–1900 hours). Egg numbers were recorded using a dissecting microscope by container and time. The experiment was replicated in January 2003 during the dry season, July 2003 during the rainy season, and Jan 2004 during the dry season using the same testing methodology. The experiment found that peak oviposition occurred from 1200 to 1800 hours during the dry and rainy season 2003, and around 1600 hours in January 2004. The study also found that the number of eggs laid increased with container diameter, volume, and surface area (Harrington et al., 2008).

In Ponlawat et al. (2016), a metofluthrin impregnated net using 5% metofluthrin and 95% ethylene-methyl methacrylate was used. The net measured 9 × 18 cm with a thickness of 0.2 cm. The net was hung in the treatment tent 10 cm from the ceiling. One tent attached to the tunnel had no net. A human attractant volunteer was randomly rotated inside the tent entrance area between the treatment tent and nontreatment tent. Each tent had an interception trap located at the start of the tunnel. One hundred and fifty *Ae. aegypti* were released into the midpoint of the tunnel at 0800 hours, and the interception traps were emptied every hour by the human volunteer using an aspirator. The experiment was stopped at 1000 hours and the remaining mosquitoes were killed using electric bats. The *An. dirus* mosquitoes were released into the tunnel at 1800 hours and the same procedure for emptying the interception traps was followed. The second experiment was stopped at 0600 hours and remaining mosquitoes were killed using electric bats. Experiments for *Ae. aegypti* were replicated 18 times, and the experiments for *An. dirus* were replicated 22 times. Two sets of experiments were done using the above procedures, a human vs human attractant assay with two volunteers in each tent, and a presence vs absence assay using only one human volunteer. A monitor assay was also done to test the spatial impacts of the treated nets. For this assay, six cups with 15 mosquitoes were placed in the tunnel at 10 m intervals. The cups were checked for mortality after 24 hours, and this was replicated for five weeks for each mosquito species. The overall finding of the experiments showed that the metofluthrin nets only had a repellency at close ranges to the mosquitoes and did not provide true spatial repellency. There also was no effect of the human volunteer attractant at any combination of treatment and control (Ponlawat et al., 2016).

11.3.2.9 North America

11.3.2.9.1 Jacksonville, Florida, Naval Entomology Center of Excellence (NECE)

The SF enclosure at NECE measures 6.1 × 10.7 × 3.4 m (W × L × H) and is constructed of a metal frame sitting on a concrete base surrounded by mosquito screen, Fig. 11.16. Liverpool and Orlando

FIG. 11.16 NECE semi-field enclosures. Semi-field enclosures at the Naval Entomology Center of Excellence; photo courtesy of Jim Cilek. *NECE*, Naval Entomology Center of Excellence.

FIG. 11.17 Inside NECE SFE. Tent Inside of the NECE semi-field enclosure. Photo courtesy of Jim Cilek. *NECE*, Naval Entomology Center of Excellence.

strains of *Ae. aegypti* obtained from the USDA Agricultural Research Service Center for Medical and Veterinary Entomology (CMAVE) in Gainesville, FL were used in the experiments by McPhatter et al. (2017) at this site. The study tested two TF SR devices in repelling mosquitoes from military-style three-person tents.

Potted plants were placed inside of the SFE to better represent the outdoor environment. One three-man tent was placed into each SFE, Fig. 11.17, and the treatment and control tent were chosen at random. For the treatment tent, the SR device was placed on the opening to the three-person tent, and a BGS trap was placed inside the tent. The two SR devices consisted of a plastic strip impregnated with TF, one commercial device contained 0.40% TF, and another noncommercially available device contained 1% TF. Prior to testing, one of the two SFEs was selected to be the treatment, at random and the assays were repeated five times. For each replication, 200 mosquitoes were released simultaneously into the control and treatment enclosures, and then collected (both inside and outside the tents) at the end of the test.

The study found that the presence of either TF device significantly reduced the number of mosquitoes collected from inside the three-person tents when compared to the control tents. The commercially available device reduced the collection of mosquitoes by 66%, and the noncommercially available device reduced the collections by 88% (McPhatter et al., 2017).

11.3.2.9.2 St. Augustine, Florida, Anastasia Mosquito Control District of St. Johns Florida

The SF enclosures measure 12.19 × 4.57 m × 6.1 m (L × W × H) and are surrounded by mosquito screening, Fig. 11.18. The enclosures have a center-covered pavilion that measures 10.92 m^2. In the experiments done by Khater et al. (2019) gravid mosquito traps with insecticide (0.7% permethrin), and larvicide (0.01% pyriproxyfen) were tested in two configurations for trapping gravid female *Ae. aegypti*. The Orlando strain, laboratory reared *Ae. aegypti* were provided by the USDA Center for Medical and Veterinary Entomology. One configuration of the trap contained a sticky paper sheet, and the other did not. Three traps were

FIG. 11.18 AMCD semi-field enclosures. Semi-field enclosures at the Anastasia Mosquito Control District of St. Johns Florida; photo courtesy of Kai Blore AMCD. *AMCD*, Anastasia Mosquito Control District of St. Johns Florida.

placed inside of the enclosure. The traps with sticky sheets were placed in the enclosure and rotated by position for three replications of the experiment, and the traps without the sticky sheets were utilized in the same way for three replications. In each iteration, 300 gravid female mosquitoes were released, and then the traps were collected after 48 hours. The number of adult mosquitoes was enumerated, and any eggs and immature mosquitoes were also collected from the traps. The eggs were placed in clean water in small mosquito breeders for adult emergence. The results of the experiment showed that the traps with the sticky sheets captured an average of 44.16% of the adults while those without the sticky sheet capturing 0.83%. The difference in the emergence of adults from the eggs in the sticky vs nonsticky sheet trap was not significant (Khater et al., 2019).

11.4 Gainesville, Florida, USDA center for medical and veterinary entomology

USDA Semi-field Environments: The first SF enclosure at CMAVE was constructed in 1989. The enclosure (548.6 m^2) measured 9.1 m wide × 18.3 m long, with sides measuring 4.9 m high, pitched to 5.5 m with a metal frame covered with mosquito mesh screening. The enclosure was oriented with the length oriented in the north–south direction; prevailing winds are out of the south. A door was located on the southeast side of the enclosure that allows entry. The enclosure was landscaped with an inside perimeter of a variety of planted shrubs, surrounding an expanse of St. Augustine grass, to simulate a typical suburban yard found in north-central Florida, United States. The cage was constructed to conduct efficacy studies on various experimental slow release/extended duration formulations of DEET intended for military use. These efficacy studies were conducted to determine the level of personal protection obtained from mosquito bites from laboratory-reared *An. quadrimaculatus*. These topical repellent studies were followed by an evaluation of the level of personal protection obtained by test subjects wearing permethrin treated uniforms either in combination with or without the extended duration DEET topical formulation to protect exposed areas of skin from *An. quadrimaculatus* bites selected in the first study. The results of these SF trials were used to design the successful field studies which followed (Schreck and Kline, 1989a, 1989b).

After these initial SF trials, the enclosure was used to evaluate the efficacy of various types of mosquito traps. It was then that we realized the need for replication. Our initial solution was to mark the mosquitoes released each day with a different color of fluorescent powder. Since requests to conduct studies in this enclosure were numerous, a different solution was required which resulted in the construction of four additional screened enclosures with the same dimensions and landscaping. Currently, we have five semi-field environments used for testing at CMAVE, Fig. 11.19.

When we began our evaluations on the efficacy parameters of SRs, we included protection of people and animals both in an open environment and those inside temporary shelters, such as tents, Fig. 11.20. Three enclosures were dedicated to open environment evaluations and two to tent studies. Since much of our efforts have been on the protection of military

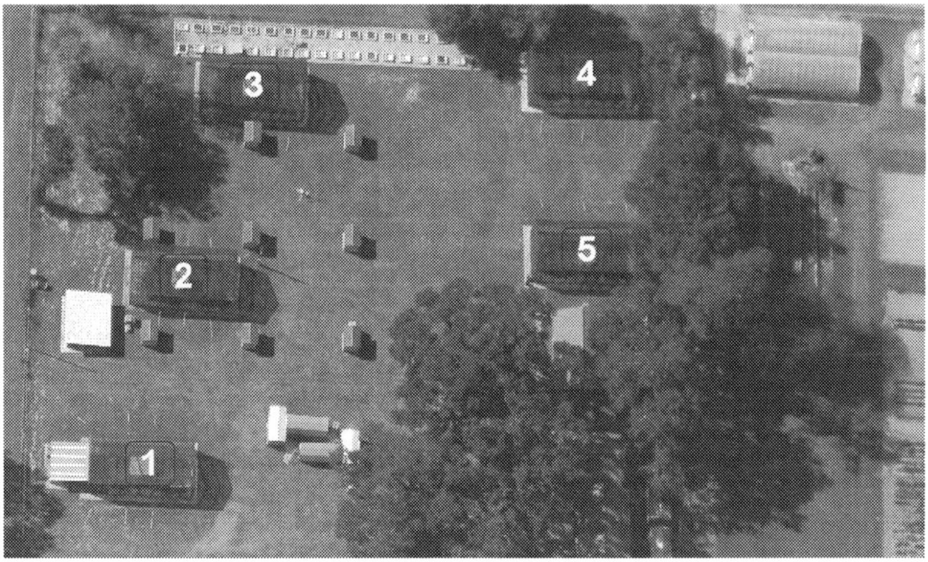

FIG. 11.19 CMAVE aerial view SFE's. Aerial view of the semi-field enclosures at the USDA Center for Medical and Veterinary Entomology (*CMAVE*). Enclosures 1–3 simulate open field environments, and 4–5 simulate an environment with a structure.

FIG. 11.20 **CMAVE semi-field enclosures**. Semi-field enclosures at CMAVE showing military tent inside of enclosure (A) and without (B). *CMAVE*, Center for Medical and Veterinary Entomology.

personnel from biting arthropods for the tent studies, HDT Base X Model 305 Shelter tents, with a floor space of 5.5 m wide x 7.6 m long, pitched to 3.1 m at the tent roof pinnacle, Fig. 11.21, were utilized to evaluate the efficacy of a candidate SR product to prevent entry of various species of biting arthropod (mosquitoes and stable flies) into the tent, or to expel them if present within the structure.

11.4.1 Initial spatial repellent research studies

Our initial evaluations of SR efficacy parameters focused on commercially available products, obtained locally or through international travel, or on experimental products/technologies reported in the literature. A diversity of both active- and passive-emission devices was evaluated. Each

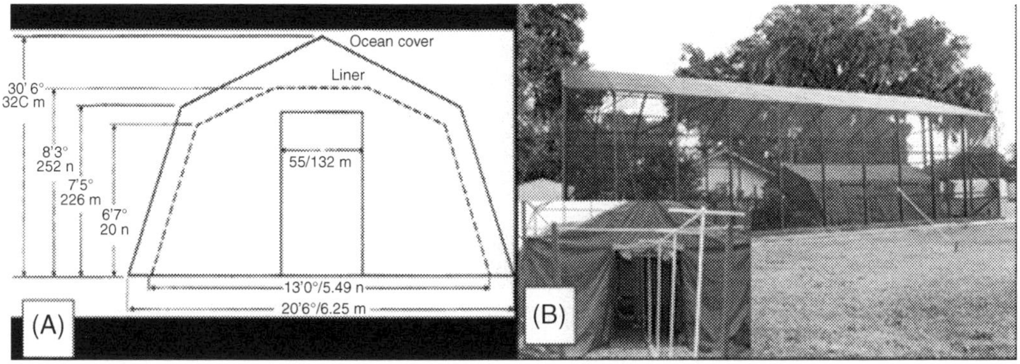

FIG. 11.21 **Schematic of CMAVE SFE**. Schematic of military tent on left (A), and views of the tent inside of the semi-field enclosure on right (B). Tent is oriented so that length is north–south direction. Inset of (B) shows a north–south facing view of the tent entrance. *CMAVE*, Center for Medical and Veterinary Entomology.

type of active ingredient/delivery system was evaluated to determine mortality and repellency. Many of these data are summarized in (Kline and Urban, 2018).

These efficacy evaluations were conducted either with free-flying mosquitoes, or mosquitoes confined in bioassay cages. In addition to determining toxicity effects and repellency effects of these products, we determined the dispersion pattern of the active ingredient and the impact on this pattern due to delivery platform. We believed both mortality effects and dispersion patterns could be determined by using caged insects. Laboratory-reared mosquitoes and stable flies (*Ae. albopictus, Ae. aegypti, Ae. taeniorhynchus, An. albimanus, An. quadrimaculatus, Cx. quinquefasciatus, Cx. tarsalis,* and *Stomoxys calcitrans*) were used in the two enclosures with the tents. The type of bioassay cages we use has evolved over time. Initially, we used custom made triangular pouches, Fig.11. 22A, constructed of gray fiberglass screen with a mesh of 22 × 22 mm. Each pouch contained 10, 4–8-day-old adult female mosquitoes or mixed sex stable flies. In later studies and currently, we are using bioassay cages constructed from cardboard hoops (materials obtained from Multi Packaging Solutions, Chicago, IL, United States) which consist of an inner ring (15.2 × 3.8 cm OL), outer ring (15.9 × 1.6 cm OL) enclosed with tulle fabric, Fig.11. 22B, in which 25 mosquitoes or stable flies are placed. We switched to using the hoop-style bioassay cages because the materials to construct them are commercially available and the activity of the mosquitoes within the cages is easily observed.

Setups to monitor the effective vertical and horizontal movement within the large screen enclosures (i.e., distribution of active ingredient within the tent) have also been evolving over time. Original studies consisted of 57 pouches distributed at 19 sites horizontally located throughout the tent. Each site consisted of a wire suspended vertically from the ceiling of the tent each site with three positions located along the vertical axis (30.5 cm from the floor, 107 cm from the floor, and 30.5 cm from the ceiling). Up to four pouches, each containing a different species of insect, were hung from hooks at each position. Whereas a repellent device can be placed anywhere within the tent or at the entrance, our standard initial evaluation of a new device is to place it in the center of the tent 107 cm above the floor, Fig. 11.23.

FIG. 11.22 **Mosquito cages. Two styles of hanging mosquito cages used at CMAVE.** Pouch style (A), and hoop style (B). *CMAVE*, Center for Medical and Veterinary Entomology.

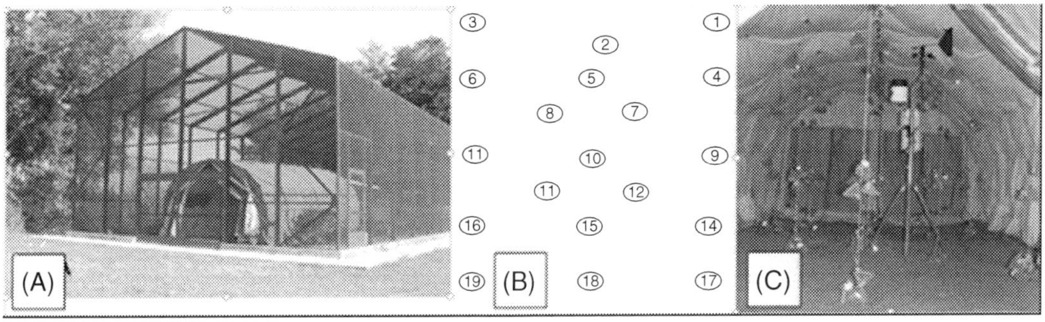

FIG. 11.23 CMAVE SFE Mosquito Cage Layout. Leftmost figure (A) showing previous style of military tent. Center schematic (B) showing mosquito hanging cage layout, and rightmost figure (C) showing bioassay setup inside of the military tent. *CMAVE*, Center for Medical and Veterinary Entomology.

In our standard protocol, the candidate SR device is activated immediately after all the bioassay cages have been hung. In our initial studies, the pouches were retrieved 18 hours post exposure and mortality data recorded was used to determine the spatial distribution pattern of a candidate device's efficacy in delivery of a lethal concentration of airborne insecticide. As previously stated, a problem with this protocol was that it was difficult to make observations through the gray mesh screen, on mosquito activity such as knockdown, thus our decision to switch to the hoop bioassay cages which allow us to make frequent observations on mosquito activity. Therefore, our current protocol calls for observing knockdown/mortality at 15-minute intervals for the first 2 hours, at 4 and 18 hours postexposure. Two enclosures are used during each test period, one serving as the untreated control. Treatment and control enclosures are rotated each test period. A minimum of 48 hours is used to allow the tents to "air out" between test periods. A replicate consists of two test nights, i.e., the same device tested in each enclosure for one night to minimize location effects. A minimum of three replicates is conducted for each candidate device or compound. Mortality data are arc sine transformed, analyzed by ANOVA, and followed by an appropriate means separation test. Efforts are made to utilize spatial statistics to determine the spatial effects of the candidate chemicals on mortality.

In our latest iteration of the bioassay cage tests, we are using a modular design to distribute the bioassay cages, which consists of a central platform with four arms radiating from it in north, south, east, and west directions. Bioassay stations are constructed from Polyvinyl chloride (PVC) pipe along each horizontal arm. Stations are established at 0.5, 1.0, 1.5. 2.0, 2.5, 3.0, 3.5, and 4.0 m from the center of the platform. At each horizontal station, bioassay cages are hung vertically at 0.5, 1.0, and 1.5 m above ground level, which are approximately the heights of the senior author's knee, waist, and shoulders. This modular bioassay cage is currently being utilized in both types of enclosures (with/without tents), Fig. 11.24.

Since we also desire to determine the potential of the candidate compound/product to repel the target species from entering a designated space, the second step in our evaluations consists of releasing free-flying insects to determine if given the freedom of movement, they will avoid the zone containing the airborne insecticide. In these types of tests, a trap is located on the center platform (within the tent or open environment

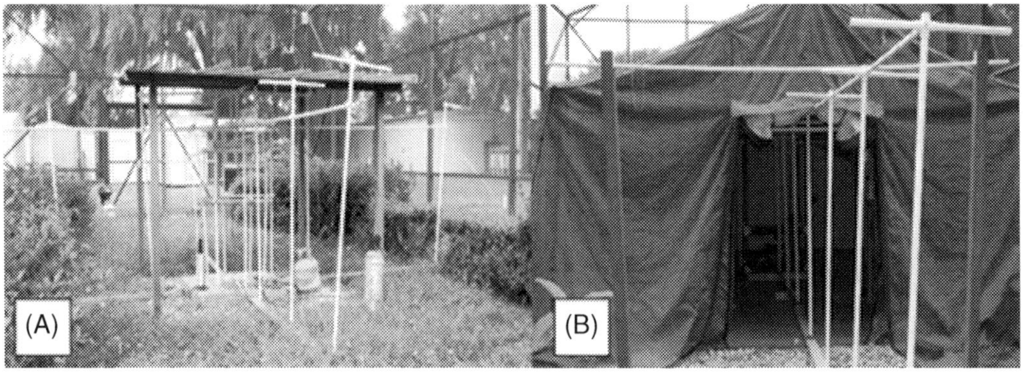

FIG. 11.24 CMAVE SFE bioassay. Bioassay hangar design shown on the left (A) within the open semi-field enclosure and on the right (B) in the semi-field enclosure with military tent. Note that bioassay hangars extend into the tent and encompasses the interior in the rightmost figure. *CMAVE*, Center for Medical and Veterinary Entomology.

enclosure; the trap simulates a host by producing carbon dioxide, heat and water vapor through the combustion of propane; 1-octenol-3-ol (octenol) is added as an additional attractant. We have been using a Mosquito Magnet Independence (MM-I) trap as a surrogate host. Mosquitoes attracted to the trap are caught in a net, which is collected 18 hours postrelease of the target insects. If stable flies are used, a Knight Stick trap, located near the MM-I trap, is used to capture the stable flies. Stable flies drawn into the tent by the attractants emitted by the MM-I trap are visually attracted to the Knight Stick trap and get stuck on the sticky sheet. As in the case of the bioassay cage study, two screened enclosures are used, one serving as an untreated control. This test provides data on the efficacy of a candidate device to prevent entry of free flying mosquitoes into an enclosed structure/area designated for protection.

In the tent studies, variations of this test include placing the SR device at different locations within the tent. As our studies have progressed, the importance of placing these SR delivery devices at the tent entrance have become the point of emphasis. In these repellency tests, the candidate is activated approximately 1 hour before release of the target insect species, from the north end of the large screen enclosure. Five hundred adults of three insect species (two mosquito species, [*Ae. aegypti*, *Ae. taeniorhynchus*] and a stable fly [*S. calcitrans*]) are released. The released mosquitoes are 4–6-day-old colony reared females. Mixed sexes of 4—6-day old colony reared stable flies are released.

We have also conducted similar tests with free-flying and mosquitoes confined in bioassay cages in the other three large outdoor cages without a tent. This was done to simulate open field conditions. Under these conditions, the host is more exposed to free-flying insects since they have no structure to protect them. These conditions can be used to simulate a scenario in which SR devices might protect an individual or a group of people in an area designated for protection. Devices tested under these conditions might be a wearable device used by an individual located at strategic places on the body, or a device strategically placed to protect a designated area. The MM-I traps have been used as the surrogate host.

The results obtained by these studies indicated that active-emission systems were more effective than passive-emission systems with the exception of the TF-treated Hessian strip, which was developed by Ogoma et al. (2012). General findings were that delivery systems containing the same active ingredient do perform differently; dispersion patterns were different; impacts on mortality and repellency were different. Target species differed in their responses to active ingredient type and concentration. Environmental factors, such as wind (air movement patterns), temperature, and relative humidity all had significant impacts (Ogoma, et al., 2012).

11.4.2 Current research

Our current research emphasizes the development of novel passive delivery devices utilizing TF as the active ingredient to protect civilian and military personnel. This research is focused on the development and evaluation of two novel types of controlled release passive SR delivery devices: TF-treated military bootlaces and custom designed tent entrance protective devices constructed from TF-containing multi-lumen polymer tubes. The TF-containing multi-lumen polymer tubes are being designed for placement at the tent entrance. The plan is to optimize these technologies under semi-field against laboratory reared mosquitoes and then confirm our findings against natural populations utilizing our modular evaluation system at local field study sites during different environmental conditions and target species to define the limits of efficacy of these technologies. Various formulations and concentrations of TF are being evaluated for immediate and long-term efficacy. A major objective is to characterize the impact of these novel delivery devices on key behavioral host-seeking and biting actions of target species to TF to optimize the release dynamics of the active ingredient which will lead to development of effective and practical products. In research conducted so far, we have observed behavioral avoidance responses by adult host-seeking females from several of our laboratory strains of mosquitoes (*Ae. aegypti, Ae. taeniorhynchus, An. quadrimaculatus,* and *Cx. quinquefasciatus*) and stable flies, *S. calcitrans.* Other colonized species, e.g., *Ae. albopictus* and *An. albimanus* and possibly insecticide resistant strains of *Ae. aegypti* and *Cx. quinquefasciatus* will also be included in future efficacy evaluations.

The overall hypothesis driving our SR research is that a sufficient concentration of an active ingredient can be continuously and effectively released from a passive device which will repel biting arthropods, thus limiting host-vector contact. Individual hypotheses which will be evaluated are as follows:

Hypothesis 1. There exists a critical concentration of airborne active ingredient that elicits repellent behavior not mortality. When biting arthropods detect this critical concentration, they will attempt to escape to an area free of the active ingredient.

Hypothesis 2. The quantity of airborne active ingredient that elicits repellency will vary according to target species, but there will be overlap in the region between repellency and mortality among the relevant species.

Hypothesis 3. The type and placement of SR delivery devices can be customized to act against different target species and protect people in a variety of localized scenarios.

A list of technical tasks and questions that will help guide us to accomplish the specific aims of our current research using TF is:

Determine if indeed there exists a critical airborne concentration of TF that will cause mosquitoes to be repelled rather than die before they make physical contact with the host.

Determine if we can produce and sustain the critical concentration through the use of passive delivery devices such as TF- treated bootlaces and multilumen polymer tubing designs containing TF. Although we intend to use different operational functions for these devices (TF-treated bootlaces to protect an individual and the TF- containing multilumen tubing devices as entrance devices to exclude mosquitoes from entering a tent and thus preventing bites to the occupants of the tent), we will design a protocol to directly compare the efficacy of the two technologies.

Determine what formulation(s) and concentrations of TF are needed to achieve the desired results. We hope that TF- treated bootlaces will reduce mosquito bites by >90% to a standing individual from at least ankle to waist for a period of at least a month.

Determine the duration of efficacy of these devices under operational conditions. We plan to wear boots with treated laces as part of our daily dress code to simulate what an infantry soldier might encounter. The senior author normally spends at least three days per week at research sites at a wildlife refuge that includes routinely walking through freshwater swamps and high salt marsh areas.

Determine the impact that different habitat/environmental conditions have on duration of efficacy (e.g., hot humid areas like Thailand versus dry hot areas like the Coachella Valley, California, United States).

Determine how much active ingredient needs to be released to create the airborne concentrations of TF that promote repellent behaviors and mortality in the target species.

Link airborne concentrations of TF to specific sets of kinematic flying behaviors in mosquitoes.

Determine how frequently TF devices will need to be replaced.

Work with appropriate regulatory agencies and industry partners to get these technologies approved for use.

11.4.3 Future Center for Medical and Veterinary Entomology studies

We plan to evaluate the impact of SRs on blood-feeding and sugar feeding behaviors of *Ae. aegypti* by utilizing sentinel cages (18 × 18 × 18 cm BioQuip cages) containing 50–100, 3—4-day-old female and male mosquitoes. Blood meals will be offered by means of heated condoms containing defibrinated cattle blood placed on top of the cage. Similarly, 10% sucrose saturated cotton pads will be placed on top of the cages at various times during the efficacy evaluation process of different SR products. These studies will be conducted in the SF enclosures without tents using the modular set-up. The sentinel mosquito cages will be placed 1.0 m high in the center of each quadrant created by the directional arms radiating from the central platform of the modular evaluation set-up. Further studies could be conducted in which these sentinel cages are placed and removed at various time intervals of exposure to determine if exposed mosquitoes can recover after exposure to various concentrations of active ingredient.

Based on the laboratory findings of Bibbs et al. (2018), we also plan to evaluate SR effects on the oviposition behavior of *Ae. albopictus* and *Ae. aegypti* in a SF environment. Oviposition containers will be distributed throughout one of the SF environments (without tents). One hundred gravid females of each species (different times) will be released into a SFE treated with a SR active ingredient and an untreated control cage. Various aspects of the oviposition process will be studied, such as skip oviposition behavior, overall number of eggs laid, hatchability, and larval development (Bibbs et al., 2018).

Efficacy evaluations of both active and passive emanators with essential oils, like linalool and candidate repellent compounds supplied by collaborators, will be conducted against three species of mosquitoes (*Ae. aegypti, Ae. quadrimaculatus,* and *Cx. quinquefasciatus*). Baited traps (either Mosquito Magnet Executive or BGS) will be used as surrogate human hosts. Only free flying mosquitoes will be used in these studies, which will allow us to compare the comparative protective efficacy obtained by using synthetic pyrethroids like TF and metofluthrin with that obtained with essential oils.

Spatial repellency of active ingredients against insecticide-resistant strains. The objective of this study is to determine any difference in spatial repellency between susceptible and insecticide-resistant strains to provide information for disease protection and control strategies.

11.5 Conclusion

Semi-field testing methodologies can bridge the gap between pure laboratory and pure field studies of both topical and SRs as published SF and subsequent field testing has been shown to yield similar results. For example, Sangoro et al. (2014) found that when testing efficacy of topical DEET against *An. arabiensis* in a SF assay, the results were similar to testing DEET against *An. gambiae* in a field study (Sangoro et al., 2014). Additional evidence that the SF environment is a useful step in evaluating the efficacy of SRs, is the increasing number, and variety of SF environments being utilized worldwide. Based on the results being generated in these SF environments, SRs may indeed play an increased role in developing future biting arthropod control interventions. However, much research is still needed to develop delivery systems that are affordable and effective. This may require the discovery of additional active ingredients that provide sustainable protection over a long period of time. Equally important is the continued improvement of the SF as an evaluation system that takes into consideration an understanding of all the various behavioral responses of biting arthropod vectors to different types of active ingredients and delivery systems, which might influence their efficacy. Based on our SF enclosure, SR efficacy studies, and those conducted at other institutions worldwide we believe that SRs can be used to create a vector-free space, thereby preventing contact between host (human or animal) and vector, thus preventing vector-borne disease transmission.

References and further readings

Abdel-Mohdy, F.A., Moustafa, M.M.G., Rehan, M.F., Aly, A.S, 2008. Repellency of controlled-release treated cotton fabrics based on cypermethrin and prallethrin. Carbohydr. Polym. 73, 92–97. https://doi.org/10.1016/j.carbpol.2007.11.006.

Achee, N.L., Bangs, M.J., Farlow, R., Killeen, G.F., Lindsay, S., Logan, J.G., Moore, S.J., Rowland, M., Sweeney, K., Torr, S.J., Zwiebel, L.J., Grieco, J.P., 2012. Spatial repellents: from discovery and development to evidence-based validation. Malar. J. 11, 1–9. doi:10.1186/1475-2875-11-164.

Achee, N.L., Sardelis, M.R., Dusrour, I., Chauhan, K.R., Grieco, J.P., 2009. Characterization of spatial repellent, contact irritant, and toxicant chemical actions of standard vector control compounds. J. Am. Mosq. Control Assoc. 25, 156–167.

Achee, N.L., Masouka, P., Smith, P., Martin, N., Chareonviryiphap, T., Polsomboon, S., Hendarto, J., Grieco, J.P., 2012a. Identifying the effective concentration for spatial repellency of the dengue vector *Aedes aegypti*. Parasit. Vectors 5, 1–9.

Achee, N.L., Grieco, J.P., 2018. Chapter 3: Current evidence, new insights, challenges and future outlooks to the use of spatial repellents for public healthAdvances in the Biorational Control of Medical and Veterinary Pests. ACS Publications, Washington, DC, pp. 25–42. American Chemical Society Symposium Series 1289.

Adanan, C.R., Zairi, J., Ng, K.H., 2005. Efficacy and sublethal effects of mosquito mats on *Aedes aegypti* and *Culex quinquefasciatus* (Diptera: Culicidae). In: Lee, C.Y., Robinson, W.H. (Eds.), Fifth International Conference on Urban Pests. Singapore, 11–13 July 2005, pp. 265–269 ref.11.

Amelia-Yap, Z.H., Chen, C.D., Sofian-Azirun, M., Low, V.L., 2018. Pyrethroid resistance in the dengue vector *Aedes aegypti* in Southeast Asia: present situation and prospects for management. Parasit. Vectors 11 (1), 332, 1–9. doi:10.1186/s13071-018-2899-0.

Anastasia Mosquito Control District of St. Johns County. AMCD 2019 Report. no. 904, 2020. St. Augustine, FL 2019.

Avicor, S.W., Wajidi, M.F.F., Jaal, Z, 2015. Laboratory evaluation of three commercial coil products for protection efficacy against *Anopheles gambiae* from southern ghana: a preliminary study. Trop. Biomed. 32 (2), 386–389. http://www.msptm.org/files/386_-_389_Avicor_SW.pdf.

Ball, T.S., Ritchie, S.R., 2010a. Sampling biases of the BG-sentinel trap with respect to physiology, age, and body size of adult *Aedes aegypti* (Diptera: Culicidae). J. Med. Entomol. 47, 649–656.

Ball, T.S., Ritchie, S.R., 2010b. Evaluation of BG-sentinel trap trapping efficacy for *Aedes aegypti* (Diptera: Culicidae) in a visually competitive environment. J. Med. Entomol. 47, 657–663.

Banks, S.D., Murray, N., Wilder-Smith, A., Logan, J.G., 2014. Insecticide-treated clothes for the control of vector-borne diseases: a review on effectiveness and safety. Med. Vet. Entomol. 28 (1), 14–25. https://doi.org/10.1111/mve.12068.

Barnard, D.R., 2000. Repellents and Toxicants for Personal Protection: A WHO Position Paper. World Health Organization, Geneva.

Barnard, D.R., Posey, K.H., Smith, D., Schreck, C.E., 1998. Mosquito density, biting rate and cage size effects on repellent tests. Med. Vet. Entomol. 12 (1), 39–45. https://doi.org/10.1046/j.1365-2915.1998.00078.x.

Barnard, D.R., Xue, R.D., 2004. Laboratory evaluation of mosquito repellents against *Aedes albopictus*, *Culex nigripalpus*, and *Ochlerotatus triseriatus* (Diptera: Culicidae). J. Med. Entomol., 41, 726–730.

Bernier, U., Clark, G., Gurman, P., Elman, N., 2015. The use of microdispensers with spatial repellents for personal protection against mosquito biting. J. Med. Entomol 53, 470–472.

Bernier, U.R., Kline, D.L., Posey, K.H., 2007. Human emanations and related natural compounds that inhibit mosquito host-finding abilities. In: Debboun, M., Frances, S.P., Strickman, D (Eds.), Insect Repellents: Principles, Methods and Uses. CRC Press, Taylor and Francis Group, Boca Raton, FL, pp. 77–100.

Bibbs, C.S., Fulcher, A.P., Xue, R.D., 2015. Allethrin-based mosquito control device causing knockdown, morbidity, and mortality in four species of field-caught mosquitoes (Diptera: Culicidae). J. Med. Entomol. 52 (4), 739–742. https://doi.org/10.1093/jme/tjv065.

Bibbs, C.S., Hahn, D.A., Kaufman, P.E., Xue, R.D., 2018a. Sublethal effects of a vapour-active pyrethroid, transfluthrin, on *Aedes aegypti* and *Ae albopictus* (Diptera: Culicidae) fecundity and oviposition behavior. Parasit. Vectors 11 (1), 1–9. doi:10.1186/s13071-018-3065-4.

Bibbs, C.S., Kaufman, P.E., 2017. Volatile pyrethroids as a potential mosquito abatement tool: a review of pyrethroid-containing spatial repellents. J. Integr. Pest Manag. 8 (1), 1–10. https://doi.org/10.1093/jipm/pmx016.

Bibbs, C.S., Xue, R.D., 2016. OFF! Clip-on repellent device with metofluthrin tested on *Aedes aegypti* (Diptera: Culicidae) for mortality at different time intervals and distances. J. Med. Entomol. 53 (2), 480–483. https://doi.org/10.1093/jme/tjv200.

Bibbs, C.S., Tsikolia, M., Bernier, U.R., Bloomquist, J.R., Xue, R.D., Kaufman, P.E., 2018b. Vapor toxicity of five volatile pyrethroids against *Aedes aegypti*, *Ae albopictus*, *Culex quinquefasciatus*, and *Anopheles quadrimaculatus* (Diptera: Culicidae). Pest Manag. Sci. 74, 2699–2706.

Bohbot, J.D., Dickens, J.C., 2012. Odorant receptor modulation: ternary paradigm for mode of action of insect repellents. Neuropharmacology 62 (5–6), 2086–2095. https://doi.org/10.1016/j.neuropharm.2012.01.004.

Briassoulis, G., Narlioglou, M., Hatzis, T., 2001. Toxic encephalopathy associated with use of DEET insect repellents: a case analysis of its toxicity in children. Hum. Exp. Toxicol. 20, 8–14. https://doi.org/10.1191/096032701676731093.

Buhagiar, T.S., Devine, G.J., Ritchie, S.A., 2017a. Effects of sublethal exposure to metofluthrin on the fitness of *Aedes aegypti* in a domestic setting in Cairns. Queensland. Parasit. Vectors, 10 (1), 1–7. doi:10.1186/s13071-017-2220-7.

Buhagiar, T.S., Devine, G.J., Ritchie, S.A., 2017b. Metofluthrin: investigations into the use of a volatile spatial pyrethroid in a global spread of dengue, chikungunya and Zika viruses. Parasit. Vec. 10, 1–11.

CDC, 2016. United States Centers for Disease Control and Prevention. Responding to local mosquito-borne transmission of Zika virus. Operational risk communication and community engagement plan Atlanta: United States Department of Health and Human Services, 1–22.

Chen-Hussey, V., Behrens, R., Logan, J.G., 2014. Assessment of methods used to determine the safety of the topical insect repellent N, N-diethyl-m-toluamide (DEET). Parasit. Vec. 7, 1–7.

Choi, D.B., Grieco, J.P., Apperson, C.S., Schal, C., Ponnusamy, L., Wesson, D.M., Achee, N.L., 2016. Effect of spatial repellent exposure on dengue vector attraction to oviposition sites. PLoS Negl.Trop. Dis. 10 (7), 1–11. doi:10.1371/journal.pntd.0004850.

Cohnstaedt, L.W., Allan, S.A., 2011. Effects of sublethal pyrethroid exposure on the host-seeking behavior of female mosquitoes. J. Vector Ecol. 36 (2), 395–403. https://doi.org/10.1111/j.1948-7134.2011.00180.x.

Collier, B.W., Perich, M.J., Boquin, G.J., Harrington, S.R., Francis, M.J., 2006. Field evaluation of mosquito control devices in Southern Louisiana. J. Am. Mosq. Control Assoc. 22 (3), 444–450. https://doi.org/10.2987/8756-971X(2006)22[444:FEOMCD]2.0.CO;2.

Cook, S.M., Khan, Z.R., Pickett, J.A., 2007. The use of push-pull strategies in integrated pest management. Annu. Rev. Entomol. 52, 375–400. https://doi.org/10.1146/annurev.ento.52.110405.091407.

Corbel, V., N'Guessan, R., 2013. Distribution, mechanisms, impact and management of insecticide resistance in malaria vectors: a pragmatic review. In: Manguin, S. (Ed.), Anopheles Mosquitoes: New Insights into Malaria Vectors. Intech Open Publ., London, pp. 579–633.

Cui, L., Yan, G., Sattabongkot, J., Cao, Y., Chen, B., Chen, X., Fan, Q., Fang, Q., Jongwutiwes, S., Parker, D., Sirichaisinthop, J., Kyaw, M.P., Su, X.z., Yang, H., Yang, Z., Wang, B., Xu, J., Zheng, B., Zhong, D., Zhou, G., 2012. Malaria in the greater mekong subregion: heterogeneity and complexity. Acta Tropica, 121 (3), 227–239. https://doi.org/10.1016/j.actatropica.2011.02.016.

Curtis, C.F., 1976. Population replacement in *Culex fatigans* by means of cytoplasmic incompatibility. II. Field cage experiments with overlapping generations. Bull. World Health Organ. 53 (1), 107–119.

Dame, D.A., Meisch, M.V., Lewis, C.N., Kline, D.L., Clark, G.G., 2014. Field evaluation of four spatial repellent devices against arkansas rice-land mosquitoes. J. Am. Mosq. Control Assoc. 30 (1), 31–36. https://doi.org/10.2987/13-6379.1.

Davis, E.E., Sokolove, P.G., 1976. Lactic acid-sensitive receptors on the antennae of the mosquito, *Aedes aegypti*. J. Comp. Physiol. 105 (1), 43–54. https://doi.org/10.1007/BF01380052.

Debboun, M., Strickman, D., 2013. Insect repellents and associated personal protection for a reduction in human disease. Med. Vet. Entomol. 27, 1–9.

Debboun, M., Frances, S.P., Strickman, D., 2007. Insect Repellents: Principles, Methods, and Uses. CRC Press, Boca Raton, FL.

Diaz, J.H., 2016. Chemical and plant-based insect repellents: efficacy, safety, and toxicity. Wilderness Environ. Med. 27 (1), 153–163. https://doi.org/10.1016/j.wem.2015.11.007.

Facchinelli, L., Valerio, L., Bond, J.G., Wise de Valdez, M.R., Harrington, L.C., Ramsey, J.M., Casas-Martinez, M., Scott, T.W., 2011. Development of a semi-field system for contained field trials with *Aedes aegypti* in southern Mexico. Am. J. Trop. Med. Hyg. 85, 248–256.

Facchinelli, L., Valerio, L., Ramsey, J.M., Gould, F., Walsh, R.K., Bond, G., Robert, M.A., Lloyd, A.L., James, A.A., Alphey, L., Scott, T.W., 2013. Field cage studies and progressive evaluation of genetically-engineered mosquitoes. PLoS Neglect. Trop. Dis., e2001.

Ferguson, H.M., Ng'habi, K.R., Walder, T., Kadungula, D., Moore, S.J., Lyimo, I., Russell, T.L., Urassa, H., Mshinda, H., Killeen, G.F., 2008. Establishment of a large semi-field system for experimental study of African malaria vector ecology and control in Tanzania. Malar. J. 7, 1–15.

Fornadel, C.M., Norris, L.C., Glass, G.E., Norris, D.E., 2010. Analysis of *Anopheles arabiensis* blood feeding behavior in southern zambia during the two years after introduction of insecticide-treated bed nets. Am. J. Trop. Med. Hyg. 83 (4), 848–853. https://doi.org/10.4269/ajtmh.2010.10-0242.

Fradin, M.S., Day, J.F., 2002. Comparative efficacy of insect repellents against mosquito bites. N. Engl. J. Med. 347 (1), 13–18. https://doi.org/10.1056/NEJMoa011699.

Frances, S.P., 1987. Effectiveness of deet and permethrin, alone, and in a soap formulation as skin and clothing protectants against mosquitoes in Australia. J. Am. Mosq. Control Assoc. 3 (4), 648–650.

Frances, S.P., Eamsila, C., Pilakasiri, C., Linthicum, K.J., 1996. Effectiveness of repellent formulations containing deet against mosquitoes in northeastern Thailand. J. Am. Mosq. Control Assoc. 12 (2), 331–333.

Frances, S.P., MacKenzie, D.O., Rowcliffe, K.L., Corcoran, S.K., 2009. Comparative field evaluation of repellent formulations containing DEET and IR3535 against mosquitoes in Queensland. Australia. J. Am. Mosq. Cont. Assoc. 25, 511–513.

Goodyer, L., Behrens, R., 1998. Short report: the safety and toxicity of insect repellents. Amer. J. Med. Hyg. 59, 323–324.

Gouck, H., McGovern, T.P., Beroza, M., 1967. Chemicals tested as space repellents against yellow-fever mosquitoes I. Esters1. J. Econ. Entomol. 60, 1587–1590. doi:10.1093/jee/60.6.1587.

Govella, N., Ogoma, SB., Paliga, J., Chaki, P.P., Killeen, G., 2015. Impregnating hessian strips with the volatile pyrethroid transfuthrin prevents outdoor exposure to vectors of malaria and lymphatic flariasis in urban Dar esSalaam,Tanzania.Para.Vec.8(1),1–9.doi:10.1186/s13071-015-0937-8.

Grieco, J.P., Achee, N.L., Chareonviriyaphap, T., Suwonkerd, W., Chauhan, K., Sardelis, M.R., Roberts, D.R., 2007. A new classification system for the actions of IRS chemicals traditionally used for malaria control. PLoS One 2 (8), e716. doi:10.1371/journal.pone.0000716.

Gryseels, C., Uk, S., Sluydts, V., Durnez, L., Phoeuk, P., Suon, S., Set, S., Heng, S., Siv, S., Gerrets, S., Tho, S., 2015. Factors influencing the use of topical repellents: implications for the effectiveness of malaria elimination strategies. Sci. Rep. 5, 1–14.

Hackett, L.W., Bates, M., 1939. The laboratory for mosquito research in AlbaniaActa Conventus Tertii de Tropicis atque Malariae Morbis, 2. Societas Neerlandica Medicinae Tropicae, Amsterdam, pp. 113–123.

Harrington, L.C., Ponlawat, A., Edman, J.D., Scott, T.W., Vermeylen, F., 2008. Influence of container size, location, and time of day on oviposition patterns of the dengue vector, *Aedes aegypti*, in Thailand. Vect. Borne Zoon. Dis. 8, 415–424.

Heng, S., Durnez, L., Gryseels, C., Van Roey, K., Mean, V., Uk, S., Siv, S., Grietens, K.P., Sochantha, T., Coosemans, M., Sluydts, V., 2015. Assuring access to topical mosquito repellents within an intensive distribution scheme: a case study in a remote province of Cambodia. Malar. J. 14, 1–13.

Hill, N., Lenglet, A., Arnéz, A.M., Carneiro, I., 2007. Plant based insect repellent and insecticide treated bed nets to protect against malaria in areas of early evening biting vectors: double blind randomised placebo controlled clinical trial in the Bolivian Amazon. Brit. Med. J. 335 (7628), 1023–1025. https://doi.org/10.1136/bmj.39356.574641.55.

Hill, N., Zhou, H.N., Wang, P., Guo, X., Carneiro, I., Moore, S.J., 2014. A household randomized, controlled trial of the efficacy of 0.03% transfluthrin coils alone and in combination with long-lasting insecticidal nets on the incidence of *Plasmodium falciparum* and *Plasmodium vivax* malaria in Western Yunnan Province, China. Malaria Journal 13 (1), 1–8. doi:10.1186/1475-2875-13-208.

Katsuda, Y., Leemingsawat, S., Thongrungkiat, S., Prummonkol, S., Samung, Y., Kanzaki, T., Watanabe, T., 2009. Control of mosquito vectors of tropical infectious diseases, Susceptibility of *Aedes aegypti* to pyrethroid and mosquito coils. Southeast Asian J. Trop. Med. Public Health 40 (5), 929–936. https://seameotropmednetwork.org/publication_current_issue.html.

Khan, A.A., Maibach, H.I., Skidmore, D.L., 1975. Insect repellents: effect of mosquito and repellent-related factors on protection time. J. Econ. Entomol. 68 (1), 43–45. https://doi.org/10.1093/jee/68.1.43.

Khater, E., Zhu, D., Xue, R.D., 2019. Semi-field evaluation of modified 00ZZZero® traps with sticky paper to increase the collection efficacy of gravid *Aedes aegypti*. J. Am. Mosq. Control Assoc. 35 (2), 137–139. https://doi.org/10.2987/19-6818.1.

Killeen, G.F., 2014. Characterizing, controlling and eliminating residual malaria transmission. Malar. J. 13, 1–22.

Killeen, G.F., Moore, S.J., 2012. Target product profiles for protecting against outdoor malaria transmission. Malar. J. 11, 1–6. doi:10.1186/1475-2875-11-17.

Kiszewski, A.E., Darling, S.T., 2010. Estimating a mosquito repellent's potential to reduce malaria in communities. J. Vector Borne Dis. 47 (4), 217–221. http://www.mrcindia.org/journal/issues/474217.pdf.

Kitau, J., Pates, H., Rwegoshora, T.R., Rwegoshora, D., Matowo, J., Kweka, E.J., Mosha, F.W., McKenzie, K., Magesa, S.M., 2010. The effect of mosquito Magnet® liberty plus trap on the human mosquito biting rate under semi-field conditions. J. Am. Mosq. Control Assoc. 26 (3), 287–294. https://doi.org/10.2987/09-5979.1.

Kline, D.L., Strickman, D., 2014. Spatial or area repellentsInsect Repellents Handbook. 2nd ed., CRC Press, Boca Raton, FL, pp. 239–252. https://doi.org/10.1201/b17407.

Kline, D.L., Urban, J., 2018. Chapter 13. Potential for utilization of spatial repellents in mosquito control interventions. In: Norris, E.J., Coats, J.R., Gross, A.D., Clark, J.M. (Eds.), Advances in the Biorational Control of Medical and Veterinary Pests, vol. 1289. American Chemical Society, ACS Symposium Series, Washington, D.C., pp. 237–248.

Kline, D.A., Strickman, D.A., 2014. Spatial or area repellents. In: Debboun, M., Frances, S.P., Strickman, D.A. (Eds.), Insect Repellents Handbook2nd Edition. CRC Press, Boca Raton, FL, pp. 239–251.

Knols, B.G., Njiru, B.N., Mathenge, E.M., Mukabana, W.R., Beier, J.C., Killeen, G.F., 2002. MalariaSphere: a greenhouse-enclosed simulation of a natural *Anopheles gambiae* (Diptera: Culicidae) ecosystem in western Kenya. Malar. J. 1, 1–13.

Kröber, T., Kessler, S., Frei, J., Bourquin, M., Guerin, P.M., 2010. An in vitro assay for testing mosquito repellents employing a warm body and carbon dioxide as a behavioral activator. J. Am. Mosq. Cont. Assoc. 26, 381–386.

Kweka, E.J., Munga, S., Mahande, A.M., Msangi, S., Mazigo, H.D., Adrias, A.Q., Matias, J.R., 2012. Protective efficacy of menthol propylene glycol carbonate compared to N, N-diethyl-methylbenzamide against mosquito bites in Northern Tanzania. Parasit. Vectors 5 (1), 1–10. doi:10.1186/1756-3305-5-189.

Lee, D.K., 2007. Lethal and repellent effects of transfluthrin and metofluthrin used in portable blowers for personal protection against *Ochlerotatus togoi* and *Aedes albopictus* (Diptera: Culicidae). Entomol. Res. 37 (3), 173–179. https://doi.org/10.1111/j.1748-5967.2007.00109.x.

Lee, M.Y., 2018. Essential oils as repellents against arthropods. Hindawi BioMesd Res. Inter. 2018, 1–9.

Lindsay, Ewald, J.A., Samung, Y., Apiwathnasorn, C., Nosten, F, 1998. Thanaka (Limonia acidissima) and deet (di-methyl benzamide) mixture as a mosquito repellent for use by Karen women. Med. Vet. Entomol. 12 (3), 295–301. https://doi.org/10.1046/j.1365-2915.1998.00115.x.

Lindsay, L.R., Surgeoner, G.A., Heal, J.D., Gallivan, G.J., 1996. Evaluation of the efficacy of 3Vo citronella candles and 5Vo citronella incense for protection against field populations of Aedes mosquitoes. J. Am. Mosq. Cont. Assoc. 12, 293–294.

Liu, W., Todd, R.G., Gerberg, E.J., 1986. Effect of three pyrethroids on blood feeding and fecundity of *Aedes aegypti*. J. Am. Mosq. Control Assoc. 2 (3), 310–313.

Lloyd, A.M., Farooq, M., Diclaro, J.W., Kline, D.L., Estep, A.S., 2013. Field evaluation of commercial off-the-shelf spatial repellents against the Asian tiger mosquito, *Aedes albopictus* (Skuse), and the potential for use during deployment. US Army Med. Dept. J. 80–86. 23584913.

Logan, J., Chen-Hussey, V., O'Halloran, L., Greaves, C., Due, C., & Macdonald, M. (2020). An expert review of spatial repellents for mosquito control. Report. IVCC. https://www.ivcc.com/wp-content/uploads/2020/08/An-Expert-Review-of-Spatial-Repellents-for-Mosquito-Control.pdf.

Maia, M.F., Moore, S.J., 2011. Plant-based insect repellents: a review of their efficacy, development and testing. Malar. J. 10, 1–15.

Maia, M.F., Onyango, S.P., Thele, M., Simfukwe, E.T., Turner, E.L., Moore, S.J., 2013. Do topical repellents divert mosquitoes within a community? –Health equity implications of topical repellents as a mosquito bite prevention tool. PLoS One 8, 1–7.

Masalu, J.P., Finda, M., Killeen, G.F., Ngowo, H.S., Pinda, P.G., Okumu, F.O., 2020. Creating mosquito-free outdoor spaces using transfluthrin-treated chairs and ribbons. Malar. J. 19 (1), 1–13. doi:10.1186/s12936-020-03180-1.

Masalu, J.P., Finda, M., Okumu, F.O., Minja, E.G., Mmbando, A.S., Sikulu-Lord, M.T., Ogoma, S.B., 2017. Efficacy and user acceptability of transfluthrin-treated sisal and hessian decorations for protecting against mosquito bites in outdoor bars. Parasit. Vectors 10 (1), 1–8. doi:10.1186/s13071-017-2132-6.

McGready, R., Hamilton, K.A., Simpson, J.A., Cho, T., Luxemburger, C., Edwards, R., Looareesuwan, S., White, N.J., Nosten, F., Lindsay, S.W., 2001. Safety of the insect repellent N, N-diethyl-m-toluamide (DEET) in pregnancy. Am. J. Trop. Med. Hyg. 65 (4), 285–289. https://doi.org/10.4269/ajtmh.2001.65.285.

McPhatter, L.P., Mischler, P.D., Webb, M.Z., Chauhan, K., Lindroth, E.J., Richardson, A.G., Debboun, M., 2017. Laboratory and semi-field evaluations of two (transfluthrin) spatial repellent devices against *Aedes aeĝypti* (L.) (Diptera: Culicidae). US Army Med. Dep. J. 13, 1–17. 28511270.

Menger, D.J., Otieno, B., Rijk, Mukabana, W.R., Loon, Takken, W, 2014. A push-pull system to reduce house entry of malaria mosquitoes. Malar. J 13, 1–8.

Mmbando, A.S., Ngowo, H., Limwagu, A., Kilalangongono, M., Kifungo, K., Okumu, F.O., 2018. Eave ribbons treated with the spatial repellent, transfluthrin, can effectively protect against indoor-biting and outdoor-biting malaria mosquitoes. Malar. J. 17, 1–14.

Moore, S.J., Davies, C.R., C.R., Hill, N., Cameron, 2007. Are mosquitoes diverted from repellent-using individuals to non-users? Results of a field study in Bolivia. Trop. Med. Intern. Heal. 12, 532–539.

Moore, S.J., 2016. A new perspective on the application of mosquito repellents. Lanc. Infect. Dis. 16, 1093–1094.

Msangi, S., Mwang'onde, B., Mahande, A., Kweka, E., 2010. Field evaluation of the bio-efficacy of three pyrethroid based coils against wild populations of anthropophilic mosquitoes in northern Tanzania. J. Glob. Infect. Dis. 2, 116–120.

Mshinda, H., Killeen, G.F., 2008. Establishment of a large semi-field system for experimental study of African malaria vector ecology and control in Tanzania. Malar. J. 7, 1–15.

Muller, G., Junnila, A., Butler, J., Kravchenko, V.D., Revay, E.E., Weiss, R.W., Schlein, Y., 2009. Efficacy of the botanical repellents geraniol, linalool, and citronella against mosquitoes. J. Vect. Ecol. 34, 2–8.

Muller, G., Junnila, A., Kravchenko, V., Revay, E., Butler, J., Weiss, R., Schlein, Y., 2008. Ability of essential oil candles to repel biting insects in high and low biting pressure environments. J. Am. Mosq. Cont. Assoc. 24, 154–160.

Nentwig, G., Frohberger, S., Sonneck, R., 2017. Evaluation of clove oil, picaridin, and transfluthrin for spatial repellent effects in three tests systems against the *Aedes aegypti* (Diptera: Culicidae). J. Med. Entomol. 4, 150–158.

Nerio, L.S., Olivero-Verbel, J., Stashenko, E., 2010. Repellent activity of essential oils: a review. Biores. Tech. 101, 372–378.

Ng'habi-Kija, R.N. Behavioural, ecological, and genetic determinants of mating and gene flow in African malaria mosquitoes. PhD Thesis, Wageningen University, 2010.

Ng'habi, K.R., 2010. PhD Thesis. Wageningen University.

Ng'habi, K.R., Lee, Y., Knols, B.G., Mwasheshi, D., Lanzaro, G.C., Ferguson, H.M., 2015. Colonization of malaria vectors under semi-field conditions as a strategy for maintaining genetic and phenotypic similarity with wild populations. Malar. J. 14, 1–11.

Nolen, J.A., Bedoukian, R.H., Maloney, R.E., Kline, D.L. (2002). Inventors; US Department of Agriculture, owner. 2002 Mar 26. Method, apparatus and compositions for inhibiting the human scent tracking ability of mosquitoes in environmentally defined three dimensional spaces. United States patent US 6,362,235.

Obermayr, U., Ruther, J., Bernier, U.R., Rose, A., Geier, M., 2015. Evaluation of a push-pull approach for *Aedes aegypti* (L.) using a novel dispensing system for spatial repellents in the laboratory and in a semi-field environment. PLoS One 10 (6), e0129878. doi:10.1371/journal.pone.0129878.

Ogoma, S.B., Lorenz, L.M., Ngonyani, H., Sangusangu, R., Kitumbukile, M., Kilalangongono, M., Simfukwe, E.T., Mseka, A., Mbeyela, E., Roman, D., Moore, J., 2014a. An experimental hut study to quantify the effect of DDT and airborne pyrethroids on entomological parameters of malaria transmission. Malar. J. 13, 1–17.

Ogoma, S.B., Mmando, A.S., Swai, J.K., Horstmann, S., Malone, D., Killeen, G.F., 2017. A low technology emanator treated with the volatile pyrethroid transfluthrin confers long term protection against outdoor biting vectors of lymphatic flariasis, arboviruses and malaria. PLoS Negl Trop. Dis 11, 1–22.

Ogoma, S.B., Moore, S.J., Maia, M.F., 2012a. A systematic review of mosquito coils and passive emanators: defining recommendations for spatial repellency testing methodologies. Parasit. Vect. 5, 1 10.

Ogoma, S.B., Ngonyani, H., Simfukwe, E.T., Mseka, A., Moore, J., Killeen, G.F., 2012b. Spatial repellency of transfluthrin-treated hessian strips against laboratory-reared *Anopheles arabiensis* mosquitoes in a semi-field tunnel cage. Parasit. Vectors 5 (1), 1–5. doi:10.1186/1756-3305-5-54.

Ogoma, S.B., Ngonyani, H., Simfukwe, E.T., Mseka, A., Moore, J., Maia, M.F., Moore, S.J., Lorenz, L.M., 2014b. The mode of action of spatial repellents and their impact on vectorial capacity of *Anopheles gambiae* sensu stricto. PLoS One 9 (12), 1–21. doi:10.1371/journal.pone.0110433.

Okumu, F.O., Titus, E., Mbeyela, E., Killeen, G.F., Moore, S.J., 2009. Limitation of using synthetic human odours to test mosquito repellents. Malar. J. 8 (1), 1–7. doi:10.1186/1475-2875-8-150.

Onyango, S., Moore, S., 2015. Evaluation of repellent efficacy in reducing disease incidence. In: Debboun, M., Frances, S., Strickman, D. (Eds.), Insect Repellents Handbook, 2nd ed. CRC Press, Boca Raton, FL, pp. 117–156.

Pates, H.V., Lines, J.D., Keto, A.J., Miller, J.E., 2002. Personal protection against mosquitoes in Dar es Salaam, Tanzania, by using a kerosene oil lamp to vaporize transfluthrin. Med. Vet. Entomol. 16 (3), 277–284. https://doi.org/10.1046/j.1365-2915.2002.00375.x.

Paz-Soldan, V.A., Plasai, V., Morrison, A.C., Rios-Lopez, E.J., Guedez-Gonzales, S., Grieco, J.P., Mundal, K.,

Chareonviriyaphap, T., Achee, N.L., 2011. Initial assessment of the acceptability of a push-pull *Aedes aegypti* control strategy in Iquitos, Peru and Kanchanaburi, Thailand. Am. J. Trop. Med. Hyg. 84, 208–217.

Ponlawat, A., Kankaew, P., Chanaimongkol, S., Pongsiri, A., Richardson, J.H., Evans, B.P., 2016. Semi-field evaluation of metofluthrin-impregnated nets on host-seeking *Aedes aegypti* and *Anopheles dirus*. J. Am. Mosq. Control Assoc. 32 (2), 130–138. https://doi.org/10.2987/moco-32-02-130-138.1.

Qualls, W.A., Day, J.F., Bowers, D.F., 2012. Altered behavioral responses of Sindbis virus-infected *Aedes aegypti* (Diptera: Culicidae) to DEET and non-DEET based insect repellents. Act. Trop. 122, 284–290.

Rapley, L.P., Russell, R.C., Montgomery, B.L., Ritchie, S.A., 2009. The effects of sustained release metofluthrin on the biting, movement, and mortality of *Aedes aegypti* in a domestic setting. Am. J. Trop. Med. Hyg. 81 (1), 94–99. https://doi.org/10.4269/ajtmh.2009.81.94.

Revay, E.E., Junnila, A., Xue, R.D., Kline, D.L., Bernier, U.R., Kravchenko, V.D., Qualls, W.A., Ghattas, N., Müller, G.C., 2013a. Evaluation of commercial products for personal protection against mosquitoes. Acta Trop. 125 (2), 226–230. https://doi.org/10.1016/j.actatropica.2012.10.009.

Revay, E., Junnila, A., Xue, R., Kline, D., Bernier, U., Kravchenko, V., Qualls, W., Ghattas, N., Muller, G., 2013a. Evaluations of commercial products for personal protection against mosquitoes. Acta. Trop. 125, 226–230.

Revay, E.E., Kline, D.L., Xue, R.D., Qualls, W.A., Bernier, U.R., Kravchenko, V.D., Ghattas, N., Pstygo, I., Muller, G.C., 2013b. Reduction of mosquito biting-pressure: spatial repellents or mosquito traps? A field comparison of seven commercially available products in Israel. Acta. Trop. 127, 63–68.

Ritchie, S.A., Devine, G.J., 2013. Confusion, knock-down and kill of *Aedes aegypti* using metofluthrin in domestic settings: a powerful tool to prevent dengue transmission? Parasit. Vectors 6 (1), 1–9. doi:10.1186/1756-3305-6-262.

Ritchie, S.A., Johnson, P.H., Freeman, A.J., Odell, R.G., Graham, N., Dejong, P.A., Standfield, G.W., Sale, R.W., O'Neill, S.L, 2011. A secure semi-field system for the study of *Aedes aegypti*. PLoS Negl.Trop. Dis. 5 (3), e988. https://doi.org/10.1371/journal.pntd.0000988.

Roberts, D.R., Alecrim, W.D., Tavares, A.M., McNeil, K.M., 1984. Influence of physiological condition on the behavioral responses of *Anopheles darlingi* to DDT. Mosq. N 44, 357–361.

Rosenberg, R., Lindsey, N.P., Fischer, M., Gregory, C.J., Hinckley, A.F., Mead, P.S., Paz-Bailey, G., Waterman, S.H., Drexler, N.A., Kersh, G.J., Hooks, H., 2018. Vital signs: trends in reported vectorborne disease cases—United States and Territories, 2004–2016. Morb. Mort. Week. Rep 67, 496.

Rowland, M., Downey, G., Rab, A., Freeman, T., Mohammad, N., Rehman, H., Durrani, N., Reyburn, H., Curtis, C., Lines, J., Fayaz, M., 2004. DEET mosquito repellent provides personal protection against malaria: a household randomized trial in an Afghan refugee camp in Pakistan. Trop. Med. Int. Health 9 (3), 335–342. https://doi.org/10.1111/j.1365-3156.2004.01198.x.

Russell, P.F., Ramachandra-Rao, T., 1942. On the swarming, mating, and ovipositing behavior of *Anopheles culicifacies*. Am. J. Trop. Med. Hyg. 22, 417–427.

Sangoro, Lweitojera, D., Simfukwe, E., Ngonyani, H., Mbeyela, E., Lugiko, D., Kihonda, J., Maia, M., Moore, S., 2014. Use of a semi-field system to evaluate the efficacy of topical repellents under user conditions provides a disease exposure free technique comparable with field data. Malar. J. 13, 1–11.

Sangoro, O.P., Gavana, T., Finda, M., Mponzi, W., Hape, E., Limwagu, A., Govella, N.J., Chaki, P., Okumu, F.O., 2020. Evaluation of personal protection afforded by repellent-treated sandals against mosquito bites in south-eastern Tanzania. Malar. J. 19, 1–12.

Sathantriphop, S., White, S.A., Achee, N.L., Sanguanpong, U., Chareonviriyaphap, T., 2014. Behavioral responses of *Aedes aegypti*, *Aedes albopictus*, *Culex quinquefasciatus*, and *Anopheles minimus* against various synthetic and natural repellent compounds. J. Vector Ecol. 39 (2), 328–339. https://doi.org/10.3376/i1081-1710-39-328.

Schaffner, F., Fontenille, D., Mathis, A., 2014. Autochthonous dengue emphasizes the threat of arboviruses in Europe. Lanc. Infec. Dis 14, 1–3.

Schreck, C.E., Kline, D.L., 1989a. Personal protection afforded by controlled-release topical repellents and permethrin-treated clothing against natural populations of *Aedes taeniorhynchus*. J. Am. Mosq. Cont. Assoc. 5, 77–80.

Schreck, C.E., Kline, D.L., 1989b. Repellency of two controlled-release formulations of deet against *Anopheles quadrimaculatus* and *Aedes taeniorhynchus* mosquitoes. J. Am. Mosq. Cont. Assoc. 5, 91–94.

Shen, Y., Xue, R.D., Bibbs, C.S., 2017. Relative Insecticidal Efficacy of Three Spatial Repellent Integrated Light Sources Against *Aedes aegypti*. J. Am. Mosq. Cont. Assoc. 4, 348–351.

Sluydts, V., Durnez, L., Heng, S., Gryseels, C., Canier, L., Kim, S., Van Roey, K., Kerkhof, K., Khim, N., Mao, S., Uk, S., 2016. Efficacy of topical mosquito repellent (picaridin) plus long-lasting insecticidal nets versus long-lasting insecticidal nets alone for control of malaria: a cluster randomised controlled trial. Lanc. Infect. Dis. 16, 1169–1177.

Stanczyk, N.M., Brookfield, J.F.Y., Field, L.M., Logan, J.G, 2013. *Aedes aegypti* mosquitoes exhibit decreased repellency by DEET following previous exposure. PLoS One 8 (2), 1–6. doi:10.1371/journal.pone.0054438.

Stevenson, J.C., Simubali, L., Mudenda, T., Cardol, E., Bernier, U.R., Vazquez, A.A., Thuma, P.E., Norris, D.E.,

Perry, M., Kline, D.L., Cohnstaedt, L.W., Gurman, P., D'hers, S., Elman, N.M, 2018. Controlled release spatial repellent devices (CRDs) as novel tools against malaria transmission: a semi-field study in Macha, Zambia. Malaria J. 17 (1), 1–16. https://doi.org/10.1186/s12936-018-2558-0.

Strickman, D., 2007b. Older synthetic active ingredients and current additives. In: Debboun, M., Frances, S.P., Strickman, D. (Eds.), Insect Repellents: Principles, Methods and Uses. CRC Press, Boca Raton, FL.

Strickman, D., 2007a. Chapter 23. Area Repellents. In: Debboun, M., Frances, S.P., Strickman, D. (Eds.), Principles of Insect Repellents: Principles, Methods, and Uses. CRC Press, Boca Raton, FL.

Sukkanon, C., Chareonviriyaphap, T., Doggett, S.L., 2019. Topical and spatial repellent bioassays against the Australian paralysis tick, Ixodes holocyclus (Acari: Ixodidae). Aust. Entomol. 58, 866–874.

Sukkanon, C., Nararak, J., Bangs, M.J., Hii, J., Chareonviriyaphap, T., 2020. Behavioral responses to transfluthrin by *Aedes aegypti*, *Anopheles minimus*, *Anopheles harrisoni*, and *Anopheles dirus* (Diptera: Culicidae). PLoS One 15, 1–21.

Tambwe, M.M., Mbeyela, E.M., Massinda, B.M., Moore, S.J., Maia, M.F., 2014. Experimental hut evaluation of linalool spatial repellent agar gel against *Anopheles gambiae* sensu stricto mosquitoes in a semi-field system in Bagamoyo, Tanzania. Parasit. Vectors 7 (1), 1–6. doi:10.1186/s13071-014-0550-2.

Ten-Bosch, Q.A., Wagman, J.M., CastroLlanos, F., Achee, N.L., Grieco, J.P., Perkins, T.A., 2020. Community-level impacts of spatial repellents for control of diseases vectored by *Aedes aegypti* mosquitoes. PLoS Comput. Biol. 16, 1–23.

Thanispong, K., Achee, N.L, Bangs, M.J, Grieco, J.P., Suwonkerd, W., Prabaripai, A., Chareonviriyaphap, T., 2009. Irritancy and repellency behavioral responses of three strains of *Aedes aegypti* exposed to DDT and alpha-cypermethrin. J. Med. Entomol. 46, 1407–1414.

Thavara, U., Tawatsin, A., Chompoosri, J., Suwonkerd, W., Chansang, U.R., Asavadachanukorn, P., 2001. Laboratory and field evaluations of the insect repellent 3535 (ethyl butylacetylaminopropionate) and debt against mosquito vectors in Thailand. J. Am. Mosq. Control Assoc. 17 (3), 190–195.

Tong, F., Bloomquist, J.R., 2013. Plant essential oils affect the toxicities of carbaryl and permethrin against *Aedes aegypti* (Diptera: Culicidae). J. Med. Entomol. 50, 826–832.

Trung, H.D., Bortel, W.V., Sochantha, T., Keokenchanh, K., Briët, O.J., Coosemans, M., 2005. Behavioural heterogeneity of *Anopheles species* in ecologically different localities in Southeast Asia: a challenge for vector control. Trop. Med. Inter. H. 10, 251–262.

United States Environmental Protection Agency, 1998. Reregistration eligibility decision DEET. United States Environmental Protection Agency Office of Pesticide Programs Special Review and Reregistration Division, Washington, DC.

Uniyal, A., Tikar, S.N., Mendki, M.J., Singh, R., Shukla, S.V., Agrawal, O.P., Veer, V., Sukumaran, D., 2016. Behavioral response of *Aedes aegypti* mosquito towards essential oils using olfactometer. J. Arthropod-Borne Dis. 10 (3), 372–382. http://jad.tums.ac.ir/index.php/jad/article/download/319/279.

van den Bercken, J., Vijverberg, H.P., 1988. Mode of action of pyrethroid insecticides. Recent Advances in Nervous System Toxicology. Springer, Boston, MA, pp. 91–105.

Van Roey, K., Sokny, M., Denis, L., Van den Broeck, N., Heng, S., Siv, S., Sluydts, V., Sochantha, T., Coosemans, M., Durnez, L., 2014. Field evaluation of picaridin repellents reveals differences in repellent sensitivity between southeast Asian vectors of malaria and arboviruses. PLoS Negl.Trop. Dis. 8 (12). doi:10.1371/journal.pntd.0003326.

Wagman, J.M., Achee, N.L., Grieco, J.P., 2015a. Insensitivity to the spatial repellent action of transfluthrin in *Aedes aegypti*: a heritable trait associated with decreased insecticide susceptibility. PLoS Negl.Trop. Dis. 9 (4), 1–18. doi:10.1371/journal.pntd.0003726.

Wagman, J.M., Grieco, J.P., Bautista, K., Polanco, J., Briceo, I., King, R., Achee, N.L., 2015b. The field evaluation of a push-pull system to control malaria vectors in Northern Belize. Central America. Malar. J. 14, 1–11.

WHO, 2012. Guidelines for field testing spatial repellents. https://www.who.int/whopes/. (Accessed December 2020).

Wilson, A.L., Chen-Hussey, V., Logan, J.G., Lindsay, S.W., 2014. Are topical insect repellents effective against malaria in endemic populations? A systematic review and meta-analysis. Malar. J. 13, 1–9.

World Health Organization, 2013. Guidelines for Efficacy Testing of Spatial Repellents. WHO.

WHO, 2013. Guidelines for field testing spatial repellents. WHO. http://www.who.int/whopes.

WHO, 2015. World Malaria Report 2015. WHO Press, World Health Organization, 20 Avenue Appia, 1211 Geneva 27, Switzerland.

Xue, R-D., Ali, A., Barnard, D.R., 2012a. Mating status and body size in *Aedes albopictus* (diptera: culicidae) affect host finding and DEET repellency. Fl. Entomol. 95, 268–272.

Xue, R.D., Qualls, W.A., Smith, M.L., Gaines, M.K., Weaver, J.H., Debboun, M., 2012b. Field evaluation of the OFF! Clip-on mosquito repellent (Metofluthrin) against *Aedes albopictus* and *Aedes taeniorhynchus* (Diptera: Culicidae) in northeastern Florida. J. Med. Entomol. 49 (3), 652–655. https://doi.org/10.1603/ME10227.

Yap, H.H., 1986. Effectiveness of soap formulations containing deet and permethrin as personal protection against outdoor mosquitoes in Malaysia. J. Am. Mosq. Control Assoc. 2 (1), 63–67.

Yayo, A.M., Ado, A., Habib, A.G., Hamza, M., Iliyasu, Z., Sadeeq, I.A., Musa, K.A., Barodo, M.M., Inuwa, M.B., Ibrahim, S.S., 2016. Effectiveness of transfluthrin-impregnated insecticide (paper rambo) and mechanical screening against culicine and anopheline mosquito vectors in Kumbotso. Nigeria. Molec. Entomol 20, 1–10.

Yoon, J.K., Kim, K.C., Cho, Y.D., Cho, H.S., Lee, Y.W., Kim, M., Choi, B.K., Oh, Y.K., Kim, Y.B., 2014. Development and evaluation of a semi-field test for repellent efficacy testing. J. Med. Entomol. 51 (1), 182–188. https://doi.org/10.1603/ME13081.

CHAPTER

12

Human subject studies of arthropod repellent efficacy, at the interface of science, ethics, and regulatory oversight

Shawn B. King[a], *Cassandre H. Kaplinsky*[a], *Ralph Washington, Jr.*[a], *Scott P. Carroll*[a,b]

[a]Carroll-Loye Biological Research, Davis, CA, United States, [b]Department of Entomology and Nematology, University of California, Davis, CA, United States

12.1 Introduction

People have used arthropod repellents for personal protection for millennia (Peterson and Coats, 2001; Moore and Debboun, 2007). Concerted studies to understand and improve topical mosquito repellents began early in the 20th Century (Smith, 1901), and were energized by the vulnerability of deployed warfighters to both malaria and typhus in WWI (Bacot and Talbot, 1919; D'Ormea, 1919) and thereafter (Bunker and Hirschfelder, 1925; Moore, 1934, among others). With only modestly effective drug treatments available for arthropod-borne diseases until mid-century, early researchers were also driven by concerns for the large numbers of people who would remain beyond the reach of mosquito population control efforts (Granett, 1940).

With their deep ecological understanding of personal exposure to blood-feeding vectors, and their pioneering experimental methods, these scientists laid the foundations for modern repellent science. Their drive for answers often led them to test novel formulations on themselves as well as associates, using both West Africa-derived *Ae. aegypti* in the laboratory and wild mosquitoes in nature. Then as now, botanical derivatives and petroleum distillates were the focus of the search for active ingredients. The ideal arthropod repellent product would be long-lasting and broad-spectrum while being generally safe under normal use, harmless to

Advances in Arthropod Repellents.
DOI: https://doi.org/10.1016/B978-0-323-85411-5.00013-3

clothing and other materials, easy to use, aesthetically attractive, and affordable (Bacot and Talbot, 1919; Granett, 1940). Those authors also correct perceived that the potential tradeoff between efficacy and those other key attributes of user experience was at the heart of the repellent development challenge. From that basis emerged the great repellent search during WWII by the US Department of Agriculture, in response to military needs. That program ultimately screened tens of thousands of compounds and culminated in the discovery of N,N-diethyl-3-methylbenzamide (DEET) in 1953 (Moore and Debboun, 2007).

The rapid preeminence of DEET among the topical repellent active ingredients did not quiet the continuing pursuit of efficacy data from human subject for other actives, however. Strickman (2007) assessed the outcomes of dozens of reports on human subject tests of synthetic actives that developed around the same time as DEET that were less repellent but still of interest to investigators, published between ca. 1960 and 1995. In much the same time frame, researchers also examined thousands of botanical prospective active ingredients using human subject trials (Sukumar et al., 1991). Since the 1990s, newer, highly efficacious active ingredients Icaridan ("Picaridin"), IR3535 and Oil of Lemon Eucalyptus (OLE) have gained acceptance among regulatory agencies and consumers. Human tests of these new actives have proliferated (reviewed by Carroll 2007), as have those comparing the growing number of EPA-exempt and international botanical repellents (reviewed by Maia and Moore, 2011; see also Fradin and Day, 2002; Carroll, 2008; Rodriguez et al., 2015, 2017 for examples of empirical approaches).

It is during this most recent phase of repellent discovery and commercialization that government regulation has begun to profoundly influence the practice of human subject repellent efficacy studies for the first time. These regulatory developments are increasing the relevance of human subject repellent studies to public health while simultaneously elevating their public profile. The dynamics are playing out in a diversity of ways. In a departure from the prior century, these changes include mandated science and ethics oversight of human subject efficacy studies for registration (Box 12.1). At the same time, as more "minimum risk" repellents are marketed, skirting the augmented regulatory requirements of the US Environmental Protection Agency (EPA), individual states are responding by requiring sounder and more ethically obtained efficacy data for such products on their own accord. Technological developments in our understanding of arthropod olfaction and molecular modeling are bringing ever more active ingredient candidates to the fore. Meanwhile, the requirements for human field studies are being debated by both US and European Union (EU) regulators (Moreno-Gómez et al., 2021).

In consequence, human subject studies have entered a dynamic new epoch. Under the demands of global change, commercial, military, and public health entities are striving to deliver personal protection from biting arthropods (Debboun and Strickman, 2013). Emerging and introduced arthropod-vectored pathogens for which vaccines are not available will continue to elevate the role of repellents in personal protection (Diaz, 2016). At its root, however, the same broad classes of repellent actives and the same product development objectives that were articulated by the first repellent scientists continue to be investigated by today's researchers, utilizing a diversity of human subject study methods.

In order to review these developments and examine their broader implications for public health practices, we focus on the following topics:

1. The current state of regulatory oversight.
2. The ethics of human subject study design and conduct.
3. Alternatives to human subject studies.

BOX 12.1

A new era develops for human subject efficacy testing potential and regulatory oversight

1. At the regulatory level, in 1996, California became the first administrative district to require Institutional Review Board oversight of human subject repellent efficacy studies conducted for product registration purposes.
2. Around that same time efforts began to register the first highly effective DEET-alternative repellents (IR3535, PMD, Picaridin) through US EPA, opening the door to wider general interest in new product development.
3. Also during this time, US EPA was also moving to revise repellent efficacy testing guidance, resulting in the draft 1999 OPPTS 810.3700 guideline, which was not finalized.
4. In adherence with legislation passed by Congress in 2005, in 2006 EPA promulgated a rule titled Protections for Subjects in Human Research, which provided for an unprecedented level of supervision ((EPA)U.S. Environmental Protection Agency, 2006).
5. Recognition of the public health importance of repellents was furthered in 2006 when the US Centers for Disease Control and Prevention (CDC) included Picaridin and Oil of Lemon Eucalyptus along with DEET as active ingredients offering protection from arthropod-borne diseases.
6. New focus on gene expression and olfaction in mosquitoes—molecular mechanisms of repellent action (e.g., Ditzen et al., 2008).
7. Computational approaches to modeling repellent molecule discovery (e.g., Miszta et al., 2013).
8. Growing recognition and acknowledgment of the importance of arthropod repellents in disease control—CDC, others in Debboun and Strickman (2013); role of different classes of repellent applications (ITNs, etc.).
9. State EPAs starting to require human subject efficacy data for US minimum risk products.
10. EU developments, e.g., emphasizing arm in cage laboratory studies instead of field studies (e.g., ECHA 2019).

4. Human subject considerations for tests of area repellents.
5. Human subject considerations for tests of the treated fabric.
6. Key considerations for the future of regulatory oversight of repellent efficacy studies.

We mainly examine studies of mosquitoes because they are the most actively studied biting arthropods and the primary target of arthropod repellents. At the same time, while the details will differ to varying degrees, the scientific and ethical practices that underlie mosquito repellent testing apply in force to studies of other blood-feeding arthropods.

12.2 Repellent testing in the context of pesticide regulation

We provide an overview of regulatory oversight of human subject testing and its influence on the ethics and science of repellent testing, product development, and registration. What follows is based on relevant literature, including governmental guidelines, as well as our own professional experience as repellent efficacy researchers for industry at our home institution. In practice, testing repellents for registration involves balancing scientific rigor, human subject protection, costs of development and registration, and the regulatory need to protect and inform consumers.

In many countries, repellents are considered and regulated as a class of pesticides. To obtain permission to market a skin-applied product with arthropod repellent claims, applicants must submit safety data for the product's active ingredient(s), as well as both safety and efficacy data for the product itself. Regulations vary among regions but are led by the US EPA (as authorized by the Federal Insecticide, Fungicide, and Rodenticide Act), the EU European Chemicals Agency (ECHA; as authorized by the Biocidal Products Directive), and Canada Pest Management Regulatory Agency (PMRA; as authorized by the Pest Control Products Act). The practices of each of these regulatory bodies have been interdependent, but more broadly determined by the political, economic, and cultural priorities of the regions each serves. From a practical standpoint, each constitutes a regulatory region with individual standards to be met by any product marketed within.

As with all pesticidal products, regulatory attention on repellents focuses ultimately on the content of the product label, which is the primary medium to inform consumers and influence their behavior around safe and effective product use (e.g., US EPA Label Review Manual https://www.epa.gov/pesticide-registration/label-review-manual). The label includes application instructions, estimated duration of protection, and any certifications for protections against specific vectors. Product duration is important both for maintaining protection and for minimizing frequency of application, because each application constitutes a pesticide exposure for the end-user. Duration of protection cannot be estimated from reduction in biting percentage, otherwise known as "percent repellency," unless a target percentage is predefined as the threshold of failure. Instead, product performance may be quantified as complete protection time (CPT), which is based on the median or mean time between product application and the first "confirmed" bites (or other similar specified measure of protection failure) observed across all test subjects. An initial bite is confirmed if followed by another within a specified time interval. Contemporary regulators typically use CPT and its variance to estimate duration of protection in actual use scenarios; historically, they used the time until percent repellency declines below the predefined threshold.

12.2.1 Harmonizing efficacy and safety testing

Region-specific requirements for testing may constrain international distribution and global consumer access to products. Global data applications require harmonization efforts aimed at preserving or enhancing data quality while reducing costs and shortening timelines for getting effective products to market. The Organization for Economic Cooperation and Development, with its Pesticide Working Group (http://www.oecd.org/env/ehs/pesticides-biocides/agriculturalpesticidesprogramme.htm), leads the effort to harmonize pesticide registration and some of its primary components: toxicity and physical chemistry testing guidelines, reporting requirements, and product labeling. An added driver behind harmonization is to reduce the use of laboratory animals for superfluous toxicological tests that add little or no value to regulatory determinations.

Missing is concerted harmonization of efficacy testing guidelines among nations and agencies. Guidelines typically cover similar testing details and topics but often arrive at different requirements or solutions for test scope and design (Appendix). One important initiative in this direction is from the World Health

Organization (WHO). Via its Pesticide Evaluation Scheme (WHOPES) and in part through collaboration with other agencies, WHO has developed standardized guidelines for human subject efficacy testing of arthropod repellents (World Health Organization, 2009). These guidelines harmonize and make comparable efficacy studies and their results across regulatory regions, and support effective product labeling and registration. By nature, the guidelines also offer a study design template for tests in regions with limited regulatory oversight. The WHO's primary objective is to protect consumers by encouraging effective vetting of repellent products worldwide. As an additional benefit, the guidelines have the potential to increase the inherent value of human subject studies by increasing the comparability of results (Maia and Moore, 2011).

12.2.2 Growing regulatory interest in streamlining registration for reduced-toxicity pesticides

Regulators and lawmakers recognize the opportunity to reduce the environmental impact of pesticides by providing shorter and less expensive pathways to less-toxic products. To this end, EPA created the reduced risk pesticide category within the Food Quality Protection Act and designed its regulatory process within the EPA's Conventional Reduced Risk Pesticide Program (Fishel, 2013). Another category, Biopesticide, was created to allow plant- and other-naturally derived compounds to be registered at lower cost and with reduced requirements for toxicological data. Similar efforts are underway in Canada and the EU.

These regulatory shifts may grant repellent users earlier and expanded access to plant-derived and other naturally occurring active ingredients and their formulations. Historically, regulatory oversight has favored larger established business entities over smaller ones, such as venture-capital-funded start-ups (Ollinger and Fernandez-Cornejo, 1995), because of the typical high cost and protracted timeline for meeting registration requirements. Therefore, reduced time and costs to register encourage the discovery and development of effective repellents by a wider business community. Accordingly, the increase in the number of viable repellent active ingredients and their derivatives will increase demand for registration-compliant human subject-based efficacy testing. Moreover, enforcement actions and data call-ins by EPA will also require additional test data from companies sponsoring compliant efficacy studies.

It is important to distinguish the general requirement to register a pesticide from the specific requirement for compliant labeling. In the United States, the EPA has determined that even repellents comprised of registration-exempt ingredients, i.e., minimum risk pesticides, may not claim efficacy against specific vector species without meeting data criteria similar to those applied to all other repellents. Specifically, "The product must not bear claims either to control or mitigate organisms that pose a threat to human health, arthropods or rodents carrying specific diseases" (https://www.epa.gov/minimum-risk-pesticides/conditions-minimum-risk-pesticides, accessed November 1, 2020). Business entities and State-level EPAs administer registration and human subject efficacy testing of minimum-risk pesticides, but the scope and scale of data requirements are dwarfed by registration for Federal EPA.

The EPA is limited in its ability to identify and pursue enforcement actions against products with unsupported label claims about efficacy against disease vectors (Mintz, 2012), which has inspired movement toward requiring efficacy data for registration of any product that claims to repel or otherwise protect a consumer from biting arthropods, regardless of a product's exemptions from registration. Regulatory agencies in other countries could follow suit because of the growing public health importance of managing and mitigating arthropod-borne diseases.

12.3 Human subjects versus surrogates for efficacy testing in wild mosquito populations

Although alternative surrogate methodologies may obviate the need for human subjects in a variety of disciplines including vertebrate toxicology, human subject testing remains a cornerstone methodology in efficacy tests of arthropod repellents. Nonetheless, there is growing pressure to limit and closely regulate human exposure in such studies. For area repellents (rather than skin-applied repellents), such as pyrethroid emanators for protection from mosquitoes, EPA already disallows human subject data, due to a perceived heightened bite risk to participants (Strickman, 2009). Instead, mechanical traps are used as human surrogates in field efficacy trials. However, variation in trap design, conditions of use, and biology of the target arthropods limit how well traps function as surrogates. Therefore, measurements of efficacy may depend, to an unknown degree, on both the choice of trapping system, as well as the response of target arthropods to interacting environmental stimuli and the properties of the test material (e.g., Laurent, 2018).

Performance characterizations are important for both accurate labeling and the ability of public health officials to predict a product's effectiveness against important pathogen vectors. Regulatory agencies require that claims of efficacy against mosquitoes be supported by data from studies on the main genera important to human health and quality of life, specifically *Aedes*, *Anopheles*, and *Culex*. Additionally, these studies must incorporate scientific understanding of the interaction between mosquito biology and host attractiveness, as well as disparities between human landing counts and surrogate testing methods (Tangena, 2015). Mosquitoes respond to a variety of host cues, including shape, color, light, heat, movement emanated chemicals in the air column (Lima et al., 2014). Far from a host, the primary attractant is carbon dioxide (CO_2) exhalations, and host-specific olfactory cues become more important as the mosquito orients to and approaches a host. Lactic and carboxylic acids are considered the primary attractive skin emanations, but hundreds of other emanated compounds likely play a role, including those generated by skin and intestinal microbiota (Verhulst et al., 2010).

Mosquito taxa also vary in their relative attraction to olfactory cues. In some species, host orientation is more influenced by CO_2, and others are more influenced by host-specific cues. For example, *Ae. aegypti* is quite responsive to lactic acid, a molecule relatively diagnostic for humans, while the similarly anthropophilic malaria vector *An. gambiae* is often comparatively unresponsive (Zwiebel and Takken, 2004). Additional compounds in commercial lures (fatty acids, ammonia) may attract the Asian Tiger mosquito, *Ae. albopictus,* but other species less reliably (Unlu et al., 2016). *Aedes* and *Culex* are more closely related to each other than *Anopheles,* so the latter is expected to exhibit a more distinct olfactory response.

The principal trap types used in research and vector control include various attractant modalities. The CDC light trap, developed by the US CDC, employs an incandescent black lightbulb often paired with a CO_2 source, typically in the form of dry ice. The BioGents BG-Sentinel trap series is baited with CO_2 from a gas canister, does not have lights, and was configured for efficacy against *Ae. aegypti,* though it also captures *Anopheles* and *Culex,* sometimes in large numbers (personal observation). BioGents also supplies optional lures containing a combination of human emanants including ammonia, lactic acid, and caproic acid. BioGents states that this lure renders the trap particularly effective for certain species, as substantiated for *Ae. albopictus* but less clearly so for other species (Unlu et al., 2016).

Several studies have examined how trap captures match simultaneous human landing counts, in this case, to measure fidelity to human landing counts in the context of vector surveillance. BG-Sentinel-2 traps have been reported to catch substantially more mosquitoes than human landing counts and other CO_2-baited traps (L' Ambert et al., 2012; Harwood et al., 2015). However, BG-Sentinel-2 trapped fewer kinds of mosquitoes, in one case almost completely lacking an *Anopheles* species present in human landing counts (Hiwat et al., 2011). The CDC and related traps used for surveillance in malaria control programs typically lack CO_2. Like the BG-Sentinel-2, their match to human landing counts also varies widely within and among studies, species, and regions (Mathenge et al., 2005). One study showed that CDC traps in homes in Suriname captured ≤10% of *Anopheles* mosquitoes with human-landing collections (Hiwat, 2011), a discrepancy with serious implications for vector control.

In summary, while there is an abundance of research on specific mosquito species' host cue preferences in certain conditions, places and contexts, there is no generalizable quantification of host cue preference (Zwiebel and Takken, 2004). This challenge makes it difficult to quantitatively relate nonhuman counts to human landing counts (Overgaard et al., 2012). Tests of new area repellents represent novel conditions with further unknown influences on how mosquitoes respond to trap stimuli. From a scientific perspective, this circumstance suggests that human landing counts may be the only reasonable method to evaluate product performance, as well as to quantify or sample important disease-vectoring species.

That is not to suggest that human subjects represent a panacea for data relevance and quality, however, including in cases where human subjects can be ethically used. Individuals vary in attractiveness to mosquitoes for both genetic and environmental reasons, reviewed by Fernández-Grandon et al. (2015). Neither of these variables can or will be strongly controlled in human efficacy studies. It is likewise notable that individual attractiveness to a particular mosquito species is not a strong predicter of attractiveness to other mosquitoes, nor to how well a person is protected by a particular repellent (reviewed in Carroll, 2007). Hence a person highly attractive to a specific species may be substantially better protected than less attractive people when all are wearing the same repellent. To ensure the utility of trial data, this variability and suite of interactions is best addressed by employing a large number of human participants in a variety of environmental conditions. For these same reasons, it also makes sense to both develop protocols that expand the realm of objectives and settings for which human subject trials can be ethically and economically utilized, and to improve the validity of *in vitro* methods where they cannot.

12.4 Reducing reliance on human subject efficacy testing

12.4.1 Alternative efficacy testing methods

Arm-in-Cage (AIC) testing has long been a staple of iterative mosquito repellent product development. As indicated elsewhere in this chapter, direct exposure of mosquitoes to a human host with product applied to the skin integrates a range of factors that influence efficacy. The AIC tests are by definition human subject tests akin to a clinical trial, where a human subject is exposed to a product's properties and the risks associated with it. As such, AIC tests are appropriate to conduct after the study protocol has been approved by an Institutional Review Board (IRB) and the subjects have undergone a consenting process vetted by that IRB.

The ethical and scientific review services provided by IRBs increase the cost, complexity, and time involved in conducting an AIC study. For new product development and, especially for screening prospective active ingredients, *in vitro* methods may often be preferable in advance of any human testing. That includes iterative developmental testing for formula optimization, including active ingredient concentration and the choice of inert ingredients.

One such method relevant for modeling candidate performance is a no-choice system in which repellent is applied to a fine screen within an enclosed flight chamber with controllable airflow rate and direction. Upwind of the screen, toward an air intake point, lures such as an inert warm body, CO_2 release, or a researcher's arm provide cues to mosquitoes released at the far downwind margin of the enclosure. Attractant and repellent molecules through the treated screen, crossing the length of the chamber. Mosquitoes must successfully orient and fly up-gradient until they encounter the screen. Hypothetically, an untreated screen will become the landing and resting point for foraging mosquitoes as it is the location in the chamber with the highest concentration of host cues. If the screen is treated with repellent and the treatment is effective, mosquitoes may still be observed moving up-gradient toward the host end of the chamber, but few or none will alight on the screen.

Such an apparatus can be used to rapidly assess candidate actives in a variety of concentrations and formulations. It was shown to provide repeatable dose-response measures and to match AIC test repellency within an order of magnitude (Sharpington et al., 2000). Ideally, such a system could be calibrated to available field repellency trial data to get a sense of how repellency observed in the chamber might translate into performance in the field. We know of no such effort to date.

Another method is to stretch a membrane over a vessel of mammalian blood maintained at approximate body temperature, then treat the exposed membrane surface with a test compound (Bar-Zeev and Smith, 1959). Various membrane materials, including animal epithelium, collagen, and paraffin film, allow female mosquitoes to draw blood through them. This approach has compared well to human subject outcomes in laboratory testing, including in comparative assays of new active ingredients formulations as well as existing commercial products (Klun et al., 2005). This general method has also been shown to work for black flies (Bernardo and Cupp, 1986). Like the chamber system described earlier, a membrane system can potentially be calibrated to field test results, but this work has yet to be done.

12.4.2 Regulatory interest in transitioning to alternative methods

Currently, regulatory agencies do not accept data from nonhuman efficacy tests for registration of skin-applied repellent products. However, there is interest in reducing reliance on field tests.

Because of the risks of contracting a mosquito-vectored disease during a field test and the costs associated with conducting them, there is an effort within the EU Biocides regulatory community to phase out field efficacy testing requirements and rely instead on AIC alone (Workshop on Repellents, 2016). However, there is no scientific evidence that such tests relate consistently with the outcomes of field studies. Colucci and Müller (2014) reviewed 871 publications and found none that provided data sufficient to quantitatively compare results with those from field studies. Colucci (2017) compared AIC with field outcomes experimentally in tests following EPA and WHO efficacy testing guidelines. At field sites with low landing rates, both p-Menthane-3,8-diol (PMD) and DEET provided CPTs of 6 h. In contrast, in laboratory, by-species protection times varied 3- or 4-fold and were much lower, 1.5 h for the DEET product and 0.75 h for the PMD product. The AIC tests neither reflected real-world performance nor provided data clearly interpretable into meaningful protection times for a product label. It also appears the estimates provided by AIC testing may underestimate protection time. Therefore, if used to estimate duration for a label, there may be excessive reapplication and thus over exposure of end users to active ingredient and other potentially toxic components of formulation. In Carroll-Loye Biological Research's own library of unpublished studies, we have likewise found variable correspondence between AIC outcomes and field outcomes. Depending on the material used, the former can either under- or over-estimate field performance.

For repellency claims against mosquitoes and biting flies, the US EPA requires (and WHO continues to recommend) data from field-based human subject tests to demonstrate product efficacy. The public health importance of repellents underscores the value of accuracy in characterizing realized protection. Without reliable AIC-based metrics, regulators face trading study accuracy and the consumer protection it affords for a reduction in risks to human subjects. The balance of such costs and benefits is discussed in the next section.

12.5 Regulation, ethics, and efficacy study design—historical overview and current conditions

Ethical considerations are always primary in the design and execution of human subject studies involving exposure to pesticides, including arthropod repellents. Moreno (2013) reviewed the history and evolution of human subject protections within the context of US government agencies and their activities. The primacy of ethical treatment of human subjects in the conduct of scientific experiments was established during the Trials at Nuremburg in the wake of WWII, when there was an international effort to codify in law and establish norms to prevent the wartime atrocities, including those committed by researchers. As a fraction of overall research, human trials are only occasionally used in the field of medical entomology despite their great scientific utility, and it is helpful to place it within the broader context and history for the ethics working with human subjects. Table 12.1 presents a timeline of key report up to and including current oversight of arthropod repellent efficacy studies (National Research Council, 2004; (EPA)U.S. Environmental Protection Agency, 2006).

The Federal Insecticide, Fungicide and Rodenticide Act, as amended by the Federal Environmental Pesticide Control Act, gives the US EPA authority to regulate the sale and distribution of pesticides, including arthropod repellents. Federal law thus requires scientific evaluation of a product's safety and efficacy prior to its registration for legal sale. For the law to be applied and enforced effectively, demonstration of efficacy must be clearly defined, and EPA maintains that skin-applied repellent efficacy is best assessed with human subjects. The agency also requires that data provide an accurate basis for label language that informs consumers how long they can expect the product to protect them from biting arthropods. The EPA is thereby both a promoter of and stakeholder in the quality of human subject efficacy tests.

12.6 Risks vs benefits: study oversight and informed consent

12.6.1 Independent and regulatory review

12.6.1.1 Creating the conditions for informed consent

To give informed consent, a candidate volunteer must clearly understand in degree and likelihood the potential risks and benefits of participating, and the nature of each study procedure that bears upon their direct experience while participating. All procedures involving

TABLE 12.1 Societal context and primary regulatory actions regarding use of human subjects in research in the United States, over time.

1946	American Medical Association establishes its first code of research ethics
1949	The Nuremburg Code of ethics is established
1953	US Department of Defense initial policy on human research, the first US Government policy on human research
1953	US National Institute for Health policy requiring independent review of human subject protocols (research and consent)
1960s	Investigative reporting uncovers multiple violations, making the general public aware
1971	Department of Health and Human Services establishes "Yellow Book" detailing guidance and justification for review committees responsible for independent review of human subject studies
1972	New York Times reports on the Tuskegee Syphilis Study
1972	National Institute of Health (NIH) creates Office for Protection from Research Risks (OPRR)
1974	The National Research Act is passed by congress (21 CFR Part 50) establishing the National Commission for the Protection of Human Subjects of Biomedical and Behavioral Research (National Commission)
1979	The National Commission issues the Belmont Report summarizing ethical principles that should guide conduct of human subject studies and government oversight of such studies
1981	The President's Commission for the Study of Ethical Problems in Medicine and Biomedical and Behavioral Research was founded. It recommended harmonization of human subject research rules and guidelines among agencies (what would become known as the Common Rule).
1991	15 US Federal agencies, including the EPA, joint-publish The Common Rule which harmonizes human subject research-related requirements among signatory agencies.
2006	US EPA issues a rule called Protection for Subjects in Human Research requiring IRB oversight and EPA ethical and scientific review of proposed study protocols by third parties intending to submit resultant data to EPA, and bans intentional pesticide exposures to pregnant or nursing women or to children in such studies. These standards apply to all arthropod repellent efficacy testing related to Federal registration of repellent products.

communication with a candidate volunteer, including how they are contacted through advertisements and follow-up communications, and both the study protocol and the consent forms detailing subject-relevant procedures, must be reviewed and approved by an Institutional Review Board (IRB) for ethical soundness prior to initiating recruitment of volunteers. For human subject studies of repellent efficacy for registration data, agencies in the United States, Canada, and the EU subsequently review the study protocol and consenting documents for ethical and scientific merit prior to granting permission to conduct the study.

12.6.1.2 The relationship of scientific quality to ethics evaluation

Considerations of study ethics necessarily include an evaluation of scientific soundness and societal relevance. If a study is unlikely to yield accurate or useful data, human subject use becomes purposeless, and all risks associated with participation in the study, no matter how minor and carefully managed, become correspondingly unacceptable. Thus, regulators evaluate both the scientific soundness and the ethics of the procedures of a proposed study protocol. Moreover, there is an inherent ethical tradeoff between precise, reliable inference, and exposure of human subjects to chemicals, ectoparasites, and vector-borne diseases. Statistical precision is heavily influenced by margin of error, which is a function of sample size. For a study design with fewer than 10 subjects, any reductions will greatly increase the margin of error, and may gravely reduce precision in estimating efficacy. Far more subjects may be needed in cases where individual variation may be important, though this latter concern will usually be more relevant to biomedical studies than studies of the sorts of topical repellents in use today.

The standard statistical approach to estimating the sample size is an *a priori* formula-based power analysis. However, the standard formula does not accommodate the multiple sources of random variation often observed in data from ecological studies such as field tests of arthropod repellents (Johnson et al., 2015). Sample size for these studies should be estimated via simulation-based power analysis, as per EPA recommendations. The EPA also recommends the use of median, rather than mean CPT, because the median is less sensitive to skewed or censored data, and also the Kaplan-Meier Survival analysis, because it accommodates non-normally distributed CPTs in generating confidence intervals.

12.6.2 Managing risks to subjects

12.6.2.1 *Inclusion, exclusion, and stopping rules to minimize risks*

Many potential subject vulnerabilities may be readily ascertained. These include exploitation (undue influence on their consent) and risks of participation (allergic to bites or physically challenged by field conditions). To prevent the enrollment of participants vulnerable to such risks, standard clinical practice involves the use of specific rules both for inclusion and exclusion: criteria that automatically disqualify a person, and the criteria each individual candidate must meet to enroll in the study.

Ethical safeguards must remain in place during all research activities, and experimental conditions may change once research is initiated. For this reason, standard clinical practice includes the use of "Stop Rules." A Stop Rule is a condition which, if observed, automatically triggers the cessation of all research activities involving either individual subjects, or all subjects if the condition applies to all. A Stop Rule functions to prevent a subject or researcher, either of whom may have emotional, financial, or professional stakes in continuing the research activities, from continuing when the activities undermine the scientific soundness of the research or have entered the realms of unacceptable risk or consequence.

12.6.2.2 *Exposure to toxic substances and potential health impacts*

In considerations of toxic exposure to a repellent product, study ethics and soundness are also directly entwined. Risks of inhalation or physical contact with a repellent product can be its acute and sub-chronic toxicological characteristics. However, it is not possible to perfectly derive human exposure risks from the results of toxicological studies on tissue samples or non-human animals. As a result, a multiplication factor, called the margin of exposure (MOE), is often applied to conservatively account for possible errors of estimation. The MOE provides a measure of confidence in the safety of a proposed or hypothetical exposure to a potentially toxic substance. For human subject arthropod repellent studies, EPA requires a minimum MOE of 100, meaning that any proposed exposure to the product must be 1/100th the level at which no toxic effects were observed in an animal or *in vitro* studies.

During efficacy studies, restricting exposures to limited skin areas, e.g., to a single forearm, is regarded as suitable for data collection while reducing risk of subject exposure to the repellent product and biting arthropods. However, for formulated end-use products, consumers can expect to apply repellent on much more of their skin, such as most of the skin on arms, legs, and necks in the case of someone wearing shorts and short sleeves outdoors. If the MOE for that scenario is less than 100, then the study itself is not ethical to conduct, as the product will not be considered safe in real-world use. Thus, considerations of the ethics of a repellent efficacy study may encompass end-use scenarios well outside the confines of study procedures and their execution.

12.6.2.3 *Exposure to biting arthropods and potential physiological or psychological reactions*

By design, a study protocol can exclude individuals who are phobic of arthropods or arthropod bites or known to be strongly allergic to bites from any biting arthropods, subjects will interact with during the study. There remains the possibility of rare or unknown reactions to an arthropod bite. A subject may have manifested an unknown allergy, a novel phobia, or experience tick paralysis, despite no previous history with the condition. The risks and impacts of such events are reduced by candidate subject screening, study medical management, and the persistent right of any subject to withdraw at his/her discretion.

12.6.2.4 *Exposure to disease agents transmissible by biting arthropods*

Worldwide, arthropod-borne diseases comprise 17% of all infectious disease in humans and cause more than 700,000 deaths annually. In tropical climates, malaria alone kills more than 400,000 people every year (World Health Organization Fact Sheet, 2020). In regions where malaria has not been recently present, less prevalent diseases remain or have emerged, such as West Nile virus, and various Encephalitides. Additionally, Lyme Disease and other serious tick-borne diseases, sicken, incapacitate, or kill many people every year.

Anyone living in an area with arthropod-borne pathogens risks disease when their actions expose them to vectors. Participants in efficacy test may therefore experience a baseline level of arthropod-vectored risk similar to what they would normally experience in nature. Careful field study design aims to reduce risks for treated subjects well below this baseline, and also controls risks for the untreated subjects. Mosquitoes are collected from candidate field sites in advance for pathogen screening, which allows confirmation that diseases agents are rare or absent in populations of competent vectors. Treated subjects are primarily protected from bites through the limited skin area exposure and limited exposure time. Protective coverings prevent bites to all parts of the body except the intentionally exposed skin. Skin-applied repellents are applied by technicians who ensure complete coverage at a specific dosing rate per skin surface area. Instead of requiring bites as data points, repellent failure is defined as a landing mosquito that ceases locomotion on the skin and lowers its proboscis toward the skin, a behavior that indicates imminent biting if not interrupted (Castilho et al., 2019). Subjects are trained and supplied with mechanical aspirators and taught to observe mosquito behavior to remove and collect mosquitoes prior to biting and without contacting the repellent. Between exposure periods, subjects shelter in mosquito-free screen houses or other structures.

Because the ultimate study datum is the time of first confirmed repellent failure, rather than percent protection with reference to the untreated controls (who serve only to verify ongoing adequate ambient biting pressure), the limb of any subject treated with repellent is covered as soon as repellent failure is verified. Finally, mosquitoes aspirated from subjects' exposed skin are screened for mosquito-borne disease. This allows for follow-up communications with subjects about disease risks that might have occurred despite the precautions listed earlier.

Repellency efficacy tests with ticks and chiggers are done in the laboratory with laboratory-reared organisms known to be disease-free.

Biting fly and midge field tests are most commonly conducted in contexts where wild mosquitoes are also active, so the same risks and precautions apply to such tests as apply to tests focused on mosquitoes alone.

12.6.2.5 *Psychological risks*

The EPA excludes pregnant people from participating, due to heightened pest and pesticide exposure concerns. As a result, adult female subjects of child-bearing potential must present supervised, negative pregnancy test results before any repellent treatment. Those women may face a risk of psychological distress associated with learning the results of a pregnancy test and the possibility that the results become known to others.

12.6.3 Beyond skin-applied arthropod repellents: special considerations specific to repellent fabrics and area arthropod repellents

Although bed nets, including repellent- or adulticide-treated ones, have been in use in malarial regions of the world for decades, repellent-treated clothing is a relatively new repellent technology. The efficacy of repellent-treated fabric has been typically assessed via AIC with laboratory-reared mosquitoes. Research has not clearly determined the degree to which repellent-treated fabric reduces vector-borne disease incidence, with study results ranging from 0% to 75% protection against malaria (Banks et al., 2014). Fabrics treated with durable pyrethroids prevent neither landing nor probing due to the slow onset of repellent action. Because of this, repellency is evaluated by comparing the proportions of blood-fed mosquitoes on treated vs untreated fabric sleeves. A probing female mosquito first injects saliva into a host to numb the local area before finding and feeding on blood (Gurera et al., 2018). This means probing females who do not take a blood meal may nonetheless inject allergens and, if present, pathogens into a host. Thus, repellency evaluated by blood-feeding rates cannot be equated with a label claim of protection from disease vectors or the diseases they carry. This is true even if the net effect of the repellent clothing is a reduction in the disease transmission, which has been observed in some cases (Banks et al., 2014). Before field testing is permitted, more research is needed about the mechanisms of treated fabric-based repellency to inform study design and ethical considerations.

Prospective field tests of area repellents with human subjects would be similar to current EPA-approved study designs but with the substitution of humans for traps. Area repellents may emanate passively from a treated substrate or actively from a device. If applied to a substrate, an area repellent product may exhibit gradual failure within a single study day. If a repellent is actively emitted, efficacy may remain stable until the point at which the test material is exhausted, which might be hours or days, depending on the type of emanator. Thus, to properly determine the efficacy of area repellents with human subjects, a single study design may need to accommodate the assessment of both relative protection and decline of repellency which may increase subject risk via additional exposure or limited protection. Finally, because repellent action occurs in the volume of air over a treated area, protection does not move with a subject, and there is no direct way to make certain repellent molecules are present in the immediate air volume around a subject is in any given moment. For these reasons, study design and consent form language must presume equal risks apply to subjects in repellent treated versus untreated conditions.

12.6.4 The future of human subject efficacy testing for arthropod repellents

As global warming drives changes in the distributions and abundance of disease-vectoring arthropods and their associated pathogens, both regulators and industry will be under pressure to flexibly meet the altered needs of consumers for effective repellents. The sudden appearance of Zika virus in the southeast United States provoked widespread news coverage and resulted in political pressure to speed up the process of registering new repellent products as well as new incentives for industry to develop novel active ingredients and formulations. Although Zika virus has left the news cycle, the phenomenon of arthropod-vectored diseases changing and expanding their ranges and encountering newly vulnerable human populations will continue.

In particular, emerging malaria should prompt coordinated strategies among nations to protect their vulnerable citizens. In North American and European regulatory regions, agencies and commercial markets have both emphasized single-product protection, and human subject testing for product registration has reflected this. The choice of CPT as the fundamental measure of efficacy, rather than percent protection, in part reflects the assumption that a consumer will be using and expecting sufficient protection from a single product or strategy (skin-applied vs area arthropod repellent). In current malaria-endemic regions world-wide, effective strategies are usually understood to require the management of combined, integrated measures by consumers. For example, bed nets, indoor area repellents, and skin-applied repellents may be used together by one person or group of people working or living inside a structure. In this scenario, products and techniques that yield short duration or incomplete protection alone can be useful and worth permitting and marketing when combined with other products and strategies. In the case of arthropod repellents, consumers might combine treated clothing with an area repellent to supplement targeted use of skin-applied repellents, thereby optimizing personal protection (Debboun and Strickman, 2013).

Increased use of pesticides worldwide drives the evolution of resistance in arthropod populations including those that focus on human hosts. Rapid resistance evolution can leave communities vulnerable to vector-borne disease outbreaks for which the flexible personal protection of products like repellents are the interventions of first resort (Debboun and Goodyer, Chapter 1 of this book). Regulatory interest in managing resistance development is likely to increase, including for repellents. Since the evolution of resistance is a response to the active ingredient of a product, the discovery of active ingredients with new modes of action will become increasingly important to transform the chronic exposure of wild populations of biting arthropods to one or few repellent molecules into exposure to a wide variety of active ingredients over space and time. More fundamentally, integrated vector management and adaptive, societally informed individual and community behaviors will be critical to sustaining control over arthropod-borne disease risks. "Soft" power against vectors, exemplified by repellents that protect but do not directly kill, remains an underappreciated approach to personal, family, and community health while helping to sustain the susceptibility of vector and pathogen populations to other methods of control (Jørgensen et al., 2020).

12.7 Conclusion

Advancements in methodologies for studies of new arthropod repellents have created a landscape of options and alternatives to human subject testing, but reliance upon human subject studies remains the standard practice for determining repellent efficacy. During the past 15 years, increased oversight of human subject studies by regulatory agencies has made human subject studies highly regulated and risk-minimized, and increasing harmonization between agencies may continue to curb the frequency with which human subject studies are required for regulatory purposes. In a world of changing climates with increasingly unpredictable ecological outcomes, cooperation between countries, regulatory agencies, and the private sector will be critical for helping mitigate the impact of endemic and spreading disease vectors through the use of personal-protection arthropod repellents, in concert with other public health initiatives.

Appendix

Tables A1–A5: Summary of primary regulatory requirements for human subject efficacy testing of skin-applied repellents across major regulatory agencies.

TABLE A1 General requirements and recommendations across all arthropods.

Study design element	WHO	US EPA	EU	PMRA Canada
Ethics review required?	As per national and/or regional laws	Yes, IRB and EPA-convened HSRB	Yes, not specified	Yes, IRB or REB and PMRA
Positive control	Recommended DEET 20% w/v in ethanol	Optional DEET 20% w/v in ethanol	Refer to WHO or US EPA	Recommended DEET 20% w/v in ethanol
Dosing	Standard 1mL/600 cm^2 skin area; preapplication accepted	Justify any dosage determination study or standard per 600 cm^2: Pump spray 0.5 mL; Aerosol, lotion 1 mL	Refer to WHO or US EPA	1 g/600 cm^2 unless other dose data indicate other; preapplication accepted; 1 formula/subject
Subject gender balance	Recommend 50/50	Not specified	Refer to WHO or US EPA	Recommend 50/50
Randomization & blinding	Completely randomized	Completely randomized and blind if possible	Refer to WHO or US EPA	Completely randomized and blind if possible

TABLE A2 General requirements and recommendations for mosquitoes (and black flies where indicated).

Study Design Element	WHO	US EPA	EU	PMRA Canada
Field study required?	Recommended	Required	No	Required
Field trials	≥2 studies in distinct habitats, *Aedes*, *Culex*, *Ano-pheles;* human landing pressure; black flies similar	≥2 studies in distinct habitats, *Aedes*, *Culex*, *Ano-pheles;* human landing pressure; 1 site black flies	Refer to WHO or US EPA	2 of ≥3 efficacy studies must in field with *Aedes*, *Culex* and *Anopheles;* same for black flies
Laboratory trials		Supplemental to field studies	Refer to WHO or US EPA	Not required

(continued)

TABLE A2 (Cont'd)

Study Design Element	WHO	US EPA	EU	PMRA Canada
Number of subjects required	Predicted min for stat significance, power analysis	Predicted min for significance under power analysis; recommend 13 subjects, same for black flies	Not specified	Predicted min for stat significance, power analysis
Negative control subjects	"… a suitable number of untreated volunteers for field…"	2 for CPT endpoints, based on stat design for RP endpoints; none serve both roles at once; same black flies	Untreated control present; Refer to WHO or US EPA for detail	Opposite limb same subject permitted, ≥1 separate control preferred (same for black flies)
Minimum biting pressure on control	Not specified	5 landings/ 5 min in field; 5/min lab	5 landings /5min	5 landings/5 min field (same black flies); 10/min arm-in-cage; 5/min black flies
Exposure duration and frequency	Periodic intervals, duration and frequency appropriate for target species	Periodic duration and frequency according to study design	5 min every h	3-min minimum exposure at least every 30 min, study duration until ≥half fail
End point	First landing or probing; observe both behaviors	First confirmed landing; remove before biting	First confirmed landing	First confirmed landing (bite preferred for black flies)
CPT or Relative Protection (RP)?	CPT	CPT most cases, some cases RP	CPT	CPT or 95% repellency duration, depending on label claim
Minimum acceptable repellency	Not specified	Not specified	Not specified	Duration of 30 min
Other criteria	Field subjects spaced ≥20 m; same black flies	None significant	None significant	None significant

TABLE A3 General requirements and recommendations for ticks (and Trombiculidae where indicated).

Study design element	WHO	US EPA	EU	PMRA Canada
Field Study Required?	No	No	No	No
Field Trials	Not recommended	Not recommended	Refer to WHO or US EPA	Three studies in any combination of field and lab
Laboratory Trials	Not Specified	*A. americanum*, *I. scapularis, and D. variabilis*—or—*R. sanguineus*, growing emphasis on adults	Refer to WHO or US EPA	≥1 species, esp. *A. americanum*. Others *I. scap* and *D. variabilis*. Chiggers: at least one species

TABLE A3 (Cont'd)

Study design element	WHO	US EPA	EU	PMRA Canada
Number of subjects required	Predicted min for stat significance, power analysis	Predicted min for significance under power analysis; recommend 25	≥10	Predicted min for stat significance, power analysis
Negative control	Not specified	Each subject uses their untreated forearm as a control	Untreated control condition present; Refer to WHO or US EPA for detail	Opposite limb same subject permitted
Minimum crossing activity on control	Not specified	Individual tick crosses on a subject's untreated arm	5 crossings/ 5 min	Not specified
Exposure duration and frequency	Periodic intervals, duration, frequency appropriate for target species	Periodic duration and frequency according to study design	Hourly after application, 5 ticks for 5 min	Study duration until ≥half fail
End point	Not specified	Crossing, confirmed	Crossing, confirmed	Crossing
CPT or Relative Protection (RP)?	CPT	CPT most cases, some cases RP	% repellency	CPT or 95% repellency duration, depending on label claim
Minimum acceptable repellency	Not specified	Not specified	≥90% repellency for claimed period	Duration of 30 min

TABLE A4 General requirements and recommendations for other blood-feeding fly species, specifically ceratopogonids ("no-see-ums, punkies, biting midges"), tabanids (e.g., deer fly, horse fly), and stable flies.

Study design element	WHO	US EPA	EU	PMRA Canada
Field Study Required?	No	No	No	No
Field Trials	May be modified from mosquito recommendations	Not recommended	Refer to WHO or US EPA; mi- mic real-world use scenarios	One field or lab study/ taxon, ≥1 taxon important native
Laboratory Trials	May be modified from mosquito recommendations	Recommended	Refer to WHO or US EPA	One field or lab study/ taxon, ≥1 taxon important native
Number of subjects required	Predicted min for stat significance, power analysis	Predicted min for significance under power analysis	≥10	Predicted min for stat significance, power analysis
Negative control subjects	Unspecified number, justified in the study protocol	Two for CPT endpoints, based on stat design for RP endpoints; none serve both roles at once	Not specified, may be modeled after current US EPA or WHO standards	Opposite limb same subject permitted, ≥1 separate control preferred

(continued)

TABLE A4 (Cont'd)

Study design element	WHO	US EPA	EU	PMRA Canada
Minimum biting pressure on control	May be modified from mosquito recommendations	1 landing or bite /5 min	Not specified	1 bite/5 min field, 5/min lab
Exposure duration and frequency	Periodic intervals, duration and frequency appropriate for target species	Periodic duration and frequency according to study design	Not specified	Study duration until ≥half fail
End point	Not specified	bite	bite	bite
CPT or Relative Protection (RP)?	CPT	CPT most cases, some cases RP	% repellency	CPT or duration of 95% repellency, depending on label claim
Minimum acceptable repellency	Not specified	Not specified	>90%	Duration of 30 min
Other criteria	Field subjects spaced ≥20 m	None significant	None significant	None significant

TABLE A5 General requirements and recommendations across all regulatory regions.

Use human subjects
Random allocation of treatments to subjects
Blinding (if possible)
Testing on specified genera/species of biting arthropods
Attractiveness assay for each candidate subject
Pre-exposure period allowed to enable aligning exposures with peak pest activity in the field
Study duration at least as long as label claim

Appendix references

Pest Management Regulatory, 2017. Value Guidelines for New Personal Insect Repellent Products and Label Amendments. https://www.canada.ca/en/health-canada/services/consumer-product-safety/reports-publications/pesticides-pest-management/policies-guidelines/value-new-personal-insect-repellent-products-label-amendments.html. (Accessed March 20, 2021).

ECHA, 2016. Transitional Guidance on Efficacy Assessment for Product Type 18, Insecticide, Acaricides & other Biocidal Products against Arthropods and Product Type 19, Repellents & Attractants. https://echa.europa.eu/documents/10162/23492134/tg_efficacy_pt18pt19_superseded_en.pdf/3e7f9fce-60db-3646-9c85-edde1105e39c. (Accessed March 20, 2021).

US Environmental Protection Agency Office of Chemical Safety and Pollution Prevention, 2010. Product Performance Test Guidelines OPPTS 810-3700 Insect Repellents to be Applied to Human Skin. https://tinyurl.com/wnye7x36. (Accessed March 20, 2021).

World Health Organization, 2009. Guidelines for efficacy testing of mosquito repellents for human skin (No. WHO/HTM/NTD/WHOPES/2009.4). World Health Organization, Genera.

Pest Management Regulatory, 2017.

References and further readings

Bacot, A., Talbot, P.G., 1919. The comparative effectiveness of certain culicifuges under laboratory conditions. *Parasitology* 11 (2), 221–236.

Banks, S.D., Murray, N., Wilder-Smith, A., Logan, J.G., 2014. Insecticide-treated clothes for the control of vector-borne diseases: a review on effectiveness and safety. Med. Vet. Entomol. 28 (S1), 14–25.

Bar-Zeev, M., Smith, C.N., 1959. Action of repellents on mosquitoes feeding through treated membranes or on treated blood. J. Econ. Entomol. 52 (2), 263–267.

Bernardo, M.J., Cupp, E.W., 1986. Rearing black flies (Diptera: Simuliidae) in the laboratory: mass-scale in vitro membrane feeding and its application to collection of saliva and to parasitological and repellent studies. J. Med. Entomol. 23 (6), 666–679.

Bunker, C.W., Hirschfelder, A.D., 1925. Mosquito repellents. Am. J. Trop. Med. Hyg. 1 (5), 359–383.

Carroll, S.P., 2007. Evaluation of topical insect repellents and factors that affect their performance. In: Debboun, M., Frances, S.P., Strickman, D. (Eds.), Insect Repellents: Principles, Methods and Uses. CRC Press, Boca Raton, FL, pp. 245–259.

Carroll, S.P., 2008. Prolonged efficacy of IR3535 repellents against mosquitoes and blacklegged ticks in North America. J. Med. Entomol. 45 (4), 706–714.

Castilho, C.J., Li, D., Liu, M., Liu, Y., Gao, H., Hurt, R.H., 2019. Mosquito bite prevention through graphene barrier layers. Proc. Natl. Acad. Sci 116 (37), 18304–18309.

Colucci, B., P.Müller, 2014. Evaluation of Topical mosquito repellents and interpretation of efficacy data: a systematic literature review. In: Müller, G., Pospischil, R., Robinson, W.H. (Eds.), Proceedings of the 8th International Conference on Urban Pests. OOK-PressKft, Hungary, pp. 163–167.

Colucci, B., 2017. *Evaluation of laboratory and field methods for measuring mosquito repellent efficacy.* Doctoral dissertation. University of Basel, Basel, Switzerland.

Colucci, B., Müller, P., 2018. Evaluation of standard field and laboratory methods to compare protection times of the topical repellents PMD and DEET. Sci. Rep. *8* (1), 1–11.

Curtis, C.F., Lines, D., Ijumba, J., Callaghan, A., Hill, N., Karimzad, M.A., 1987. The relative efficacy of repellents against mosquito vectors of disease. Med. Vet. Entomol. 1 (2), 109–119.

D'Ormea, G., 1919. Note on the employment of a pomade of thymol as a culicifuge for troops in malarial districts. Giorn. de Med. Milit. 67 (2), 296–300.

Debboun, M., Strickman, D., 2013. Insect repellents and associated personal protection for a reduction in human disease. Med. Vet. Entomol. 27 (1), 1–9.

Diaz, J.H., 2016. Preparing the United States for Zika Virus: pre-emptive vector control and personal protection. Wilderness Enviro. Med. 27, 350–359.

Ditzen, M., Pellegrino, M., Vosshall, L.B., 2008. Insect odorant receptors are molecular targets of the insect repellent DEET. Science 319 (5871), 1838–1842.

ECHA 2016. Transitional guidance on efficacy assessment for product type 18, insecticide, acaricides & other biocidal products against arthropods and product type 19, repellents & attractants. https://echa.europa.eu/documents/10162/23492134/tg_efficacy_pt18pt19_superseded_en.pdf/3e7f9fce-60db-3646-9c85-edde1105e39c. (Accessed March 20, 2021).

ECHA. 2019c. Biocidal products regulation guidance structure. (https://echa.europa.eu/guidance-documents/guidance-on-biocides-legislation). (Accessed March 20, 2021).

(EPA)U.S. Environmental Protection Agency, 2006. Protections for subjects in human research; final rule. 40 Code of Federal Regulations Parts 9 and 26. Federal Regist. 71, 6138–6176.

Fernández-Grandon, G.M., Gezan, S.A., Armour, J.A., Pickett, J.A., Logan, J.G., 2015. Heritability of attractiveness to mosquitoes. PLoS One 10 (4), e0122716.

Fishel, F.M., 2013. The EPA conventional reduced risk pesticide program. PI-224, University of Florida Institute of Food and Agricultural Sciences, Gainesville, FL.

Fradin, M.S., Day, J.F., 2002. Comparative efficacy of insect repellents against mosquito bites. N. Engl. J. Med. 347 (1), 13–18.

Granett, P., 1940. Studies of mosquito repellents, II. Relative performance of certain chemicals and commercially available mixtures as mosquito repellents. J. Econ. Entomol. 33 (3).

Gurera, D., Bhushan, B., Kumar, N., 2018. Lessons from mosquitoes' painless piercing. J. Mech. Behav. of Biomed. Mat 84, 178–187.

Harwood, J.F., Arimoto, H., Nunn, P., Richardson, A.G., Obenauer, P.J., 2015. Assessing carbon dioxide and synthetic lure-baited traps for dengue and chikungunya vector surveillance. J. Am. Mosq. Control Assoc. 31 (3), 242–247.

Hiwat, H., Andriessen, R., de Rijk, M., Koenraadt, C.J.M., Takken, W., 2011. Carbon dioxide baited trap catches do not correlate with human landing collections of Anopheles aquasalis in Suriname. *Memorias do Instituto Oswaldo Cruz* 106 (3), 360–364.

Johnson, P.C.D., Barry, S.J.E., Ferguson, H.M., Müller, P., 2015. Power analysis for generalized linear mixed models in ecology and evolution. Methods Ecol. Evol. 6, 133–142.

Jørgensen, P.S., Folke, C., Henriksson, P.J., Malmros, K., Troell, M., Zorzet, A., 2020. Coevolutionary governance of antibiotic and pesticide resistance and the Living with Resistance Project. Trends Ecol. Evol. 35 (6), 484–494.

Klun, J.A., Kramer, M., Debboun, M., 2005. A new in vitro bioassay system for discovery of novel human-use mosquito repellents. J. Am. Mosq. Control Assoc. 21 (1), 64–70.

L'Ambert, G., Ferré, J-B., Schaffner, F., Fontenille, D., 2012. Comparison of different trapping methods for surveillance of mosquito vectors of West Nile virus in Rhône Delta. France. J. Vector Ecol. 37 (2), 269–275.

Laurent, B.S., Sukowati, S., Burton, T.A., Bretz, D., Zio, M., Firman, S., Sudibyo, H., Safitri, A., Asih, P.B., Kosasih, S., Hawley, W.A., 2018. Comparative evaluation of anopheline sampling methods in three localities in Indonesia. Malar. J. 17 (1), 1–11.

Lima, J.B.P., Rosa-Freitas, M.G., Rodovalho, C.M., Santos, F., Lourenço-de-Oliveira, R., 2014. Is there an efficient trap or collection method for sampling *Anopheles darlingi* and other malaria vectors that can describe the essential parameters affecting transmission dynamics as effectively as human landing catches?—A review. Memorias do Instituto Oswaldo Cruz 109 (5), 685–705.

Maia, M.F., Moore, S.J., 2011. Plant-based insect repellents: a review of their efficacy, development and testing. Malar. J. 10 (1), S11.

Mathenge, E.M., Misiani, G.O., Oulo, D.O., Irungu, L.W., Ndegwa, P.N., Smith, T.A., Killeen, G.F., Knols, B.G., 2005. Comparative performance of the Mbita trap, CDC light trap and the human landing catch in the sampling of *Anopheles arabiensis, An. funestus* and culicine species in a rice irrigation in western Kenya. Malar. J. 4 (1), 1–6.

Mintz, J.A., 2012. *Enforcement at the EPA: High Stakes and Hard Choices*. University of Texas Press, Austin, Texas.

Miszta, P., C Basak, S., Natarajan, R., Nowak, W., 2013. How computational studies of mosquito repellents contribute to the control of vector borne diseases. Curr. Comput. Aided Drug. Des. 9 (3), 300–307.

Moore, S.J., Debboun, M., 2007. History of insect repellents. In: Debboun, M., Frances, S.P., Strickman, D. (Eds.), Insect Repellents: Principles, Methods and Uses. CRC Press, Boca Raton, FL, pp. 3–29.

Moore, W., 1934. Esters as repellents. J. NY Entomol. Soc. 42 (2), 185–192.

Moreno, J.D., 2013. *Undue Risk: Secret State Experiments on Humans*. Routledge, New York, NY.

Moreno-Gómez, M., Bueno-Marí, B., Drago, A., Miranda, M.A., 2021. From the Field to the Laboratory: Quantifying Outdoor Mosquito Landing Rate to Better Evaluate Topical Repellents. J. Med. Entomol. 58 (3), 1287–1297.

National Research Council, 2004. *Intentional Human Dosing Studies for EPA Regulatory Purposes: Scientific and Ethical Issues*. National Academies Press, Washington, D.C.

Ollinger, M., Fernandez-Cornejo, J., 1995. *Regulation, Innovation, and Market Structure in the US Pesticide Industry*. US Department of Agriculture, Economic Research Service, Washington, DC.

Overgaard, H.J., Sæbø, S., Reddy, M.R., Reddy, V.P., Abaga, S., Matias, A., Slotman, M.A., 2012. Light traps fail to estimate reliable malaria mosquito biting rates on Bioko Island, Equatorial Guinea. Malar. J. 11, 1–14.

Pest Management Regulatory Agency, 2017. Value guidelines for new personal insect repellent products and label amendments. https://www.canada.ca/en/health-canada/services/consumer-product-safety/reports-publications/pesticides-pest-management/policies-guidelines/value-new-personal-insect-repellent-products-label-amendments.html. (Accessed March 20, 2021).

Peterson, C., Coats, J., 2001. Insect repellents: past, present, and future. Pestic. Outlook 12, 154–158.

Rodriguez, S.D., Chung, H.N., Gonzales, K.K., Vulcan, J., Li, Y., Ahumada, J.A., Romero, H.M., De La Torre, M., Shu, F., Hansen, I.A., 2017. Efficacy of some wearable devices compared with spray-on insect repellents for the yellow fever mosquito, *Aedes aegypti* (L.)(Diptera: Culicidae). J. Insect Sci. 17 (1), 24.

Rodriguez, S.D., Drake, L.L., Price, D.P., Hammond, J.I., Hansen, I.A., 2015. The efficacy of some commercially available insect repellents for *Aedes aegypti* (Diptera: Culicidae) and Aedes albopictus (Diptera: Culicidae). J. Insect Sci. 15 (1), 140.

Schofield, S., Plourde, P., 2012. Statement on personal protective measures to prevent arthropod bites: an Advisory Committee Statement (ACS) Committee to Advise on Tropical Medicine and Travel (CATMAT). Canada Commun. Dis. Rep. 38 (ACS-3), 1.

Sharpington, P.J., Healy, T.P., Copland, M.J., 2000. A wind tunnel bioassay system for screening mosquito repellents. J. Am. Mosq. Control Assoc. 16 (3), 234–240.

Smith, J.B., 1901. Report of the Entomological Dep. N. J. Agric. Coll. Exp. Sta. 542.

Strickman, D., 2009. Answers to EPA HSRB questions on spatial repellents. US EPA Archives. https://archive.epa.gov/osa/hsrb/web/pdf/hsrb_questions_on_spatial_repellents.pdf. (Accessed March 20, 2021).

Strickman, D., 2007. Older synthetic active ingredients and current additivesInsect Repellents: Principles, Methods, and Uses. CRC Press, Boca Raton, FL, pp. 361–383.

Sukumar, K., Perich, M.J., Boobar, L.R., 1991. Botanical derivatives in mosquito control: a review. J. Am. Mosq. Control Assoc. 7 (2), 210.

Tangena, J.A.A., Thammavong, P., Hicox, A., Lindsay, S.W., Brey, P.T., 2015. The human-baited double net trap: an alternative to human landing catches for collecting outdoor biting mosquitoes in Lao PDR. *PLoS One* 10 (9), e0138735.

Unlu, I., Faraji, A., Indelicato, N., Rochlin., I., 2016. TrapTech R-Octenol lure does not improve the capture rates of aedes albopictus (Diptera: Culicidae) and other container-inhabiting species in biogents sentinel traps. J. Med. Entomol. 53 (4), 982–985.

US Environmental Protection Agency Office of Chemical Safety and Pollution Prevention, 2010. Product Performance Test Guidelines OPPTS 810-3700 Insect Repellents to be Applied to Human Skin. https://tinyurl.com/wnye7x36. (Accessed March 20, 2021).

Verhulst, N.O., Takken, W., Dicke, M., Schraa, G., Smallegange, R.C., 2010. Chemical ecology of interactions between human skin microbiota and mosquitoes. FEMS Microb. Ecol. 74 (1), 1–9.

Workshop on Repellents, 2016. Ministry for Agriculture, Forestry, Environment and Water Management. Vienna. https://www.bmk.gv.at/dam/jcr:61c4c2fb-407d-4476-9f1a-271b9d30faf0/ResultsWorkshopOnRepellents22-23June2016.pdf. (Accessed March 20, 2021).

World Health Organization Fact Sheet, 2020. Vector-borne diseases. https://www.who.int/news-room/fact-sheets/detail/vector-borne-diseases. (Accessed March 20, 2021).

World Health Organization, 2009. Guidelines for efficacy testing of mosquito repellents for human skin. In: No, WHO/HTM/NTD/WHOPES/2009.4, World Health Organization, Geneva.

Zwiebel, L.J., Takken, W., 2004. Olfactory regulation of mosquito-host interactions. Insect Biochem. Molec. 34 (7), 645–652.

CHAPTER

13

Arthropod repellent research in Northwest Florida, United States

John P. Smith

Public Health Entomology Services, LLC, Panama City Beach, FL, United States

13.1 Introduction

Mosquitoes have existed before dinosaurs roamed the planet (Reidenbach et al., 2009) and now occupy all continents of the world except Antarctica (Lounibos, et al., 1985). These insects have a tremendous reproductive capacity and are one of the most successful life forms, with over 3500 species described. Most can replicate to incredibly large numbers within a couple of weeks to a month, depending on temperatures and day length. Even the coldest winters in the most northern climes cannot quell their growth. Given the diversity of habitats utilized by fresh, salt, and brackish water mosquitoes, mosquito control can be quite daunting (Becker et al., 2010).

Control agencies manage mosquitoes by integrating methods, including surveillance, source production, and site elimination through draining and ditching, propagating, and releasing indigenous predators, applying biorational and chemical insecticides, and public education (Barker et al., 2017). Chief among the latter is the use of arthropod repellents as the first line of defense (Debboun et al., 2014). In locations where there is no organized mosquito and vector control, arthropod repellents are often the best products for abating mosquitoes, as well as the discomfort and disease potential they transmit.

Arthropod repellent use is one of the most effective means of personal protection against arthropod bites. Numerous topical and spatial arthropod repellents are commercially available. Earlier chapters detail these products and their differences. The Centers for Disease Control and Prevention (CDC) recommends topical arthropod repellents containing N,N-diethyl-3-methyl benzamide (DEET), 2-(2-hydroxyethyl)-1-piperidine carboxylic acid 1-methylpropyl ester (Picaridin), p-methane-3,8 diol (PMD), ethyl butylacetylaminopropionate (IR3535), or methyl nonyl ketone (2-undecanone) for reasonably long-lasting protection (CDC, 2020). There are no CDC recommendations for spatial arthropod repellents because there is still a need for more efficacy studies. New arthropod repellent products are continuously being manufactured to satisfy

Advances in Arthropod Repellents.
DOI: https://doi.org/10.1016/B978-0-323-85411-5.00017-0

growing public demand for nonsynthetic alternatives. Interest in botanicals containing essential oils has become very popular as topical and spatial arthropod repellents. Regardless of the arthropod repellents used, there are ongoing needs for additional scientific evaluation. This chapter provides an overview of arthropod repellent testing conducted in Northwest Florida. The goal of this research is to evaluate protocols that can be easily and efficiently implemented under replicated and controlled conditions with minimal resources. Several alternative arthropod repellents with encouraging results are suggested for further research.

13.2 Regulations

In the United States, all arthropod repellents except those classified as minimum risk 25(b) are required to be federally registered by the US Environmental Protection Agency (EPA) as explained in the Federal Insecticide Fungicide and Rodenticide Act (FIFRA) (40 CFR 152.40-152.55). Most EPA registered arthropod repellents are synthetic, although some are botanicals such as eucalyptus oil and p-methane-3,8 diol (PMD). To qualify for 25(b) status, arthropod repellents must have active and inert ingredients as identified by the EPA (40 CFR 125.25(f) (1,2) and 40 CFR 180.950(a), (b), (c), and (e). Through the authority granted by EPA in FIFRA, state regulatory officials in the Association of American Pesticide Control Officials (AAPCO) have implemented significant changes in registering 25(b) minimum risk insecticides. Although these insecticides are exempted from EPA regulation, each state can adopt its own label and/or data requirements. Guidelines to help companies comply with these requirements have been published (Association of American Pesticide Control Officials (AAPCO), 2019). In short, all product claims must be supported by replicated scientific data. A minimum efficacy of ≥90% is required for topical arthropod repellents and ≥75% for spatial arthropod repellents. In addition, EPA has elevated requirements for human testing to safeguard against potential health risks. Documentation detailing the research protocol, human subject recruitment/informed consent, Institutional Review Board application, correspondence, and approval must be submitted and approved by the institution and by EPA. Health issues such as potential allergic reactions and arthropod-borne diseases, as well as lengthy regulatory requirements, have led to the development of new techniques for screening arthropod repellents without using human subjects.

13.3 Topical arthropod repellent bioassays

Field and laboratory bioassays help determine repellent efficacy and duration, although there are pros and cons to each technique. There are two field and four published laboratory bioassay techniques (Barnard, 2005). Field bioassays with feral mosquitoes are usually conducted in a more natural environment; however, they are difficult to perform because of numerous uncontrollable variables that can affect results. These include differential human attractiveness, varying weather conditions, changing arthropod species, densities, and physiologies prevalent at the time of testing. Reliance on human subjects in field tests can also be problematic as discussed previously. It is possible through laboratory studies to control most of these variables, however, at the expense of foregoing a more natural setting.

Mosquito repellent research in Northwest Florida has evolved from semifield tests where researchers take biting counts inside screen enclosures filled with thousands of laboratory-reared *Ae. taeniorhynchus* and *Cx. quinquefasciatus* mosquitoes (Fig. 13.1A and B) (Smith et al., 2000), to simple hand-in-cage laboratory assays utilizing as few as 50–100 female mosquitoes/cage (Fig. 13.2A and B) (Smith, 2001a, 2001b, 2002,

FIG. 13.1 (A, B) Biting counts conducted in semifield outdoor testing screen enclosures.

Cilek et al., 2004). With the development of the Klun and Debboun (K&D) *in vivo* bioassay, more treatments and replications could be performed with fewer human subjects and under more controlled conditions (Klun and Debboun, 2000). The technique entails exposing five female *Ae. aegypti* or *Cx. quinquefasciatus* in each of six chambers contained in a Plexiglass module measuring approximately 30 cm L × 7 cm W × 4 cm H to arthropod repellent treatments applied on the upper legs of three human subjects (Fig. 13.3A and B). Biting counts are taken for 1–1.5 minutes depending on the species, and environmental conditions are recorded with data loggers. Results from these studies are more repeatable and less affected by external variables experienced in the field (Smith, 2003a, 2003b, 2004a, 2004b, 2005a, 2005b, 2005c, 2006a, 2006b, 2007a, 2007b, 2007c).

Human volunteers were eliminated when *in vitro* methods utilizing stored blood in the K&D system were developed (Klun et al., 2005). In this procedure, preserved blood is supplied in a collagen-covered Plexiglass assembly placed beneath the K&D mosquito module serving as a surrogate host. Water heated and circulated through the lower Plexiglass assembly warms the blood making it attractive for mosquito bites. By inserting thin cloth strips treated with candidate arthropod repellents between the mosquito and blood-holding modules, timed biting counts can be made by opening sliding doors beneath the K&D module. It is possible to replicate several tests simultaneously using up to nine modules interconnected to a water bath fitted with a temperature inversion circulator (Fig. 13.4). This procedure can be reliably

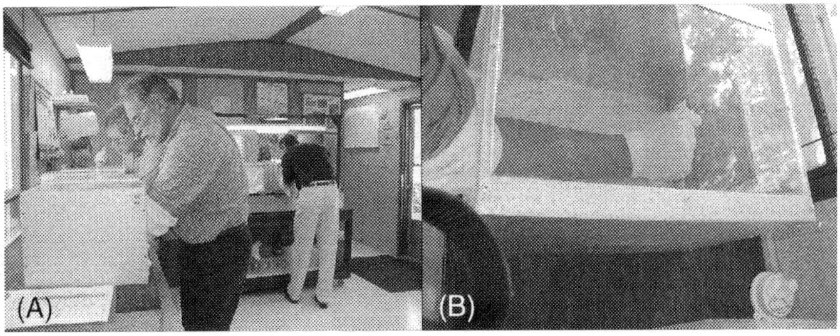

FIG. 13.2 (A, B) Hand-in-cage arthropod repellent bioassay.

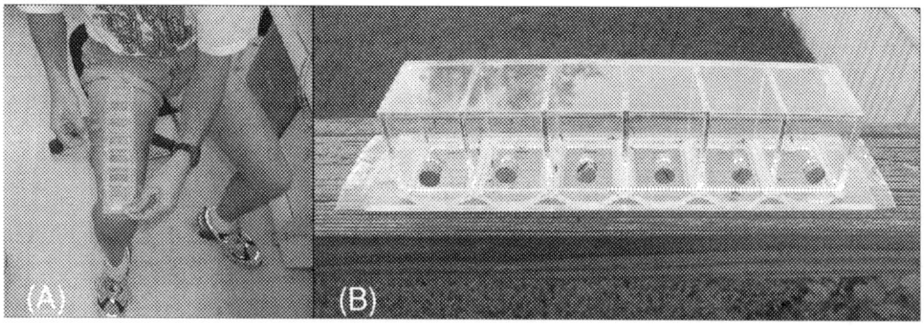

FIG. 13.3 (A, B) *In vivo* Klun and Debboun (K&D) arthropod repellent bioassay with close-up of Plexiglas mosquito module.

performed by two trained persons. Biting data can be acquired over several posttreatment time intervals by replenishing the K&D modules between intervals.

The development of a blood substitute composed of adenosine triphosphate, citrate, phosphate, dextrose, and adenine to replace human or animal blood was another major advancement (Klun et al., 2008, 2013). Using this in the K&D bioassay has become the standard technique for topical arthropod repellent screening in Northwest Florida (Smith, 2013, 2014a, 2014b, 2015a, 2015b, 2016, Smith and Thrall, 2017, Smith and Smith, 2019a). It takes considerable experience and practice to operate the methodologies correctly and consistently. Laboratory bioassays are much less costly, quicker, and effective screening methods for initial evaluation and development of arthropod repellents. Field assays with human volunteers remain a necessary component in validating arthropod repellent efficacy.

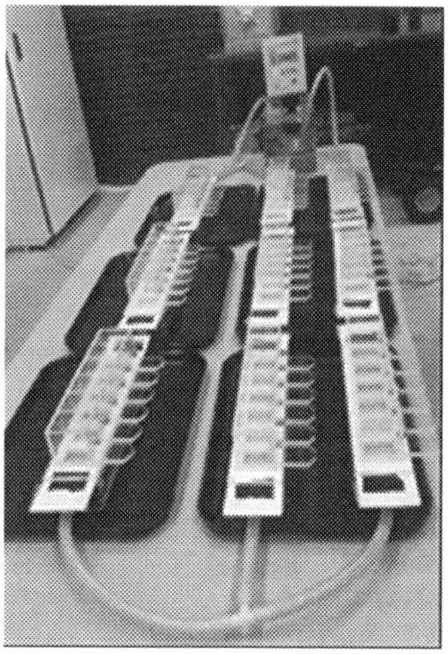

FIG. 13.4 *In vitro* K&D arthropod repellent bioassay connected to a temperature-controlled water bath and replicated nine times.

13.4 Spatial arthropod repellent bioassays

Over the last 10 years, there has been increased emphasis on spatial arthropod repellents, particularly the essential oils. This is because there is the considerable public interest in natural botanical alternatives to synthetic repellents such as DEET and preference for avoiding direct application to the skin. Semifield studies evaluating formulated spatial arthropod repellents are typically performed in outdoor screened enclosures to control variables, such as mosquito species, densities, age, and physiological state (World Health Organization and WHO Pesticide Evaluation Scheme, 2013). Similar studies can also be conducted indoors in screened enclosures or free-flight rooms to control environmental conditions. In these studies, female *Ae. aegypti* or other species are released in the enclosures or rooms. In Northwest Florida, the studies are

FIG. 13.5 Outdoor screen house testing of spatial arthropod repellents.

performed in two 3.7 m(W) × 7.4 m(L) × 2.4 m(H) or 54.5 m^3 ShelterLogic enclosures covered with 20 × 20 (0.841 mm) mesh no-see-um screen (Smith and Smith, 2018, 2019b, 2019c). One enclosure is designated for the formulated arthropod repellent treatment and the other for a negative control (Fig. 13.5). Fifteen minutes after starting the arthropod repellent treatment, 100 3-5-day-old mosquitoes are released in each enclosure and allowed to acclimate for 15 minutes. One-minute biting counts are made thereafter at 0, 1, and 2 hours by exposing a hand through a zipper porthole in the enclosure (Fig. 13.6). Two human volunteers are required for these studies, i.e., one for each enclosure. After a test is completed, mosquitoes are removed by aspiration, and air in the repellent enclosure is exhausted for 15 minutes with a fan. The test is rerun with the repellent and control switched to the opposite enclosure in a 2 × 2 Latin-square design. Two tests are conducted per day and the entire study is repeated under similar weather conditions on three separate days for a total of six tests. Temperature, humidity, wind direction, and speed are recorded with data loggers. The positioning of the treatment is dependent on the number of devices per unit area specified on the product label. Most are placed in the center of the enclosure halfway between the mosquito release cage on one end and the evaluator on the other or can be placed nearer to the evaluator.

Several other methods are used to evaluate spatial arthropod repellents in the laboratory and field. The use of modular cylinders (Grieco et al., 2005) and Y-tube olfactometers (Bernier et al., 2007) are commonly deployed for laboratory studies accessing arthropod repellency and host attraction-inhibition, respectively. These are usually employed to determine effective dosage of an active ingredient. Field trials using human volunteers are frequently conducted to evaluate a formulated product in a natural

FIG. 13.6 Investigator taking hand-in-cage biting count through porthole in outdoor testing screen house.

environment against indigenous species. These studies are encumbered with issues like those discussed previously with topical arthropod repellent testing. However, field studies using human subjects remain the ultimate test in assessing product efficacy.

13.5 Promising arthropod repellents

Although synthetic arthropod repellents containing DEET and Picaridin are considered the "gold standards," there has been much interest in the development of more natural-based arthropod repellents. The most promising botanical arthropod repellents are those containing mixtures of essential oils, as shown in Table 13.1. Because of the proprietary nature of our research, percentages, inert ingredients, and product trade names are not mentioned. We found these repellent mixtures to perform as well as the low concentration of DEET (7%) providing at best, efficacious repellency for up to 6 hours. One would expect outdoor results would vary depending on the arthropod species tested and environmental conditions. Further testing with human volunteers is certainly warranted.

13.6 Conclusion

The general objectives of this chapter were to (1) review regulatory requirements for marketing arthropod repellents in the United States; (2) assess methods for collecting and evaluating efficacy data on topical and spatial arthropod repellents; and (3) recommend natural repellent alternatives with efficacy comparable to synthetics for further research.

Numerous arthropod repellents are marketed each year by private companies in the United States and internationally. All synthetic arthropod repellents are required to be registered by the EPA before being sold in the United States. Although exempted from EPA registration, botanical arthropod repellents marketed as alternatives to traditional synthetics such as DEET are also regulated in many states. Most contain EPA listed 24(b) active and inert ingredients. The AAPCO has published guidelines to assist companies in satisfying requirements for selling these arthropod repellent products in several US states. Aside from minimum efficacy of 90% for topical and 75% for spatial arthropod repellents, additional scientific evidence must be submitted to support all labeled product claims.

TABLE 13.1 Botanical repellents mixtures.

Active ingredients	% Repellency	Species tested[a]
Jojoba, citronella, aloe vera, and eucalyptus oils	100% for 4 hours and >90% at 6 hours	*Cx. quinquefasciatus*
Lemongrass, peppermint, thyme, geranium, and rosemary oils	90% through 4 hours	*Cx. quinquefasciatus*
Lemongrass, peppermint, thyme, geranium, rosemary, soybean, and wheat germ	90% through 6 hours	*Cx. quinquefasciatus*
Citronella, geranium, peppermint, lemongrass, thyme, rosemary, and limonene	90% through 6 hours	*Cx. quinquefasciatus*
Oil of lemon eucalyptus (p-methane-3,8 diol)	100% for 2 hours and >90% up to 6 hours	*Ae. aegypti*
Geraniol, soybean, clove, peppermint, and rosemary oils	≥90% through 6 hours	*Ae. aegypti*
Rosemary and cedarwood oils	≥90% through 6 hours	*Ae. aegypti*

[a] *Laboratory tests.*

There are many laboratory and field techniques developed for efficacy testing. Although laboratory bioassays can be performed more quickly and at less cost, field testing with wild mosquitoes on human volunteers in a natural environment remains the gold standard. Human subject testing is encumbered with many federal requirements and typically entails considerably more time and cost than can be afforded by many manufacturers. As such, the methods presented herein potentially provide alternatives.

Several 25(b) essential oil mixtures provide ≥90% efficacy compared to a negative control for up to 6 hours postapplication (Table 13.1). The efficacy and duration of these blends compare favorably with DEET and other synthetics.

References

Association of American Pesticide Control Officials (AAPCO) FIFRA 25(B) Workgroup., 2015. https://aapco.org/2015/07/02/fifra-25b-workgroup/(accessed October 8, 2020).

Barker, C., Collins, C., Colon, J., Rutledge-Connelly, C., Debboun, M., Dormuth, E., Faraji, A., Fujioka, K., Lesser, C., Michaels, S., Schankel, B., Smith, K., Unlu, I., White, G., 2017. Best practices for integrated mosquito management: a focused update. Am. Mosq. Control Assoc. https://cdn.ymaws.com/www.mosquito.org/resource/resmgr/docs/Resource_Center/Training_Certification/12.21_amca_guidelines_final_.pdf (accessed February 13, 2021).

Barnard, D.R., 2005. Biological assay methods for mosquito repellents. J. Am. Mosq. Control Assoc. 21 (Suppl. 4), 12–16.

Becker, N., Petric, D., Zgomba, M., Boase, C., Madon, M., Dahl, C., Kaiser, A., 2010. Mosquitoes and Their Control, 2nd ed. Springer-Verlag Berlin, Heidelberg and Dordrecht and London and New York, NY. ISBN 978-3-540-92873-7.

Bernier, U.R., Kline, D., Posey, K.H., 2007. Human emanations and related natural compounds that inhibit mosquito host-finding abilities. In: Debboun, M., Frances, S.P., Strickman, D. (Eds.), Insect Repellents: Principles, Methods and Uses, Boca Raton, Florida. CRC Press and Taylor & Francis Group, Boca Raton, FL, pp. 31–46.

CDC, 2020.Ch. 3: Health information for international travel. Environmental Hazards & Other Noninfectious Health Risks. CDC Yellow Book 2020. Mosquitoes, Ticks, & Other Arthropods, https://wwwnc.cdc.gov/travel/yellowbook/2020/noninfectious-health-risks/mosquitoes-ticks-and-other-arthropods (accessed February 1, 2021).

Cilek, J.E., Petersen, J.L., Hallmon, C.F., 2004. Comparative efficacy of IR3535 and DEET as repellents against adult *Aedes aegypti* and *Culex quinquefasciatus*. J. Am. Mosq. Control Assoc. 20 (3), 299–304.

Debboun, M., Frances, S.P., Strickman, D.A., 2014. Insect Repellents Handbook, 2nd ed. CRC Press, Taylor & Francis Group, Boca Raton, FL. ISBN-13: 978-1-4665-5358-3.

Grieco, J.P., Achee, N.L., Sardelis, M.R., Chauhan, K.R., Roberts, D.R., 2005. A novel high-throughput screening system to evaluate the behavioral response of adult mosquitoes to chemicals. J. Am. Mosq. Control Assoc. 21, 404–411.

Klun, J.A., Debboun, M., 2000. A new module for quantitative evaluation of repellent efficacy using human subjects. J. Med. Entomol. 37, 177–181.

Klun, J.A., Kramer, M., Debboun, M., 2005. A new in vitro bioassay system for discovery of novel human-use mosquito repellents. J. Am. Mosq. Control Assoc. 21, 64–70.

Klun, J.A., Kramer, M., Zhang, A., Wang, S., Debboun, M., 2008. A quantitative in vitro assay for mosquito deterrent activity without human blood cells. J. Am. Mosq. Control Assoc. 24, 508–512.

Klun, J.A., Kramer, M., Debboun, M., 2013. Four simple stimuli that induce host-seeking and blood-feeding behavior in two mosquito species, with a clue to DEET's mode of action. J. Vector Ecol. 38, 1–11.

Lounibos, L.P., Rey, J.R., Frank, J.H., 1985. Ecology of mosquitoes: proceedings of a workshop. ISBN-13: 978-0961522407.

Reidenbach, K.R., Cook, S., Bertone, M.A., Harbach, R.E., Wiegmann, B.M., Besansky, N.J., 2009. Phylogenetic analysis and temporal diversification of mosquitoes (Diptera: Culicidae) based on nuclear genes and morphology. BMC Evol. Biol. 9 (1), 298. doi:10.1186/1471-2148-9-298.

Smith, J.P., Floore, T.G., Petersen, J., Shaffer, K., 2000. Good housekeeping mosquito repellent study. The Good Housekeeping Institute. https://www.mosquitoresearchlab.com/repellent-research (accessed October 19, 2020).

Smith, J.P., 2001a. Evaluation of the BugOFF! biting insect repellent wrist band. KAZ, Inc. https://www.mosquitoresearchlab.com/repellent-research (accessed October 19, 2020).

Smith, J.P., 2001b. IBI repellent test. Insect Biotechnology Inc. https://www.mosquitoresearchlab.com/repellent-research (accessed October 19, 2020).

Smith, J.P., 2002. CitroGuard repellent test. CitroLux Corporation. https://www.mosquitoresearchlab.com/repellent-research (accessed October 19, 2020).

Smith, J.P., 2003a. Lewey's Eco-Blends, Inc. repellent study. Lewey's Eco-Blends, Inc. https://www.mosquito researchlab.com/repellent-research (accessed October 19, 2020).

Smith, J.P., 2003b. Dr. Homola's repellent evaluation study–II. Citrolux Corporation. https://www.mosquitoresearchlab.com/repellent-research (accessed October 19, 2020).

Smith, J.P., 2004a. XLTG repellent evaluation study I. XL Tech Group, LLC. https://www.mosquitoresearchlab.com/repellent-research (accessed October 19, 2020).

Smith, J.P., 2004b. HOMS Bio Block repellent evaluation studies. HOMS, LLC. https://www.mosquitoresearchlab.com/repellent-research (accessed October 19, 2020).

Smith, J.P., 2005a. XLTG repellent evaluation studies. XL Tech Group, LLC. https://www.mosquitoresearchlab.com/repellent-research (accessed October 19, 2020).

Smith, J.P., 2005b. XLTG repellent evaluation studies II-IV. XL Tech Group, LLC. https://www.mosquitoresearchlab.com/repellent-research (accessed October 19, 2020).

Smith, J.P., 2005c. XLTG repellent evaluation studies V-VII. XL Tech Group, LLC. https://www.mosquitoresearchlab.com/repellent-research (accessed October 19, 2020).

Smith, J.P., 2006a. TyraTech repellent evaluation studies VIII & IX. TyraTech, LLC. https://www.mosquitoresearchlab.com/repellent-research (accessed October 19, 2020).

Smith, J.P., 2006b. TyraTech repellent evaluation studies X & XI. TyraTech, LLC. https://www.mosquitoresearchlab.com/repellent-research (accessed October 19, 2020).

Smith, J.P., 2007a. TyraTech repellent evaluation studies (Amendment 4 & 5). TyraTech, LLC. https://www.mosquitoresearchlab.com/repellent-research (accessed October 19, 2020).

Smith, J.P., 2007b. TyraTech repellent evaluation studies (Amendment VI). TyraTech, LLC. https://www.mosquitoresearchlab.com/repellent-research (accessed October 19, 2020).

Smith, J.P., 2007c. TyraTech repellent evaluation studies (Amendment VII). TyraTech, LLC. https://www.mosquitoresearchlab.com/repellent-research (accessed October 19, 2020).

Smith, J.P., 2013. EverSafe All natural mosquito control product evaluation: Phase I – Trial #2. Matrix24 Laboratories, LLC. https://www.mosquitoresearchlab.com/repellent-research (accessed October 19, 2020).

Smith, J.P., 2014. In vitro laboratory evaluation of EverSafe repellent. Matrix24 Laboratories, LLC. https://www.mosquitoresearchlab.com/repellent-research (accessed October 19, 2020).

Smith, J.P., 2014. In vitro laboratory evaluation of Natureza repellent. Natureza, Inc. https://www.mosquitoresearchlab.com/repellent-research (accessed October 19, 2020).

Smith, J.P., 2015a. MOSQUITO PAQ in vitro repellent bioassay phase I – laboratory trial. Mosquito Paq, LLC. https://www.mosquitoresearchlab.com/repellent-research (accessed October 19, 2020).

Smith, J.P., 2015b. MOSQUITO PAQ Secure Organic No-Bite proof of concept study phase I—laboratory trial. Mosquito-Paq, LLC. https://www.mosquitoresearchlab.com/repellent-research (accessed October 19, 2020).

Smith, J.P., 2016. MOSQUITO PAQ Secure NOW proof of concept study phase I—laboratory trial. Mosquito Paq, LLC. https://www.mosquitoresearchlab.com/repellent-research (accessed October 19, 2020).

Smith, J.P., Thrall T.J., 2017. Penta5USA in vitro repellent bioassay phase I—laboratory trial. Penta5USA, LLC. https://www.mosquitoresearchlab.com/repellent-research (accessed October 19, 2020).

Smith, J. P., Smith, M.A., 2018. Semi-field trials of Mainstays® outdoor citronella candle against *Aedes aegypti*. Shanghai Daisy. https://www.mosquitoresearchlab.com/repellent-research (accessed October 19, 2020).

Smith, J.P., Smith, M.A., 2019a. Penta5USA Phase I—In Vitro repellent bioassay final report. Penta5USA, LLC. https://www.mosquitoresearchlab.com/repellent-research (accessed October 19, 2020).

Smith, J. P., Smith, M.A., 2019b. Outdoor screen house tests of Packaging Service Co., Inc. tiki fuel against *Aedes aegypti*. Packaging Service Co. Inc. https://www.mosquitoresearchlab.com/repellent-research (accessed October 19, 2020).

Smith, J. P., Smith, M.A., 2019c. Semi-field trials of Hunter's Kloak™ Bugg Off Mist against *Aedes aegypti*. The Gyde Group, LLC. https://www.mosquitoresearchlab.com/repellent-research (accessed October 19, 2020).

World Health Organization and WHO Pesticide Evaluation Scheme, 2013. Guidelines for efficacy testing of spatial repellents. World Health Organization. https://apps.who.int/iris/handle/10665/78142 (accessed October 21, 2020).

CHAPTER

14

Current status of spatial repellents in the global vector control community

Nicole L. Achee, John P. Grieco

Department of Biological Sciences, Eck Institute for Global Health, University of Notre Dame, Notre Dame, IN, United States

14.1 The public health problem

In 2018, a total of 228 million cases of malaria were reported from 91 countries, a decrease of 3 million cases over the previous year (WHO, 2019b). The global tally of malaria deaths reached 405,000 deaths, similar to the number reported in 2017 (WHO, 2018). Although malaria case incidence has fallen globally since 2010, the rate of decline has stalled and even reversed in some regions since 2014. Mortality rates have followed a similar pattern. The World Health Organization (WHO) African region continues to account for about 93% of malaria cases and deaths worldwide. Nineteen countries—all but one in sub-Saharan Africa—carry 85% of the global malaria burden. Remaining challenges to reducing malaria morbidity and mortality include residual transmission (Killeen, 2014), mosquito biting during times when insecticide treated nets (ITNs) are not in use as well as accommodating varied housing structures where ITNs and indoor residual spraying (IRS) may be constrained (Hemingway et al., 2016).

Aedes aegypti, a day-biting mosquito, is a primary vector for dengue, Chikungunya, Zika, and urban Yellow Fever viruses. Each of these viruses can be associated with explosive epidemics, where high disease incidence and public fear combine to overwhelm health systems (Wilder-Smith et al., 2016). Dengue has become the most important human arthropod-borne viral infection worldwide (Brady et al., 2012; Bhatt et al., 2013). Dengue is a public health risk for over 3.9 billion people across 129 countries where the mosquito-borne disease is actively transmitted, with recent estimates of 390 million dengue infections per year (WHO, 2020a). Remaining challenges to controlling *Aedes*-borne viruses include constraints in day-time access to homes in urban environments for ultra-low volume (ULV)-spray for vector control. Although

Advances in Arthropod Repellents.
DOI: https://doi.org/10.1016/B978-0-323-85411-5.00009-1

this emergency control strategy appears to mitigate the impact of outbreaks, the economic costs are high, sometimes leading health departments to abandon campaigns after 1–2 cycles that are far less likely to impact disease transmission, and their success is linked to high coverage rates that are difficult to achieve logistically.

14.2 Market shortcomings

14.2.1 Malaria

Malaria control strategies that have demonstrated success include treatment of infected individuals with drugs, application of insecticide to reduce mosquito populations through IRS, and reduction of human contact with infected mosquitoes via ITNs, including long-lasting insecticidal nets (LLINs) (D'Acremont et al., 2010; O'Meara et al., 2010). However, further achievements toward malaria elimination using seasonal malaria chemoprophylaxis (SMC), intermittent preventive treatment of malaria in pregnancy, LLINs, and/or IRS are constrained.

Primary challenges for SMC, similarly to other mass drug administration, are that the strategy uses the routine health service delivery system for a mass administration thus overburdening the existing health system for providing routine services. On intermittent preventive treatment of malaria in pregnancy, primary challenges include frequent stock-outs of sulfadoxine-pyrimethamine and delayed identification of pregnancy, delaying early uptake of sulfadoxine-pyrimethamine. Variations in anopheline mosquito bionomics, regarding day-time and early evening biting and patterns, can reduce their contact with current intra-domiciliary vector control (ITNs/LLINs and IRS), thus reducing the efficacy of these interventions. Lastly, these same interventions are inappropriate for some eco-epidemiological contexts and are facing challenges of insecticide resistance.

Between 2016 and 2018, a total of 578 million ITNs—most of which were LLINs—were reported by manufacturers as having been delivered globally in malaria-endemic countries (WHO, 2019b). In 2018, about 197 million ITNs were delivered, 87% of which were to countries in sub-Saharan Africa. Across sub-Saharan Africa, household ownership of at least one ITN increased from 50% in 2010 to 72% in 2018. However, the proportion of households with sufficient nets (i.e., one net for every two people) remains inadequate, at 40% in 2018 (WHO, 2019b). Use of ITNs by household members, measured as the percentage of people who slept under an ITN the night before the survey, was 61% in 2018 compared with 36% in 2010 for both pregnant women and children aged under 5 years and was 50% in 2018 compared with 29% in 2010 for the overall population. These indicators represented impressive progress from 2010, but no significant change since 2016.

Fewer people at risk of malaria are being protected by IRS (WHO, 2019b). Globally, IRS protection declined from a peak of 5.8% in 2010 to 2.4% in 2018, with decreases seen across all WHO regions. Coverage dropped from 180 million people globally in 2010 to about 93 million in 2018, a decrease of 13 million compared with 2017. The decline in IRS coverage is occurring for varied reasons to include countries changing or rotating to insecticides that are more expensive, an overall decrease in at-risk populations in those countries in malaria pre-elimination phase, as well as lack of sufficient government infrastructure and/or technical capacity to implement and fund the operational expenses of IRS.

The role of spatial repellent (SR) products in malaria public health vector control can be highlighted when considering temporary shelters that are not conducive to ITN or IRS implementation and/or against daytime, early evening biting anopheline mosquitoes (malaria mosquitos)—areas of transmission where traditional interventions have incomplete effectiveness.

One use-case scenario is during emergency situations. Forcibly displaced persons, including those internally displaced, and refugees now total more than 65 million individuals (UNHCR, 2020). Many individuals suffer or are at increased risk for vector-borne diseases, e.g., malaria throughout much of Africa, leishmaniasis in Syria, Turkey, and Iraq, dengue in Yemen. Relief agencies are struggling to meet the challenges of vector-borne diseases due to the limited availability of resources and limited vector control options. While some displaced persons may be settled into camps where standard vector control tools can be deployed, many others are mobile, in makeshift shelters and situations where IRS or LLINs are not practical, but where tools under development can play a life-saving role.

14.2.2 *Aedes*-borne viruses

Aedes-borne virus control has traditionally focused on larval source reduction (to include larvicides and container removal) and predominantly outdoor and limited indoor ULV space sprays, the latter restricted to public health emergencies. Despite the lack of a well-informed evidence base (Bowman et al., 2016), these strategies, widely practiced by public health programs, are considered ineffective (Valdez et al., 2011; Simmons et al., 2012; Reiter, 2014; Andersson et al., 2015).

There is increasing recognition that programs lacking interventions specifically directed at adult mosquitoes are insufficient for suppression of *Aedes*-borne diseases (Morrison et al., 2008; Achee et al., 2015). Recently, evidence from 10 years of city-wide emergency indoor ULV-spray campaigns showed that when control interventions were applied during the first half of the dengue virus transmission season, fewer dengue cases were detected and the transmission season was shorter (Stoddard et al., 2014). Although these emergency control strategies appear to mitigate the impact of outbreaks, the economic costs are high, sometimes leading health departments to abandon campaigns after 1–2 cycles that are far less likely to impact disease transmission, and their success is linked to high coverage rates that are difficult to achieve logistically. Furthermore, these methods require strict technical controls (properly calibrated machines, skilled operators) and should be applied early in the morning or late afternoon, further complicating access to households.

The role for SR products in *Aedes*-borne virus control is envisioned to address constraints in day-time access to homes in urban environments for ULV-spray—a challenge SR product ease-of-use and distribution may overcome. The SR products could potentially provide the same level of protection with a single deployment overcoming many of the household coverage and technical problems associated with space sprays. They also have the potential to limit the spread and/or emergence of insecticide resistance alleles due to low/no selection pressure when considering nonlethality of effect.

There are thousands of registered SR products already on market, adopted, and used for protection from nuisance biting. There are sophisticated, expensive products that are used in the United States and Europe (liquid vaporizers); there are inexpensive and simpler products (e.g., mosquito coils) that are widely used throughout Africa and Asia. However, there is presently no public sector use of mosquito coils (or any other SR product format) for disease control, due to insufficient evidence for WHO policy recommendation. There is building but limited data in the scientific literature regarding factors that influence human use of SR products, such as, preferences for and perceived efficacy of different SR products, and procurers' perceptions about when and where SR products are needed to supplement (or serve as an alternative to) more commonly used vector control strategies (ITNs, IRS, and larvicides) where those may not be appropriate or able to be implemented (Paz-Soldan et al., 2011; Avicor et al., 2017; Liverani et al., 2017; Masalu et al., 2017; Jumbam et al., 2020).

14.3 The spatial repellent product class

14.3.1 Behavioral effects of vector control chemicals

By definition, an insecticide (insecticide or insecticidal) is a chemical that kills insects. This single term is not adequate for meaningful discourse about chemicals, chemical actions, insect responses to chemicals, and the different ways in which chemicals are used. However, this single response is the foundation for an old paradigm that classifies chemicals sprayed on house walls for malaria control based solely on their killing action despite other chemical actions which also exist, such as spatial repellency (prevention of entry) and contact irritancy (causing escape) (Grieco et al., 2007).

More than 70 years ago, excitant and repellent effects on mosquitoes of sublethal contacts with dichloro-diphenyl-trichloroethane (DDT) were recognized as a potential benefit to malaria control (Kennedy, 1947). As early as 1953, Muirhead-Thomson (1953) concluded chemicals could disrupt contact between humans and malaria-transmitting mosquitoes and stop disease transmission without killing the mosquitoes. Dethier et al. (1960) showed that chemicals elicit multiple actions and that insects respond to those actions through a variety of behaviors. He noted that if we were to take a closer look at modes of action, we could find a much more diverse set of terms for oriented movements of insects toward or away from a chemical source.

In a simple conception, chemical actions can elicit arthropod behavior that prevents entry into a space occupied by a human host (to include spatial repellency), whereas contact irritancy elicits an escape response once the arthropod is in a treated space and toxicity follows uptake of an insecticide occurring either before or after human biting (Grieco et al., 2007). In addition, there may be no impact on entry, feeding success or survival (Fig. 14.1).

Regardless, the search for alternative compounds has focused almost entirely on toxicity. Insecticides recommended for IRS and LLINs continue to be evaluated almost entirely on mosquito mortality (WHO, 2006) and laboratory evaluations continue to use toxicity as the primary measure of success (Schreck and McGovern, 1989; WHO, 1998; Brogdon and McAllister, 1998).

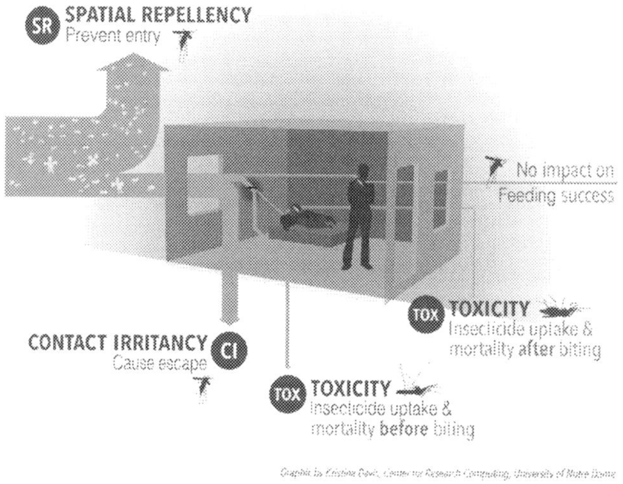

FIG. 14.1 Early probability model demonstrating the impact of behavior modifying properties of chemicals on house entering mosquitoes.

14.3.2 Mode of action of spatial repellent products

Spatial repellents are products designed to release volatile chemicals into the air and prevent human-vector contact within the treated space thus reducing pathogen transmission (Achee et al., 2012). This mode of action holds the potential to have an impact against anopheline mosquitos (carrying malaria parasites) and aedine mosquitos (carrying dengue, Chikungunya, Zika or other *Aedes*-borne viruses).

Thresholds exist for when and how insects respond to chemical actions (Grieco et al., 2007) (Fig. 14.2). These thresholds are governed by intrinsic and extrinsic factors such as inherent strength of a chemical action, chemical volatility, environmental temperature, humidity, proximity, and length of exposure, and a mosquito's sensitivity to a compound, among others. The dose-dependent order in which thresholds are exceeded determines whether the primary mode of chemical action is repellent, irritant, or toxicant.

The term "spatial repellency" is used here to refer to a range of arthropod behaviors induced by airborne chemicals that result in a reduction in human-vector contact and therefore personal protection. The behaviors can induce arthropods to move away from a chemical stimulus, interfere with their host detection (attraction-inhibition) and/or feeding response, and consequently can operate on all adult behaviors that incorporate movement (WHO, 2013). These effects have been measured in laboratory studies, in semifield testing under experimentally controlled conditions, and in open field settings against *Aedes* spp., *Anopheles* spp., and *Culex* spp. of varying insecticide resistance profiles. However, the full extent of behaviors elicited and/or modified by SRs is still unclear.

14.3.3 Laboratory studies

(Achee et al., 2009) in a suite of laboratory assays demonstrated the range of spatial repellency, contact irritancy, and toxicity chemical actions elicited from common active ingredients used for both malaria and *Aedes*-borne virus vector control. Recent laboratory studies have continued to demonstrate the ability of SRs to target a range of vector life traits that impact behaviors required for vector survival and propagation (Achee, Grieco, 2018). In addition, experiments evaluating effects of pre-exposure of mosquitoes to SR

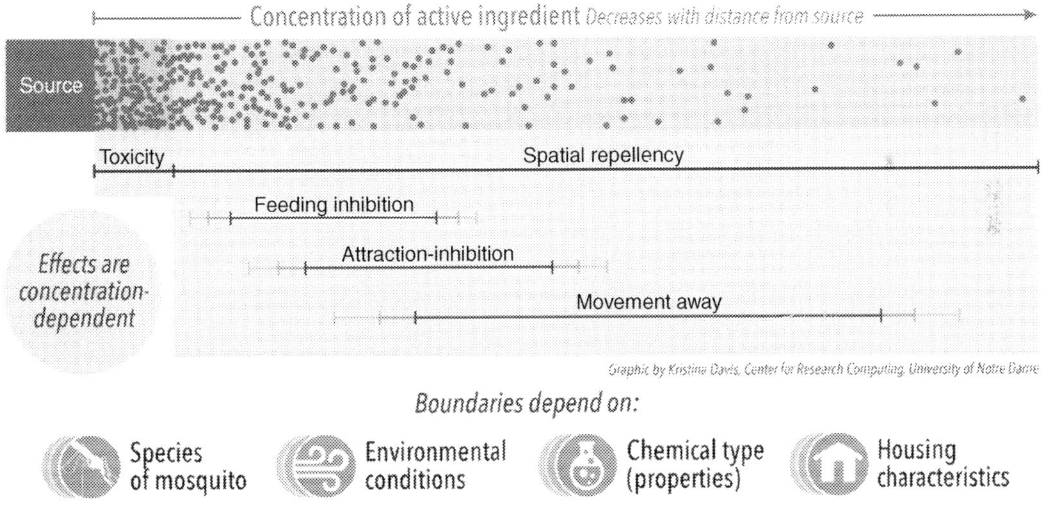

FIG. 14.2 Illustration of the concentration dependent activity of chemicals exhibiting spatial repellent properties and factors that contribute to their effectiveness.

product active ingredients have demonstrated increased attraction to oviposition cues that could intervene in the vector life-cycle or enhance combination interventions (i.e., push–pull) (Choi et al., 2016). Still other laboratory studies have demonstrated effects against insecticide-resistant vector species linked to malaria and *Aedes*-borne virus transmission (Horstmann and Sonneck, 2016; Sonneck and Horstmann, 2017).

14.3.4 Semifield and experimental hut studies

Evaluation of both passive and active emanating SR products in varied semifield environments implementing experimental hut/house study designs have reported a wide range of mosquito behaviors affected by airborne chemical actives such as deterrence and reduced entry (Grieco et al., 2000; Wagman et al., 2015) blood feeding inhibition and reduced survival (Ogoma et al., 2012; Ritchie and Devine, 2013), as well as reduced fecundity following exposure (Ogoma et al., 2014; Kawada et al., 2014; Bibbs et al., 2018). In addition, evaluations of a novel controlled-release spatial repellent device have been conducted in Zambia to demonstrate that devices reduce indoor densities of *An. gambiae* (Stevenson et al., 2018). Through these evaluations, it is evident that volatile semiochemicals are able to modify epidemiologically relevant mosquito behaviors and have the potential to disrupt human pathogen transmission.

14.3.5 Open field settings

Evaluations of SR products in numerous open field settings have demonstrated the prevention of human-vector contact (i.e., fewer mosquito bites) using vapor-active pyrethroids. Outdoor field trials in North America have demonstrated paper emanators impregnated with metofluthrin to reduce landing rates of *Aedes vexans* by up to 95% compared to pre-treatment conditions and *A. canadensis* by 85%–100% compared to untreated controls; both species are nuisance biters and incriminated in vectoring West Nile virus (Lucas et al., 2005; Lucas et al., 2007). Kawada et al. (2004, 2005a, 2005b) reported a reduction in *Anopheles* spp. mosquitoes when using metofluthrin-impregnated paper strips as compared to untreated sites in Lombok, Indonesia (Kawada et al., 2004; Kawada et al., 2005a; Kawada et al., 2005b). Evaluations of transfluthrin have also shown up to 90% bite protection from *Anopheles* spp. inside houses of Dar es Salaam, Tanzania compared to control after volatizing using a kerosene oil lamp (Pates et al., 2002). A randomized field trial in Mexico has recently indicated a 57.2% reduction in indoor *Ae. aegypti* blood fed females using a SR product containing metofluthrin (Devine et al., 2021). The range of mosquitoes affected—major genera (e.g., *Anopheles* spp. and *Aedes* spp.) and species—as well as the range of insecticide susceptibility statuses, suggests the added value of SR products in public health. However, the full extent of mosquito behaviors elicited and/or modified by SRs is still unclear.

14.3.6 Existing evidence of public health impact from clinical trials

Large-scale trials with epidemiological endpoints have demonstrated SR impact on pathogen transmission. A trial in China using 0.3% transfluthrin coils alone or in combination with LLINs demonstrated up to 77% reduction in *Plasmodium falciparum* and 80% reduction in *P. vivax* prevalence rates in a four-treatment arm study design (Hill et al., 2014). In addition, a proof-of-principle study in Indonesia showed that communities (village clusters) with high coverage (90%) of metofluthrin coils had lower biting pressure from mosquitoes and reduced malaria transmission by 52% (Syafruddin et al., 2014).

Most recently, two prospective double-blinded, large-scale controlled randomized cluster trials (cRCTs) using a matched transfluthrin passive emanator were conducted in Sumba Island, Indonesia, to evaluate efficacy against new malaria infections in children ≤ 6 mo-≥ 5 years (Syafruddin et al., 2020) and in Iquitos, Peru, to evaluate efficacy against new *Aedes*-borne viruses (WHO, 2020c). In Indonesia, the 24-cluster protective effect was 27.7% and 31.3%, for time to first-event and overall (total new) infections, respectively, although it was not statistically significant due in part to zero to low incidence in some clusters, undermining the ability to detect a protective effect (Syafruddin et al., 2020). A subgroup analysis of 19 clusters where at least one infection occurred during baseline showed 33.3% and 40.9%, statistically significant protective effect to first infection and overall infections, respectively. And among 12 moderate- to high-risk clusters, a statistically significant decrease in infection by intervention was detected (60% PE). In Peru, the same transfluthrin passive emanator significantly reduced *Aedes*-borne virus infection by 34.1% in association with a significant reduction in adult female *Ae. aegypti* abundance by 28.6% and blood fed rate by 12.4% (WHO, 2020c).

While both of these current cRCTs support evidence of SR efficacy, the Indonesia trial was underpowered and the Iquitos, Peru cRCT represented only one of two RCTs required by WHO Vector Control Advisory Group (VCAG) thus, WHO VCAG requirements for conclusively assessing public health value of the intervention class for malaria control remain partially satisfied (WHO, 2019a).

14.3.7 Global assessment of spatial repellent public health value

The WHO assessment of SR product public health value has been ongoing for the past decade, starting with a position statement regarding the repellent (reduced house entry) mode of action of DDT in 2011 (WHO, 2011). Formal national and international meetings have been convened to bring together academics, industry, funders, and global public health experts, including representatives from the WHO, to discuss the role of SR products in the reduction of arthropod-borne diseases based on existing evidence. A critical aspect of these meetings and subsequent efforts have been to establish a critical path of development for SR products based on expert advice. This includes measures related to scientific, regulatory, and social parameters. In part, these criteria outline the endpoints of preferred product characteristics (WHO, 2020b) and a target product profile (WHO, 2017a) for the SR intervention class.

The SR intervention class has been under WHO VCAG assessment for public health value since 2014. Despite existing clinical trial outcomes, there remains a lack of sufficient epidemiological evidence needed to demonstrate public health impact across a range of eco-epidemiological settings to confidently inform a potential WHO policy recommendation for the incorporation of SR products into current multilateral disease control programs. In 2017, the WHO VCAG recommended additional clinical trials to evaluate SR against malaria in Africa and against *Aedes*-borne viruses in Asia (WHO, 2017b). These knowledge gaps must be addressed to inform WHO SR policy recommendation, to have national disease control programs adopt a SR policy, and to incentivize SR product research and development.

The VCAG concluded that the cRCT conducted in Indonesia (Syafruddin et al., 2020) investigating the impact of a SR product on malaria showed promising results but was underpowered for demonstrating clear protective efficacy (WHO, 2019a). Consequently, VCAG concluded that the trial was inconclusive with respect to the protective efficacy of the intervention, and as such there is a pressing need for further evidence. While more evidence will be required to determine whether or not a WHO recommendation for SR products as a malaria control intervention is warranted, VCAG highlighted that the Indonesia trial generated useful data that can contribute to a future assessment. Lastly, given that the entomological results and subgroup analyses were promising, they strongly supported the continued evaluation of the potential epidemiological impact of this tool.

Regarding the outcomes from the cRCT conducted in Peru (WHO, 2020c), VCAG noted the results were promising given the significant and conclusive 34.1% protective effect of the SR product against arbovirus infection in persons susceptible to Zika virus and/or seronegative or monotypic to dengue virus. Given the significant evidence gap on vector control for *Aedes*-borne viruses, and the urgent need for new tools, the successful completion of the trial was marked by WHO as a major milestone.

14.4 Status in closing the knowledge gap

In 2019, funding for a multicenter program entitled Advancing Evidence for Global Implementation of Spatial repellents (AEGIS) (Unitaid, 2019) has provided the opportunity to close the knowledge gap on SR public health value and barriers to implementation under operational use conditions. The ultimate goal being to validate an SR mode of action across ecological settings and complete the historical timeline of achieving a full WHO global policy recommendation for the inclusion of SR products in vector control programs (Fig. 14.3).

The AEGIS project has four major components evaluating a single SR product: two cRCTs, one each in Kenya and Mali, designed to evaluate protection against new human malaria infections in residual transmission settings, and a cRCT in Sri Lanka to quantify the protective effect of the same SR product against *Aedes*-borne viruses. The Kenya trial will also inform community/diversion effect against insecticide-resistant vector populations. Based on a stage-gated process, an operational study will be conducted in a rural African environment linked to displaced persons where LLINs and/or IRS are constrained to achieve universal coverage. This study is designed to inform SR effectiveness under operational conditions and quantify differences

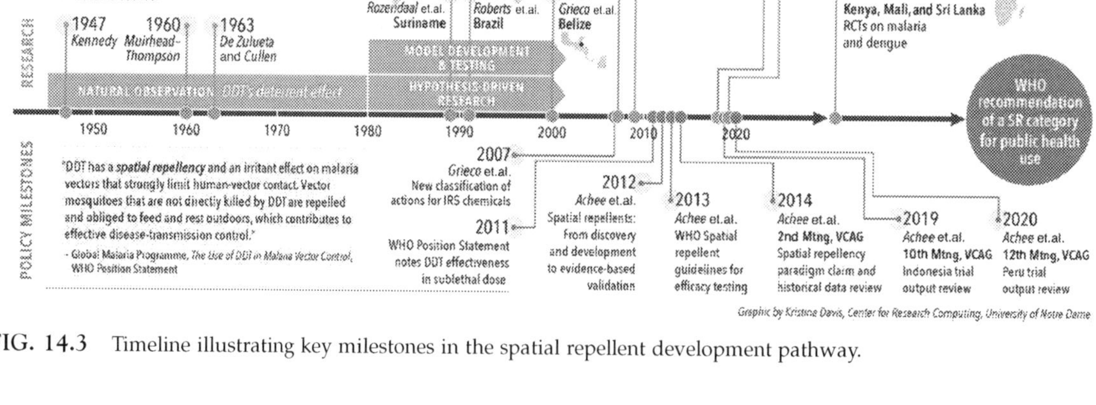

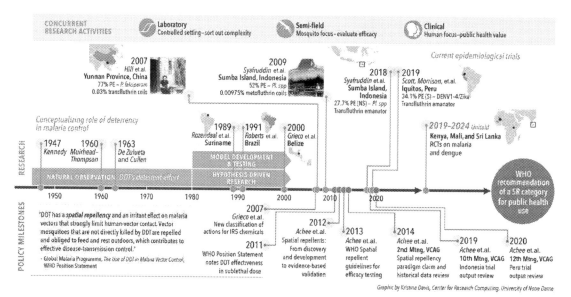

FIG. 14.3 Timeline illustrating key milestones in the spatial repellent development pathway.

in impact using different distribution channels in preparation for donor market introduction.

14.5 Conclusion

Spatial repellent products are envisioned to complement and enhance existing vector control strategies in integrated approaches to disease control. Current evidence from laboratory, semi-field, and clinical trials building over 70 years since spatial repellency was first recognized suggest the potential of SR products to impact anopheline mosquitos carrying malaria parasites and aedine mosquitos carrying dengue and Zika viruses. The added benefit of SR products in the arsenal for vector control is the continual release of volatile active ingredients which precludes the requirement of mosquito contact with a treated surface (i.e., ITN, IRS, ULV-sprays) thus providing protection against varied vector biting patterns and housing construction. This unique mode of action facilitates the development of a range of formulations to address context-specific application requirements thereby facilitating health systems strengthening and coverage of at-risk populations.

Progress toward achieving a WHO global health policy for inclusion of SR products in national programs is advancing, with demonstrated efficacy in public health across a range of epidemiological studies. The remaining gaps in evidence required by the WHO VCAG are anticipated to be addressed through clinical trials planned over the next few years through a recently funded multicenter program. While data outcomes from those trials are pending, detailed entomological small-scale studies evaluating more granular product performance continue to be performed across investigative teams around the world. Through these combined scientific and policy-shaping efforts, the SR product class will be well-positioned for design, manufacturing, and implementation scale-up following a positive WHO product class endorsement.

References

Achee, N.L., Grieco, J.P., 2018. Current evidence, new insights, challenges and future outlooks to the use of spatial repellents for public health. in: Norris, E., Coats, J. (Eds.), Advances in the Biorational Control of Medical and Veterinary Pests ACS Books, Washington, D.C., pp. 25–42.

Achee, N.L., Bangs, M.J., Farlow, R., Killeen, G.F., Lindsay, S., Logan, J.G., Moore, S.J., Rowland, M., Sweeney, K., Torr, S.J., Zwiebel, L.J., Grieco, J.P., 2012. Spatial repellents from discovery and development to evidence-based validation. Malar. J. 11, 164.

Achee, N.L., Gould, F., Perkins, T., Reiner, R.C., Morrison, A.C., Ritchie, S.A., Gubler, D.J., Teyssou, R., Scott, T.W., 2015. A Critical Assessment of Vector Control for Dengue Prevention. PLoS Negl. Trop. Dis. 9 (5), e0003655.

Achee, N.L., Sardelis, M.R., Dusfour, I., Chauhan, K.R., Grieco, J.P., 2009. Characterization of spatial repellent contact irritant and toxicant chemical actions of standard vector control compounds. J. Am. Mosq. Control Assoc. 25 (2), 156–167.

Andersson, N., Nava-Aguilera, E., Arosteguí, J., Morales-Perez, A., Suazo-Laguna, H., Legorreta-Soberanis, J., Hernandez-Alvarez, C., Fernandez-Salas, I., Paredes-Solís, S., Balmaseda, A., Cortés-Guzmán, A.J., Serrano de los Santos, R., Coloma, J., Ledogar, R.J., Harris, E., 2015. Evidence based community mobilization for dengue prevention in Nicaragua and Mexico (camino verde, the green way): Cluster randomized controlled trial. BMJ, 351, h3267.

Avicor, S.W., Wajidi, M.F.F., Owusu, E.O., 2017. To coil or not to coil application practices perception and efficacy of mosquito coils in a malaria-endemic community in Ghana. Environ. Sci. Pollut. Res. Int. 24 (26), 21138–21145.

Bhatt, S., Gething, P.W., Brady, O.J., Messina, J.P., Farlow, A.W., Moyes, C.L., Drake, J.M., Brownstein, J.S., Hoen, A.G., Sankoh, O., Myers, M.F., George, D.B., Jaenisch, T., Wint, G.R.W., Simmons, C.P., Scott, T.W., Farrar, J.J., Hay, S.I., 2013. The global distribution and burden of dengue. Nature 496 (7446), 504–507.

Bibbs, C.S., Hahn, D.A., Kaufman, P.E., Xue, R., 2018. Sublethal effects of a vapour-active pyrethroid, transfluthrin on *Aedes aegypti* and *Ae. albopictus* (Diptera: Culicidae) fecundity and oviposition behaviour. Parasit. Vectors 11 (1), 486.

Bowman, L.R., Donegan, S., McCall, P.J., 2016. Is dengue vector control deficient in effectiveness or evidence systematic review and meta-analysis. PLoS Negl. Trop. Dis. 10 (3), e0004551.

Brady, O.J., Gething, P.W., Bhatt, S., Messina, J.P., Brownstein, J.S., Hoen, A.G., Moyes, C.L., Farlow, A.W., Scott, T.W., Hay, S.I., 2012. Refining the global spatial limits of dengue virus transmission by evidence-based consensus. PLoS Negl. Trop. Dis. 6 (8), e1760.

Brogdon, W.G., McAllister, J.C., 1998. Simplification of adult mosquito bioassays through use of time-mortality determinations in glass bottles. J. Am. Mosq. Control Assoc. 14 (2), 159–164.

Choi, D.B., Grieco, J.P., Apperson, C.S., Schal, C., Ponnusamy, L., Wesson, D.M., Achee, N.L., 2016. Effect of spatial repellent exposure on dengue vector attraction to oviposition sites. PLoS Negl. Trop. Dis. 10 (7), e0004850.

D'Acremont, V., Lengeler, C., Genton, B., 2010. Reduction in the proportion of fevers associated with *Plasmodium falciparum* parasitaemia in Africa—a systematic review". Malar. J. 9, 240.

Dethier, V.G., Browne, B.L., Smith, C.N., 1960. The designation of chemicals in terms of the responses they elicit from insects. J. Econ. Entomol. 53 (1), 134–136.

Devine, G.J., Vazquez-Prokopec, G.M., Bibiano-Marin, W., Pavia-Ruz, N., Che-Mendoza, A., Medina-Barreiro, A., Villegas, J., Gonzalez-Olvera, G., Dunbar, M.W., Ong, O., Ritchie, S.A., 2021. The entomological impact of passive metofluthrin emanators against indoor *Aedes aegypti*: A randomized field trial. PLoS Neglected Tropical Diseases 15. doi:10.1371/journal.pntd.0009036.

Grieco, J.P., Achee, N.L., Andre, R.G., Roberts, D.R., 2000. A comparison study of house entering and exiting behavior of *Anopheles vestitipennis* (Diptera: Culicidae) using experimental huts sprayed with DDT or deltamethrin in the Southern District of Toledo Belize C.A. J. Vector Ecol. 25 (1), 62–73.

Grieco, J.P., Achee, N.L., Chareonviriyaphap, T., Suwonkerd, W., Chauhan, K., Sardelis, M.R., Roberts, D.R., 2007. A new classification system for the actions of IRS chemicals traditionally used for malaria control. PLoS One 2 (8), e716.

Hemingway, J., Shretta, R., Wells, T.N.C., Bell, D., Djimdé, A.A., Achee, N., Qi, G., 2016. Tools and strategies for malaria control and elimination what do we need to achieve a grand convergence in malaria. PLoS Biol. 14 (3), e1002380.

Hill, N., Zhou, H.N., Wang, P., Guo, X., Carneiro, I., Moore, S.J., 2014. A household randomized, controlled trial of the efficacy of 0.03% Transfluthrin coils alone and in combination with long-lasting insecticidal nets on the incidence of *Plasmodium falciparum* and *Plasmodium vivax* Malaria in Western Yunnan Province China. Malar. J. 13, 208.

Horstmann, S., Sonneck, R., 2016. Contact bioassays with Phenoxybenzyl and Tetrafluorobenzyl pyrethroids against target-site and metabolic resistant mosquitoes. PloS One 11 (3), e0149738.

Jumbam, D.T., Stevenson, J.C., Matoba, J., Grieco, J.P., Ahern, L.N., Hamainza, B., Sikaala, C.H., Chanda-Kapata, P., Cardol, E.I., Munachoonga, P., Achee, N.L., 2020. Knowledge attitudes and practices assessment of malaria interventions in rural Zambia. BMC Public Health 20 (1), 216.

Kawada, H., Maekawa, Y., Takagi, M., 2005a. Field trial on the spatial repellency of metofluthrin-impregnated plastic strips for mosquitoes in shelters without walls (Beruga) in Lombok Indonesia. J. Vector Ecol. 30 (2), 181–185.

Kawada, H., Maekawa, Y., Takagi, M., 2005b. Field trial on the spatial repellency of metofluthrin-impregnated resin strip against mosquitoes in shelters without walls (Beruga) in Lombok Indonesia. JSMEZ 56, 63.

Kawada, H., Maekawa, Y., Tsuda, Y., Takagi, M., 2004. Laboratory and field evaluation of spatial repellency with metofluthrin-impregnated paper strip against mosquitoes in Lombok Island Indonesia. J. Am. Mosq. Control Assoc. 20 (3), 292–298.

Kawada, H., Ohashi, K., Dida, G.O., Sonye, G., Njenga, S.M., Mwandawiro, C., Minakawa, N., 2014. Insecticidal and repellent activities of pyrethroids to the three major pyrethroid-resistant malaria vectors in Western Kenya. Parasit. Vectors 7, 208.

Kennedy, J.S., 1947. The excitant and repellent effects on mosquitos of sub-lethal contacts with DDT. Bull. Entomol. Res. 37 (4), 593–607.

Killeen, G.F., 2014. Characterizing, controlling and eliminating residual malaria transmission. Malar. J. 13, 330.

Liverani, M., Charlwood, J.D., Lawford, H., Yeung, S., 2017. Field assessment of a novel spatial repellent for malaria control a feasibility and acceptability study in Mondulkiri Cambodia. Malar. J. 16 (1), 412.

Lucas, J.R., Shono, Y., Iwasaki, T., Ishiwatari, T., Spero, N., 2005. Field efficacy of Metofluthrin—a new mosquito repellent. In: Lee, C.-Y., Robinson, W.H. (Eds.), Proceedings of the Fifth International Conference on Urban Pests. Perniagaan Ph'ng. P&Y Design Network, George Town, Malaysia.

Lucas, J.R., Shono, Y., Iwasaki, T., Ishiwatari, T., Spero, N., Benzon, G., 2007. U.S. laboratory and field trials of metofluthrin (SumiOne) emanators for reducing mosquito biting outdoors. J. Am. Mosq. Control Assoc. 23 (1), 47–54.

Masalu, J.P., Finda, M., Okumu, F.O., Minja, E.G., Mmbando, A.S., Sikulu-Lord, M.T., Ogoma, S.B., 2017. Efficacy and user acceptability of transfluthrin-treated sisal and hessian decorations for protecting against mosquito bites in outdoor bars. Parasit. Vectors 10 (1), 197.

Valdez, M.R.W.de, Nimmo, D., Betz, J., Gong, H.-F., James, A.A., Alphey, L., Black, W.C., 2011. Genetic elimination of dengue vector mosquitoes. PNAS 108 (12), 4772–4775.

Morrison, A.C., Zielinski-Gutierrez, E., Scott, T.W., Rosenberg, R., 2008. Defining challenges and proposing solutions for control of the virus vector *Aedes aegypti*. PLoS Med. 5 (3), e68.

Muirhead-Thomson, R.C., 1953. Mosquito Behaviour in Relation to Malaria Transmission and Control in the Tropics. Edward Arnold & Co., London.

Ogoma, S.B., Moore, S.J., Maia, M.F., 2012. A systematic review of mosquito coils and passive emanators: Defining recommendations for spatial repellency testing

methodologies. Parasit Vectors, 5, 287. Parasites & Vectors 5 (1), 1–10.
Ogoma, S.B., Lorenz, L.M., Ngonyani, H., Sangusangu, R., Kitumbukile, M., Kilalangongono, M., Simfukwe, E.T., Mseka, A., Mbeyela, E., Roman, D., Moore, J., Kreppel, K., Maia, M.F., Moore, S.J., 2014. An experimental hut study to quantify the effect of DDT and airborne pyrethroids on entomological parameters of malaria transmission. Malar. J. 13, 131.
O'Meara, W.P., Mangeni, J.N., Steketee, R., Greenwood, B., 2010. Changes in the burden of malaria in sub-Saharan Africa. Lancet Infect. Dis. 10 (8), 545–555.
Pates, H.V., Line, J.D., Keto, A.J., Miller, J.E., 2002. Personal protection against mosquitoes in Dar es Salaam, Tanzania by using a kerosene oil lamp to vaporize transfluthrin. Med. Vet. Entomol. 16 (3), 277–284.
Paz-Soldan, V.A., Plasai, V., Morrison, A.C., Rios-Lopez, E.J., Guedez-Gonzales, S., Grieco, J.P., Mundal, K., Chareonviriyaphap, T., Achee, N.L., 2011. Initial assessment of the acceptability of a push-pull *Aedes aegypti* control strategy in Iquitos Peru and Kanchanaburi Thailand". Am. J. Trop. Med. Hyg. 84 (2), 208–217.
Reiter, P., 2014. Surveillance and control of urban dengue vectors. In: Gubler, D.J., Ooi, E.E., Kuno, G., Vasudevan, S., Farrar, J. (Eds.), Dengue and Dengue Hemorrhagic Fever2nd edn. Cabi, Wallingford, pp. 481–516.
Ritchie, S.A., Devine, G.J., 2013. Confusion, knock-down and kill of *Aedes aegypti* using metofluthrin in domestic settings: a powerful tool to prevent dengue transmission? Parasites & Vectors 6 (1), 1–9.
Schreck, C.E., McGovern, T.P., 1989. Repellents and other personal protection strategies against *Aedes albopictus*. J. Am. Mosq. Control Assoc. 5 (2), 247–250.
Simmons, C.P., Farrar, J.J., van Vinh Chau, N., Wills, B., 2012. Dengue. N. Engl. J. Med. 366 (15), 1423–1432.
Sonneck, R., Horstmann, S., 2017. The impact of resistance in mosquitoes to the efficacy of transfluthrin and other pyrethroids. Int. Pest Control 59 (1), 36–39.
Stevenson, J.C., Simubali, L., Mudenda, T., Cardol, E., Bernier, U.R., Vazquez, A.A., Thuma, P.E., Norris, D.E., Perry, M., Kline, D.L., Cohnstaedt, L.W., Gurman, P., D'hers, S., Elman, N.M., 2018. Controlled release spatial repellent devices (CRDs) as novel tools against malaria transmission a semi-field study in Macha Zambia. Malar. J. 17 (1), 437.
Stoddard, S.T., Wearing, H.J., Reiner, R.C., Morrison, A.C., Astete, H., Vilcarromero, S., Alvarez, C., Ramal-Asayag, C., Sihuincha, M., Rocha, C., Halsey, E.S., Scott, T.W., Kochel, T.J., Forshey, B.M., 2014. Long-term and seasonal dynamics of dengue in iquitos Peru. PLoS Negl. Trop. Dis. 8 (7).
Syafruddin, D., Asih, P.B.S., Rozi, I.E., Permana, D.H., Nur Hidayati, A.P., Syahrani, L., Zubaidah, S., Sidik, D., Bangs, M.J., Bøgh, C., Liu, F., Eugenio, E.C., Hendrickson, J., Burton, T., Baird, J.K., Collins, F., Grieco, J.P., Lobo, N.F., Achee, N.L., 2020. Efficacy of a spatial repellent for control of malaria in Indonesia a cluster-randomized controlled trial. Am. J. Trop. Med. Hyg. 103 (1), 344–358.
Syafruddin, D., Bangs, M.J., Sidik, D., Elyazar, I., Asih, P.B.S., Chan, K., Nurleila, S., Nixon, C., Hendarto, J., Wahid, I., Ishak, H., Bøgh, C., Grieco, J.P., Achee, N.L., Baird, J.K., 2014. Impact of a spatial repellent on malaria incidence in two villages in Sumba Indonesia. Am. J. Trop. Med. Hyg. 91 (6), 1079–1087.
UNHCR, 2020. Figures at a glance. http://www.unhcr.org/en-us/figures-at-a-glance.html (Accessed October 28, 2020).
Unitaid, 2019. Innovative repellents for disease-carrying mosquitoes. https://unitaid.org/project/innovative-repellents-for-disease-carrying-mosquitoes/. (Accessed July 10, 2019).
Wagman, J.M., Grieco, J.P., Bautista, K., Polanco, J., Briceño, I., King, R., Achee, N.L., 2015. The field evaluation of a push-pull system to control malaria vectors in Northern Belize, Central America. Malar. J. 14, 1–11.
WHO, 1998. Test Procedures for Insecticide Resistance Monitoring in Malaria Vectors Bio-Efficacy and Persistence of Insecticides on Treated Surfaces. World Health Organization, Geneva.
WHO, 2006. Malaria Vector Control and Personal Protection Report of a WHO Study Group. World Health Organization, Geneva.
WHO, 2011. The Use of DDT in Malaria Vector Control WHO Position Statement. World Health Organization, Geneva.
WHO, 2013. Guidelines for Efficacy Testing of Spatial Repellents. World Health Organization, Geneva.
WHO, 2017a. The Evaluation Process for Vector Control Products. World Health Organization, Geneva.
WHO, 2017b. Seventh Meeting of the Vector Control Advisory Group (VCAG). World Health Organization, Geneva.
WHO, 2018. World Malaria Report 2018. World Health Organization, Geneva.
WHO, 2019a. Tenth Meeting of the WHO Vector Control Advisory Group (VCAG). World Health Organization, Geneva.
WHO, 2019b. World Malaria Report 2019. World Health Organization, Geneva.
WHO, 2020a. Dengue and severe dengue. www.who.int/mediacentre/factsheets/fs117/en/ (Accessed October 30, 2020).
WHO, 2020b. Public consultation on preferred product characteristics for malaria vector control interventions. https://www.who.int/news-room/articles-detail/public-consultation-on-preferred-product-characteristics-for-malaria-vector-control-interventions (Accessed October 30, 2020).
WHO, 2020c. Twelfth Meeting of the WHO Vector Control Advisory Group (VCAG). World Health Organization, Geneva.
Wilder-Smith, A., Gubler, D.J., Weaver, S.C., Monath, T.P., Heymann, D.L., Scott, T.W., 2016. Epidemic arboviral diseases priorities for research and public health. Lancet Infect. Dis. 17 (3), e101–e106.

CHAPTER

15

Repellent semiochemical solutions to mitigate the impacts of global climate change on arthropod pests

Agenor Mafra-Neto[a], Mark Wright[b], Christopher Fettig[c], Robert Progar[d], Steve Munson[e], Darren Blackford[e], Jason Moan[f], Elizabeth Graham[g], Gabe Foote[h], Rafael Borges[a], Rodrigo Silva[a], Revilee Lake[a], Carmem Bernardi[a], Jesse Saroli[a], Stephen Clarke[i], James Meeker[j], John Nowak[k], Arthur Agnello[l], Xavier Martini[m], Monique J. Rivera[n], Lukasz L. Stelinski[m]

[a]ISCA, Inc., Riverside, CA, United States; [b]University of Hawaii at Manoa, College of Tropical Agriculture and Human Resources, Honolulu, HI, United States, [c]USDA Forest Service, Pacific Southwest Research Station, Davis, CA, United States, [d]USDA Forest Service, Sustainable Forest Management Research (SFMR), Washington, DC, United States, [e]USDA Forest Service, Forest Health Protection, Ogden, UT, United States, [f]Alaska Department of Natural Resources, Division of Forestry, Anchorage, AK, United States, [g]USDA Forest Service, Forest Health Protection, Juneau, AK, United States, [h]University of California, Department of Entomology and Nematology, Davis, CA, United States, [i]USDA Forest Service, Forest Health Protection, Lufkin, TX, United States, [j]USDA Forest Service, Forest Health Protection, Pineville, LA, United States, [k]USDA Forest Service, Forest Health Protection, Asheville, NC, United States, [l]Cornell University, College of Agriculture and Life Sciences, Ithaca, NY, United States, [m]University of Florida, Institute of Food and Agricultural Sciences, Gainesville, FL, United States, [n]University of California at Riverside, Department of Entomology, Riverside, CA, United States

Advances in Arthropod Repellents.
DOI: https://doi.org/10.1016/B978-0-323-85411-5.00010-8

15.1 Introduction

In the millennia-long rivalry for resources between humanity and pestiferous arthropods, the arthropods appear to have gained a significant advantage in present times from the same phenomenon at the heart of so many of today's greatest challenges: the effects of global climate change. Populations of arthropods harmful to human interests, such as pests that feed on food and fiber crops and vectors that transmit disease, have responded to increasing average temperatures in a variety of ways, many of which exacerbate negative impacts. The mosquitoes responsible for the transmission of blood-borne pathogens such as malaria and dengue—the world's deadliest and most common mosquito-transmitted diseases—are steadily expanding their ranges into regions that were once ecologically unsuitable for them (Ellis, 2019; Ryan et al., 2019). In the United States, *Ae. aegypti* (L.), the yellow fever mosquito (which also transmits a variety of other pathogens, including the causal agents of dengue, chikungunya, and Zika) has spread north at an annual rate of approximately 240 km over the last half-decade, while the Asian tiger mosquito, *Ae. albopictus* (Skuse), has expanded its range northward in Europe by around 150 km per year (Kraemer et al., 2019; Yale School of the Environment, 2019). Should these trends continue, by 2050, populations of *Ae. aegypti* could be found in such northerly cities as Shanghai and Chicago, placing nearly half (49%) of the global human population at risk from the diseases carried by these vectors (Kraemer et al., 2019; Yale School of the Environment, 2019). Mosquitoes are also likely to move into higher altitudes as average temperatures rise, opening opportunities for mosquito-borne diseases like malaria to reemerge in urban environments such as Mexico City, where malaria was eradicated decades ago (Reiter, 2001). In addition to these expansions in range, warmer temperatures and higher levels of carbon dioxide in the atmosphere contribute to increased health threats from mosquitoes by increasing rates of evolution, and thus speciation, as well as changes to seasonal dynamics such as extending breeding seasons, creating more robust and diverse populations with higher density (Tang et al., 2018; Pope, 2019). According to Ryan et al. (2019), the net increase in people threatened by exposure to *Aedes*-transmitted pathogens over the next century could be as high as 1 billion, with most new cases likely to occur in Europe and tropical and subtropical high-elevation regions.

Productivity and sustainability of agriculture and forestry resources also face increasing threats from arthropod pests as the world grows warmer. North American forests, already threatened by more frequent, severe, and larger wildfires, such as those consuming the state of California at the time of this writing (October 2020) (Higuera and Abatzoglou, 2021), are suffering increasing tree mortality due to bark beetle outbreaks of unprecedented scale in some forest systems (Vose et al., 2018). Warmer temperatures foster longer, more intense periods of drought, which in turn reduce trees' vigor and capacity to defend themselves against many wood-boring insects (Freeman, 2017; Mietkiewicz et al., 2018; Negrón and Popp, 2018), while simultaneously promoting conditions favorable to wood-boring beetle outbreaks, accelerating insect development (Negrón and Popp, 2018), extending flight periods (Lill, 2019; Puikkonen, 2019), increasing overwintering survival (Miller and Werner, 1987; Hart et al., 2017), and even producing larger beetles and larger populations of fertile female beetles (Puikkonen, 2019). One of the world's most valuable tropical crops, coffee, is expected to suffer higher infestation rates from its chief insect pest, the coffee berry borer (CBB), *Hypothenemus hampei* Ferrari, as global temperatures rise, expanding the range of the insect and increasing the number of generations produced over each coffee crop season (Jaramillo et al., 2009, 2011).

At the same time that these challenges are emerging, effective management tools for destructive arthropods are becoming scarce. Conventional chemical insecticides have formed the backbone of arthropod control for much of the last century, for both public health and agricultural applications. In recent decades, however, ample evidence has arisen to indicate that unrestricted pesticide use is technically, environmentally, and culturally unsustainable. Because arthropods reproduce so rapidly and in such large numbers, they are readily capable of responding to selective pressures in their environment, including evolving tolerances to chemical agents applied to suppress their populations. Applications of even the most effective pesticides lose their effectiveness over time as natural selection confers physiological or behavioral resistance to the chemical active ingredient (AI) (Martini et al., 2012; Mota-Sanchez and Wise, 2021). One particularly alarming instance of this phenomenon is the increasing frequency of mosquito resistance to pyrethroids, a group of synthetic insecticides designed to mimic the biological activity of pyrethrins, found naturally in chrysanthemum flowers (Bond et al., 2014). Pyrethroids have played a critical role in vector control since the 1970s (Kupferschmidt, 2016). They are the only insecticides approved for use in long-lasting insecticide-treated bed nets and the primary agent used in indoor residual insecticide sprays, both of which tactics are pillars of the World Health Organization's (WHO) global strategy to eliminate malaria (World Health Organization, 2015; Larson et al., 2016). In 2012, the WHO estimated that if pyrethroids were to lose most of their efficacy, 55% of the benefits of vector control would be lost (World Health Organization, 2012). Pyrethroid resistance was rarely observed until the last decade when cases began to appear across Africa with alarming speed (Kupferschmidt, 2016); in 2018, the WHO reported widespread resistance in the sub-Saharan countries most impacted by malaria (World Health Organization, 2018).

Public acceptance of conventional pesticide use has declined over recent decades. Many insecticidal chemicals have been linked to a range of health and environmental risks, particularly older classes with neurotoxic modes of action, such as organochlorine, organophosphate, and carbamate insecticides. Regulations regarding the use of these chemicals on crops have generally tightened globally in the last few decades, with lower tolerances for maximum pesticide residues exported produce, mandated delays between pesticide application and entry of workers into treated fields, and even total bans on some chemistries. For example, the chlorinated cyclodiene pesticide, endosulfan, was once one of the most effective tools available to coffee growers seeking to manage CBB (Ingram, 1968; Vega et al., 2015) but has since been banned in 70 countries, including Brazil, the United States, and the member states of the European Union, on account of its negative environmental impacts and the threat its continued use would pose to agricultural workers (U.S. Environmental Protection Agency, 2010; European Commission, 2016). Nonselective chemical insecticide applications also can lead to the emergence of other pest problems. Thacker (2002) estimated that approximately half of the arthropod species that are currently considered pests only emerged as pests as a result of unintended effects of insecticide usage, such as the removal of predators and other natural enemies that would otherwise have acted as a natural check against such secondary outbreaks.

Considering the hazards and limitations of traditional pest control methods and the growing challenges presented by a warming world, there is a clear need to develop more lastingly effective, safe, and sustainable strategies to mitigate the impacts of harmful insects on human health and prosperity. This chapter will discuss three cases of a method of insect control that does not rely on traditional pesticides to reduce impacts: pheromone- or semiochemical-based repellents to protect agricultural and forestry resources. While repellents have been broadly used to protect people from hematophagous arthropods (ticks, mosquitoes, other biting flies), their commercial use in agriculture and forestry is comparatively rare. The technologies described herein demonstrate the feasibility of agricultural and/or forestry pest management by means of semiochemical-based repellents in situations made more challenging or likely to become more frequent as climate change continues to reshape ecosystems around the world. *CBB Repel* is a repellent designed to deliver effective, long-lasting control of CBB in coffee plantations. CBB Repel can be applied alone or in combination with strategic insecticide applications, where the volatiles of the spatial repellent flush the insects out of the interior of the coffee berries where they feed and breed, making them more susceptible to pesticide treatments. Adoption of CBB Repel could help coffee growers to cope with the increasing populations of CBBs fostered by warming climates in tropical and subtropical regions. *SPLAT Verb* contains the bark beetle antiaggregation pheromone verbenone in a viscous thixotropic fluid matrix that increases in firmness following application, which when applied to the bark of pine trees protects them and their neighboring trees from colonization by host-seeking mountain pine beetle (MPB), *Dendroctonus ponderosae* Hopkins. This product has been tested in multiple pine species and multiple locations in the Western United States, including high-altitude stands of whitebark pine, *Pinus albicaulis* Engelmann, a critical tree species in North American mountain ecosystems that is being threatened by climate-driven changes. *Beetle Guard*, utilizing semiochemical repellents, has applications in both forestry and agriculture, as it has been developed to prevent attack by multiple species of ambrosia beetles known to colonize a broad range of tree species that include endangered or ecologically important species and specialty food crops. The redbay ambrosia beetle (RAB), *Xyleborus glabratus* Eichhoff, for example, is a serious threat to the native Lauraceae species that characterize coastal forests of the Southeastern United States and has also severely impacted the avocado industry in Florida.

15.2 Coffee berry borer, *Hypothenemus hampei* (Ferrari)

15.2.1 Background

Coffee is the world's most widely consumed beverage and the most valuable export crop produced in the tropics (Davis et al., 2012). Globally, the coffee industry is worth over 100 billion USD a year and supports the livelihoods of roughly 100 million people (Bunn et al., 2015; Davis et al., 2019; Schiffman, 2019). Two main species are cultivated: Arabica coffee, *Coffea arabica* (L.), and robusta coffee, *C. canephora* (Pierre ex A. Froehner), with Arabica being the chief crop grown, accounting for 60%–70% of the world's commercial production (Davis et al., 2012). Both species originated and flourished in the humid forests of Africa (Davis et al., 2006) but are currently grown in 80 countries over an area of >10 million ha (FAOSTAT, 2014). A third species, *C. liberica* (Bull ex Hiern), is grown around the world, but this crop is used almost exclusively as grafting rootstock for Arabica and robusta coffee rather than for actual bean production (Davis et al., 2019). The majority of commercially traded coffee, around 70%, is produced on farms ≤2 ha in size, partly because of the high-altitude environment required for Arabica beans to thrive, which renders larger-scale growing operations impossible (Davis et al., 2019; Knowledge@Wharton, 2019).

In terms of crop productivity, coffee has proven to be highly sensitive to changes in climate. The ideal mean annual temperature range for Arabica has been estimated at 18–24°C, depending on the location (Alègre, 1959; Teketay, 1999; Davis et al., 2012). Temperatures above 30°C, if sustained for too long, can severely stress the crop, slowing its growth and creating such defects as stem tumors and yellowing of the leaves (Franco, 1958; Davis et al., 2012).

Accelerated fruit development also occurs at higher than optimal temperatures, reducing the quality of the beverage produced from the brewing of the ground beans (Camargo, 1985; Davis et al., 2012). This presents an uncertain future for the coffee industry in many locations, as the Intergovernmental Panel on Climate Change estimates that average global temperatures will increase by at least 1.8°C and as much as 4°C by 2100 (IPCC, 2007; Davis et al., 2012). Such an increase will impact coffee production in a variety of ways, including causing more frequent extreme weather events (i.e., heat waves or frosts, floods, or droughts) that are detrimental to crop quality and productivity (Conde et al., 2006; Gay et al., 2006). In Veracruz, Mexico, for example, where coffee is grown on over 153,000 ha by 67,000 producers, winter frosts led to nearly complete destruction of coffee plants in 1970 (Gay et al., 2006). Because of the limited range of geographies and conditions over which coffee can be cultivated properly, climate-driven challenges may affect large proportions of the market simultaneously, limiting or disabling supply chains and driving coffee prices up (Worland, 2018). Wild *Coffea* species are also threatened by these changes, with an estimated 60% (75 out of 124 identified species) currently at risk of extinction (Knowledge@Wharton, 2019). While these species are not commercially cultivated for coffee production, coffee growers rely on the genetic diversity of wild plants for cross-breeding, to make their crops more resistant to drought, pests, or disease, which is ever more critical in the face of growing variability in conditions (Knowledge@Wharton, 2019). The drying up of this genetic reservoir could therefore make Arabica and robusta crops more susceptible to these stresses, while at the same time warmer temperatures enable fungal diseases such as stem rust to thrive (Knowledge@Wharton, 2019). These collective impacts are projected to cut the global land area capable of producing coffee by half across all scenarios (Schiffman, 2019). In some areas, such as the Colombian mountain region, these changes are already underway. Cultivated coffee area in this region has decreased by 20% since the 1990s, with almost 40,500 ha (4% of the total area) lost in an 18-month span (2018–2019) (Schiffman, 2019). Daylight hours have been shortened by 19% as cloud cover has increased, which in some cases has hindered coffee productivity, and extremes in precipitation and outbreaks of insect pests and fungal diseases have become more common (Schiffman, 2019).

The CBB, is the most damaging pest on coffee worldwide; indeed, it is the only organism in the world known to feed solely on coffee beans (Fig. 15.1A–D). This feeding behavior is enabled by the presence of gut bacteria that detoxify caffeine (1,3,7-trimethylxanthine), a purine alkaloid that is both distasteful and toxic to most herbivorous insects, and also break down the large carbohydrate molecules that comprise coffee berries (Ceja-Navarro, 2015; Vega et al., 2015). Infestation of a coffee berry is typically initiated by a single individual, a founder female beetle, which bores into the endosperm of the berry starting around 8 weeks after crop flowering and continuing to the harvest (Centre for Agriculture and Biosciences International, 2016). She creates galleries within the berry in which she lays 2–3 eggs daily for a period of 20 days, but she does not leave the berry after this oviposition period has ended, remaining within the fruit to tend her brood (University of Hawaii, 2018). After hatching, CBB larvae also feed on the endosperm and at maturity practice sibling mating. Female beetles, more common than males at a ratio of 10:1 (Waterhouse and Norris, 1989; Baker et al., 1992), leave their berry of origin ready to infest other berries and begin rearing their own young, while males never leave the berry (Centre for Agriculture and Biosciences International, 2016). Each infested berry may contain up to 100 CBB spanning three generations, and reproduction continues even after berries have dropped to the ground, making these berries a major source of continuing infestation of the crop (Centre for

FIG. 15.1 (A) Coffee beans susceptible to the attack of (B) adult female coffee berry borers, *Hypothenemus hampei*, that penetrate the bean and (C) oviposit, creating new foci of CBB infestation in the field (D). Frequently, each infected coffee bean hosts a multigenerational colony of CBB that is well protected from predators, external pesticide applications, and potential desiccation. The use of CBB Repel in the field induces female CBB to leave infested coffee beans, exposing themselves to these dangers in the environment. Furthermore, CBB Repel seems to reduce the likelihood that CBB females will deposit eggs in the protected field. Although CBB Repel has been designed to be applied as dollops to the base of the plant and or the soil, the tests described herein used 3-g dollops of the formulation applied inside containers. CBB Repel, a gray paste, was placed as 3-g dollops inside (E) prototype plastic orange containers clipped to the coffee branches or (F) black plastic containers placed on the ground at the base of the plant. Credits: (A and D) Mafra-Neto; (B and C) Wright; (E and F) Borges. *CBB*, coffee berry borer.

Agriculture and Biosciences International, 2016; University of Hawaii, 2018). One study found that 1.5–2 million CBB per ha emerged from fallen berries in Colombian coffee plantations after pruning (Byers, 1992; Baker, 1999).

Damage caused by CBB feeding on coffee berries renders the infested beans unsaleable, reduces the quality and yield of the remaining crop, and can lead to bacterial or fungal infestations and premature drop of immature berries (Sponagel, 1994; Damon, 2000; Duque, 2000; Duque-Orrego, 2002; Duque and Baker, 2003; Vega et al., 2009). Left uncontrolled, CBB infestation rates can reach 50%–100%, essentially eliminating crop yield, enabled by the pest's rapid generation time (Le Pelley, 1968; Mathieu et al., 1999; Vega et al., 2015; Centre for Agriculture and Biosciences International, 2016). Under favorable temperatures, adult CBB can emerge in as little as 30 days, producing multiple generations in a single season (Baker et al., 1992). Global losses due to CBB infestations have generally been estimated at 500 million USD each year (Vega et al., 2003), but recent evaluations conducted in Brazil place the pest's annual impact in that country alone at 215–358 million USD (Oliveira et al., 2013; Yong, 2015), suggesting that the half-billion USD toll is likely an underestimation of CBB's real global impact. Native to Central Africa, CBB has since been introduced to every country in the world where coffee is produced except Nepal and China, with Puerto Rico and Hawaii being the most recently infested areas (2007 and 2010, respectively) (Jaramillo et al., 2011). In the first 5 years after the pest's introduction in Puerto Rico, coffee values dropped by approximately 12.6 million USD, a 33% decrease, and 1200 coffee farms were abandoned (U.S. Department of Agriculture (USDA), 2014; Mariño et al., 2017). CBB's first appearance in Hawaii occurred on Kona Island, but it has since spread to Hawaii Island, Maui, and Oahu, putting at risk a coffee industry valued at >49 million USD in the 2015–2016 season (Hawaii Department of Agriculture website, 2017a, 2017b).

Climate change is expected to worsen the impacts of CBB on coffee. Jaramillo et al. (2009, 2011) projected that an increase in average temperatures of 1–2°C would alter CBB's geographic range, quicken the insect's development process, and increase its number of generations per season. These predictions were followed only 2 years later by reports from Tanzania that CBB had been detected at altitudes 300 m above the limit where they were detected 10 years previously (Le Pelley, 1968; Mangina et al., 2007; Jaramillo et al., 2011). The warm, dry conditions created by a La Niña event occurring in East Africa in 2011 contributed further to the problem, sparking a series of CBB infestations across the area (Jaramillo et al., 2011).

Control measures for CBB rely on cultural practices such as replacement of traditional burlap coffee bags with bags made of synthetic fiber (which are less likely to be infested by CBB), varying storage methods for harvested berries, timely pruning and removal of all unharvested berries from coffee trees and the ground, and proper sanitation in processing mills (Department of Agriculture, State of Hawaii, 2014; Centre for Agriculture and Biosciences International, 2016; University of Hawaii, College of Tropical Agriculture & Human Resources, 2018). Like many other bark beetles, CBB is a challenging pest to manage using traditional chemical insecticides due to its internal feeding behavior. As these beetles spend most or all of their lives within the coffee berry where they hatched, they are shielded from contact by insecticide sprays, and are only susceptible to these toxins during a limited window of time, before adult females have burrowed into the endosperm (Centre for Agriculture and Biosciences International, 2016).

Until recently, growers had two reliable weapons in their pesticidal arsenal against CBB, the organochlorine insecticide, endosulfan (6,7,8,9,10,10-hexachloro-1,5,5a,6,9,9a-hexahydro-6,9-methano-2,4,3-benzodioxathiepine-3-oxide) and the organophosphate, chlorpyrifos [0,0-diethyl 0-(3,5,6-trichloro-2-pyridinyl)-phosphorothioate] (Baker et al., 2002). Introduced in the 1950s, endosulfan possesses characteristics that make it ideal for use against CBB and led to its rapid adoption by coffee growers around the world. Rather than inducing toxicity on physical contact, this chemical exerts fumigant effects, improving its efficacy against internal-feeding CBB (Vega et al., 2015). However, the use of endosulfan for CBB management has since fallen victim to the twin banes of conventional pesticides: deregistration and development of resistance. CBB resistance to endosulfan was first reported in 1989 (Brun et al., 1989), and as data accumulated regarding its negative health and environmental impacts, governments have acted to restrict or eliminate its use (Cone, 2010). Endosulfan has been banned in 70 countries, including the United States and the member states of the EU (Lubick, 2010; U.S. Environmental Protection Agency, 2010; Janssen, 2011; European Commission, 2016). Chlorpyrifos is in the process of meeting the same fate. The pesticide was banned in Europe in April 2016, and the United States was prepared to follow suit later that year—proposing to revoke tolerances for chlorpyrifos on any crop (U.S. Environmental Protection Agency, 2016)—before reversing itself following the 2017 change in the US federal administration, though the compound has been banned for use in California by 2021. The loss of these chemicals leaves coffee growers with a critical shortage of effective tools to manage CBB, which in some cases has translated into substantial economic losses. The ban on endosulfan in Brazil in 2013 brought an end to four decades of the pesticide's use against CBB in that country. Compounded by a heavy rain season and improperly maintained sanitation programs, this ban facilitated the largest CBB infestation in recent history in Brazil, resulting in coffee crop losses of 5%–30% (Szal, 2017).

The confluence of factors threatening the coffee industry—a changing climate creating conditions harmful to the crop and favorable to its pest, emergence of resistance to previously effective pesticides, and removal of entire classes of control agents by regulatory agencies—serves as a prime example of the challenges faced by agricultural producers in the modern age and highlights the need for more integrated, sustainable methods of arthropod pest control. Rather than continuing to rely on conventional chemical toxins to suppress or eradicate pest populations, many pest control developers have focused on the use of behavior-modifying chemicals, or semiochemicals, to reduce crop damage. Diverse semiochemical-based strategies for agricultural pest control have been developed and tested in the field: the use of attractant lures for mass trapping; the application of toxic baits in an attract-and-kill method; push–pull strategies, employing arthropod repellents to push pests away from crops and attractants to pull them toward a trap; and mating disruption, a technique that consists of the deployment of sex pheromones to interfere with male arthropods' ability to detect and orient to female mating partners. These techniques have been applied successfully in several cropping systems and against many insect pests including codling moth, *Cydia pomonella* L., tephritid fruit flies (e.g., oriental fruit fly, *Bactrocera dorsalis* Hendel, melon fruit fly, *Bactrocera cucurbitae* Coquillett, and Mediterranean fruit fly, *Ceratitis capitata* Wiedemann), and carob moth, *Ectomyelois ceratoniae* Zeller, but the use of insect semiochemicals as repellents in agriculture is comparatively rare.

Several compounds have been evaluated for potential repellency against CBB. A blend of two volatiles produced by *Coffea* plants, (*Z*)-3-hexenyl acetate and (*Z*)-3-hexenol, was found to repel CBB adults, as was (*E*,*E*)-α-farnesene, a sesquiterpene produced by CBB-infested coffee berries (Borbón et al., 2000; Castro et al., 2017; Vega et al., 2017). A field trial in Hawaii revealed that application of a bubble capsule formulation of (*E*,*E*)-α-farnesene reduced the number of CBB captured in traps baited with standard methanol/ethanol lures by 80%, compared with traps baited with the lure alone (Vega et al., 2017). However, the compound is highly reactive and can be degraded by photochemical and air oxidation, limiting its usefulness in coffee-growing regions, which are generally characterized by intense sunlight and frequent rainfall (Anet, 1969; Spicer et al., 1993; Vega et al., 2017). Nonhost compounds from several plants, including *Lantana camara* (L.), *Allium sativum* (L.), *Capsicum frutescens* (L.), *Moringa oleifera* (Lam.), *Piper* spp., and *Tilesia baccata* (L.) Pruski, have also demonstrated some repellency to CBB (Giraldo and Valencia, 2000; Henao, 2008; Santos et al., 2010; Santoro et al., 2011; Benavides and Góngora, 2015; Castro et al., 2017), but the extracts used are expensive and have generally failed to maintain their efficacy under field conditions (Schmutterer, 1990; Gurr et al., 2004; Nicholls, 2009; Castro et al., 2017).

Verbenone [4,6,6-trimethylbicyclo [3.1.1]hept-3-en-2-one], an antiaggregation pheromone of some species of *Dendroctonus* bark beetles has been developed as a patented alternative semiochemical-based repellent for the management of CBB in coffee (Mafra-Neto et al., 2016). Antiaggregation pheromones function to prevent the influx of new colonizing beetles on a particular substrate to a degree that would be unsustainable and thus harmful to the colony's members. In the case of CBB, higher concentrations of semiochemicals produced by larvae or emerged adults may signal overcrowding within an infested coffee berry, causing founder females to avoid that fruit (Vega et al., 2011). The antiaggregation semiochemicals of other bark beetles, including frontalin, (Z)-3-hexenol, methylcyclohexenone (3-methylcyclohex-2-en-1-one) and verbenone, have shown some degree of repellency toward CBB in several trials (Borbón et al., 2000; Vega et al., 2015; Vega, et al., 2017). A 2014 field trial in a Kenyan coffee plantation revealed that frontalin reduced CBB attraction to methanol-ethanol lures. Traps containing both the lure and frontalin captured 77% fewer CBB than traps with the lure alone (Njihia et al., 2014; Vega et al., 2017). CBB also has been shown to avoid verbenone and α-pinene, leading Jaramillo et al. (2013) to recommend intercropping coffee plants with those that produce repellent conifer monoterpenes (Vega et al., 2015). In laboratory and field trials, verbenone has been shown to act as a repellent to CBB and *Xylosandrus compactus* (Eichhoff), another beetle pest of coffee in Hawaii (Burbano et al., 2011; Jaramillo et al., 2013). To date, however, semiochemical-based repellent strategies have yet to be successfully commercialized or incorporated into area-wide control programs for CBB. The utility of insect repellents in general has often been limited by practical shortcomings, including relatively limited residual effectiveness in the field (a result of the volatility of most repellent AIs), loss of efficacy to environmental degradation, and the resulting need for frequent reapplications, challenges that would be especially daunting to coffee growers given the already high costs of their operations.

15.2.2 Repellent trials

To create a technically and commercially viable repellent for CBB in coffee, Mafra-Neto et al. (2016) incorporated verbenone into a wax-based controlled-release emulsion technology, specialized pheromone & lure application technology (SPLAT) manufactured by ISCA, Inc., Riverside, CA. The base SPLAT emulsion, which has been discussed at length in previous publications (Fan and Singh, 1989; Stelinski et al., 2006; Teixeira et al. 2010; Lapointe and Stelinski, 2011; Mafra-Neto et al., 2013, 2014), is a shear-thinning, thixotropic fluid that becomes thinner and less viscous when agitated by stirring or vibration but thickens when at rest. This characteristic makes the formulation amenable to the application using a broad range of manual or mechanized equipment and allows the user to designate a desired application rate or quantity. After the point source is applied, the aqueous component of the formulation evaporates, leaving the active components (AIs, additives incorporated to stabilize the formulation and/or refine release rates, etc.) securely attached to the application surface. Most SPLAT formulations become secure and resistant to rain within 3 h of application. The release rate of an AI from SPLAT can be modulated by altering one or more of several factors of the formulation, AI, or application process, including size of the AI molecule, AI concentration, composition, proportion, and order of addition of nonactive components of the matrix, temperature at varying stages of the formulation process, and size of the point source. Most SPLAT-based products developed to date have an active field life ranging from 2 weeks to 6 months.

Trials with CBB Repel (Fig. 15.1E and F) began with a series of laboratory assays to assess the response of CBB adults to varying doses of a prototype SPLAT formulation containing 20% verbenone: 0.1, 0.25, 0.5, 1.0, and 3.0 g (four replicates per dose). Ten coffee berries infested with a known number of beetles were placed in a Petri dish or watch glass in the center of a 50-L plastic container filled with soapy water to a depth of 5 cm. Containers were placed ≥20 m away from each other to prevent cross-contamination of any volatiles emerging from the prototype repellent, and each container was covered with a fabric net to prevent insects from escaping. Point sources of CBB Repel were applied in a small plastic holder affixed to the side of the 50-L container. Each container was checked for emergence of CBB from infested berries at 5, 15, and 30 min and 2, 5, 10, and 22 h after the start of the bioassay. Exposure to CBB Repel caused CBB to dislodge from the inside of the bean; after 22 h, significantly more adults emerged from fresh berries exposed to 0.1 g and 0.25 g doses than the control.

These results suggested an interesting mechanism by which CBB Repel might be used to reduce the impacts of CBB on coffee crops. The repellent could be applied at an appropriate interval before a planned insecticide application to flush the beetles out from the berries, increasing their exposure to the toxins and making pest suppression interventions more efficacious and efficient. Even in the absence of such insecticide treatment, exiting coffee berries would likely reduce the survival of female adult beetles simply by exposing them to the hazards of the environment (e.g., predation, microbial pathogenic infection, dehydration, or starvation) as a result of exposure to CBB Repel. Such a repellent might also dissuade female CBB from ovipositing in fruit in an area treated with verbenone, further reducing crop damage. To determine whether the effects observed in the lab could be induced at larger scales outside a contained environment, preliminary field trials were conducted in areas of Brazil experiencing high CBB pressure. In the first of these trials, a set of 10 × 10-m plots were randomly assigned to receive either no treatment (control) or a single application of CBB Repel. Rates of damage to coffee berries were assessed in each plot, revealing that treatment with CBB Repel significantly lowered berry damage in treated plots after 60 and 90 days compared to untreated control to a similar degree as treatment with chlorpyrifos.

However, because these data showed relatively low infestation rates across the experimental area (both treatment and control plots), limiting the certainty of any conclusions regarding the efficacy of the repellent, a second field trial was performed in coffee plantations (Red Catuai variety) at Alegre/ES region (20° 46′ 03″ S, 41° 35′ 23″W, elevation 687 m). Replicates were set up in a block design with six 10 m × 10 m^2 plots per block, and each plot spaced at least 25 m apart within the block. Four treatments were installed: (1) CBB Repel, (2) CBB Repel + chlorpyrifos, (3) chlorpyrifos, and (4) control. Ten 3-g point sources of CBB Repel were placed in the top half of the coffee plant once at the beginning of the trial. Positions of each treatment within a block were randomly selected. To evaluate the efficacy of each treatment, a monitoring trap (baited with 200 mL of 1:1 methanol: ethanol) was placed in the center of each plot and checked every 2 weeks for CBB captures. Chlorpyrifos was applied via backpack sprayer 1–3 days after trap installation and reapplied every 2 weeks until the end of the trial. Coffee berries were sampled 15 days after experimental setup and collected on a monthly basis thereafter. One hundred randomly selected coffee berries from 20 randomly selected plants were collected in each plot and evaluated for larval and adult infestation. Treatments remained in place until the end of the season when the crop was harvested. No significant difference was observed in CBB infestation rate (number of insects detected in damaged fruit) or percentage of berry damage among any of the evaluated treatments for the first, second, and third months of the trial. In the fourth month, however, coffee berries in plots treated with CBB Repel with or without chlorpyrifos had significantly fewer insects within damaged fruit than control plots. There were no significant differences among CBB Repel, CBB Repel + chlorpyrifos, and chlorpyrifos alone in the number of insects within damaged fruit. Coffee berries in plots containing CBB Repel, CBB Repel + chlorpyrifos, and chlorpyrifos alone displayed significantly less damage than control plots. No significant differences were observed in damage rate among CBB Repel, CBB Repel + chlorpyrifos, and chlorpyrifos alone. These data suggest that the CBB Repel prototype applied a single time at the beginning of the season has the same suppressive effect on CBB field populations and coffee berry damage as a biweekly chlorpyrifos spray regimen sustained throughout the season and that CBB Repel retains its capacity to alter female beetles' behavior for a period of at least 90 days in the field.

Following some adjustments to the formulation to improve flowability, rainfastness, repellency, and field longevity, Borges et al. proceeded to evaluate CBB Repel in a series of field trials representing a range of severity levels in CBB infestation, with or without additional insecticide treatments (personal communication, unpublished results). Two experimental areas, 2.7 and 4.5 ha in size, were selected in Cabo Verde, Minas Gerais, Brazil, to represent midlevel infestations based on monitoring for CBB infestation in fruit, while a third (Serrana, São Paulo; 1.8 ha) and fourth area (Arceburgo, Minas Gerais; 5.25 ha) represented low and high infestation scenarios. Three treatments were evaluated at each site: (1) CBB Repel alone, with no additional insecticide applications; (2) CBB Repel with a single application of chlorpyrifos, 3–5 days after the repellent application; and (3) control, represented by the grower's standard insecticide regime for CBB. Four replicates were made per treatment, with one replicate in each experimental area (Area 1 plots: 0.9 ha; Area 2 plots: 1.5 ha; Area 3 plots: 0.6 ha; Area 4 plots: 1.8 ha). For both treatments using CBB Repel, 3 kg of the formulation was applied per ha, distributed in point sources of 3 g each (1000 point sources per ha), contained within a plastic dispenser pinned on a branch in the middle of the coffee plant. This was done to preclude any damage to the crop, as some trials with CBB Repel prototypes have shown some phytotoxic effects when applied directly to plant leaves. CBB Repel was applied in November during the period between the blooming of the crop and the greening of the coffee beans (*'chumbinho'* stage). No reapplications were made throughout the trial period, which continued until the harvest. The effect of each treatment was assessed by dissecting 200 fruits randomly sampled from each plot and inspecting them for signs of CBB-induced damage, starting 15 days after treatment application and continuing monthly. Sampling took place on December 19, 2016, January 3, February 6, March 6, April 10, and May 6, 2017.

In Experimental Area 3, characterized by the lowest CBB-infestation rate of any site used in this trial, no significant difference was observed in the percentage of damaged fruit among the three treatments. Fruit damage was relatively low at this site, likely correlated to the sparse presence of the pest, ranging from 0.5% to 1.5%. Treatment effects were clearer at the sites experiencing higher infestation rates. In Areas 1 and 2, characterized by midlevel pest pressure, and in Area 4, the high-level infestation site, treatment with CBB Repel suppressed crop damage below the rate of damage detected in plots treated with the grower's standard insecticide regime. No significant difference was observed in the level of crop damage reduction between plots treated with CBB Repel alone and those treated with a combination of CBB Repel and chlorpyrifos, except in Area 4, where the latter treatment appeared more effective in reducing damage. This suggests that interventions using CBB Repel alone effectively suppress CBB crop damage in the field. The largest reduction in damage observed with the use of CBB Repel over the course of the trial was 8%, which at the contemporaneous price of raw coffee fruit (Hawaii, 2016/2017), would translate into a gross financial gain for the grower of nearly 2300 USD per ha.

In addition to the trials conducted in Brazil, CBB Repel has also been tested in Hawaiian coffee plantations in collaboration with Dr. Mark Wright's Insect Ecology and Integrated Pest Management Laboratory (University of Hawaii at Manoa). In June of 2019, a field trial was installed on two orchards at Dole Coffee

Plantations, Haleiwa, Oahu, comprising a total of 50.6 ha, separated by a distance of 1 km. These orchards had rows running in rough alignment with an N–S parallel, with standard planting distances. Trees were 2–3 m tall at the start of the trial. CBB Repel was applied on a grid pattern, with one 25-g point source applied to the trunk of a tree (1.5 m above the ground) every second row, with five treated trees in each row. This produced a treated area of approximately 40 × 38 m. Control plots under grower's best CBB pest control practices were established at the same size, located at least 11 m upwind from the treatment blocks. Eight CBB traps baited with a methanol–ethanol lure were installed at even spacing in each treated area (four traps per plot, two on one border, and two in the middle row). Thirty trees in the middle three rows of each treatment or control block were randomly selected for sampling, and one fruit-bearing branch per tree was subsampled. For each of these branches, all coffee fruit present were counted and classified as "attacked" or "not attacked" by CBB. Assessments of CBB trap captures and fruit evaluations were made every 2 weeks, with traps emptied and all captured insects transported to the laboratory for counting.

Although there was no significant difference in the numbers of CBB captured in traps (Treatment mean = 214.8 per trap; Control mean = 453.9 per trap, $F_{7,8} = 1.21$, $P = .39$) there were differences in the rate of berry infestation across the treatments. As with the previous trials, CBB Repel-treated plots exhibited significantly lower rates of berry infestation than control plots ($F_{1;1} = 171.36$, $P < .0001$). Rate of CBB infestation increased significantly over the duration of the trial ($F_{6;6} = 13.52$, $P < .0001$) but remained consistently lower in the CBB Repel treatment blocks than in the control. Some caution should be exercised in interpreting these results, as individual trees in each block were treated as replicates, rather than using means of infestation levels per block, which may have resulted in an underestimation of error. However, the patterns appear to be robust, with 2021 field results corroborating these trends.

15.2.3 Future directions

The effects of global climate change are making profitable coffee production more challenging on multiple fronts. Higher temperatures, if sustained for too long during the growing season, can stress coffee plants and reduce the quality of the crop. The warming climate is driving wild *Coffea* species—which provide cross-breeding stock vital to maintaining the health and resilience of *Coffea arabica* crops—to extinction and creating favorable conditions for diseases such as stem rust and insect pests such as CBB. The ban on endosulfan in many geographies and the tenuous regulatory status of chlorpyrifos has exacerbated the latter threat by creating a shortage of viable control options for CBB in coffee. These challenges will be particularly difficult to deal with for the small-holder farms (≤2 ha) that comprise the majority (~70%) of the coffee industry, a fact that is illustrative of a general trend in the socio-economic consequences of climate change: The worst of these impacts tend to fall heaviest on those of lower economic status who have the fewest resources to draw on to cope with them.

CBB has been shown to respond well to certain spatial repellent compositions. The studies reported here demonstrate that a strategy using CBB Repel, alone or in combination with traditional pesticide sprays, can effectively reduce crop damage inflicted by CBB. While the application of a "flush and kill" strategy—applying CBB Repel to drive the beetles out of the coffee berries where they live, feed, and breed, to make them more vulnerable to insecticide treatment—requires further evaluation in the field, this approach holds the potential not only to improve the efficacy of CBB control but also to enable more precise, targeted pesticide applications, reducing the amount of toxic chemical applied in the field over the course of a growing season. As the agronomic use of CBB Repel is being assessed under different conditions and geographies, its developers have commenced the registration of the product with the US EPA and in other countries of interest. Product launch of CBB Repel is expected to take place in key coffee-growing geographies starting in the next few years.

15.3 *Dendroctonus* bark beetles: mountain pine beetle, southern pine beetle, Douglas-fir beetle, and spruce beetle

15.3.1 Background

North American forests have had a difficult time of it lately, to put it mildly. At the peak of the fire season in the late summer and early fall of 2020, there were >8200 wildfires burning across California alone, consuming over 1.6 million ha of forest—more than double the previous record set in 2018—killing 31 people, forcing >96,000 to evacuate, and destroying >8400 structures (as of October 7, 2020) (Peñaloza, 2020; Stelloh, 2020). Factors contributing to these destructive events include the years-long drought affecting the area, a particularly dry 2019/2020 winter, an abundance of fuels resulting from 100 years of fire suppression programs, record heat waves, and a series of freak lightning storms accompanied by little to no rain (Rogers, 2020; Stelloh, 2020). All available data indicate that these challenges to forest health, diversity, and productivity are expected to continue for the foreseeable future, suggesting that the severity of the 2020 fire season across the American West may become more of the new norm (Vose et al., 2018). Since global temperatures began to be monitored and recorded in 1880, the 10 hottest years on record have all occurred since 1998 (Rogers, 2020). As this pattern continues, heat waves are expected to become more intense and more common, and drought risk and severity are projected to increase across the Western United States, particularly in the Southwest, even as heavy precipitation and flooding become more frequent in other regions (Cook et al., 2018; Gray and Merzdorf, 2019).

In addition to these direct impacts on forest ecosystems, climate change is facilitating another threat to North American coniferous forests in the form of mass outbreaks of tree-killing bark beetles. While most bark beetle species (Coleoptera: Curculionidae) feed on dead or dying trees, several are aggressive tree killers that colonize living trees to feed on their nutritive phloem tissues and to gain a place of shelter to lay eggs and rear their brood. Conifer trees have evolved a number of natural defense mechanisms to deter attacks by wood-boring insects, specifically by producing large volumes of resin to drown or encapsulate attacking insects; and by altering the chemical composition of their tissues to release insecticidal compounds (Franceschi et al., 2005; Hain et al., 2011). Systemic resistance can also be induced so that future attacks are more easily defended against. Aggressive bark beetles, such as MPB, southern pine beetle (SPB), *Dendroctonus frontalis* Zimmerman, Douglas-fir beetle (DFB), *Dendroctonus pseudotsugae* Hopkins, and spruce beetle (SB), *Dendroctonus rufipennis* Kirby, have in turn evolved a sequence of semiochemical-mediated mass attack behaviors to overwhelm these defenses by sheer numbers.

Following a similar trend as forest wildfires, outbreaks of aggressive bark beetles have become more common in recent decades, resulting in the deaths of billions of conifer trees over millions of ha of North American forests (Bentz et al., 2009, 2010; Mattson et al., 2019; Foote et al., 2020). Recent outbreaks of MPB (Fig. 15.2A), which attacks several North American pine species (lodgepole pine, *Pinus contorta* Douglas ex Loudon, ponderosa pine, *Pinus ponderosa* Lawson & C. Lawson, sugar pine, *Pinus lambertiana* Douglas, and whitebark pine, among others) (Furniss and Carolin, 1977; Keane, 2001; Wood

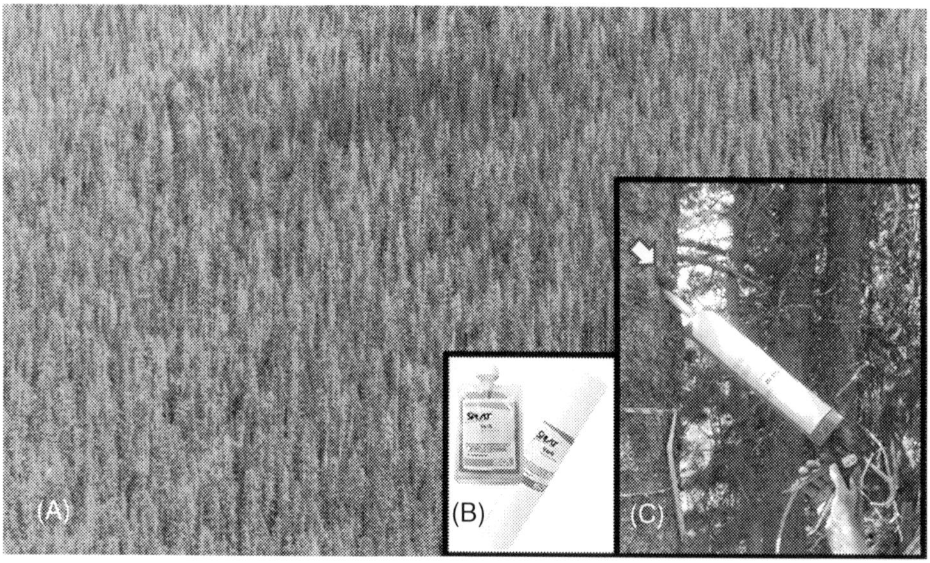

FIG. 15.2 (A) Example of a study area that has been devastated by bark beetles. SPLAT Verb can be used to protect single trees or stands of trees from attack by bark beetles repelled by verbenone. The product is sold in (B) multidose caulking gun cartridges or in a single application squeeze-bag. (C) A dollop of SPLAT Verb (indicated by white arrow) is applied directly to a tree trunk using a caulking gun. The dollop eventually biodegrades, eliminating the need for crews to return to the site to retrieve plastic pouches. Credits: (A, B) ISCA Technologies, Inc.; (C) Fettig. *SPLAT*, Specialized Pheromone & Lure Application Technology.

et al., 2003; Schwandt and Kegley, 2004), have been responsible for the loss of billions of trees across over 27 million ha of western North American forests (Fettig et al., 2020). Such outbreaks typically begin with the colonization of trees that are physically compromised, stressed by injury, overcrowding, disease, or old age, but as the density of the beetle population grows, even healthy trees may be infested. The most severe outbreak of MPB on record, occurring in Western Canada, affected over 20 million ha of pine forests (McIntosh and Macdonald, 2013; Walton, 2013; Dhar et al., 2016).

Southern pine beetle (SPB) is the most destructive pest of southern yellow pine species, particularly in forests across the southeastern United States, Mexico, and Central America (Fig. 15.3A) (Billings, 2011b). Outbreaks of this species are not only destructive on a large scale—according to USDA Forest Service estimates, over the course of eight outbreaks from 1973 to 2004, SPB caused a total of 3.57 billion USD in damage—but are also unpredictable (Waldron, 2011). SPB populations can remain low for long periods, then explode in outbreaks that kill thousands of trees and can last 2–3 years, only coming to an end when there are no suitable host trees left or environmental conditions render the affected region inhospitable to the beetles (Hain et al., 2011). Currently, the most severe impacts of SPB outbreaks are felt in the Southern United States, where ~80.9 million ha of forest is nearly a third (30%) composed of pine species susceptible to infestation by SPB. However, outbreaks have recently occurred much farther north (Guldin, 2011). Since 2010, outbreaks in New Jersey, where SPB-susceptible pitch pines, *Pinus rigida* Mill., are the dominant pine species (Crocker, 2012), have impacted ~11,000 ha of forests (Ferguson et al., 2013; Strom et al., 2015). In 2015, SPB was detected along a front >320 km long, stretching from

FIG. 15.3 (A) A pine forest infested with southern pine beetle, *D. frontalis*. The infestation is directional, expanding from the center to the borders. Attacked trees in the center are gray, dead, and defoliated, whereas canopies fade and trees show different senescent colors, from brown, red, and yellow to light green approaching the edges of the infestation. (B) SPB Repel dollop (white arrow) being applied to a pine tree. Credits: (A) Tanner; (B) Fettig. *SPB*, Southern pine beetle.

Long Island to Martha's Vineyard and Cape Cod (Schlossberg, 2016). Additional infestations have been found in Bear Mountain State Park in New York, in Rhode Island, Massachusetts, and Connecticut (Schlossberg, 2016).

DFB and SB are the most destructive pests of the trees they are named for, Douglas-fir (*Pseudotsuga menziesii* Mirb.) and spruce (*Picea* spp.) (Fig. 15.4A) (Furniss and Carolin, 1977). Occurring from Southern Canada to Northern Mexico, DFB was once fairly limited in its impact on North American forests, its populations maintained at low levels by the availability of stressed, dying, or dead trees in a given landscape and only reaching outbreak levels in the aftermath of destructive events such as windstorms or wildfires (State of Colorado, 2016). DFB-induced tree mortality is now on the rise in many locations, even in some of the healthier forests of the United States (Dickson, 2019; Harper, 2019). In the Pacific Northwest, for example, historical annual tree mortality rates associated with DFB infestations fluctuated between 1% and 5% but have now risen to 30%–50% (Bible, 2001; Larson et al., 2008; Freeman, 2017). SB is the most broad-ranging bark beetle in North America and the leading cause of death for mature spruces (Holsten et al., 1999; Raffa et al., 2008; Mietkiewicz et al., 2018), colonizing all spruce species within their range, including white spruce, *Picea glauca* Moench, Sitka spruce, *Picea sitchensis* Bong., and Engelmann spruce, *Picea engelmannii* Parry ex Engelm., and destroying 333–500 million board feet of timber each year (Oatman, 2013). As with DFB, research suggests that current SB outbreaks are causing unusually high rates of spruce mortality, nearly 90%, compared to the historically typical 20%–30% (Salvail, 2019). SB has infested more than 49,370 ha of spruce forests in Utah (3 million trees killed) and caused losses of >100 million board feet of timber in Arizona, 2 billion in Alaska, 25 million in Montana, and 31 million in Idaho (Holsten et al., 1999). Outbreaks continue in British Columbia, Alaska, Colorado, Idaho, New Mexico, Utah, and Wyoming (Hansen et al., 2019; Murkowski, 2019). According to USDA Forest Service estimates, >364,000 ha of Alaskan spruce forest has become infested with SB on the Kenai Peninsula and in the Susitna River Valley, and nearly 204,000 ha in Matanuska-Susitna, in

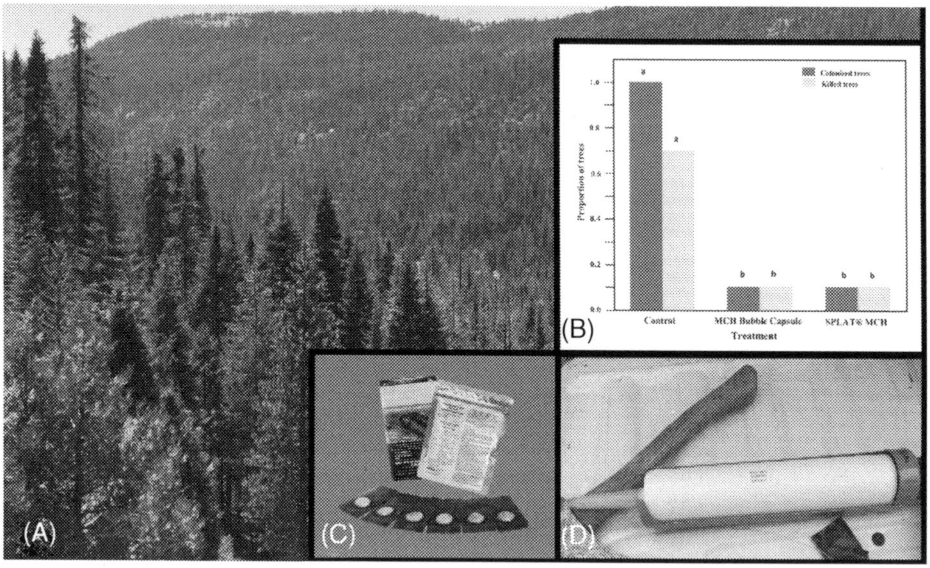

FIG. 15.4 (A) Douglas-fir forests devastated by bark beetles. (B) A higher proportion of Douglas-fir trees were colonized or killed when untreated (control), than when treated either with MCH bubble capsules, or the prototype SPLAT MCH formulation. (C) Commercial ISCA/Contech plastic MCH bubble capsules. (D) Prototype biodegradable MCH Repel. Credits: (A–C) ISCA Technologies, Inc.; (D) Fettig. *MCH*, 3-methylcycolhex-2-en-1-one.

addition to the 1.2 million ha of spruce forest damaged in the last major SB outbreak, which ended in 2004 (Murkowski, 2019; O'Malley, 2019).

The impacts of bark beetle outbreaks are not limited to the destruction of timber. They also complicate long-term forest stewardship by disrupting logging plans and change tree stocking levels and age distributions. Forests enduring beetle infestations also may be more likely to burn, as beetle-killed trees have higher emissions of flammable terpenes from foliage and create an abundance of combustible fuels (State of Colorado, 2016; Giunta et al., 2016). Widespread tree mortality increases soil erosion, reduces soil stability, and impacts water yields by eliminating transpiration from dying and dead trees and by exposing snow to rapid melting in the spring, consequently increasing water inputs to rivers and lakes and causing flooding (Holsten et al., 1999; State of Colorado, 2016).

As is the case with the effects of global climate change in general, the influence of rising temperatures on bark beetle outbreaks and their impact on forest health are complicated, varying with geography and tree species. In some cases, prolonged drought increases the risks of severe bark beetle outbreaks by placing trees under chronic stress, reduces their general vigor, and impedes their capacity to repel beetle attacks through resin outflow (Waldron, 2011; Kolb et al., 2016; Freeman, 2017; Mietkiewicz et al., 2018; Negrón and Popp, 2018). Shorter, warmer winters may increase overwintering survival in more locations, allow beetle populations to move into higher elevations or more northerly territories as shown by SPB, and also may lead to an additional generation each year (Miller and Werner, 1987; Ungerer et al., 1999; Bentz et al., 2010; Hain et al., 2011; Hart et al., 2017). Other impacts of current climate trends include accelerated insect development, longer, and more unpredictable flight periods, larger populations of fertile female beetles, and larger insect body size (Negrón and Popp, 2018; Lill, 2019; Puikkonen, 2019). These changes may also alter host selection in a destructive way, as unusually large beetle infestations may result in attacks on younger, smaller-diameter trees than would ordinarily not be targeted, reducing the number and size of the next generation of mature trees (Raffa et al., 2008; Hart et al., 2014; Mietkiewicz et al., 2018).

Attempts to control bark beetle pests are hampered by the scale over which outbreaks occur, as well as by the remote, sensitive natural environments constituting susceptible forests. Removal of infested material, stand thinning, creation of felled-tree buffer zones, and preventative insecticide treatments may be effective against some species if implemented properly, but these techniques may be too costly and labor-intensive to be applied at the scale of a bark beetle outbreak (Billings, 1980; Mitchell et al., 1983; Fiddler et al., 1989; Amman and Logan, 1998; Johnstone, 2002; Billings, 2011a; Fettig et al., 2014; Fettig and Hilszczański, 2015). Insecticide treatments have a further disadvantage in terms of the negative impacts they can have on the environment and their negative connotations in the minds of the public. One of the most commonly employed chemical agents for bark beetle control, the carbamate insecticide, carbaryl (1-naphthyl methylcarbamate), must be applied as a bark drench over each treated tree to the point of runoff, to a height where the bole diameter is too small to sustain beetle larvae; this level of coverage typically requires 15–30 L of insecticide solution per tree (DeGomez et al., 2006; Fettig et al., 2006; Fettig et al., 2008; Fettig et al., 2013). Because of the risks of exposure to this toxic chemical (headaches, memory loss, muscle weakness, and cramps, vomiting, bronchoconstriction, blurred vision, convulsions, coma, and respiratory failure), carbaryl applications require all application personnel to wear personal protective gear and use special application equipment that may be difficult or impossible to move across rugged forest terrain (Bond et al., 2016). Carbaryl treatments are not effective against all bark beetle species. SPB, for instance, can rapidly break down the insecticide into its metabolites and then excrete them (Ragenovich and Coster, 1974; Zhong et al., 1994; Fettig et al., 2009; Cancelliere, 2016a, 2016b).

The search for more effective, sustainable methods of bark beetle control has found fertile ground in the same set of chemical compounds that enable these insects to cause such widespread and substantial damage: the semiochemicals that mediate mass attack behavior (Seybold et al., 2018; Gillette and Fettig, 2021). To successfully colonize a living tree, aggressive bark beetles must recruit enough insects (hundreds to thousands, depending on the species, size, and vigor of the host tree) to the selected host to overwhelm its defenses and penetrate the phloem beneath the outer bark (Gara et al., 1965; Hain et al., 2011). This beetle battle cry, as it were, takes the form of aggregation pheromones released by the insects, such as (*Z*)- and (*E*)-verbenol, frontalin, and *exo*-brevicomin (Pitman et al., 1969; Conn et al., 1983; Miller and LaFontaine, 1991; Billings, 2011b; Sullivan, 2011), amplified by the release of certain kairomones from the tissues of the plant being colonized (e.g., α-pinene, myrcene) (Miller and Lindgren, 2000; Billings, 2011b, Sullivan, 2011). Toward the end of a successful mass attack, once the tree is fully colonized, beetles begin releasing antiaggregation pheromones to deter incoming beetles, which might otherwise result in overcrowding that would be harmful to the larval brood within the tree. Antiaggregation pheromones are interpreted by host-seeking beetles as an indication that a target tree is already fully occupied and therefore unsuitable for further colonization (Seybold et al., 2018; Gaylord et al., 2020).

The functionality of antiaggregation pheromones offers a valuable opportunity to manipulate the behavior of bark beetles in such a way as to reduce the destruction they inflict on forests. Verbenone is the most commonly deployed

FIG. 15.5 (A) Plastic pouch containing verbenone nailed to a tree. (B) A verbenone pouch being nailed to a tree trunk. (C) A verbenone pouch stapled to a young fir tree. Credit: Fettig.

antiaggregation pheromone in bark beetle management, particularly against MPB, and has been used by forest scientists and managers for decades to protect trees and forest stands in high-value or high-visibility locations, such as residential or recreational areas and seed orchards (Amman et al., 1989; Ross and Daterman, 1995; Seybold et al., 2018; Fettig et al., 2020). Multiple verbenone-based repellents are registered in the United States and Canada, including plastic bubble capsules, pouches, and pheromone-impregnated flakes. While these products have delivered effective control in many cases, treatment failures are common (Progar et al., 2014; Seybold et al., 2018), attributed to inconsistent or insufficient release of the pheromone (Bentz et al., 1989), photoisomerization of verbenone to chrysanthenone (2,7,7-trimethylbicyclo[3.1.1]hept-2-en-6-one, a compound with no known impact on bark beetle behavior, therefore of no use as a repellent) (Kostyk et al., 1993), and a limited range of repellency, especially when beetle populations are high (Miller, 2002). In addition to these technical shortcomings, device-type repellents like pouches and bubble caps (Fig. 15.5) present practical challenges, as they must be removed from many forested areas after the beetle flight period ends. Plastic pheromone flakes may be broadcast more quickly over larger areas than discrete devices and are less conspicuous in the environment, but they tend to have short release periods (Seybold et al., 2018; Gillette and Fettig, 2021) while introducing plastic pollutants to treated environments.

15.3.2 Repellent trials

An alternative verbenone-based repellent was developed starting in 2011: SPLAT Verb (Fig. 15.2B and C), a flowable controlled-release formulation designed to improve the efficiency of application and ensure a more consistent and effective release rate of the pheromone AI, so that a single application of the repellent would be sufficient to protect treated trees from bark beetle colonization for an entire flight period (Fettig et al., 2015). Because SPLAT Verb is comprised entirely of food-safe ingredients and biodegrades within a year of application, point sources of this formulation do not require retrieval from the field as do plastic verbenone products (Fettig et al., 2020). SPLAT Verb was first tested for its capacity to protect lodgepole pine from attack by MPB, beginning with a pilot study conducted in July 2011 in Bridger-Teton National Forest, Wyoming. A prototype SPLAT formulation containing verbenone was applied as four large point sources spaced evenly around the bole of 21 randomly selected healthy, non-MPB infested lodgepole pines. Twenty-one pines were treated with roughly 533 g of the SPLAT formulation (32 g verbenone/tree), and six were treated with 650 g of formulation (39 g verbenone/tree). Thirty additional healthy lodgepole pines were left untreated and used as controls. All trees used in the experiment, both SPLAT Verb-treated and untreated control, received a commercial MPB tree bait, manufactured by Contech Inc. (Delta, BC, Canada, since acquired by ISCA, Inc. Riverside, CA, US), to ensure sufficient pest pressure to evaluate any treatment effect. One year after the treatment, control and treated trees were assessed for crown fade, an indication of a successful mass attack and tree death. Results of the assessment indicated that the SPLAT Verb prototype delivered a high degree of protection against MPB. While 28 out of 30 untreated control trees were killed by MPB, a mortality rate of 93.3%, all SPLAT Verb-treated trees remained alive and healthy 1 year after treatment.

While the level of efficacy demonstrated by the initial Bridger-Teton study was encouraging, the high AI loading rate was cost-prohibitive for a viable commercial product. The prototype formulation, therefore, underwent further refinement to reduce its content of verbenone and ensure a steady, consistent release rate, followed by an additional study in the same area. This second trial was conducted according to a similar protocol as the initial study but with an AI application rate equivalent to the dose of verbenone used in pouch formulations: 7 g of verbenone AI per tree. The reduction in AI loading rate showed no significant impact on the level of tree protection provided. As with the first trial, 100% of SPLAT Verb-treated trees survived with no signs of MPB colonization 1 year after application of the repellent, while 93.3% of untreated trees died (Fettig et al., 2015). Assessments of trees growing near SPLAT Verb-treated trees also displayed a reduced frequency of MPB attack. No trees growing within 11 m of SPLAT Verb-treated trees were found to have been successfully colonized, compared to 61 infested trees growing within 11 m of control trees (Fettig et al., 2015).

A small stand-level trial was performed on a series of 0.4-ha plots in the Caribou Targhee National Forest, Idaho. A SPLAT Verb formulation containing 10% verbenone was compared to Contech verbenone pouches (7 g AI) applied according to manufacturers' instructions along a 9.1 × 9.1 m grid, and untreated control. SPLAT Verb was applied in such a manner as to achieve an equivalent AI rate/area as the verbenone pouch treatment. At this equivalent rate, SPLAT Verb delivered more effective protection of treated stands than verbenone pouches compared to the untreated control (Fettig et al., 2015). When a final assessment of tree mortality on the experimental plots was made in July 2013, eight trees in SPLAT Verb-treated plots had died as a result of the MPB attack, compared to 15 trees killed in pouch-treated plots and 49 trees killed in untreated plots. In another small-scale stand study in the Beaverhead-Deerlodge National Forest, Montana, fewer trees were colonized by MPB on 0.41-ha experimental plots treated with 2.5 kg and 3.5 kg of SPLAT Verb [250 g and 350 g of verbenone] and the 7-g verbenone pouch [50 pouches/plot, 350.0 g of verbenone/plot] compared to the untreated control. No significant difference was observed between 1.5 kg of SPLAT Verb [150 g of verbenone] and the untreated control. Fewer trees were killed by MPB on experimental plots treated with 1.5, 2.5, and 3.5 kg of SPLAT Verb and the verbenone pouch compared to the untreated control. No other significant differences were observed among treatments. Collectively, these data suggest that lower doses of SPLAT Verb can be used for tree protection than previously considered and at a substantial cost savings that could make this strategy economically viable for area-wide management of MPB (Fettig et al., 2020). A third study on SPLAT Verb (7 g verbenone/tree) was conducted in Darby, Montana, on ponderosa pine but was confounded by the presence of two additional bark beetle pests at the trial site, western pine beetle, *Dendroctonus brevicomis* LeConte, and pine engraver beetle, *Ips pini* Say (Fettig et al., 2016).

SPLAT Verb was registered for use against MPB in lodgepole pine in 2014 and went on to become the first semiochemical-based control product for bark beetles to be successfully commercialized in the past 20 years. It has since been tested in several other pine species, including sugar pine in Stanislaus National Forest and Yosemite National Park, California—where SPLAT Verb provided complete or nearly complete protection of treated trees (<5% mortality in Stanislaus, 0% mortality in Yosemite compared to ≥60% mortality in untreated, baited control trees (Fettig et al., 2016)—and whitebark pine, a foundational species that provides multiple crucial functions in subalpine ecosystems, including wildlife habitat and watershed protection (Progar et al., 2021). Whitebark pine densities have declined by as much as 90% in some locations as a result of MPB infestations and white pine blister rust to the extent that it has been considered for inclusion under the Endangered Species Act (Progar et al., 2021). Small stand studies were conducted in whitebark pine in the Southern Cascade Range in northern California and in the Strawberry Mountain Range in central Oregon (Progar et al., 2021). Survival rates were 4%–7% higher in plots treated with SPLAT Verb, compared to 2%–3% higher with verbenone pouches. SPLAT Verb and verbenone pouches delivered roughly equivalent levels of protection in small and medium trees at the Oregon trial site, but for large trees, survival was significantly higher in plots treated with SPLAT Verb (93%) than in those treated with pouches (82%) (Progar et al., 2021).

The developers of SPLAT Verb advanced similar repellents for other bark beetle pests, including SPB, DFB, and SB. SPB has proven a particularly difficult pest to manage by semiochemical means, as most studies have indicated it does not respond to verbenone in the same way as MPB. The reason for this difference can likely be found in the chemical nature of verbenone, which occurs in two enantiomeric forms. The "R" enantiomer of verbenone, also written as (–)-verbenone, is the more common of the two and is easier and less expensive to synthesize than its opposite, "S" or (+)-verbenone but has not been shown to be an effective repellent for SPB. Multiple studies have demonstrated that SPB responds more readily to verbenone blends containing higher proportions of the (+) isomer, which costs tens of thousands of USD per kg, compared to low hundreds for a kg of the (–) isomer (Mafra-Neto 2020, unpublished data). Commercial high-volume synthesis of verbenone typically produces only 9% (±3%) of the (+) isomer, while most verbenone repellent formulations contain about 12% (+)-verbenone (Bunt et al., 1980; Salom et al., 1995), far short of the 30% that the current literature suggests is the minimum necessary for an effective SPB repellent (Strom and Clarke, 2011). In their efforts to develop an analogous product to SPLAT Verb for SPB (Fig. 15.3B), ISCA scientists first developed a novel synthetic method for production of high purity (+)-verbenone that reduced the cost of the final product to 1200 USD/kg, 29x less expensive than the current price of purchase (35,000 USD/kg). Further optimization of the synthetic pathways has allowed ISCA to produce (+)-verbenone at costs similar to that of (-)-verbenone.

Work to develop a semiochemical repellent for SPB is ongoing, but several field trials with

SPLAT formulations containing (+)-verbenone have been completed. A trapping study in the Bienville National Forest in Mississippi showed significant inhibitory effects of the (+)-verbenone repellent on the SPB population. Significantly fewer SPB were trapped in baited traps treated with the repellent than in control traps (baited with no repellent). Three spot-suppression trials, evaluating the repellent's capacity to disrupt growing SPB infestations, were also conducted in Mississippi, the first in the Bienville National Forest and the second and third in the Homochitto National Forest. At Bienville, five SPB infested spots were treated with SPLAT containing (+)-verbenone [17.5 g/tree applied to uninfested trees in a buffer creating a barrier (avg. 3 treated trees deep) separating the area of spot growth from the area without infested trees, where the growth of SPB infection was predicted to progress], while another five spots were left untreated. Three out of five repellent-treated spots exhibited significantly reduced infestations. Two treated spots were considered completely successful (0 untreated trees attacked beyond the buffer zone), and the third was considered partially successful (<10 trees colonized). More than 10 trees were infested in the remaining two spots, but in both cases, the breach of the barrier that led to treatment failure occurred at its narrowest points, where there was a gap in treated trees or the barrier was only one or two trees deep. This result can be interpreted as an indication that the cause of the treatment failure at these sites may be attributable to faulty application rather than an ineffective repellent formulation.

Similar problems were encountered in the first trial conducted at the Homochitto Forest, where only one repellent treatment was determined to have been completely successful in SPB spot disruption (three considered partially successful, and one considered a treatment failure); as with the Bienville trial, all breaches of the repellent barrier occurred at its narrowest points. A second trial at Homochitto addressed the application difficulty by increasing the depth of the barrier to a ubiquitous minimum of three trees around the spot head and all currently infested trees. Of the three SPB infestations treated, two achieved partially successful spot disruption, while a third achieved completely successful disruption. According to a mathematical model developed by one of ISCA's collaborators on this project, Dr. Stephen Clarke at the USDA Forest Service, to predict the likelihood and direction of expansion of a given SPB spot, 60% of the spots treated in these trials were either slowed or contained by the presence of the (+)-verbenone barrier.

To develop an effective repellent for DFB and SB, a different semiochemical AI was required, 3-methylcycolhex-2-en-1-one (MCH), which serves an analogous function in the chemical ecology surrounding a mass attack on trees in these two species as verbenone does in MPB and SPB infestations. Repellent bubble capsules that release MCH have been commercially available for tree protection against DFB since 1999 and are used to treat several thousand ha of high-value forest annually at a cost of 200 USD per ha (2 USD per tree) (Ross et al., 2001, 2015; Brookes et al., 2016; Hansen et al., 2019). The cost of this repellent treatment is significantly lower than for carbaryl treatments and significantly reduces environmental impact (Gillette and Munson, 2009; Strom and Clarke, 2011). However, as with verbenone pouches, MCH bubble caps are not suitable for larger-scale management programs, as their application and post-season removal are labor-intensive (Fig. 15.4C). They are also not consistently effective against SB (Zogas, 2001). MCH Repel (Fig. 15.4D), a controlled-release formulation similar to SPLAT Verb, incorporating MCH, was developed to correct these issues and field tested against DFB in Idaho and New Mexico (Foote et al., 2020) and against SB in Wyoming and Alaska. A field trial performed in the Boise National Forest in Idaho demonstrated that both MCH Repel and MCH bubble capsules significantly reduced DFB attacks on Douglas-fir trees and subsequent levels of tree mortality compared to the untreated control and that there was no significant difference between MCH bubble capsules and MCH Repel (Fig. 15.4B) (Foote et al., 2020).

MCH Repel was also tested in two stand-level trials in the Boise National Forest and the Cibola National Forest in New Mexico. Results from the New Mexico trial indicated that MCH Repel was as effective as MCH bubble capsules. Both MCH treatments significantly reduced the proportion of trees attacked and killed by DFB compared to untreated control plots, with no significant difference between MCH treatments (Foote et al., 2020). The Idaho trial found that neither MCH treatment significantly reduced DFB attack and that only MCH bubble capsules significantly reduced tree mortality due to DFB attack compared to untreated control plots, though no significant difference in tree mortality was observed between MCH Repel and bubble capsule treatments (Foote et al., 2020). As this contradicts the findings of multiple studies on MCH bubble capsule efficacy against DFB, the authors of the study attributed this result to the use of a strong three-component attractant in the trial, which has not been used in previous studies (Ross and Daterman, 1994, 1995; Ross et al., 1996, 2002; Ross and Wallin, 2008; Brookes et al., 2016; Foote et al., 2020).

Repellents combining MCH Repel with nonhost volatiles were tested against SB in 2019 in white spruce in Alaska and in Engelmann spruce in Wyoming. Preliminary results from Wyoming suggested that MCH Repel + acetophenone + (*E*)-2-hexen-1-ol and (*Z*)-2-hexen-1-ol, and MCH Repel + AKB [linalool, β-caryophyllene and (*Z*)-2-hexen-1-ol] were effective for protecting individual trees and small groups of trees from mortality attributed to SB (Hansen et al., 2019, Fettig et al., USDA Forest Service, unpublished data). The best performing MCH Repel prototype with nonhost volatiles, applied at 3.5 and 7 g MCH per tree, reduced spruce mortality due to SB attack in Wyoming to 0 and 8%, compared to 100% mortality in untreated trees at the same site. However, no evaluated MCH Repel treatment was effective in Alaska (Fettig et al., USDA Forest Service, unpublished data). In the summer of 2020, the manufacturers of SPLAT Verb and all other bark beetle repellents described in this section, ISCA, Inc., obtained its latest Phase II grant from the USDA Small Business Innovation Research (SBIR) program to continue research and development on SPLAT-based MCH repellents for DFB and SB, which will include continued field trials to assess the capacity of these semiochemical techniques to deliver effective protection of North American forests at the individual tree, small stand, and area-wide scale.

15.3.3 Future directions

As with CBB in commercial coffee, the warming global climate is creating conditions more challenging to the health of North American forests and more favorable to their most damaging insect pests, bark beetles in the *Dendroctonus* genus. More frequent and severe droughts and other extreme weather events place conifer trees under chronic stress, leaving them more vulnerable to insect attacks. Aggressive *Dendroctonus* species are expanding their territories northward and into higher altitudes, enabled by higher average temperatures to survive the winter at higher rates. Warmer temperatures also accelerate insect development and foster larger, more fertile populations of female beetles, allowing outbreaks to multiply faster. Some species are altering their attack behaviors as they move into new areas. Infestations of SPB in the northern United States do not progress with the same directionality as in their native southern range, making them less predictable and harder to treat effectively. As these shifts continue and an ever-increasing area of forest biomass is lost to bark beetle outbreaks, deforestation, and widespread wildfires, some forests may be converted from carbon sinks (systems that absorb more carbon than they release) to carbon sources. A critical resource for mitigating the effects of anthropogenic carbon emissions could therefore become a worsening factor for climate-related problems.

Forests represent both a challenge and an opportunity for semiochemical-based pest management technologies. As most pheromones and other semiochemicals used for insect control are nature-identical and selective in their mode of action, they are considered safer for people and other nontarget species, and therefore better suited for use in sensitive, diverse forest environments. This is in contrast with agricultural pest control applications, where semiochemical techniques must compete directly with cheaper, faster-acting broad-spectrum pesticides. However, the field environment over which bark beetle outbreaks take place poses a significant challenge for a pest control application of any kind. Previous bark beetle repellents have failed to deliver consistent tree protection due to difficulties achieving a sufficient release rate and period, degradation of the AI through environmental exposure, and insufficient range of active repellency.

The repellent products described here, SPLAT Verb, SPB Repel, and MCH Repel, have completed field trials demonstrating the resolution of many of these previous shortcomings in bark beetle control products. SPLAT Verb was registered in 2014 following field trials demonstrating effective repellency in lodgepole pine and has since been tested in multiple additional pine species, including the high-altitude foundational species, whitebark pine. Mafra-Neto et al. are also engaged in field trials exploring aerial application methods for SPLAT Verb using fixed-wing aircraft, helicopters, or unmanned aerial vehicles to apply the repellent to larger and/or more remote areas than can be treated effectively using manual application. Evaluations of SPB Repel demonstrated that a repellent using (+)-verbenone partly or completely disrupted SPB infestations, though in several cases, the treatment buffer zone proved too narrow to entirely prevent spot expansion. Continuing development on SPB Repel will focus on determining how to most effectively disrupt the linear expansion of SPB infestations in the southern regions of their range and evaluation of new methods of application to combat the atypical and unpredictable SPB outbreaks observed in the north. MCH Repel achieved an equivalent level of tree protection against DFB infestation as MCH bubble capsules in field trials conducted in New Mexico and Idaho, though a stand-level trial in the Boise National Forest revealed no significant treatment effect on tree mortality, likely due to the potent lure deployed in the trial. A series of prototype MCH Repel formulations containing nonhost volatiles delivered effective protection of Engelmann spruce trees in Wyoming but not of white spruce trees in Alaska.

Because the MCH Repel formulation is biodegradable, breaking down within 1–2 years, it will not require postflight removal, giving it a decided advantage where such retrieval is costly and labor-intensive. Continuing trials on MCH Repel will include field studies assessing the efficacy of MCH Repel against DFB and SB at larger scales and in a wider range of habitats across the lower 48 states, as well as field trials in Alaska with lower SB infestation rates. Mafra-Neto et al. expect full registration and product launch of MCH Repel as a control product for DFB with the US EPA within the next year.

15.4 Ambrosia beetles: redbay ambrosia beetle, black stem borer, and polyphagous shot hole borer

15.4.1 Background

The first section of this chapter discussed the use of a semiochemical-based repellent against a major pest of commercial agriculture, the CBB, whose impacts are expected to increase as a result of global climate change, while the second deals with a group of insects exerting expanding influence on North American forest ecosystems. The subjects of this section, ambrosia beetle pests, have harmful impacts on both forestry and commercial agriculture. Ambrosia beetles are in the same subfamily as bark beetles (Scolytinae), and display some of the same attack behaviors and ecology, boring beneath the bark of host trees to create galleries for themselves and their offspring. However, unlike bark beetles, ambrosia beetles do not directly feed on the wood tissues themselves; instead, they use these galleries as a substrate for the cultivation of the symbiotic "ambrosia" fungi they rely upon for sustenance (Batra, 1966, 1967; Beaver, 1989; Castrillo et al., 2011; Ploetz et al., 2013). Adult ambrosia beetles introduce the fungi to the interior tissues of their host tree, providing them with a suitable sheltered environment and an abundant food supply. Typically, ambrosia beetle species in their original native habitats limit their attacks to dead or dying trees, serving as nature's recyclers. For reasons not completely understood, some invasive ambrosia beetles also may infest healthy living trees (Kuhnholz et al., 2001; Reding et al., 2010; Castrillo et al., 2011; Ploetz et al., 2013; Wilson et al., 2014). In these instances, the transmitted fungus disrupts the transport of water and nutrients throughout the plant's vascular system, inflicting serious and, in some cases, lethal damage. Some species also transmit secondary pathogenic fungi or bacteria (Beaver, 1989; Whitney, 1982; Hijii et al., 1991; Castrillo et al., 2011).

The RAB is an example of an introduced ambrosia beetle species attacking living and healthy hosts in its new environment (Fig. 15.6B and C). RAB transmits the deadly fungal pathogen *Raffaelea lauricola* (Harr., Fraedrich & Aghayeva) (Fig. 15.6D), which causes laurel wilt disease in trees in the family Lauraceae. Infection with the disease can result from as few as 100 fungal spores and disrupt xylem tissue function in as little as 3 days after the initial inoculation (Inch and Ploetz, 2012; Inch et al., 2012; Ploetz et al., 2013). Symptoms of laurel wilt include withering of leaves, branch death, and blackening of sapwood; in some cases, infection with *R. lauricola* can be lethal to the tree. Most species of Lauraceae native to the United States are susceptible to infection with laurel wilt (Fig. 15.6A and E), including redbay, *Persea borbonia* L., swamp bay, *Persea palustris* Rafinesque, avocado, *Persea americana* Mill., sassafras, *Sassafras albidum* Nutt. (Fraedrich et al., 2008; Mayfield et al., 2008; Smith et al., 2009; Hughes et al., 2011), and several other commercially, culturally, and ecologically important species in the Coastal Plain forests of the southeastern states (Fraedrich et al., 2011; Ploetz et al., 2011; Ploetz et al., 2013).

Since RAB was first detected in the United States in 2002 in Georgia (Haack, 2002), it has spread rapidly throughout the South, killing millions of redbay and swamp bay trees (Cameron et al., 2015). Laurel wilt was detected in Northern Florida in 2005 and in the southern regions of the state in 2011, causing widespread mortality in redbay growing across Florida, Georgia, and South Carolina (Fraedrich et al. 2008; Kendra et al., 2013; Hughes et al., 2015). An estimated one-third of redbay in the United States (>300 million trees) have already been killed by infestation with *R. lauricola* (Hughes et al., 2017b). RAB attacks multiple *Persea* spp., including California laurel, swampbay, sassafras, and endangered Lauraceae species like pondspice, *Litsea aestivalis* L., and pondberry, *Lindera melissifolia* Walter (Fraedrich, 2008; Fraedrich et al., 2011; Hughes et al., 2011; Peña et al., 2012). Lauraceae trees are key riparian species protecting rivers, lakes, and other bodies of water across continental North America. The reduction of protective vegetation or the sudden degradation of the woody key species in riparian areas greatly increases the risk of habitat degradation and soil erosion and loss through fire, wind, and rain effects. Laurel wilt is also of concern to the avocado industry. The fungus *R. lauricola* has already been reported on avocado

FIG. 15.6 (A) The native Lauraceae species that characterize coastal forests of the southeastern United States, including redbay, as well as agricultural crops such as (E) avocado, have been severely impacted by the invasive redbay ambrosia beetle (*RAB*), *X. glabratus*. (B) Adult RAB females oviposit preferentially onto Lauraceae trees, where (C) the larvae tunnel beneath the bark. During oviposition, RAB inoculates its host tree with (D) the deadly fungal pathogen *R. lauricola*, which invariably causes a widespread fungal infection that overwhelms the tree defenses, ultimately killing it. Credits: (A and E) Martini; (B and C) Buss; (D) Hughes.

in Florida, the second-largest producer of avocado in the country. Avocados were grown on roughly 2350 ha in Florida in the 2018–2019 season, producing a total yield of 13,900 tons (13,620 tons utilized, with an estimated value of 15.3 million USD) (USDA National Agricultural Statistics Service (NASS), 2019). In the case of avocado, the fungus is transmitted to the host through other ambrosia beetle species that acquire fungal spores through lateral transfer from an infected laurel tree (Carrillo et al., 2014), so in this crop, control of laurel wilt needs to focus on the ambrosia beetle group rather than on a specific species (Rivera et al., 2020). If the pest and its pathogenic symbiont were to spread to California, the economic impacts would be far larger. According to estimates by the USDA National Agricultural Statistics Service, a total of 171,000 tons of avocados were produced in California over 19,000 ha, with a utilized production value of 383 million USD (169,100 tons) (USDA National Agricultural Statistics Service (NASS), 2019). Laurel wilt disease distribution has been moving westwards, and it has been confirmed in Arkansas, Louisiana, and Texas (USDA, 2015).

Black stem borer (BSB), *Xylosandrus germanus* (Blandford), an ambrosia beetle native to eastern Asia that was introduced to the United States in 1932, has become one of the most serious pests of nurseries and landscapes across the Southern, Midwest, Upper northwestern, and Northeastern United States (Gill et al., 1998; Mudge et al., 2001; Rabaglia et al., 2006). BSB preys on more than 200 species in 51 families, including beech, *Fagus*, oak, *Quercus*, *Magnolia*, maple, *Acer*, black walnut, *Juglans nigra* L., ash, *Fraxinus*, cedar, *Cedrus*, grape, *Vitis*, pecan, *Carya illinoinensis* Wangenh, plum, *Prunus*, willow, *Salix*, dogwood, *Cornus*, chestnut, *Castanea*, apple, *Malus domestica* Borkh., cherry, *Prunus*, peach, *Prunus persica* L., pear, *Pyrus*, and rose, *Rosa* (Weber and McPherson, 1983; Oregon State University, 2007; Ranger et al., 2010; Centre for Agriculture and Biosciences International, 2020). Virtually any woody plant is vulnerable to infestation by BSB, though it does show some preference for deciduous species over conifers (Weber and McPherson, 1983; Oregon State University, 2007; Ranger et al., 2010; Centre for Agriculture and Biosciences International, 2020). It also displays a preference for saplings and small-diameter trees, making it a particular threat to ornamental nursery and orchard industries. BSB is also highly mobile, capable of flying distances of 2 km or more, and can travel even farther if humans provide assistance by transporting infested wood (Grégoire et al., 2001; Centre for Agriculture and Biosciences International, 2020). Because this species practices sibling mating, only a few mobile female beetles are required to produce an active infestation (Centre for Agriculture and Biosciences International, 2020). Infestations by BSB, along with its symbiotic fungal partner, *Ambrosiella grosmanniae* (McNew, Mayers, and Harr) produce discoloration and blistering in the bark of infested trees and disruption of the vascular system, leading to wilting, dieback, and eventually mortality (Fig. 15.7A and B). BSB also transmits pathogenic *Fusarium* spp. fungi, known to cause dieback, stem cankers, sprouting, and may lead to tree death (Anderson and Hoffard, 1978; Weber, 1980; Centre for Agriculture and Biosciences International, 2020). In recent years, BSB has expanded its impacts in orchard crops, including apples, apricots, and plums in Michigan, and apples in Ohio, North Carolina, and New York (Oregon State University, 2007; Lehnert, 2015). In 2013, apple growers in Western New York began to see significant numbers of trees (up to 30%) collapsing in their orchards. By the end of 2013, hundreds of apple trees had to be removed and destroyed, and 50 more infestation sites have since been identified, extending all the way to

FIG. 15.7 The black stem borer (*BSB*), *X. germanus*, is an invasive species that preys on more than 200 plant species in 51 families, including important crops like pome and stone fruit, nuts, and grapes. BSB creates (A) galleries hosting adults and brood, and in the process inoculates the host tree with a symbiotic fungus, *A. grosmanniae*, which accelerates the senescence of the infested plant. (B) An infested apple tree shows signs of decline during early summer, with dead branches and brown leaves. Field trials reported here used potted apple trees with trunk applications of (C) SPLAT Verb (indicated by white arrow), or (D) Verbenone Disrupt Micro-Flakes VBN (indicated by white arrow). (E) A group of potted trees treated with individual ethanol lures (indicated by red arrows) plus a central verbenone dispenser (indicated by white arrow). Credits: (A, C, D, and E) Agnello; (B) Biltonen. *SPLAT*, Specialized Pheromone & Lure Application Technology.

Long Island (Agnello et al., 2017). The pest had apparently been present in the forests around these orchards for several years before colonizing the orchard trees (Lehnert, 2015). The reason for this shift in behavior is currently unknown.

In 2003, an additional invasive ambrosia beetle species was found in Whittier, California (Boland, 2016; Seybold et al., 2016; Chen et al., 2017). It was initially misidentified as tea shot hole borer, *Euwallaecea fornicatus* (Eichhoff), only to be revealed as a new species in the same genus through genetic analysis (Eskalen et al., 2013). Seven years later, this species—now assigned the moniker of polyphagous shot hole borer—was implicated in the deaths of a large number of box elder trees in Long Beach (University of California; Pest and Diseases of Southern California Oaks, 2014). The beetle's range expanded to four other counties in Southern California, Ventura, San Bernardino, Riverside, and Orange County, becoming a serious pest of multiple hardwood species in urban and wild forests (Chen et al., 2017). Over 200 species are known to be susceptible to infestation by polyphagous shot hole borer, with 41 species classified as true reproductive hosts (in which beetles are successful in creating galleries and cultivating their fungal gardens), including coast live oaks, *Quercus agrifolia* Née, willow, big leaf maple, *Acer macrophyllum* Pursh, sycamore, *Platanus occidentalis* L., California box elder, *Acer negundo* L., Castor bean, *Ricinus communis* L., English oak, *Quercus robur* L., and blue palo verde, *Parkinsonia florida* Benth. ex A. Gray S. (Oregon State University, 2007; University of California; Pest and Diseases of Southern California Oaks, 2014; Aguilar, 2016; Boland, 2016). Like BSB, polyphagous shot hole borer transmits *Fusarium* spp. fungi, causing a condition called Fusarium dieback, characterized by damage to the xylem and phloem tissues, discoloration of the leaves and wood, wilting, branch dieback, and possible mortality (Oregon State University, 2007; Eskalen et al., 2012, 2013; Freeman et al., 2013; Eatough Jones and Paine, 2015). According to an estimate by the Orange County Parks staff, polyphagous shot hole borer has killed 1.4 million USD worth of trees. This pest's broad host range could make it a substantial threat in California, where major agricultural centers, wildland forests, and urban landscapes—many of them dominated by tree species that serve as reproductive hosts for the borer—often all exist in the same region (Eskalen et al., 2013). Nearly half (48%) of street trees in Southern California are species vulnerable to Fusarium dieback, and six of these, including American sweetgum, *Liquidambar styraciflua* L., one of the most common street trees, are confirmed reproductive hosts for polyphagous shot hole borer (Eskalen et al., 2013).

Management options for ambrosia beetles are limited. Because these insects spend so much of their time shielded within their host trees, they are protected from topical insecticide treatments such as bark drenches, and because they do not feed directly on the plant tissues, most systemic treatments have little to no impact on them, either (Wilson et al., 2014; Centre for Agriculture and Biosciences International, 2020). In addition, the beetles' small size makes them difficult to detect until after they have inflicted visible damage to the tree, making proper timing of insecticide applications very difficult (Reding et al., 2010). Treatment for the fungal infestations ambrosia beetles transmit can be equally challenging. The only available treatment for laurel wilt, for example, is preventative injection with the fungicide propiconazole (1-[[2-(2,4-dichlorophenyl)-4-propyl-1,3-dioxolan-2-yl]methyl]-1,2,4-triazole), a costly and labor-intensive process that is not suitable for area-wide treatment, such as in the southeastern forests where RAB thrives (Gitau et al., 2013; Ploetz et al., 2017). There is no cure for a tree already infested with laurel wilt; infested trees typically die within weeks. Cultural control practices are currently the most commonly recommended method for preventing ambrosia beetle infestations, including removal and destruction of infested trees (typically by burning, chipping and solarization, or heat sterilization) and eliminating transport of potentially infested wood (Buffam and Lucht, 1968; Denlinger and Yocum, 1998; Wang et al., 2000; McCullough et al. 2007; Spence et al., 2013; Wilson et al., 2014). However, in the case of the RAB, this method is only applied by avocado growers, not by forest managers. Most have limited resources and lack the time and people to scout for laurel wilt trees, and to remove and dispose of them properly. In theory, the surest method of tree protection is to maintain their health, and with it their capacity to defend themselves from wood-boring insects, but this simple strategy is becoming more challenging with changing environmental conditions, such as the years-long drought in California, resulting in large swaths of severely stressed forests.

15.4.2 Repellent trials

Following on the success of ISCA's SPLAT Verb against MPB, in 2014, Mafra-Neto et al. began the development of Beetle Guard, a controlled-release repellent combining SPLAT Verb with additional repellent semiochemicals. As with bark beetles, verbenone has been shown to have an inhibitory effect on certain ambrosia beetle species. Dodge et al. (2017) demonstrated that verbenone pouches significantly reduced the number of polyphagous shot hole borers and Kuroshio shot hole borers attracted to quercivorol [(1S,4R)-4-Isopropyl-1-methyl-2-cyclohexen-1-ol], a known attractant to these species, to the extent of almost eliminating the lure's effect. Similar behaviors have also been documented in BSB (Ranger et al., 2013, 2014; Van DerLaan and Ginzel, 2013), where verbenone combined with terpinolene (1-methyl-4-propan-2-ylidene-cyclohexene) reduced attraction to ethanol, a volatile emitted by stressed trees commonly used as a monitoring lure for ambrosia beetles (Grégoire et al., 2001; Oliver and Mannion, 2001; Reding et al., 2010; Centre for Agriculture and Biosciences International, 2020). Methyl salicylate, also released by plants stressed by injury or infection (Shulaev et al., 1997), has been tested as a semiochemical repellent against a broad array of harmful insect species, including the yellow fever mosquito, (Dekker et al., 2011) and black bean aphid *Aphis fabae* Scopoli (Hardie et al., 1994). The ambrosia beetles *X. glabratus*, *Xyleborus bispinatus*, and *Xyleborus volvulus* were also repelled by methyl salicylate in laboratory conditions (Hughes et al., 2017a; Rivera et al., 2020).

Prototype formulations of Beetle Guard, including those containing methyl salicylate and both (−) and (+) verbenone isomers, have also been field-tested against BSB in New York apple orchards, under the supervision of one of ISCA's research partners, Dr. Art Agnello (Department of Entomology, Cornell University, New York, NY). Several treatments were evaluated based on trap capture rates and damage to the crop (Fig. 15.7C–E), including SPLAT Verb (commercial formulation described in the first section of this chapter), with or without chlorpyrifos treatment; two prototype formulations of Beetle Guard, combining verbenone and methyl salicylate; Verbenone Disrupt Micro-Flakes with or without chlorpyrifos treatment; chlorpyrifos treatment alone; blank flakes (no repellent AI), and untreated control. SPLAT formulations were applied to young apple trees using a caulking gun at 35 g/tree (3 testing sites/treatment × 6 trees/treatment per site = 18 trees per treatment). Trap captures in the treated areas showed a relatively normal adult flight occurrence, indicating active BSB populations at these sites, but assessments of damage indicated that the rate of infestation was low at two out of three sites, perhaps due to abundant rain during the season. Treatments utilizing SPLAT Verb, with or without chlorpyrifos, failed to provide effective protection for treated trees from beetle attack, but both Beetle Guard treatments completely prevented crop damage.

Several ISCA repellent formulations also have been field-tested against RAB. A series of large-scale experiments were conducted from November 2016 to October 2017 at three locations in North-central and Northwest Florida, selected for their recent history of laurel wilt infestation (Martini et al., 2020). The north-central trial sites were located at the Historic Haile Homestead in Gainesville and Ichetucknee Springs State Park in Fort White, while the northwestern trial was established at the Edward Ball Wakulla Springs State Park in Quincy. Each site comprised five blocks, assigned in a randomized complete block design to receive one of six treatments: (1) SPLAT Verb, single application; (2) Beetle Guard, single application; (3) SPLAT Verb double application; (4) Beetle Guard double application; (5) asymptomatic redbay without SPLAT Verb or Beetle Guard treatment, serving as a negative control; and (6) nonhost tree without SPLAT Verb or Beetle Guard treatment, also serving as a negative control (Martini et al., 2020). All SPLAT Verb and Beetle Guard treatments were applied via caulking gun to live redbay trees, as four 17.5-g point sources spaced evenly around the circumference of the trunk (70 g formulation/tree). Reapplications were made at the same rate and in the same pattern for treatments 3 and 4, totaling 140 g/tree over the course of the trial (Martini et al., 2020). At the Historic Haile Homestead and Ichetucknee Springs State Park, treatment blocks were separated by at least 20 m, and by 50 m at Edward Ball Wakulla Springs State Park (Martini et al., 2020). Treatment impacts were assessed based on trap capture rates of adult beetles, obtained through the placement of sticky card traps on all experimental trees, and monitoring of the trees for symptoms of laurel wilt. Traps were replaced and monitoring assessments were made every 2 weeks. To cope with a high frequency of 0 beetle capture rates at the two north-central locations, cumulative totals of all beetles captured over the course of the trial were used for data analysis (generalized linear model with Poisson distribution, treatment, and location as explanatory variables), while separate analyses were implemented for the northwestern site. First and second application data were also analyzed separately at the Wakulla site, using a general linear mixed model (library MASS, function glmmPQL), with treatment and time as explanatory variables and block numbers as the random variable. A Poisson distribution was used for the first application, a negative binomial distribution for the second (θ=0.160). Combined data on tree mortality from all three trial sites were subjected to Cox' survival analysis (Martini et al., 2020).

Trap monitoring revealed high RAB populations at Wakulla State Park, and low populations at Historic Haile Homestead and Ichetucknee Springs State Park (Martini et al., 2020). This was likely due to differences in pest infestation history among the different locations. While Historic Haile Homestead and Ichetucknee Springs State Park have been infested with RAB for several years, the first RAB was found in Wakulla State Park in November 2016. The Wakulla trial began in February 2017, when RAB populations were increasing, and in late September populations increased exponentially in the aftermath of Hurricane Irma. After the first application at Wakulla, traps placed on SPLAT Verb-treated trees showed significantly lower capture rates of RAB than those on control trees (t-value = 3.763, $P < .001$). Treatment with Beetle Guard showed no such effect, as there was no significant difference from control (t-value = 3.763, $P < .001$). Similar results were obtained following the second treatment applications (for treatments 3 & 4): trees treated with SPLAT Verb once or twice captured significantly fewer beetles than control (t-value = 2.257, $P = .035$; t-value = 3.194, $P = .005$), while no significant difference was observed for trees treated with Beetle Guard once or twice (t-value = 1.649, $P = .115$; t-value = 0.36, $P = .718$). SPLAT Verb's treatment effect was not significantly different between the single and double treatment. As expected, significantly fewer beetles were captured on nonhost trees compared to untreated host trees (t-value = 3.837, $P = .001$) (Martini et al., 2020). The Ichetucknee trial site had a significantly higher RAB population, as measured by trap captures, than the Haile site (1.6 ± 0.79; 0.46 ±0.10, respectively). As at the Wakulla site, significantly fewer beetles were captured in traps placed on trees treated with a single (z value = 2.331, $P = .020$) or double application of SPLAT Verb (z value = 2.120, $P = .034$) than in control traps (Martini et al., 2020). Beetle trap captures on trees treated with Beetle Guard were not significantly different from the control, whether applied once (z value = 0.343, $P = .732$) or twice ($z = 1.287$, $P = .198$) (Martini et al., 2020). By the end of the trial, 40% of redbay trees that received no repellent treatment across the three trial sites suffered fatal infestations of laurel wilt. In trees treated with single or double applications of SPLAT Verb, only 13% died from the disease. Survival analyses that included trials from Martini et al. (2020) as well as new trials conducted in 2018, showed a significant decline in mortality due to laurel wilt when trees were treated with SPLAT Verb (Martini et al., 2020).

These results contradicted previous findings by Hughes et al. (2017a) and Martini et al. (2020), which demonstrated repellency of Beetle Guard against RAB in lab tests, and in small-scale field trials when applied to cut logs. To further elucidate the impact of Beetle Guard in the field vs. in the lab, an additional field trial was conducted in May 2018 against ambrosia beetle vectors of *R. lauricola* in avocado, *X. volvulus* (Fabricius) and *X. bispinatus* (Eichhoff) (Rivera et al., 2020). A 4-ha avocado orchard in Homestead, Florida, planted with the West Indian variety of the crop, was selected as the trial for its known incidence of laurel wilt infestation. Trees at this site were treated with fungicide, and any trees showing symptoms of laurel wilt were removed from the orchard, resulting in a low infestation rate at the time of the trial. A neighboring orchard, separated by a dirt road and a distance of ~10 m, was subject to unknown control practices; trees growing therein had visible symptoms of laurel wilt. This situation allowed for an effective assessment of the efficacy of the evaluated repellents (deployed alone or as part of a push–pull strategy), as the presence of a presumably large beetle population from the neighboring orchard would ensure high pest pressure on the tested treatments. The first trial conducted at this location evaluated the effects of SPLAT Verb or Beetle Guard, each applied alone, compared to untreated control; and assessed whether their repellency extended beyond a range of 1 m. SPLAT formulations were applied as eight 1-g point sources to each tree in a 3 × 3 tree plot (0.05 ha/plot, five replicates per treatment). Four unbaited sticky traps were placed in each replicate plot, two on each of a pair of trees. One trap was placed 5–10 cm away from the applied Beetle Guard point source, and the other was placed 1–1.5 m away. Traps were collected and replaced every 10 days for 1 month, and all ambrosia beetles captured were counted.

A second trial was conducted in June 2018, testing a push–pull strategy using verbenone or methyl salicylate as a repellent (push) and ethanol as an attractant (pull) (Rivera et al., 2020). The experiment was installed in a split-plot design, with the "pull" as whole plot factor at two levels: (1) control, (2) ethanol dispensers (Synergy Semiochemicals, Burnaby, BC, Canada) and "push" as the split-plot factor at three levels: (1) control, (2) Beetle Guard, and (3) SPLAT Verb. Treatments were replicated three times. SPLAT applications were made in an identical manner as in the repellent trial, with a buffer row separating plots from each other. In the block assigned for treatment with ethanol, dispensers were applied to the buffer row separating the plots treated with SPLAT Verb and Beetle Guard. Two additional ethanol dispensers were placed on each west-facing border of the plot, to establish the "pull" whole plot factor. Unbaited sticky traps were placed at the west-facing edge of the control plots, facing the unmanaged avocado orchard neighboring the test site. As in the repellent trial, traps were placed at differing distances from SPLAT Verb or Beetle Guard point source (5–10 cm vs 1–1.5 m) and were checked for ambrosia beetles every 10 days.

Trap capture data from the first field trial (repellent only) indicated that treatments had no overall effect on the number of ambrosia beetles caught in the field, with the exception of sentinel traps placed close to SPLAT Verb and Beetle Guard point sources, in which reduced captures were observed compared to controls ($F1 = 3.37$, $P = .0682$) (Rivera et al., 2020). The number of beetles captured 20 days after treatment was significantly higher than captures at 10 or 30 days after treatment ($F2 = 3.45$, $P = .0340$). No statistical difference was observed between traps placed close or far from SPLAT Verb and Beetle Guard point sources at 10 or 30 days after treatment, but analysis within treatment times at 20 days revealed that traps placed nearer to the repellent captured fewer beetles than control traps ($F1 = 4.75$, $P = .0429$) (Rivera et al., 2020). Reduced capture rates were also observed at 20 days after treatment in traps placed close to point sources of Beetle Guard than in traps placed farther away or in control traps. No difference was observed between the "far" placed traps and control with either the SPLAT Verb or Beetle Guard formulations. In the push–pull trial, significant interaction was observed between the effects of the applied repellents (SPLAT Verb and Beetle Guard) and attractants (ethanol) in traps placed near treated trees (Rivera et al., 2020). Significantly fewer beetles were trapped in push–pull-treated plots than in the controls at 10 and 30 days after treatments were applied, using either SPLAT Verb or Beetle Guard as the "push" factor. This effect was not seen at the 20-day interval, likely because the beetle population across the treated site was lower at this time. In plots treated with the SPLAT Verb-based push–pull system, no beetles were captured. As in the repellents-only trial, no significant difference was observed in the number of beetles captured in "far"-placed traps and control traps in push- or push–pull-treated plots (Rivera et al., 2020). These results suggest that the push–pull system was more effective at reducing trap capture rates of ambrosia beetles but only when traps were placed fewer than 10 cm away from the applied repellent. At the current stage of these products' development, SPLAT Verb and Beetle Guard would need to be applied to every avocado tree in a given field to provide effective protection of the crop.

15.4.3 Future directions

Impacts of ambrosia beetle pests, while currently less well known and less visible than those of bark beetle pests, have both expanded and worsened in the last few decades. The appearance of RAB in Georgia in 2002 was followed by devastating ecological effects throughout the Lauracea-dominated forests of the American South. The fungal pathogen transmitted by this species, *R. lauricola*, or laurel wilt, has killed off an estimated one-third of the redbay trees in the United States and now threatens commercial avocado crops in Florida. BSB, presumably present for years in the forests around New York apple orchards without impacting them, has now spilled over into these orchards, introducing the deadly fungal disease, Fusarium dieback. A similar Fusarium pathogen has also appeared in Southern California with the emergence of a previously unknown ambrosia beetle species, polyphagous shot hole borer in that area, resulting in the loss of roughly 1.4 million USD in trees.

As with bark beetles, the sensitive natural environments where ambrosia beetle outbreaks occur demand a selective, ecologically sound approach to manage these pests. Beetle Guard provides a means of protecting susceptible trees from attack by ambrosia beetles without the application of toxic chemicals, instead relying on a combination of repellent semiochemicals, including the bark beetle antiaggregation pheromone, verbenone, and a kairomone produced by stressed or injured plants, methyl salicylate. Field trials with Beetle Guard are ongoing. The results obtained from these studies provide evidence of the potential of a repellent strategy to deliver effective tree protection from ambrosia beetle attack—in trials against BSB in New York apple orchards, Beetle Guard completely suppressed crop damage by the target pest (Agnello et al., 2021)—as well as illumination of the importance of rigorous field testing. Beetle Guard displayed relatively high levels of repellency toward RAB when tested in lab trials, but results in the field were less consistent. Field studies in North Central and Northwestern Florida demonstrated that treatment with SPLAT Verb significantly decreased the number of RAB adults captured in baited traps, and decreased mortality due to laurel wilt. When tested in an avocado system, both SPLAT Verb and Beetle Guard were efficient in repelling beetle species responsible of laurel wilt. Additional trials evaluating both repellents, SPLAT Verb and Beetle Guard, alone or as part of a push–pull mechanism, revealed that the latter strategy was more effective in suppressing trap capture, but that both strategies were limited to a close range of activity.

Further work on Beetle Guard will include optimization of the agronomic application of the formulation to protect trees of interest, including avocado. Additional field trials will assess Beetle Guard's capacity to prevent or reduce attack by other invasive ambrosia beetle species of concern in the United States, including BSB in apples and polyphagous shot hole borer in urban forest environments.

15.5 Conclusion

Decades of reliance on a single tactic for agricultural pest control—cover sprays of broad-spectrum chemical pesticides like DDT, organochlorines, and organophosphates—have revealed the weaknesses of any such simplistic approach. As insects are infinitely varied and adaptable, so too must be the methods employed to minimize their detrimental impacts on human health and prosperity. The resources offered by the field of arthropod chemical ecology, by the myriad chemical signals by which arthropods perceive and respond to various aspects of their environment, present an equally vast array of opportunities to investigate such means of mitigation, which has been previously underutilized, especially in agriculture and forestry applications. The repellent technologies described in this chapter, CBB Repel, SPLAT Verb, SPB Repel, MCH Repel, and Beetle Guard, have proven effective in field trials against seven beetle pest species of agriculture and forestry: *H. hampei*, *D. ponderosae*, *D. frontalis*, *D. pseudotsugae*, *D. rufipennis*, *X. glabratus*, and *X. germanus*. A similar repellent formulation is also being evaluated by Mafra-Neto et al. against darkling beetle, *Alphitobius diaperinus* Panzer, a major pest of the poultry industry. The varied applications for these semiochemical repellents in forestry, orchard crop agriculture, and animal husbandry demonstrate the largely untapped potential of insect semiochemicals in general to effect suppression of pest impacts without the introduction of large quantities of toxic chemicals.

This potential efficacy is, surprisingly, not limited to the insect world but can be deployed against other pest organisms as well. The wide-ranging utility of insect pheromones has been demonstrated by a recent collaborative project between ISCA Technologies and Wright et al. (2018), during which an alarm pheromone produced by honeybees in response to a perceived threat to the hive was examined for its impact on African bush elephants, *Loxodonta africana* Blumenbach. Elephants pose a threat to human communities in Africa by causing damage to farmland and infrastructure and may interfere with conservation priorities by trampling through preserved environments (Ngama et al., 2016; Cook et al., 2017; Wright et al., 2018), but because elephants are themselves members of a protected species, means to minimize or prevent these impacts are limited. Elephants' natural aversion to African honey bee, *Apis mellifera scutellata* Lepeletier, colonies opens an opportunity to repel these animals away from sensitive areas without harming these animals and without the time and investment required to establish complete hives around the areas to be protected. When a crude approximation of the honey bee alarm pheromone (comprising a blend of isoamyl acetate and 2-heptanone in equal proportions) was applied in field trials in the Greater Kruger Park Associated Private Nature Reserves, Limpopo Province, South Africa, incorporated into a prototype SPLAT formulation (20% AI blend by volume) and applied in 25-g point volumes in weighted white socks hung from tree branches, elephants responded not by bolting in fear, which could have created its own problems, but by stopping, turning in the direction of the semiochemical plume, and extending their trunks toward it before calmly moving away from the formulation (Wright et al., 2018).

The diversity of potential applications for repellent semiochemicals is an enormous asset in attempts to mitigate the impacts of destructive arthropods, which heretofore has not been accurately represented in the range of control applications adopted on a broad commercial scale. Arthropod repellents have been readily developed and adopted to protect people from blood-feeding arthropods such as ticks, mosquitoes, and other biting flies for much of the latter 20th and early 21st centuries, but relatively few analogous methods have been successful in agriculture or forestry applications. The studies reported herein provide strong evidence that the utility of pheromone- or semiochemical-based repellents is not limited to vector management but may also prove more broadly efficacious and cost-effective in meeting the challenges of pest control in the age of anthropogenic climate change, whether in regard to public health, agriculture, or natural resource management.

References

Agnello, A.M, Breth, D.I., Tee, E.M., Cox, K.D., Villani, S.M., Ayer, K.M., et al., 2017. *Xylosandrus germanus* (Coleoptera: Curculionidae: Scolytinae) occurrence, fungal associations, and management trials in New York apple orchards. J. Econ. Entomol. 110, 2149–2164.

Agnello, A.M., Combs, D.B., Filgueiras, C.C., Willett, D.S., Mafra-Neto, A., 2021. Reduced infestation by *Xylosandrus germanus* (Coleoptera: Curculionidae: Scolytinae) in apple trees treated with host plant defense compounds. J. Econ. Entomol. In press.

Aguilar, E., 2016. The insect that could destroy California's avocado crop. http://www.scpr.org/news/2016/04/12/59521/who-will-stop-the-shot-hole-borer-beetle/. (Accessed October 23, 2020).

Alègre, C., 1959. Climates et cafe´iers d'Arabie. Agron. Trop. 14, 23–58.

Amman, G.D., Logan, J.A., 1998. Silvicultural control of mountain pine beetle: prescriptions and the influence of microclimate. Am. Entomol. 44 (3), 166–178.

Amman, G.D., Their, R.W., McGregor, M.D., Schmitz, R.F., 1989. Efficacy of verbenone in reducing lodgepole pine infestation by mountain pine beetles in Idaho. Can. J. For. Res. 19, 60–64.

Anderson, R.L., Hoffard, W.H., 1978. Fusarium canker-ambrosia beetle complex on tulip poplar in Ohio. Plant Dis 62, 751.

Anet, E.F.L.J., 1969. Autoxidation of α-farnesene. Aust. J. Chem. 22 (11), 2403–2410.

Baker, P.S., 1999. The coffee berry borer in Colombia. CABI Bioscience, Egham ISBN 958-96554-1-6.

Baker, P.S., Barrera, J.F., Rivas, A., 1992. Life-history studies of the coffee berry borer (*Hypothenemus hampei*, Scolytidae) on coffee trees in Southern Mexico. J. Appl. Ecol. 29 (3), 656.

Baker, P.S., Jackson, J.A.F., Murphy, S.T., 2002. Natural enemies, natural allies: how scientists and coffee farmers forged new partnerships in the war against pests and low prices. CFC/ICO/02 (1998–2002), The Commodities Press, CABI commodities.

Batra, L.R., 1966. Ambrosia fungi: extent of specificity to ambrosia beetles. Science 173, 193–195.

Batra, L.R., 1967. Ambrosia fungi: a taxonomic revision and nutritional studies of some species. Mycologia 59, 976–1017.

Beaver, R.A., 1989. Insect–fungus relationships in bark and ambrosia beetles. In: Wildling, N., Collins, N.M., Hammond, P.M., Webber, J.F. (Eds.), Insect–Fungus Interactions. Academic Press, London, pp. 121–143.

Benavides, P., Góngora, C., 2015. Combination of Biological Pesticides. US Patents No. 20, 150, 359, 229. US Patent and Trademark Office, Washington, DC.

Bentz, B., Logan, J., MacMahon, J., Allen, C.D., Ayres, M., Berg, E., et al., 2009. Bark beetle outbreaks in western North America: causes and consequences, Bark Beetle Symposium; Snowbird, Utah. University of Utah Press, Salt Lake City, UT, p. 42.

Bentz, B.J., Régnière, J., Fettig, C.J., Hansen, E.M., Hayes, J.L., Hicke, J.A., et al., 2010. Climate change and bark beetles of the western United States and Canada: direct and indirect effects. BioScience 60 (8), 602–613.

Bentz, B.J., Lister, C.K., Schmid, J.M., Mata, S.A., Rasmussen, L.A., Haneman, D., 1989. Does verbenone reduce mountain pine beetle attacks in susceptible stands of ponderosa pine? RN-RM-495. U.S. Department of Agriculture, Forest Service, Ogden, UT.

Bible, K., 2001. Long-term patterns of Douglas-fir and western hemlock mortality in. In: the Cascade Mountains of Oregon and Washington. Doctoral dissertation. University of Washington, Seattle, WA.

Billings, R.F., Thatcher, R.C., Searcy, J.L., Coster, J.E., Hertel, G.D., 1980. Direct control The Southern Pine Beetle. Tech. Bull. 1631. U.S. Department of Agriculture Forest Service, Expanded Southern Pine Beetle Research and Applications Program, Washington, DC, pp. 179–192.

Billings, R.F., 2011a. Aerial detection, ground evaluation, and monitoring of the southern pine beetle: state perspectives. In: Coulson, R.N., Klepzig, K.D. (Eds.), Southern Pine Beetle II. Gen. Tech. Rep. SRS-140. U.S. Department of Agriculture Forest Service, Southern Research Station, Asheville, NC, pp. 245–261.

Billings, R.F., 2011b. Use of chemicals for prevention and control of southern pine beetle infestations. In: Coulson, R.N., Klepzig, K.D. (Eds.), Southern Pine Beetle II. Gen. Tech. Rep. SRS-140. U.S. Department of Agriculture Forest Service, Southern Research Station, Asheville, NC, pp. 367–379.

Boland, J.M., 2016. The impact of an invasive ambrosia beetle on the riparian habitats of the Tijuana River Valley. California. Peer J 4, e2141.

Bond, C., Buhl, K., Stone, D., 2014. Pyrethrins general fact sheet. National Pesticide Information Center, Oregon State University Extension Services, Corvallis, Oregon. http://npic.orst.edu/factsheets/pyrethrins.html. (Accessed October 23, 2020).

Bond, C., Cross, A., Buhl, K., Stone, D., 2016. Carbaryl general fact sheet. National Pesticide Information Center, Oregon State University Extension Services, Corvallis, OR. http://npic.orst.edu/factsheets/carbarylgen.html. (Accessed October 23, 2020).

Borbón, O., Mora, O., Mora, R., Oehlschlager, C., Gonzáles, L., Andrade, R., Alvarez, L., 2000. Attraction and inhibition of attraction of coffee berry borer, *Hypothenemus hampei* (Coleoptera: Scolytidae), Proceedings International Society of Chemical Ecology (ISCE) 17th Annual Meeting. Locos de Caldas.

Brookes, H.M., Ross, D.W., Strand, T.M., Thistle, H.W., Ragenovich, I.R., Lowrey, L.L., 2016. Evaluating high release rate MCH (3-methylcyclohex-2-en-1-one) treatments for reducing *Dendroctonus pseudotsugae* (Coleoptera: Curculionidae) infestations. J. Econ. Entomol. 109 (6), 2424–2427.

Brun, L.O., Marcillaud, C., Gaudichon, V., Suckling, D.M., 1989. Endosulfan resistance in coffee berry borer *Hypothenemus hampei* (Coleoptera: Scolytidae) in New Caledonia. J. Econ. Entomol. 82, 1311–1316.

Buffam, P.E., Lucht, D.D., 1968. Use of polyethylene sheeting for control of *Ips* spp. in logging debris. J. Econ. Entomol. 61, 1465–1466.

Bunn, C., Läderach, P., Rivera, O.O., Kirschke, D., 2015. A bitter cup: climate change profile of global production of Arabica and robusta coffee. Clim. Change 129 (1), 89–101.

Bunt, W.D., Coster, J.E., Johnson, P.C., 1980. Behavior of the southern pine beetle on the bark of host trees during mass attack. Ann. Entomol. Soc. Am. 73 (6), 647–652.

Burbano, E., Wright, M., Bright, D.E., Vega, F.E., 2011. New record for the coffee berry borer, *Hypothenemus hampei*, in Hawaii. J. Insect Sci. 11 (117), 1–3.

Byers, J.A., 1992. Attraction of bark beetles, *Tomicus piniperda*, *Hylurgops palliatus*, and *Trypodendron domesticum* and other insects to short-chain alcohols and monoterpenes. J. Chem. Ecol. 18 (12), 2385–2402.

Camargo, A.P., 1985. Florescimento e frutificação de café arábica nas diferentes regiões cafeeiras do Brasil. Pesqui. Agropecu. Bras. 20, 831–839.

Cameron, R.S., Hanula, J., Fraedrich, S., Bates, C., 2015. Progression and impact of laurel wilt disease within redbay and sassafras populations in southeast Georgia. Southeast. Nat. 14 (4), 650–674.

Cancelliere, J., 2016a. Systemic insecticides for the treatment of southern pine beetle division of lands and forests—forest health southern pine beetle response. New York Department of Environmental Conservation. Division of Lands and Forest-Forest Health, Southern Pine Beetle Response, Albany, NY.

Cancelliere, J., 2016b. Can systemic insecticides or verbenone control southern pine beetle on Long Island? New York Department of Environmental Conservation. Division of Lands and Forest-Forest Health, Southern Pine Beetle Response, Albany, NY.

Carrillo, D., Duncan, R.E., Ploetz, J.N., Campbell, A.F., Ploetz, R.C., Peña, J.E., 2014. Lateral transfer of a phytopathogenic symbiont among native and exotic ambrosia beetles. Plant Pathol. 63 (1), 54–62.

Castrillo, L.A., Griggs, M.H., Ranger, C.M., Reding, M.E., Vandenberg, J.D., 2011. Virulence of commercial strains of *Beauveria bassiana* and *Metarhizium brunneum* (Ascomycota: Hypocreales) against adult *Xylosandrus germanus* (Coleoptera: Curculionidae) and impact on brood. Biol. Control 58 (2), 121–126.

Castro, A.M., Tapias, J., Ortiz, A., Benavides, P., Góngora, C.E., 2017. Identification of attractant and repellent plants to coffee berry borer, *Hypothenemus hampei*. Entomol. Exp. Appl. 164 (2), 120–130.

Ceja-Navarro, J.A., Vega, F.E., Karaoz, U., Hao, Z., Jenkins, S., Lim, H.C., et al., 2015. Gut microbiota mediate caffeine detoxification in the primary insect pest of coffee. Nat. Commun. 6, 7618.

Centre for Agriculture and Biosciences International (CABI) Invasive species compendium: *Hypothenemus hampei* (coffee berry borer) datasheet, 2016. https://www.cabi.org/isc/datasheet/51521. (Accessed October 19, 2020).

Centre for Agriculture and Biosciences International (CABI) Invasive species compendium: *Xylosandrus germanus* (black timber bark beetle) datasheet, 2020. http://www.cabi.org/isc/datasheet/57237. (Accessed October 23, 2020).

Chen, Y., Dallara, P.L., Nelson, L.J., Coleman, T.W., Hishinuma, S.M., Carrillo, D., Seybold, S.J., 2017. Comparative morphometric and chemical analyses of phenotypes of two invasive ambrosia beetles (*Euwallacea* spp.) in the United States. J. Insect Sci. 24 (4), 647–662.

Conde, C., Vinocur, M., Gay, C., Seiler, R., Estrada, F., 2006. Climatic threat spaces as a tool to assess current and future climate risks: case studies in México and Argentina. AIACC Working Paper, p. 56.

Cone, M., 2010. Endosulfan to be banned, pesticide poses "unacceptable risks". EPA Says. Sci. Am. https://www.scientificamerican.com/article/endosulfan-banned-epa/. (Accessed October 19, 2020).

Conn, J.E., Borden, J.H., Scott, B.E., Friskie, L.M., Pierce Jr, H.D., Oehlschlager, A.C., 1983. Semiochemicals for the mountain pine beetle, *Dendroctonus ponderosae* (Coleoptera: Scolytidae) in British Columbia: field trapping studies. Can. J. For. Res. 13 (2), 320–324.

Cook, B.I., Mankin, J.S., Anchukaitis, K.J., 2018. Climate change and drought: from past to future. Curr. Clim. 4, 164–179.

Cook, R.M., Witkowski, E.T.F., Helm, C.V., Henley, M.D., Parrini, F., 2017. Recent exposure to African elephants after a century of exclusion: rapid accumulation of marula tree impact and mortality, and poor regeneration. For. Ecol. Manag. 401, 107–116.

Crocker, S.J., 2012. New Jersey's Forest Resources, 2011. Res. Note NRS-156. U.S. Department of Agriculture, Forest Service, Northern Research Station, Newtown Square, PA, p. 4.

Damon, A., 2000. A review of the biology and control of the coffee berry borer, *Hypothenemus hampei* (Coleoptera: Scolytidae). Bull. Entomol. Res. 90 (6), 453–465.

Davis, A.P., Chadburn, H., Moat, J., O'Sullivan, R., Hargreaves, S., Lughadha, E.N., 2019. High extinction risk for wild coffee species and implications for coffee sector sustainability. Sci. Adv. 5 (1), eaav3473.

Davis, A.P., Gole, T.W., Baena, S., Moat, J., 2012. The impact of climate change on indigenous arabica coffee (*Coffea arabica*): predicting future trends and identifying priorities. PloS One 7 (11), e47981.

Davis, A.P., Govaerts, R., Bridson, D.M., Stoffelen, P., 2006. An annotated taxonomic conspectus of the genus *Coffea* (Rubiaceae). Bot. J. Linn. Soc. 152 (4), 465–512.

DeGomez, T.E., Hayes, C.J., Anhold, J.A., McMillin, J.D., Clancy, K.M., Bosu, P.P., 2006. Evaluation of insecticides for protecting southwestern ponderosa pines from attack by engraver beetles (Coleoptera: Curculionidae: Scolytinae). J. Econ. Entomol. 99 (2), 393–400.

Dekker, T., Ignell, R., Ghebru, M., Glinwood, R., Hopkins, R., 2011. Identification of mosquito repellent odours from *Ocimum forskolei*. Parasit. Vectors 4 (1), 183.

Denlinger, D.L., Yocum, G.D., 1998. Physiology of heat sensitivity. In: Delinger, D.L., Hallman, G.J. (Eds.), Temperature Sensitivity in Insects and Applications for Integrated Pest Management. Westview Press, Boulder, CO, pp. 7–54.

Department of Agriculture, State of Hawaii, 2014. Coffee berry borer *Hypothenemus hampei* (Ferrari). http://hdoa.hawaii.gov/pi/files/2013/01/Hypothenemus-hampei-NPA-MASTER.pdf. (Accessed October 19, 2020).

Dhar, A., Parrott, L., Hawkins, C.D., 2016. Aftermath of mountain pine beetle outbreak in British Columbia: stand dynamics, management response and ecosystem resilience. Forests 7 (8), 171.

Dickson, C., 2019. Forestry service advising property owners to apply MCH repellent as Douglas fir beetles move on B.C. Interior. CBC News posted March 26, 2019. https://www.cbc.ca/news/canada/british-columbia/douglas-fir-beetle-repellant-1.5071390. (Accessed October 23, 2020).

Dodge, C., Coolidge, J., Cooperband, M., Cossé, A., Carrillo, D., Stouthamer, R., 2017. Quercivorol as a lure for the polyphagous and Kuroshio shot hole borers, *Euwallacea* spp. nr. fornicatus (Coleoptera: Scolytinae), vectors of Fusarium dieback. Peer J. 5, e3656.

Duque, O.H., 2000. Economics of coffee berry borer (*Hypothenemus hampei*) in ColombiaWorkshop, Coffee Berry Borer: New Bark Beetles Approaches to Integrated Pest Management. Mississippi State University, Starkville, MS, p. 14.

Duque-Orrego, H., Márquez-Q., A., Hernández-S., M., 2002. Estudios de caso sobre costos de manejo integrado de la broca del café en el Departamento de Risaralda. Cenicafé 53, 106–118.

Duque, H., Baker, P.S., 2003. Devouring Profit, the Socio-economics of Coffee Berry Borer IPM. The Commodities Press-CABI-CENICAFÉ, Chinchiná, CO.

Eatough Jones, M., Paine, T.D., 2015. Effect of chipping and solarization on emergence and boring activity of a recently introduced ambrosia beetle (Euwallacea sp., Coleoptera: Curculionidae: Scolytinae) in southern California. J. Econ. Entomol. 108 (4), 1852–1859.

Ellis, R., 2019. CHART: where disease-carrying mosquitoes will go in the future." NPR.org: https://www.npr.org/sections/goatsandsoda/2019/03/28/707604928/chart-where-disease-carrying-mosquitoes-will-go-in-the-future. (Accessed October 23, 2020).

Eskalen, A., Stouthamer, R., Lynch, S.C., Rugman-Jones, P.F., Twizeyimana, M., Gonzalez, A., Thibault, T., 2013. Host range of Fusarium dieback and its ambrosia beetle (Coleoptera: Scolytinae) vector in Southern California. Plant Dis. 97, 938–951.

Eskalen, A., Wang, D.H., Twizeyimana, M., 2012. A new pest: Fusarium sp. and its vector tea shot-hole borer (*Euwallacea fornicatus*) causing Fusarium dieback on avocado in California. Phytopathology 102, 35.

European Commission, 2016. EU Pesticides Database: Endosulfan. https://ec.europa.eu/food/plant/pesticides/eu-pesticides-database/public/?event=activesubstance.detail&language=EN&selectedID=1281. (Accessed October 24, 2020).

Fan, L.T., Singh, S.K., 1989. Controlled Release: A Quantitative Treatment. Springer-Verlag, New York, NY.

FAOSTAT, 2014. Food and agriculture organization of the United Nations, statistics division. http://faostat.fao.org/. (Accessed October 23, 2020).

Ferguson, T.L., Buccowich, M., Lueckel, R., 2013. Southern pine beetle in New Jersey: the infestation is expanding. U.S. Department of Agriculture, Forest Service, Northern Area State and Private Forestry Report, 2 p.

Fettig, C.J., DeGomez, T.E., Gibson, K.E., Dabney, C.P., Borys, R.R., 2006. Effectiveness of permethrin plus-C (Masterline®) and carbaryl (Sevin SL®) for protecting individual, high-value pines from bark beetle attack. Arboric. Urban For. 32, 247–252.

Fettig, C.J., Gibson, K.E., Munson, A.S., Negrón, J.F., 2014. A comment on "management for mountain pine beetle outbreak suppression: does relevant science support current policy?". Forests 5 (4), 822–826.

Fettig, C.J., Grosman, D.M., Munson, A.S., 2013. Advances in insecticide tools and tactics for protecting conifers from bark beetle attack in the western United StatesInsecticides—Development of Safer and More Effective Technologies. InTech, Rijeka, Croatia, pp. 472–492.

Fettig, C.J., Hilszczański, J., 2015. Management strategies for bark beetles in conifer forests. In: Vega, F.E., Hofstetter, R.W. (Eds.), Bark Beetles: Biology and Ecology of Native and Invasive Species. Academic Press, London, pp. 555–584.

Fettig, C.J., Munson, A.S., McKelvey, S.R., Bush, P.B., Borys, R.R., 2008. Spray deposition from ground-based applications of carbaryl to protect individual trees from bark beetle attack. J. Environ. Qual. 37 (3), 1170–1179.

Fettig, C.J., Munson, A.S., McKelvey, S.R., DeGomez, T.E., 2009. Deposition from ground-based sprays of carbaryl to protect individual trees from bark beetle attack in the Western United States. AZ1493, University of Arizona, College of Agriculture and Life Sciences Bulletin, Tucson, AZ.

Fettig, C.J., Munson, A.S., Reinke, M., Mafra-Neto, A., 2015. A novel semiochemical tool for protecting lodgepole pine from mortality attributed to mountain pine beetle (Coleoptera: Curculionidae). J. Econ. Entomol. 108, 173–182.

Fettig, C.J., Steed, B.E., Bulaon, B.M., Mortenson, L.A., Progar, R.A., Bradley, C.A., Munson, A.S., Mafra-Neto, A., 2016. The efficacy of SPLAT® Verb for protecting individual *Pinus contorta*, *Pinus ponderosa*, and *Pinus lambertiana* from colonization by *Dendroctonus ponderosae*. J. Entomol. Soc. B.C. 113, 11–20.

Fettig, C.J., Steed, B.E., Munson, A.S., Progar, R.A., Mafra-Neto, A., 2020. Evaluating doses of SPLAT® verb to protect lodgepole pine trees and stands from mountain pine beetle. Crop Prot. 136, 105228.

Fiddler, G.O., Hart, D.R., Fiddler, T.A., McDonald, P.M., 1989. Thinning Decreases Mortality and Increases Growth of Ponderosa Pine in Northeastern California. USDA Forest Service, Pacific Southwest Forest and Range Experiment Station, Berkeley, CA, p. 11.

Foote, G.G., Fettig, C.J., Ross, D.W., Runyon, J.B., Coleman, T.W., Gaylord, M.L., et al., 2020. A biodegradable formulation of MCH (3-Methylcyclohex-2-en-1-one) for protecting *Pseudotsuga menziesii* from *Dendroctonus pseudotsugae* (Coleoptera: Curculionidae) colonization. J. Econ. Entomol. 113, 1858–1863.

Fraedrich, S.W., 2008. California laurel is susceptible to laurel wilt caused by *Raffaelea lauricola*. Plant Dis. 92 (10), 1469.

Fraedrich, S.W., Harrington, T.C., Bates, C., Johnson, J., Reid, L., Leininger, T., Hawkins, T., 2011. Susceptibility to laurel wilt and disease incidence in two rare plant species, pondberry and pondspice. Plant Dis. 95 (9), 1056–1062.

Fraedrich, S.W., Harrington, T.C., Rabaglia, R.J., Ulyshen, M.D., Mayfield, A.E., Hanula, J.L., et al., 2008. A fungal symbiont of the redbay ambrosia beetle causes a lethal wilt in redbay and other Lauraceae in the Southeastern United States. Plant Dis. 92, 215–224.

Franceschi, V.R., Krokene, P., Christiansen, E., Krekling, T., 2005. Anatomical and chemical defenses of conifer bark against bark beetles and other pests. New Phytol. 167 (2), 353–376.

Franco, C.M., 1958. Influence of Temperature on Growth of Coffee Plant. IBEC Research Institute, New York, NY.

Freeman, M.B., 2017. The role of abiotic and biotic factors in Douglas-fir decline in the Western Cascades, Doctoral dissertation, Washington, DC.

Freeman, S., Sharon, M., Maymon, M., Mendel, Z., Protasov, A., Aoki, T., Eskalen, A., O'Donnell, K., 2013. Fusarium euwallaceae sp. nov.—a symbiotic fungus of *Euwallacea* sp., an invasive ambrosia beetle in Israel and California. Mycologia 105 (6), 1595–1606.

Furniss, R.L., Carolin, V.M., 1977. Western forest insects US Department of Agriculture, Forest Service, Washington DC, p. 654.

Gara, R.I., Vité, J.P., Cramer, H.H., 1965. Manipulation of *Dendroctonus frontalis* by use of a population aggregating pheromone. Contrib. Boyce Thompson Inst. 23 (3), 55–66.

Gay, C., Estrada, F., Conde, C., Eakin, H., Villers, L., 2006. Potential impacts of climate change on agriculture: a case of study of coffee production in Veracruz, Mexico. Clim. Change 79 (3-4), 259–288.

Gaylord, M., McKelvey, S., Fettig, C., McMillin, J.D., 2020. Verbenone inhibits attraction of *Ips pini* (Coleoptera: Curculionidae) to pheromone-baited traps in northern Arizona. J. Econ. Entomol. 113, 3017–3020.

Gill, S., Blessington, T., Dutky, E., 1998. Greenhouse Weekly IPM Report Maryland Central Maryland Research and Education Center, Ellicott.

Gillette, N.E., Fettig, C.J., 2021. Semiochemicals for bark beetle management: where do we go from here? Can. Entomol. 153, 121–135.

Gillette, N.E., Munson, A.S., 2009. Semiochemical sabotage: behavioral chemicals for protection of western conifers from bark beetles, The Western Bark Beetle Research. Group: A unique collaboration with Forest Health Protection, Proceedings of a Symposium at the 2007 Society of American Foresters Conference, October 23–28, 2007 PNW-GTR-784. U.S. Department of Agriculture, Forest Service, Pacific Northwest Research Station, Portland, Oregon.

Giraldo, N., Valencia, C.E., 2000. Evaluación de Productos Biológicos para el Manejo de la Broca del Café *Hypothenemus hampei* Ferrari (Coleoptera: Scolytidae). MSc Thesis, Facultad de Agronomía, Santa Rosa de Cabal, Colombia.

Gitau, C.W., Bashford, R., Carnegie, A.J., Gurr, G.M., 2013. Review: a review of semiochemicals associated with bark beetle (Coleoptera: Curculionidae: Scolytinae) pests of coniferous trees: a focus on beetle interactions with other pests and their associates. For. Ecol. Manag. 297, 1–14.

Giunta, A.D., Runyon, J.B., Jenkins, M.J., Teich, M., 2016. Volatile and within-needle terpene changes to Douglas-fir trees associated with Douglas-fir beetle (Coleoptera: Curculionidae) attack. Environ. Entomol. 45 (4), 920–929.

Gray, E., Merzdorf, J., 2019. Earth's freshwater future: extremes of flood and drought. NASA website: Global Climate Change. https://climate.nasa.gov/news/2881/earths-freshwater-future-extremes-of-flood-and-drought/. (Accessed October 22, 2020).

Grégoire, J.C., Piel, F., De Proft, M., Gilbert, M., 2001. Spatial distribution of ambrosia-beetle catches: a possibly useful knowledge to improve mass-trapping. Integrated Pest Manag. Rev. 6 (3–4), 237–242.

Guldin, J.M., 2011. Silvicultural considerations in managing southern pine stands in the context of southern pine beetle. In: Coulson, R.N., Klepzig, K.D. (Eds), Southern Pine Beetle II. Gen. Tech. Rep. SRS-140. US Department of Agriculture Forest Service, Southern Research Station. Asheville, NC, pp. 317–352.

Gurr, G., Wratten, S.D., Altieri, M.A., 2004. Ecological Engineering for Pest Management: Advances in Habitat Manipulation for Arthropods. CSIRO Publishing.

Haack, R.A., 2002. Intercepted bark- and wood-boring insects in the United States: 1985-2000. Newsletter Michigan Entomol. Soc. 6 (3–4), 253–282.

Hain, F.P., Duehl, A.J., Gardner, M.J., Payne, T.L., 2011. Natural history of the southern pine beetle. In: Coulson, R.N., Klepzig, K.D. (Eds.), Southern Pine Beetle II. Gen. Tech. Rep. SRS-140. US Department of Agriculture Forest Service, Southern Research Station, Asheville, NC, pp. 1–24.

Hansen, E.M., Munson, A.S., Wakarchuk, D., Blackford, D.C., Graves, A.D., Stephens, S.S., Moan, J.E., 2019. Advances in semiochemical repellents to mitigate host mortality from the spruce beetle (Coleoptera: Curculionidae). J. Econ. Entomol. 112 (5), 2253–2261.

Hardie, J., Isaacs, R., Pickett, J.A., Wadhams, L.J., Woodcock, C.M., 1994. Methyl salicylate and (–)-(1R, 5S)-myrtenal are plant-derived repellents for black bean aphid, *Aphis fabae* Scop. (Homoptera: Aphididae). J. Chem. Ecol. 20 (11), 2847–2855.

Harper, T., 2019. Douglas-fir beetle infestation is a provincial crisis: B.C. expert. Coast Mountain News. https://www.coastmountainnews.com/news/douglas-fir-beetle-infestation-is-a-provincial-crisis-expert/. (Accessed October 23, 2020).

Hart, S.J., Veblen, T.T., Eisenhart, K.S., Jarvis, D., Kulakowski, D., 2014. Drought induces spruce beetle (*Dendroctonus rufipennis*) outbreaks across Northwestern Colorado. Ecology 95 (4), 930–939.

Hart, S.J., Veblen, T.T., Schneider, D., Molotch, N.P., 2017. Summer and winter drought drive the initiation and spread of spruce beetle outbreak. Ecology 98 (10), 2698–2707.

Hawaii Department of Agriculture website: Coffee berry borer confirmed on Maui, 2017a. http://hdoa.hawaii.gov/blog/main/nr17-1-cbbonmaui/. (Accessed October 19, 2020).

Hawaii Department of Agriculture website: HDOA Quarantines coffee plants on Kauai that may have been shipped from Oahu, 2017b. http://hdoa.hawaii.gov/blog/main/nr17-04kauai-coffee-plants/. (Accessed October 19, 2020).

Henao, L., 2008. Control de la broca del café *Hypothenemus hampei* (Ferrari) con extractos vegetales de plantas de la flora regional (Tesis Tecnólogo en Química). Universidad Tecnológica de Pereira, Facultad de Tecnología, Pereira, CO.

Higuera, P.E., Abatzoglou, J.T., 2021. Record-setting climate enabled the extraordinary 2020 fire season in the western United States. Glob. Change Biol. 27, 1–2. doi:h10.1111/gcb.15388.

Hijii, N., Kajimura, H., Nishibe, Y., 1991. A note on the discoloration and fungal infiltration processes on wood tissues surrounding the gallery systems of scolytid beetles. Bull. Nagoya Univ. For. 11, 31–38.

Holsten, E.H., Thier, R.W., Munson, A.S., Gibson, K.E., 1999. The spruce beetle. Forest Insect and Disease Leaflet 127. US Department of Agriculture, Forest Service, Washington DC.

Hughes, M., Smith, J.A., Mayfield III, A.E., Minno, M.C., Shin, K., 2011. First report of laurel wilt disease caused by *Raffaelea lauricola* on pondspice in Florida. Plant Dis. 95 (12), 1588.

Hughes, M.A., Martini, X., Kuhns, E., Colee, J., Mafra-Neto, A., Stelinski, L.L., et al., 2017a. Evaluation of repellents for the redbay ambrosia beetle, *Xyleborus glabratus*, vector of the laurel wilt pathogen. J. Appl. Entomol. 141.

Hughes, M.A., Riggins, J.J., Koch, F.H., Cognato, A.I., Anderson, C., Formby, J.P., et al., 2017b. No rest for the laurels: symbiotic invaders cause unprecedented damage to southern USA forests. Biol. Invasions 19, 2143–2157.

Hughes, M.A., Smith, J.A., Ploetz, R.C., Kendra, P.E., Mayfield, A., Hanula, J., et al., 2015. Recovery plan for laurel wilt on redbay and other forest species caused by *Raffaelea lauricola* and disseminated by *Xyleborus glabratus*. Plant Health Prog. 16, 173–210.

Inch, S.A., Ploetz, R.C., 2012. Impact of laurel wilt, caused by *Raffaelea lauricola*, on xylem function in avocado, *Persea americana*. For. Pathol. 42, 239–245.

Inch, S.A., Ploetz, R.C., Held, B., Blanchette, R., 2012. Histological and anatomical responses in avocado, *Persea americana*, induced by the vascular wilt pathogen, *Raffaelea lauricola*. Botany 90, 627–635.

Ingram, W.R., 1968. Observations on the control of the coffee berry borer *Hypothenemus hampei* (Ferr.), with endosulfan in Uganda. Bull. Entomol. Res. 30, 539–547.

IPCC, 2007. Summary for policymakers. In Climate Change 2007, published for the Intergovernmental Panel on Climate Change. Cambridge University Press, Cambridge. 18 p.

Janssen, M.P.M., 2011. Endosulfan. A closer look at the arguments against a worldwide phase out. RIVM letter report 601356002/2011. National Institute for Public Health and the Environment. Ministry of Health, Welfare and Sport, The Netherlands.

Jaramillo, J., Chabi-Olaye, A., Kamonjo, C., Jaramillo, A., Vega, F.E., Poehling, H.M., Borgemeister, C., 2009. Thermal tolerance of the coffee berry borer *Hypothenemus hampei*: predictions of climate change impact on a tropical insect pest. PloS One 4 (8), e6487.

Jaramillo, J., Muchugu, E., Vega, F.E., Davis, A., Borgemeister, C., Chabi-Olaye, A., 2011. Some like it hot: the influence and implications of climate change on coffee berry borer (*Hypothenemus hampei*) and coffee production in East Africa. PloS One 6 (9), e24528.

Jaramillo, J., Torto, B., Mwenda, D., Troeger, A., Borgemeister, C., Poehling, H.-M., Francke, W., 2013. Coffee berry borer joins bark beetles in coffee klatch. PLoS One 8 (9), e74277.

Johnstone, W.D., 2002. Thinning Lodgepole Pine in Southeastern British Columbia: 46-year Results (Vol. 63). Research Branch, British Columbia, Ministry of Forests, Victoria, B.C., pp. 1–22.

Keane, R.E., 2001. Can the fire-dependent whitebark pine be saved? Fire Manag. Today 61, 17–20.

Kendra, P.E., Montgomery, W.S., Niogret, J., Epsky, N.D., 2013. An uncertain future for American Lauraceae: a lethal threat from redbay ambrosia beetle and laurel wilt disease-a review. Am. J. Plant Sci. 4, 727–738.

Knowledge@Wharton, 2019. SOCIAL IMPACT: how climate change is killing coffee. Podcast, Wharton Business Daily. https://knowledge.wharton.upenn.edu/article/coffee-climate-change/. (Accessed October 19, 2020).

Kolb, T.E., Fettig, C.J., Ayres, M.P., Bentz, B.J., Hicke, J.A., Mathiasen, R., et al., 2016. Observed and anticipated impacts of drought on forests insects and diseases in the United States. For. Ecol. Manag. 380, 321–334.

Kostyk, B.C., Borden, J.H., Gries, G., 1993. Photoisomerization of antiaggregation pheromone verbenone: biological and practical implications with respect to the mountain pine beetle, *Dendroctonus ponderosae* Hopkins. J. Chem. Ecol. 19, 1749–1759.

Kraemer, M.U., Reiner, R.C., Brady, O.J., Messina, J.P., Gilbert, M., Pigott, D.M., et al., 2019. Past and future spread of the arbovirus vectors *Aedes aegypti* and *Aedes albopictus*. Nat. Microbiol. 4 (5), 854–863.

Kuhnholz, S., Borden, J.H., Uzunovic, A., 2001. Secondary ambrosia beetles in apparently healthy trees: adaptations, potential causes and suggested research. Integr. Pest Manag. Rev. 6, 209–219.

Kupferschmidt, K., 2016. After 40 years, the most important weapon against mosquitoes may be failing. Science Magazine (online). https://www.sciencemag.org/news/2016/10/after-40-years-most-important-weapon-against-mosquitoes-may-be-failing. (Accessed October 23, 2020).

Lapointe, S.L., Stelinski, L.L., 2011. An applicator for high viscosity semiochemical products and intentional treatment gaps for mating disruption of *Phyllocnistis citrella*. Entomol. Exp. Appl. 141 (2), 145–153.

Larson, A.J., Lutz, J.A., Gersonde, R.F., Franklin, J.F., Hietpas, F.F., 2008. Potential site productivity influences the rate of forest structural development. Ecol. Appl. 18 (4), 899–910.

Larson, B.A., Ngoma, T., Silumbe, K., Rutagwera, M.R.I., Hamainza, B., Winters, A.M., et al., 2016. A framework for evaluating the costs of malaria elimination interventions: an application to reactive case detection in Southern Province of Zambia, 2014. Malar. J. 15 (1), 408.

Le Pelley, R.H., 1968. The Pests of Coffee. Longmans, London.

Lehnert, R., 2015. The growing threat of black stem borer. Good Fruit Grower. http://www.goodfruit.com/the-growing-threat-of-black-stem-borer/. (Accessed October 23, 2020).

Lill, A., 2019. The earth is warming and that means spruce beetles are getting bigger but it's not all bad news. CPR News. https://www.cpr.org/2019/08/21/the-earth-is-warming-and-that-means-spruce-beetles-are-getting-bigger-but-its-not-all-bad-news/. (Accessed October 22, 2020).

Lubick, N., 2010. Endosulfan's exit: U.S. EPA pesticide review leads to a ban. Science 328, 1466.

Mafra-Neto, A., de Lame, F.M., Fettig, C.J., Munson, A.S., Perring, T.M., Stelinski, L.L., et al., 2013. Manipulation of insect behavior with specialized pheromone and lure application technology (SPLAT®). In: Beck, J.J., Coats, J.R., Duke, S.O., Koivunen, M.E. (Eds.), Pest Management with Natural Products. ACS Symposium Series, 1141. American Chemical Society Washington, D.C., pp. 31–58.

Mafra-Neto, A., Fettig, C.J., Munson, A.S., Rodriguez-Saona, C., Holdcraft, R., Faleiro, J.R., et al., 2014. Development of specialized pheromone and lure application technologies (SPLAT®) for management of coleopteran pests in agricultural and forest systems, Biopesticides: State of the Art and Future Opportunities. American Chemical Society, pp. 211–242.

Mafra-Neto, A., Ponce, J.I., Urrutia, W.H., Bernardi, C.R., Da Silva, R.O., 2016. Compositions and methods for repelling coffee berry borer. Patent Publication 20160128324.

Mangina, F.L., Makundi, R.H., Maerere, A.P., Maro, G.P., Teri, J.M., 2007. Temporal variations in the abundance of three important insect pests of coffee in Kilimanjaro region, Tanzania. Sokoine University of Agriculture, Tanzania, p. 1. http://www.suaire.sua.ac.tz/handle/123456789/969.

Mariño, Y.A., Vega, V.J., García, J.M., Rodrigues, V., José, C., García, N.M., Bayman, P., 2017. The coffee berry borer (Coleoptera: Curculionidae) in Puerto Rico: distribution, infestation, and population per fruit. J. Insect Sci. 17 (2).

Martini, X., Conover, D., Hughes, M., Smith, J., 2020. Use of semiochemicals for the management of the redbay ambrosia beetle. Insects 11 (11), 796.

Martini, X., Kincy, N., Nansen, C., 2012. Quantitative impact assessment of spray coverage and pest behavior on contact pesticide performance. Pest Manag. Sci. 68 (11), 1471–1477.

Martini, X., Sobel, L., Conover, D., Mafra-Neto, A., Smith, J., 2020. Verbenone reduces landing of the redbay ambrosia beetle, vector of the laurel wilt pathogen, on live standing Redbay trees. Agric. For. Entomol. 22 (1), 83–91.

Mathieu, F., Brun, L.-O., Frerot, B., Suckling, D.M., Frampton, C., 1999. Progression in field infestation is linked with trapping of coffee berry borer, *Hypothenemus hampei* (Col., Scotylidae). J. Appl. Entomol. 123, 535–540.

Mattson, L.R., Coop, J.D., Battaglia, M.A., Cheng, A.S., Sibold, J.S., Viner, S., 2019. Post-spruce beetle timber salvage drives short-term surface fuel increases and understory vegetation shifts. For. Ecol. Manag. 437, 348–359.

Mayfield, A.E.I., Smith, J.A., Hughes, M., Dreaden, T.J., 2008. First report of laurel wilt disease caused by a *Raffaelea* sp. on avocado in Florida. Plant Dis. 92, 976.

McCullough, D.G., Poland, T.M., Cappaert, D., Clark, E.L., Fraser, I., Mastro, V., et al., 2007. Effects of chipping, grinding, and heat on survival of emerald ash borer, *Agrilus planipennis* (Coleoptera: Buprestidae), in chips. J. Econ. Entomol. 100, 1304–1315.

McIntosh, A.C., Macdonald, S.E., 2013. Potential for lodgepole pine regeneration after mountain pine beetle attack in newly invaded Alberta stands. For. Ecol. Manag. 295, 11–19.

Mietkiewicz, N., Kulakowski, D., Veblen, T.T., 2018. Pre-outbreak forest conditions mediate the effects of spruce beetle outbreaks on fuels in subalpine forests of Colorado. Ecol. Appl. 28 (2), 457–472.

Miller, D.R., Lafontaine, J.P., 1991. *cis*-Verbenol: an aggregation pheromone for the mountain pine beetle, *Dendroctonus ponderosae* Hopkins (Coleoptera: Scolytidae). J. Entomol. Soc. B.C. 88, 34–38.

Miller, D.R., Lindgren, B.S., 2000. Comparison of a-pinene and myrcene on attraction of mountain pine beetle, *Dendroctonus ponderosae* (Coleoptera: Scolytidae) to pheromones in stands of western white pine. J. Entomol. Soc. B.C. 97, 41–46.

Miller, D.R., 2002. Short-range horizontal disruption by verbenone in attraction of mountain pine beetle (Coleoptera: Scolytidae) to pheromone-baited funnel traps in stands of lodgepole pine. J. Entomol. Soc. B.C. 99, 103–105.

Miller, L.K., Werner, R.A, 1987. Cold-hardiness of adult and larval spruce beetles *Dendroctonus rufipennis* (Kirby) in interior Alaska. Can. J. Zool. 65 (12), 2927–2930.

Mitchell, R.G., Waring, R.H., Pitman, G.B., 1983. Thinning lodgepole pine increases tree vigor and resistance to mountain pine beetle. For. Sci. 29 (1), 204–211.

Mota-Sanchez, D., Wise, J.C., 2021. The Arthropod Pesticide Resistance Database (APRD). Michigan State University, East Lansing, MI. http://www.pesticideresistance.org.

Mudge, A.D., LaBonte, J.R., Johnson, K.J.R., LaGasa, E.H., 2001. Exotic wood-boring Coleoptera (Micromalthidae,

Scolytidae) and Hymenoptera (Xiphydriidae) new to Oregon and Washington. Proc. Entomol. Soc. Wash. 103, 1011–1019.

Murkowski, L., 2019. Spruce beetle outbreak will require all hands, all lands solutions. Anchorage Daily News. https://www.adn.com/opinions/2019/05/22/spruce-beetle-outbreak-will-require-all-hands-all-lands-solutions/. (Accessed October 22, 2020).

Negrón, J.F., Popp, J.B., 2018. Can spruce beetle (*Dendroctonus rufipennis* Kirby) pheromone trap catches or stand conditions predict Engelmann spruce (*Picea engelmannii* Parry ex Engelm.) tree mortality in Colorado? Agric. For. Entomol. 20 (2), 162–169.

Ngama, S., Korte, L., Bindelle, J., Vermeulen, C., Poulsen, J.R., 2016. How bees deter elephants: beehive trials with forest elephants (*Loxodonta africana cyclotis*) in Gabon. PLoS One 11, e0155690.

Nicholls, C.I., 2009. Bases Agroecológicas para Diseñar e Implementar una Estrategia de Manejo de Hábitat para Control Biológico de Plagas. Vertientes del Pensamiento Agroecológico: Fundamentos y Aplicaciones (ed. By MA Altieri), pp. 207–228. Sociedad Científica Latinoamericana de Agroecología SOCLA, Medellín, CO.

Njihia, T.N., Jaramillo, J., Murungi, L., Mwenda, D., Orindi, B., Poehling, H.M., Torto, B., 2014. Spiroacetals in the colonization behaviour of the coffee berry borer: a 'push-pull'system. PLoS One 9 (11), e111316.

O'Malley, J., 2019. Southcentral Alaska's spruce bark beetle infestation is spreading in Anchorage. There's only one defense. Anchorage Daily News. https://www.adn.com/alaska-news/anchorage/2019/05/11/southcentral-alaskas-spruce-bark-beetle-infestation-is-spreading-in-anchorage/. (Accessed October 22, 2020).

Oatman, M., 2013. when spruce beetles attack! http://www.motherjones.com/blue-marble/2013/10/growing-spruce-beetle-outbreak-tied-drought-colorado. (Accessed October 22, 2020).

Oliveira, C.M., Auad, A.M., Mendes, S.M., Frizzas, M.R., 2013. Economic impact of exotic insect pests in Brazilian agriculture. J. Appl. Entomol. 137 (1–2), 1–15.

Oliver, J.B., Mannion, C.M., 2001. Ambrosia beetle (Coleoptera: Scolytidae) species attacking chestnut and captured in ethanol-baited traps in middle Tennessee. Environ. Entomol. 30, 909–918.

Oregon State University, 2007. PACIFIC NORTHWEST NURSERY IPM Website, Insects: *Xylosandrus germanus* (Blandford). http://oregonstate.edu/dept/nurspest/Xylosandrusgermanus.htm. (Accessed October 23, 2020).

Peña, J.E., Carrillo, D., Duncan, R.E., Capinera, J.L., Brar, G., McLean, S., et al., 2012. Susceptibility of *Persea* spp. and other Lauracear to attack by redbay ambrosia beetle, *Xyloborus glabratus* (Coleoptera: Curculionidae: Scolytinae). Fla. Entomol 95 (3), 783–787.

Peñaloza, M., 2020. California wildfires near tragic milestone: 4 million acres burned. NPR online: https://www.npr.org/2020/10/02/919554698/california-wildfires-near-tragic-milestone-4-million-acres-burned. (Accessed October 19, 2020).

Pitman, G.B., Vité, J.P., Kinzer, G.W., Fentiman Jr, A.F., 1969. Specificity of population-aggregating pheromones in *Dendroctonus*. J. Insect Physiol. 15 (3), 363–366.

Ploetz, R.C., Harrington, T., Hulcr, J., Fraedrich, S., Smith, J.A., Inch, S., et al., 2011. Recovery plan for laurel wilt of avocado (caused by *Raffaelea lauricola*). National Plant Disease Recovery System, Homeland Security Presidential Directive Number 9 (HSPD-9). p. 22.

Ploetz, R.C., Hulcr, J., Wingfield, M.J., de Beer, Z.W., 2013. Destructive tree diseases associated with ambrosia and bark beetles: black swan events in tree pathology? Plant Dis 97 (7), 856–872.

Ploetz, R.C., Kendra, P.E., Choudhury, R.A., Rollins, J.A., Campbell, A., Garrett, K., et al., 2017. Laurel wilt in natural and agricultural ecosystems: understanding the drivers and scales of complex pathosystems. Forests 8, 1–27.

Pope, K., 2019. New research suggests climate change could enable mosquitoes to evolve more rapidly. Yale Climate Connections. https://yaleclimateconnections.org/2019/04/climate-change-could-foster-rapid-mosquito-evolution/. (Accessed October 23, 2020).

Progar, R.A., Fettig, C.J., Munson, A.S., Mortenson, L.A., Snyder, C.L., Kegley, S., et al., 2021. Comparisons of efficiency of two formulations of verbenone (4, 6, 6-trimethylbicyclo [3.1.1] hept-3-en-2-one) for protecting whitebark pine, *Pinus albicaulis* (Pinales: Pinaceae) from mountain pine beetle (Colopetera: Curculionidae). J. Econ. Entomol. 114 (1), 209–214.

Progar, R.A., Gillette, N.E., Fettig, C.J., Hrinkevich, K.H., 2014. Applied chemical ecology of the mountain pine beetle. For. Sci. 60, 414–443.

Puikkonen, K., 2019. New research shows importance of climate on spruce beetle flight. Phys.org. https://phys.org/news/2019-06-importance-climate-spruce-beetle-flight.html. (Accessed October 23, 2020).

Rabaglia, R.J., Dole, S.A., Cognato, A.I., 2006. Review of American Xyleborina (Coleoptera: Curculionidae: Scolytinae) occurring North of Mexico, with an illustrated key. Ann. Entomol. Soc. Am. 99, 1034–1055.

Raffa, K.F., Aukema, B.H., Bentz, B.J., Carroll, A.L., Hicke, J.A., Turner, M.G., Romme, W.H., 2008. Cross-scale drivers of natural disturbances prone to anthropogenic amplification: the dynamics of bark beetle eruptions. Bioscience 58 (6), 501–517.

Ragenovich, I.R., Coster, J.E., 1974. Evaluation of some carbamate and phosphate insecticides against southern pine beetle and *Ips* bark beetles. J. Econ. Entomol. 67, 763–765.

Ranger, C.M., Gorzlancyk, A.M., Addesso, K.M., Oliver, J.B., Reding, M.E., Schultz, P.B., Held, D.W., 2014. Conophthorin enhances the electroantennogram and field behavioural response of *Xylosandrus germanus* (Coleoptera: Curculionidae) to ethanol. Agric. For. Entomol. 16 (4), 327–334.

Ranger, C.M., Reding, M.E., Persad, A.B., Herms, D.A., 2010. Ability of stress-related volatiles to attract and induce attacks by *Xylosandrus germanus* and other ambrosia beetles. Agric. For. Entomol. 12 (2), 177–185.

Ranger, C.M., Tobin, P.C., Reding, M.E., Bray, A.M., Oliver, J.B., Schultz, P.B., et al., 2013. Interruption of the semiochemical-based attraction of ambrosia beetles to ethanol-baited traps and ethanol-injected trap trees by verbenone. Environ. Entomol. 42, 539–547.

Reding, M., Oliver, J., Schultz, P., Ranger, C., 2010. Monitoring flight activity of ambrosia beetles in ornamental nurseries with ethanol-baited traps: influence of trap height on captures. J. Environ. Hortic. 28 (2), 85.

Reiter, P., 2001. Climate change and mosquito-borne disease. Environ. Health Perspect. 109 (Suppl. 1), 141–161.

Rivera, M.J., Martini, X., Conover, D., Mafra-Neto, A., Carrillo, D., Stelinski, L.L., 2020. Evaluation of semiochemical based push-pull strategy for population suppression of ambrosia beetle vectors of laurel wilt disease in avocado. Sci. Rep. 10 (1) 1v12.

Rogers, P., 2020. Heat, lightning, drought create perfect storm for record California fires. Marin Independent J. https://www.marinij.com/2020/09/12/heat-lightning-drought-create-perfect-storm-for-record-california-fires/ (Accessed October 19, 2020).

Ross, D.W., Daterman, G.E., 1994. Reduction of Douglas-fir beetle infestation of high-risk stands by antiaggregation and aggregation pheromones. Can. J. For. Res. 24, 2184–2190.

Ross, D.W., Daterman, G.E., 1995. Efficacy of an antiaggregation pheromone for reducing Douglas-fir beetle *Dendroctonus pseudotsugae* Hopkins (Coleoptera: Scolytidae), infestation in high risk stands. Can. Entomol. 127, 805–811.

Ross, D.W., Wallin, K.F., 2008. High release rate 3-methylcyclohex-2-en-1-one dispensers prevent Douglas-fir beetle (Coleoptera: Curculionidae) infestation of live Douglas-fir. J. Econ. Entomol. 101, 1826–1830.

Ross, D.W., Gibson, K.E., Daterman, G.E., 2001. Using MCH to protect trees and stands from Douglas-fir beetle infestation, FHTET-2001-09. United States Department of Agriculture, Forest Service, Forest Health Technology Enterprise Team, Morgantown, WV.

Ross, D.W., Gibson, K.E., Daterman, G.E., 2015. Using MCH to protect trees and stands from Douglas-fir beetle infestation. US Department of Agriculture Forest Service, Forest Health Technology Enterprise Team, Morgantown, WV.

Ross, D.W., Gibson, K.E., Their, R.W., Munson, A.S., 1996. Optimal dose of an antiaggregation pheromone (3-methylcyclohex-2-en-1-one) for protecting live Douglas-fir from attack by *Dendroctonus pseudotsugae* (Coleoptera: Scolytidae). J. Econ. Entomol. 89, 1204–1207.

Ross, D., Daterman, G., Gibson, K., 2002. Elution rate and spacing of antiaggregation pheromone dispensers for protecting live trees from *Dendroctonus pseudotsugae* (Coleoptera: Scolytidae). J. Econ. Entomol. 95, 778–781.

Ryan, S.J., Carlson, C.J., Mordecai, E.A., Johnson, L.R., 2019. Global expansion and redistribution of *Aedes*-borne virus transmission risk with climate change. PLoS Negl. Trop. Dis. 13 (3), e0007213.

Salom, S.M., Grosman, D.M., McClellan, Q.C., Payne, T.L., 1995. Effect of an inhibitor-based suppression tactic on abundance and distribution of the southern pine beetle (Coleoptera: Scolytidae) and its natural enemies. J. Econ. Entomol. 88 (6), 1703–1716.

Salvail, A., 2019. Spruce beetle outbreak likely to hit high-country forests. Aspen Daily News. https://www.aspendailynews.com/news/spruce-beetle-outbreak-likely-to-hit-high-country-forests/article_e1c5f216-b338-11e9-b303-0f69ee053345.html. (Accessed October 22, 2020).

Santoro, P.H., Zorzetti, J., Constanski, K.C. Neves, P.M.O.J., 2011. Repência da broca-do-cafeeiro, Hypothenemus hampei (Ferrari, 1876) (Coleoptera: curculionidae) a extratos vegetais. VII Simpósio de Pesquisa dos Cafés do Brasil, Araxá.

Santos, M.R.A., Silva, A.G., Lima, RA, Lima, DKS, Sallet, LAP, et al., 2010. Atividade inseticida do extrato das folhas de *Piper hispidum* (Piperaceae) sobre a broca-do-café (*Hypothenemus hampei*). Rev. Bras. Bot. 33, 319–324.

Schiffman, R., 2019. As climate changes, Colombia's small coffee farmers pay the price. Yale Environment 360, Published at the Yale School of the Environment. https://e360.yale.edu/features/as-climate-changes-colombias-small-coffee-farmers-pay-the-price. (Accessed October 19, 2020).

Schlossberg, T., 2016. Warmer winter brings forest-threatening beetles north. New York Times. http://www.nytimes.com/2016/03/22/science/southern-pine-beetles-new-england-forests.html?_r=2. (Accessed October 23, 2020).

Schmutterer, H., 1990. Properties and potential of natural pesticides from the neem tree, *Azadirachta indica*. Annu. Rev. Entomol. 35 (1), 271–297.

Schwandt, J., Kegley, S., 2004. Mountain pine beetle, blister rust, and their interaction on whitebark pine at Trout Lake and Fisher Peak in northern Idaho from 2001 to 2003. Forest Health Protection Report. US Department of Agriculture Forest Service, Northern Region, Missoula, MT.

Seybold, S.J., Bentz, B.J., Fettig, C.J., Lundquist, J.E., Progar, R.A., Gillette, N.E., 2018. Management of western North American bark beetles with semiochemicals. Annu. Rev. Entomol. 63, 407–432.

Seybold, S.J., Penrose, R.L., Graves, A.D., 2016. Invasive bark and ambrosia beetles in California Mediterranean forest ecosystems. In: Lieutier, F., Paine, T.D. (Eds.), Insects and Diseases of Mediterranean Forest Systems. Springer, Cham, pp. 583–662.

Shulaev, V., Silverman, P., Raskin, I., 1997. Airborne signalling by methyl salicylate in plant pathogen resistance. Nature 385 (6618), 718–721.

Smith, J.A., Dreaden, T.J., Mayfield Iii, A.E., Boone, A., Fraedrich, S.W., Bates, C., 2009. First report of laurel wilt disease caused by *Raffaelea lauricola* on sassafras in Florida and South Carolina. Plant Dis. 93, 1079.

Spence, D.J., Smith, J.A., Ploetz, R., Hulcr, J., Stelinski, L.L., 2013. Effects of chipping on emergence of the redbay ambrosia beetle (Coleoptera: Curculionidae: Scolytinae) and recovery of the laurel wilt pathogen from infested wood chips. J. Econ. Entomol. 106, 2093–2100.

Spicer, J.A., Brimble, M.A., Rowan, D.D., 1993. Oxidation of α-farnesene. Aust. J. Chem. 46 (12), 1929–1939.

Sponagel, K.W., 1994. The Coffee Borer *Hypothenemus hampei* in Plantations of Robusta Coffee in Equatorial Amazonia. Wissenschaftlicher Fachverlag, Niederkleen, p. 185.

State of Colorado, 2016. Douglas-Fir Beetle, Quick Guide Series. Colorado State Forest Service, Colorado State University and Colorado Department of Natural Resources, pp. 1–8.

Stelinski, L.L., Miller, J.R., Ledebuhr, R., Gut, L.J., 2006. Mechanized applicator for large-scale field deployment of paraffin-wax dispensers of pheromone for mating disruption in tree fruit. J. Econ. Entomol. 99 (5), 1705–1710.

Stelloh, T., 2020. California exceeds 4 million acres burned by wildfires in 2020. NBC News. https://www.nbcnews.com/news/us-news/california-exceeds-4-million-acres-burned-wildfires-2020-n1242078 (Accessed October 19, 2020).

Strom, B., Clarke, S.R., 2011. Use of semiochemicals for southern pine beetle infestation management and resource protection. In: Coulson, R.N., Klepzig, K.D. (Eds.), Southern Pine Beetle II. Gen. Tech. Rep. SRS-140. US Department of Agriculture Forest Service, Southern Research Station, Asheville, NC, 381–397.

Strom, B.L., Oldland, W.K., Meeker, J.R., Dunn, J., 2015. Evaluation of general-use insecticides for preventing host colonization by New Jersey southern pine beetles. Arboric. Urban For. 41 (2), 88–102.

Sullivan, B.T., 2011. Southern pine beetle behavior and semiochemistry. In: Coulson, R.N., Klepzig, K.D. (Eds.), Southern Pine Beetle II. Gen. Tech. Rep. SRS-140. U.S. Department of Agriculture Forest Service, Southern Research Station, Asheville, NC, pp. 25–50.

Szal, A., 2017. Pesticide ban fuels beetle infestation in Brazilian coffee crop. https://www.chem.info/news/2017/07/pesticide-ban-fuels-beetle-infestation-brazilian-coffee-crop. (Acccessed October 23, 2020).

Tang, C., Davis, K.E., Delmer, C., Yang, D., Wills, M.A., 2018. Elevated atmospheric CO_2 promoted speciation in mosquitoes (Diptera, Culicidae). Commun. Biol. 1 (1), 1–8.

Teixeira, L.A.F., Mason, K., Mafra-Neto, A., Isaacs, R., 2010. Mechanically-applied wax matrix (SPLAT-GBM) for mating disruption of grape berry moth (Lepidoptera: Tortricidae). Crop Prot. 29 (12), 1514–1520.

Teketay, D., 1999. History, botany and ecological requirements of Coffee. Walia. J. Ethiopian Wildlife National Hist. Soc. 20, 28–50.

Thacker, J.R., 2002. An Introduction to Arthropod Pest Control. Cambridge University Press, Cambridge, UK.

U.S. Department of Agriculture (USDA), 2014. 2012 Census of Agriculture, vol. 1. Part 52: Puerto Rico. USDA, Beltsville, MD.

USDA, 2015. Distribution of counties with laurel wilt disease by year of initial detection. http://www.fs.usda.gov/Internet/FSE_DOCUMENTS/stelprd3840620.pdf. (Accessed October 23, 2020).

USDA National Agricultural Statistics Service (NASS), 2019. Noncitrus Fruits and Nuts 2018 Summary. 101. https://www.nass.usda.gov/Publications/Todays_Reports/reports/ncit0619.pdf. (Accessed October 29, 2020).

U.S. Environmental Protection Agency, 2016. Proposal to revoke chlorpyrifos food residue tolerances. https://archive.epa.gov/epa/ingredients-used-pesticide-products/proposal-revoke-chlorpyrifos-food-residue-tolerances.html. (Accessed October 25, 2020).

U.S. Environmental Protection Agency, 2010. Endosulfan phase-out https://archive.epa.gov/pesticides/reregistration/web/html/endosulfan-agreement.html. (Accessed October 23, 2020).

Ungerer, M.J., Ayres, M.P., Lombardero, M.J., 1999. Climate and the northern distribution limits of *Dendroctonus frontalis* Zimmermann (Coleoptera: Scolytidae). J. Biogeogr. 26, 1133–1145.

University of California: Pest and Diseases of Southern California Oaks, 2014. http://ucanr.edu/sites/socaloakpests/Polyphagous_Shot_Hole_Borer/. (Accessed October 25, 2020).

University of Hawaii, College of Tropical Agriculture and Human Resources, 2018. Extension website: the coffee berry borer is in Hawai'i; How can we manage it? https://www.ctahr.hawaii.edu/site/CBBManage.aspx. (Accessed October 19, 2020).

Van DerLaan, N., Ginzel, M., 2013. The capacity of conophthorin to enhance the attraction of two *Xylosandrus* species (Coleoptera: Curculionidae: Scolytinae) to ethanol and the efficacy of verbenone as a repellent. Agric. For. Entomol. 15, 391–397.

Vega, F.E., Rosenquist, E., Collins, W., 2003. Global project needed to tackle coffee crisis. Nature 435, 343.

Vega, F.E., Infante, F., Castillo, A., Jaramillo, J., 2009. The coffee berry borer, *Hypothenemus hampei* (Ferrari) (Coleoptera: Curculionidae): a short review, with recent findings and future research directions. Terr. Arthropod Rev. 2 (2), 129–147.

Vega, F.E., Infante, F., Johnson, A.J., 2015. The genus *Hypothenemus*, with emphasis on *H. hampei*, the coffee berry borer In: Fernando E. Vega, F.E., Hofstetter, F.W. (Eds.), Bark Beetles, Biology and Ecology of Native and Invasive Species. Academic Press, London, UK, pp. 427–494.

Vega, F.E., Kramer, M., Jaramillo, J., 2011. Increasing coffee berry borer (Coleoptera: Curculionidae: Scolytinae) female density in artificial diet decreases fecundity. J. Econ. Entomol. 104 (1), 87–93.

Vega, F.E., Simpkins, A., Miranda, J., Harnly, J.M., Infante, F., Castillo, A., et al., 2017. A potential repellent against the coffee berry borer (Coleoptera: Curculionidae: Scolytinae). J. Insect Sci. 17 (6), 122.

Vose, J.M., Peterson, D.L., Domke, G.M., Fettig, C.J., Joyce, L.A., Keane, R.E., et al., 2018. Forests. In: Reidmiller, D.R., Avery, C.W., Easterling, D.R., Kunkel, K.E., Lewis, K.L.M., Maycock, T.K., Stewart, B.C. (Eds.), Impacts, Risks, and Adaptation in the United States: Fourth National Climate Assessment, Volume II. U.S. Global Change Research Program, Washington, DC, pp. 223–258.

Waldron, J.D., 2011. Forest Restoration following southern pine beetle. In: Coulson, R.N., Klepzig, K.D. (Eds.), Southern Pine Beetle II. Gen. Tech. Rep. SRS-140. U.S. Department of Agriculture Forest Service, Southern Research Station, Asheville, NC, pp. 353–363.

Walton, A., 2013. Provincial-level projection of the current mountain pine beetle outbreak: update of the infestation projection based on the provincial aerial overview surveys of forest health conducted from 1999 through 2012 and the BCMPB Model (Year 10). BC Ministry of Forests, Lands and Natural Resource Operations: Victoria, BC. http://www.for.gov.bc.ca/ftp/hre/external/!publish/web/bcmpb/year10/BCMPB.v10.BeetleProjection.Update.pdf. (Accessed October 25, 2020).

Wang, B., Mastro, V.C, McLane, W.H., 2000. Impacts of chipping on surrogates for the long-horned beetle *Anoplophora glabripennis* (Coleoptera: Cerambycidae) in logs. J. Econ. Entomol. 93, 1832–1836.

Waterhouse, D.F., Norris, K.R., 1989. Biological control: pacific prospects–Supplement 1 (No. 435-2016-33676). Australian Centre for International Agricultural Research, Canberra.

Weber, B.C., 1980. *Xylosandrus germanus* (Blandford) (Coleoptera: Scolytidae): an ambrosia beetle pest of young hardwood plantations in the United States, Abstracts of the 16th International Congress of Entomology. Kyoto.

Weber, B.C., McPherson, J.E., 1983. World list of host plants of *Xylosandrus germanus* (Blandford) (Coleoptera: Scolytidae). Coleopt. Bull. 37, 114–134.

Whitney, H.S., 1982. Relationships between bark beetles and symbiotic organisms. In: Mitton, J.B., Sturgeon, K.B. (Eds.), Bark Beetles in North American Conifers. University of Texas Press, Austin, pp. 183–211.

Wilson, J., Irish-Brown, A., Shane, B., 2014. Black stem borer: an opportunistic pest of young fruit trees under stress. Michigan State University Extension website. http://msue.anr.msu.edu/news/black_stem_borer_an_opportunistic_pest_of_young_fruit_trees_under_stress. (Accessed October 22, 2020).

Wood, D.L., Koerber, T.W., Scharpf, R.F., Storer, A.J., 2003. Pests of the Native California Conifers (Vol. 70). Univ of California Press, Berkeley, CA.

Worland, J., 2018. Your morning cup of coffee is in danger. Can the industry adapt in time? Time. http://time.com/5318245/coffee-industry-climate-change. (Accessed October 19, 2020).

World Health Organization, 2015. Malaria: insecticide resistance. World Health Organization, Geneva. http://www.who.int/malaria/areas/vector_control/insecticide_resistance/en/ (Accessed October 25, 2020).

World Health Organization, 2018. Global report on insecticide resistance in malaria vectors: 2010-2016. World Health Organization, Geneva, pp. 1–72. https://apps.who.int/iris/bitstream/handle/10665/272533/9789241514057-eng.pdf (Accessed October 20, 2020).

World Health Organization, 2012. The Global Plan for Insecticide Resistance Management in Malaria Vectors (GPIRM). World Health Organization, Geneva. https://www.who.int/malaria/publications/atoz/gpirm/en/. (Accessed October 20, 2020.).

Wright, M.G., Spencer, C., Cook, R.M., Henley, M.D., North, W., Mafra-Neto, A., 2018. African bush elephants respond to a honeybee alarm pheromone blend. Curr. Biol. 28 (14), R778–R780.

Yale School of the Environment, 2019. Climate change will expose half of world's population to disease-spreading mosquitoes by 2050. YaleEnvironment360. https://e360.yale.edu/digest/climate-change-will-expose-half-of-worlds-population-to-disease-spreading-mosquitoes-by-2050. (Accessed October 23, 2020).

Yong, E., 2015. This beetle is ruining your coffee with the help of bacteria. National Geographic. http://phenomena.nationalgeographic.com/2015/07/14/this-beetle-is-ruining-your-coffee-with-the-help-of-bacteria/. (Accessed October 19, 2020).

Zhong, H., Hastings, F.L., Hain, F.P., Werner, R.A., 1994. Toxicity of carbaryl toward the southern pine beetle in filter paper, bark and cut bolt bioassays. J. Entomol. Sci. 29, 247–253.

Zogas, K., 2001. Summary of thirty years of field testing of MCH: antiaggregation pheromone of the spruce bark beetle and the Douglas-fir beetle. U.S. Department of Agriculture, Forest Service, Alaska Region, State & Private Forestry, Juneau, AK.

CHAPTER

16

The role of arthropod repellents in the control of vector-borne diseases

Stephen P. Frances[a], *Mustapha Debboun*[b]

[a]Australian Defence Force Malaria and Infectious Disease Institute, Gallipoli Barracks, Enoggera, Qld, Australia, [b]Delta Mosquito & Vector Control District, Visalia, CA, United States

16.1 Introduction

The use of arthropod repellents is the first line of defense and protection against arthropods. They provide protection against nuisance, biting, vectors, and vector-borne diseases. In this chapter, we discuss both topical (skin) and spatial arthropod repellent treatments that have been used to reduce biting of haematophagous arthropods, including mosquitoes, and some biting flies, as well as ticks and mites. The main focus is to discuss the studies that have tested the arthropod repellent products against vectors of vector-borne diseases, and the future role of topical and spatial arthropod repellents in reducing the incidence of vector-borne diseases. An advantage of the use of commercial arthropod repellent products is that their safety for human use has been reviewed so that countries with limited resources can purchase them and investigate field use in small or large communities.

Readers will note that none of the methods or their combination is 100% effective due to inconsistency of arthropod repellent application, and not all individuals in a community adhere to personal protective measures (Strickman and Debboun, 2013).

The use of topical arthropod repellents applied to exposed skin is the most important personal intervention against vectors of vector-borne diseases, and numerous studies have been undertaken to enumerate the protection against biting arthropods. Some of the main active ingredients used in repellent formulations for biting arthropods include N,N-diethyl-3-methylbenzamide (DEET), 2-(2-hydroxyethyl)-1-piperidine carboxylic acid 1-methylpropyl ester

Advances in Arthropod Repellents.
DOI: https://doi.org/10.1016/B978-0-323-85411-5.00006-6

(Picaridin), ethyl butylacetylaminoproprionate (IR3535), Lemon Scented Eucalyptus Oil or p-menthane-3,8-diol (PMD), methyl nonyl ketone (2-Undecanone or BIoUD), and a recent new repellent active ingredient, 4,α,5-Dimethyl-1,2,3,4,4α,5,6,7-octahydro-7-keto-3-isopropenyl naphthalene (Nootkatone) that was registered by the US Environmental Protection Agency (EPA) in August 2020. All of these active ingredients will be briefly discussed.

16.2 N,N-diethyl-3-methylbenzamide

This active ingredient (also known as *N,N*-diethyl-*m*-toluamide) was synthesized and developed in collaboration between the US Department of Agriculture and the US Department of Defense and became widely available in 1956. Since its introduction and registration by EPA for use by the public in 1957, DEET has become the most widely used active ingredient in topical arthropod repellents to protect against a variety of biting flies, especially mosquitoes, and other arthropods such as chiggers, ticks, and leeches. DEET has a broad spectrum of activity and has become the gold standard for the comparison of novel active ingredients for other arthropod repellent formulations (Fig. 16.1). It is available worldwide in a wide variety of formulations including aerosols, cream, lotions, sprays, gels, sticks, and wipes at concentrations of 5%–100%.

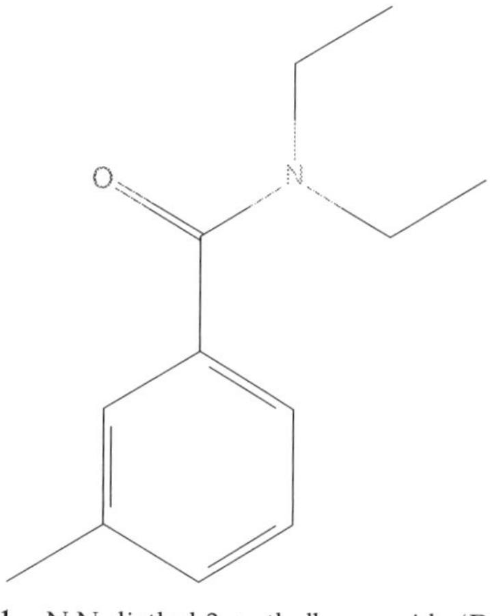

FIG. 16.1 N,N-diethyl-3-methylbenzamide (*DEET*).

Numerous formulations containing DEET have been evaluated in the field and in laboratory settings since the 1940s (Frances, 2007a). DEET provided good protection against all genera of mosquitoes, including *Culex* spp. and *Aedes* spp. Field studies over the last 20 years in Africa, Australia, Papua New Guinea, and Thailand have shown that DEET provided less protection against *Anopheles* spp. than against Culicine mosquitoes. The response of different mosquito species to DEET is variable. Field tests of arthropod repellent formulations containing DEET against biting *Culex* spp., *Aedes* spp., *Mansonia* spp., and *Verralina* spp. have shown longer protection against these mosquitoes compared to *Anopheles* spp. (Frances et al., 1998).

There have been a number of reviews of the safety of DEET (Goodyer and Behrens, 1998; Sudakin et al., 2003; Diaz, 2016), which have attested to its generally acceptable safety profile. There are only a few reports of systemic toxicity in human adults following dermal application. The use of DEET was shown to be safe in the second and third trimester of pregnancy (McCready et al., 2001), and animal models do not indicate any teratogenic effects (Schoenig et al., 1994). Recently, a 2020 study found no evidence that DEET exposure has any impact on the human biomarkers related to systemic inflammation, immune function, liver function, and kidney function (Haleem et al., 2020). The scientific evidence and its continual use for more than 64 years have shown that DEET is one of the best broad-spectrum repellents available for minimizing the bites of mosquitoes and other biting arthropods.

16.3 Picaridin

This active ingredient was developed by Bayer AG in the 1990s, and by 2005, it was registered in over 50 countries throughout the world as a 7% formulation referred to as Cutter Advanced (Fig. 16.2).

A number of Picaridin formulations have been evaluated throughout the world (Frances, 2007b). In comparative studies conducted to date, Picaridin has been shown to be as good as or better than comparable formulations of DEET. In initial field tests in Malaysia (Yap et al., 1998, 2000) against *Ae. albopictus* and *Cx. quinquefaciatus*, Picaridin provided similar protection to DEET. In subsequent field studies in the United States (Barnard et al., 2002) and Australia (Frances et al., 2002), Picaridin provided extended protection against mosquitoes. In a seminal study in Burkino Faso against *Anopheles gambiae*, the main vector of malaria in Africa, Picaridin provided the best protection compared to DEET and IR3535 (Costantini et al., 2004). In human field efficacy trials, they found that 10% and 20% concentrations of Picaridin provided high level of protection for up to 12 hours against lone star ticks, *Amblyomma americanum*.

16.4 p-Menthane-3,8-diol

The principal repellent component of Oil of Lemon Eucalyptus extract is PMD, which is the main by product after hydrodistillation (Fig. 16.3). PMD is available in pump sprays in concentrations of 10% to 40%. The repellent was known initially as Quwenling when used in China. The first field study in the United States of PMD provided poor results (Schreck and Leonhardt, 1991). However, subsequent studies in the laboratory and field have shown PMD to have equal or better efficacy as DEET (Goodyer et al., 2010, 2020). Formulations containing this active ingredient were evaluated against Anopheline mosquitoes in Tanzania (Trigg, 1996). It was found that 50% PMD provided 6–8 hours protection against *An. gambiae*, the main malaria vector, compared with 7 hours protection provided by

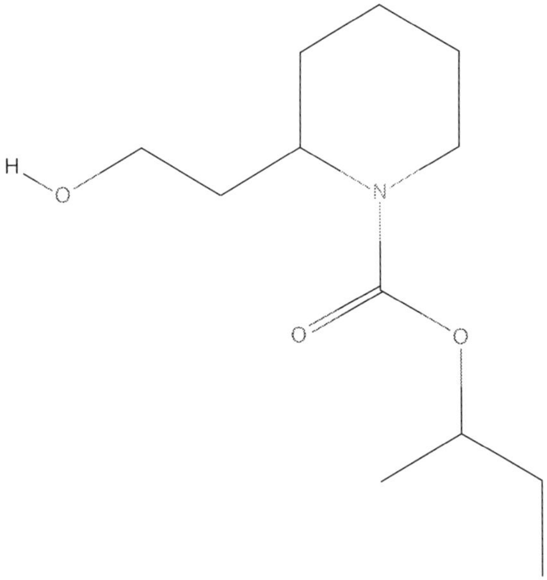

FIG. 16.2 Picaridin.

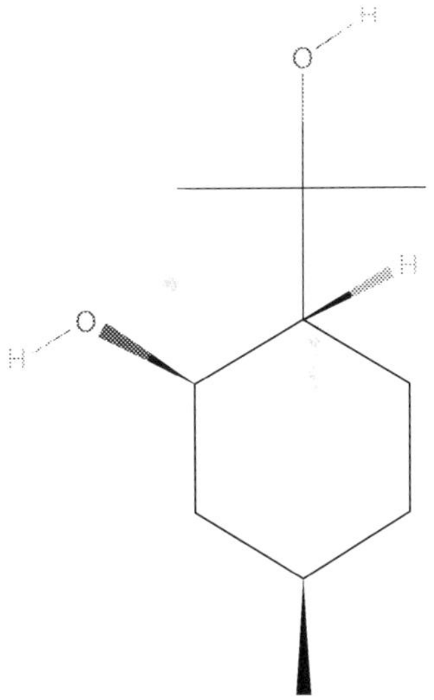

FIG. 16.3 p-menthane-3,8-diol (*PMD*).

50% DEET. Carroll and Loye (2006) showed that PMD provided excellent protection against *Ae. melanimon* and *Ae. vexans* in field trials in California. Also, PMD reduced attachment and blood-feeding by 77% against the tick vectors of Lyme disease, Rocky Mountain spotted fever, and was effective against some species of biting midges (Trigg and Hill, 1996).

16.5 IR3535

The efficacy of formulations containing this active ingredient, ethyl butyl acetylaminopropionate (IR3535) are reviewed by Puccetti (2007). IR3535 is a synthetic molecule derived from a natural amino acid, β-alanine (Fig. 16.4). It was developed in the early 1970s and is currently in more than 150 consumer products worldwide. This chemical has been used in commercial arthropod repellent products since 1999. Field tests conducted in Florida, showed 25% IR3535 in ethanol provided protection of 3–4 hours, compared with 5 hours for 25% DEET and 7 hours for 25% Picaridin (Barnard et al., 2002). Field trials in Burkino Faso showed higher concentrations of IR3535 were needed to provide similar protection as DEET and Picaridin (Costantini et al., 2004). In other field trials, IR3535 provided good protection against mosquitoes in southeast Asia (Thavara et al., 2001; Liu et al., 2003) and greater efficacy against onchocerciasis-transmitting black flies and leishmaniasis-transmitting sand flies in endemic areas than DEET (Naucke et al., 2006).

16.6 2-Undecanone

The arthropod repellent compound, methyl nonyl ketone (2-undecanone or BioUD) is a US EPA registered natural compound from leaves and stems of the wild tomato plant, *Lycopersicon hirsutum* (Farrar and Kennedy, 1987) (Fig.16.5). It is also known commercially as Bite Blocker with BioUD formula and was classified as a biopesticide product in 2007. The Bite Blocker BioUD arthropod repellent formulation is non-flammable and not a plasticizer. It is safe for use on pregnant women, children, elderly, and can be reapplied as often as needed. Witting-Bissinger et al. (2008) evaluated 7.75% 2-undecanone (Bite Blocker BioUD) in comparison to 30% DEET in North Carolina against field populations of the mosquito, *Psorophora ferox* and found that BioUD provided similar repellency

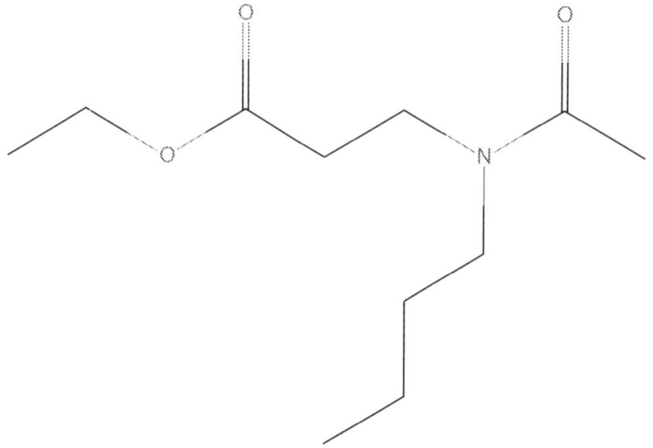

FIG. 16.4 IR 3535.

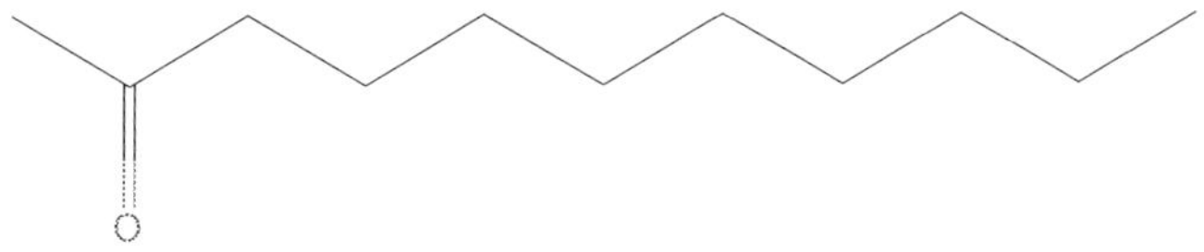

Fig. 16.5 2-Undecanone.

as 30% DEET and a 90% reduction in mosquito bites for up to 6 hours. In another study conducted by Qualls et al. (2011), they found that Bite Blocker BioUD Insect Repellent provided 140 minutes of protection against the floodwater mosquito, *P. columbiae* in St. Johns County, FL. In addition, BioUD was found to be an effective arthropod repellent against ticks (Witting-Bissinger et al., 2008; Bissinger et al., 2009a, 2009b).

16.7 Nootkatone

Nootkatone is an active ingredient that was isolated from the terpene oil of the Alaskan yellow cedar tree, *Cupressus nootkatensis* and is essentially the essence of grapefruit and has a very pleasant citrus-like odor (Dolan and Panella, 2011) (Fig.16.6). The CDC discovered its arthropod repellent properties about 25 years ago when investigating plant-based products for controlling medically important arthropods, particularly for new tick control compounds to protect from tick bites and Lyme disease. Recently, i.e., on August 10, 2020, the US EPA approved and registered Nootkatone as the newest arthropod repellent. Nootkatone has been found to be an effective repellent against deer ticks, lone star ticks, fleas, and mosquitoes (Panella et al., 2005; Dolan et al., 2009; Jordan et al., 2012).

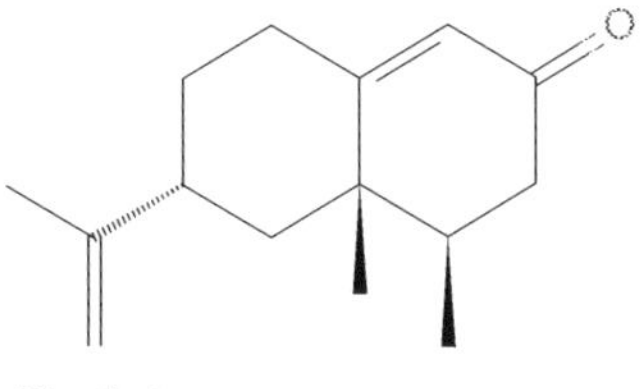

FIG. 16.6 Nootkatone.

16.8 Cost, formulation, and user acceptability

An important consideration in the use of topical arthropod repellents by individuals within communities is the cost of obtaining them (Frances and Wirtz, 2005). This is a limiting factor in the use of arthropod repellents in developing countries, where vector-borne diseases may be endemic and personal incomes are low. This has led to the development of lower-cost arthropod repellent formulations. For example, an arthropod repellent soap formulation containing a mixture of 20% DEET and 0.5% permethrin was developed in Australia in 1985 (Simmons, 1985). It was prepared in small blocks (approximate weight, 70 g) and packed in relatively low-cost greaseproof paper, and was initially priced at $US 0.25 per piece. This formulation was applied to wet skin, lathered, and the residue was left on the skin surface to dry. Several field trials showed the arthropod soap formulation provided satisfactory protection against mosquitoes in Malaysia (Yap, 1986), Papua New Guinea (Charlwood and Dagoro, 1987), Australia (Frances, 1987), India (Mani et al., 1991), and Ecuador and Peru (Kroeger et al., 1997). This formulation was commercialized and marketed as "Mosbar" in Southeast Asia. A survey of personal protective measures used by inhabitants of East Honiara, Solomon Islands, showed a variety of personal protective measures were used by people to protect themselves from potential vectors of malaria (Bell et al., 1997). The survey showed 10.3% of respondents used Mosbar and 8.4% used unidentified arthropod repellent formulations. The study also showed that respondents who used prophylactic drugs or Mortein (pyrethroid aerosol) had increased protection against malaria (Bell et al., 1997).

Another important consideration in the use of arthropod topical repellents and other personal protective measures is the user acceptability of the formulation. A number of studies have been undertaken to determine what factors affect the use of arthropod repellents by individuals and groups (Frances and Debboun, 2007).

The availability of personal protective measures against mosquitoes is variable within communities, and differences are due primarily to socio-economic reasons. In most communities, the seasonal increase in the density of mosquitoes is expected to result in an increase in the use of mosquito control and protection products. In Tanzania, Chavasse et al. (1996) showed a relationship between mosquito densities obtained from trap collections and the sale of mosquito coils in the shops of Mikocheni, Dar es Salaam. In Gambia, districts with higher mosquito densities had higher rates of bed net usage (Thomson et al., 1994). Conversely, a study in southern Tanzania showed that mosquitoes were diverted away from households where the occupants used arthropod repellents to households who did not (Maia et al., 2012). These researchers noted that policymakers should take into consideration results showing vectors diverted from privileged families to those less privileged who may be exposed to nuisance and vector mosquitoes.

The evaluation of new active ingredients and formulations of arthropod repellents continues. Many studies have tested plant extracts as topical repellents against mosquitoes. The success of these active ingredients to repel mosquitoes has been variable, and as yet, active ingredients to replace those mentioned earlier, namely DEET, Picaridin, PMD, and IR3535, have not become widely available. It is possible that new and potentially safer active ingredients may be developed in the future (Nentwig et al., 2017; Tisgratog et al., 2018).

16.9 Spatial arthropod repellents

In the last decade, arthropod repellent researchers have investigated the use of highly volatile chemicals that are dispersed around people using candles, coils (Fig. 16.7), and battery-operated dispensing devices to protect against arthropod bites by providing spatial repellency. For example, spatial arthropod repellents such as the Off! Clip-On device have been suggested as area repellent products to protect against mosquito-borne arboviruses and malaria in the future (Achee et al., 2019).

A number of recent studies have evaluated the use of battery-powered devices dispersing

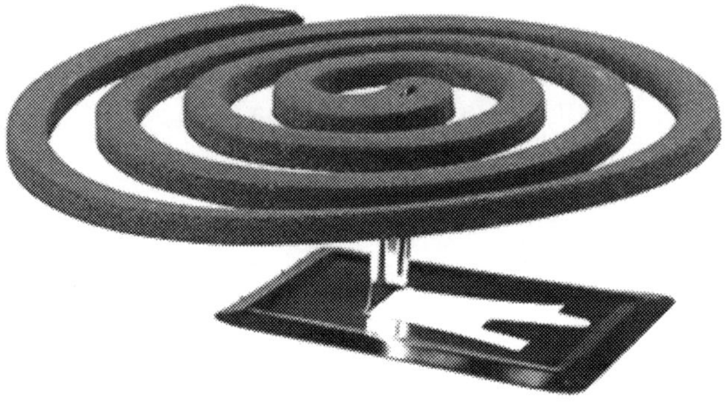

FIG. 16.7 Mosquito Coil, courtesy of Coleman® Repellents, Wisconsin Pharmacal Co., all rights reserved.

the chemical metofluthrin, and some conducted in North Queensland, Australia, have shown that they have potential to control *Aedes aegypti*, the main vector of dengue (Buhagiar et al, 2017; Darbro et al., 2017). The trial by Lucas et al. (2007) showed that these devises reduced landing rates of mosquitoes by 85% (5.5 landings per minute) to 100 % protection against *Ae. canadensis* in Pennsylvania in 2003. Bibbs and Xue (2016) evaluated a metofluthrin emanator against caged *Ae. aegypti* in the field and showed high mortality of mosquitoes at 0.3 m from the emanator, thereby showing that mortality of mosquitoes could be provided outdoors. A battery-powered chemical emanator called Off! Clip-On was evaluated in field trials in northeastern Florida by Xue et al. (2012). They showed the emanator provided 79% protection against *Ae. albopictus* and *Ae. taeniorhynchus* in the field and 79% protection against *Ae. taeniorhynchus* for 3 hours after it had remained open for more than a week. In contrast, Lucas et al. (2007) showed that devices used for 40 hours or more would be unlikely to provide good protection in the field. This was contradicted by the findings of Kawada et al. (2004a, 2004b) who demonstrated that impregnated papers remained active for a 4-week period. A more recent study in Australia showed that a battery-powered emanator emitting 31% metofluthin provided only 42.2 to 60.8% protection against *Ae. vigilax*, an important pest and arbovirus vector (Frances et al., 2020). The recent studies have shown the potential of battery-powered devices, which can be attached to clothing at times when biting mosquitoes are active. These devices may be preferred to the application of arthropod repellents topically on the skin. The overall effectiveness of these chemical emanators has been variable, but it is likely that newer, better emanators, and active ingredients will be developed in the future.

16.10 The use of arthropod repellents against vectors and vector-borne diseases

The World Health Organization requires field studies undertaken in several countries that show a positive effect on vector-borne diseases such as malaria, conducted in randomized control trials (World Health Organization, 2013). Arthropod repellents have been widely used to avoid and lessen the chance of getting vector-borne diseases (Barnard et al., 1998, 2000; Barnard and Xue, 2004; Rowland et al., 2004; Hill et al., 2007; Debboun and Strickman, 2013; Onyango and Moore, 2015).

Historically, the first report of an arthropod repellent chemical causing a significant reduction in vector-borne disease was shown in Australia during World War II (Rutledge et al., 1978). During World War II, Allied forces operating in the southwest Pacific region experienced significant morbidity and mortality from scrub typhus, an arthropod-borne disease caused by *Orientia tsutsugamushi* (Hayashi) and carried by trombiculid mite larvae, commonly called chiggers (McCulloch, 1946). Studies in Australia and Papua New Guinea in 1943 showed that dibutylphthalate applied externally to military shirts and trousers provided excellent protection against chiggers. A concerted effort was undertaken to show soldiers the best method of application and encouraged its use among the Allies, and in 1944, there was an 80% decrease in the incidence of scrub typhus (McCulloch, 1947). This method was subsequently adapted for use in civilian communities after the war for protection against chiggers and scrub typhus.

Arthropod repellent products have been studied in malaria-endemic areas as a method to reduce the risk of exposure to *Anopheles* spp. during the period between dusk and when people went to bed. Onyango and Moore (2015) showed that reducing the vectorial capacity of a mosquito vector of disease was integral to reducing the incidence of that disease. They showed that reducing human-biting rate of the vector by 50%, resulted in a consequent 75% reduction in the vectorial capacity of a mosquito population. Their use has a strong effect on the overall vectorial capacity by reducing the probability of infecting or being infected by an arthropod vector (Smith et al., 2012). Thus, the use of arthropod repellents will reduce the probability of infecting or being infected by the vector.

Between April 1995 and September 1996, McCready et al. (2001) undertook an evaluation of the use of DEET repellent in preventing malaria in pregnant Karen women, living on the Thai-Burmese border. They showed a 28% reduction in the incidence of *Plasmodium falciparum* malaria in the women using DEET compared to no disease reduction in the untreated group. The authors stated that compliance with the daily use of repellent was self-reported at 90.5% for evening use, and actively detected at 84.6% on random home visits. They concluded that malaria transmission was quite low to give a confident assessment of repellent efficacy, but that the popularity of compliance suggested that it should be evaluated in other areas of low malaria transmission (McCready et al., 2001).

A study in Pakistan showed that a repellent containing DEET was popular among Afghan refugees and provided protection from malaria in the early evening (Rowland et al., 2004). A subsequent study in Bolivia showed the incidence of malaria in adults was reduced when people used a repellent formulation containing 30% PMD in combination with sleeping under permethrin-treated bed nets (Hill et al., 2007). This is an example of a notable large field study in Bolivia that did show an important effect by regular use of an arthropod repellent to reduce the incidence of malaria, an important worldwide mosquito-borne disease.

In a field study in three locations in Cambodia, the effects of one or four slow-release emanators containing metofluthrin were compared against biting rates of Anopheline mosquitoes (Charlwood et al., 2016). The landing rates were reduced by 48% by a single emanator and 67% by four emanators in two locations. However, the emanators did not reduce mosquito biting in a third location. The authors concluded that although the emanators reduced biting, they needed further development (Charlwood et al., 2016).

In malarious countries, it is suggested that topical and spatial arthropod repellents could be used when people are outside and not under bed nets or in indoor residual sprayed areas. In southeast Myanmar, a study was proposed to assign village health workers and allow them to distribute repellent and investigate the effects on malaria (Oo et al., 2018). A subsequent recent study reported that the distribution of DEET repellent through the village health workers network did not reduce detectable infections using rapid diagnostic test kits, but appeared to reduce the odds of PCR detectable infections of *P. falciparum* malaria. The costs to conduct this widespread study in Myanmar showed that extensive field studies to evaluate arthropod repellents in multiple villages over a 15-month time frame was more than $US75K (Agius et al., 2020).

An extensive study was undertaken in Sumba Island, Indonesia, between June 2015 and March 2016 to determine the efficacy of a transfluthrin-based spatial repellent in a passive emanator that released the volatile active ingredient into the air placed at two units per 9 m^2 application rates (Syafruddin et al., 2020). The study showed that there was no significant difference between treated and placebo-treated households, possibly due to low incidence of malaria in some clusters. However, among 12 moderate to high-risk clusters, a statistically significant decrease in infection by intervention was detected. The authors concluded that the study suggested that spatial arthropod repellents prevented malaria, but additional evidence was needed to demonstrate if the arthropod repellent product could provide an operationally feasible and effective means of reducing malaria transmission (Syafruddin et al., 2020).

16.11 Conclusion

There is a need for additional studies to determine the absolute effectiveness of the use of arthropod repellents in controlling vector-borne diseases. The studies that have been done so far have shown that the use of arthropod repellents alone has reduced the incidence of disease in some of the studies, but the reduction has been small. In addition, there have been conflicting factors in some of the studies, resulting in several being inconclusive.

The use of arthropod repellents continues to serve as the primary means of personal protection against biting arthropods (Piesman and Eisen, 2008). In addition, their use against vectors of vector-borne diseases has focused on methods that are low cost and can be conducted by local communities. In the last decade, more funds have become available to pursue some of the innovative personal protective methods of protection, and progress to reduce the incidence of vector-borne diseases, especially malaria, was made. The vector control workers in these countries have contributed knowledge and expertise in using and developing new methods.

This chapter discussed the use of topical and spatial arthropod repellents to reduce vector, nuisance biting, and disease transmission among people in varied locations around the world. Despite a variety of vector control methods, mosquitoes still cause nuisance and transmit mosquito-borne diseases, and it is hoped that the development of newer, better, and more effective topical and spatial arthropod repellents in the future will further decrease the burden of vector-borne diseases in the world.

References and further readings

Achee, N.L., Grieco, J.P., Vatandoost, H., Seixsas, G., Pinto, J., Choing-Ng, L., Martins, A.J., Juntarajumnong, W., Corbel, V., Gouagna, C., David, J.P., Logan, J.G., Osbourne, J., Marois, E., Devine, G.J, Vontas, J., 2019. Alternate strategies for mosquito-borne arbovirus control. PLoS Negl. Trop. Dis. 13, 1–22.

Agius, P.A., Cutts, J.C., Oo, W.H., Thi, A., O'Flaherty, K., Aung, K.Z., Thu, H.K., Aung, P.P., Thein, M.M., Zaw, N.N., Htay, W.Y.M., Soe, A.P., Razook, Z., Barry, A.E., Htike, W., Devine, A., Simpson, J.A., Crabb, B.S., Beeson, J.G., Pasricha, N., Fowkes, F.J.I., 2020. Evaulation of the effectiveness of topical repellent distributed by village health networks against *Plasmodium* spp. infection in Myanmar: a stepped-wedge cluster randomised trial. Plos Med. 17 (8), e1003177.

Anderson, A.L., Apperson, C.S., Knake, R., 1991. Effectiveness of mist-blower applications of malathion and permethrin to foliage as barrier sprays for salt marsh mosquitoes. J. Am. Mosq. Control Assoc. 7, 116–117.

Asilan, A., Sadeghinia, A., Shariati, F., Jome, I., Ghoddusi, A., 2003. Efficacy of permethrin-impregnated uniforms in the prevention of cutaneous leishmaniasis in Iranian soldiers. J. Clin. Pharm. Therapeut. 28, 175–178.

Barnard, D.R., Bernier, U.R., Posey, K.H., Xue, R-D., 2002. Repellency of IR3535, KBR 3023, *para*-menthane -*3*-diol, and deet to black salt marsh mosquitoes (Diptera: Culicidae) in the Everglades National Park. J. Med. Entomol. 39, 895–899.

Barnard, D.R., Posey, K.H., Smith, D., Schreck, C.E., 1998. Mosquito density, biting rate and cage size effects on repellent tests. Med. Vet. Entomol. 12 (1), 39–45.

Barnard, D.R., Xue, R.D., 2004. Laboratory evaluation of mosquito repellents against *Aedes albopictus*, *Culex nigrpalpus*, and *Ochlerotatus triseriatus* (Diptera:Culicidae). J. Med. Entomol. 41, 726–730.

Barnard, D.R., 2000. Repellents and Toxicants for Personal Protection: A WHO Position Paper. World Health Organization, Geneva.

Batra, C.P., Raghavendra, K., Adak, T., Singh, O.P., Singh, S.P., Mittal, P.K., Malhotra, M.S., Sharma, R.S., Subbarao, SK., 2005. Evaluation of bifenthrin treated mosquito nets against anopheline and culicine mosquitoes. Indian J. Med. Res. 121, 55–62.

Bell, D., Bryan, J., Cameron, A., Fernando, M., Leafascia, J., Pholsyna, K., 1997. Malaria in Honiara, Solomon Islands: reasons for presentation and human and environmental factors influencing prevalence. Southeast Asian J. Trop. Med. Publ. Hlth. 28, 482–488.

Bibbs, C.S., Xue, R-D., 2016. OFF! Clip-on repellent device with metofluthrin tested on *Aedes aegypti* (Diptera: Culicidae) for mortality at different time intervals and distances. J. Med. Entomol. 53, 480–483.

Bissinger, B.W., Apperson, C.S., Sonenshine, D.E., Watson, D.W., Roe, R.M., 2009a. Efficacy of the new repellent BioUD against three species of ixodid ticks. Ex. Appl. Acarol. 48, 239–250.

Bissinger, B.W., Zhu, J., Apperson, C.S., Sonenshine, D.E., Watson, D.W., Roe, R.M., 2009b. Comparative efficacy of BioUD to other commercially available arthropod repellents against the ticks, *Amblyomma americanum* and

Dermacentor variabilis on cotton cloth. Am. J. Trop. Med. Hyg. 81, 658–690.

Buhagiar, T.S., Devine, G.J, Ritchie, S.A., 2017. Metofluthrin: investigations into the use of a volatile spatial pyrethroid in a global spread of dengue, chikungunya and Zika viruses. Parasi. Vectors 10, 270.

Burns, M., Rowland, M., N'Guessan, R., Carneiro, I., Beeche, A., Ruiz, S.S., Kamara, S., Takken, W., Carnevale, P., Allan, R., 2012. Insecticide-treated plastic sheeting for emergency malaria prevention and shelter among displaced populations: an observational cohort study in a refugee setting in Sierra Leone. Am. J. Trop. Med. Hyg. 87, 242–250.

Carroll, S.P., Loye, J., 2006. A registered botanical mosquito repellent with DEET-like efficacy. J. Am. Mosq. Control Assoc. 22, 507–514.

Chandre, F., Dabire, R.K., Hougard, J-M., Djogbenou, L.S., Irish, S.R., Rowland, M., N'Guessan, R., 2010. Field efficacy of pyrethroid treated plastic sheeting (durable lining) in combination with long lasting insecticidal nets against malaria vectors. Parasi. Vectors 3, 65–71.

Charlwood, J.D., Dagoro, H., 1987. Repellent soap for use against malaria vectors in Papua New Guinea. Papua New Guinea Med. J. 30, 301–303.

Charlwood, J.D., Nenhep, S., Protopopoff, N., Sovannaroth, S., Morgan, J.C., Hemingway, J., 2016. Effects of the spatial repellent metofluthrin on landing rates of the outdoor biting anophelines in Cambodia. Southeast Asia. Med. Vet. Entomol. 30, 229–234.

Chavasse, D.C., Lines, J.D., Ichimori, K., 1996. The relationship between mosquito density and mosquito coil sales in Dar es Salaam. Trans. R. Soc. Trop. Med. Hyg. 90, 493–495.

Coleman, R.E., Burkett, D.A., Putnam, J.L., Sherwood, V., Caci, J.B., Jennings, B.T., Hochberg, L.P., Spradling, S.L., Rowton, E.D., Blount, K., Bloch, J., Hopkins, G., Raymond, J-L.W., O'Guinn, M.L., Lee, J.S., Weina, P.J., 2006. Impact of Phlebotomine sand flies on U.S military operations at Tallil Air Base: 1. Background, Military situation, and development of a "Leishmaniasis control program". J. Med. Entomol. 43, 647–662.

Costantini, C., Badolo, A., Iboudo-Sanoo, E., 2004. Field evaluation of the efficacy and persistence of insect repellents DEET, IR3535 and KBR 3023 against *Anopheles gambiae* complex and other Afrtropical vector mosquitoes. Trans. R. Soc. Trop. Med. Hyg. 98, 644–646.

Darbro, J.M., Muzari, M.O., Giblin, A., Damczyk, R.M., Ritchie, S.A., Devine, G.J., 2017. Reducing biting rates of *Aedes aegypti* with metofluthrin: investigations in time and space. Parasi. Vectors 10, 69.

Debboun, M., Strickman, D., 2013. Insect repellents and associated personal protection for a reduction in human disease. Med. Vet. Entomol. 27, 1–9.

Deparis, X., Frere, B., Lamizana, M., N'Guessan, R., Leroux, F., Lefevre, P., Finot, L., Hougard, J-M., Carnevale, P., Gillet, P., Baudon, D., 2004. Efficacy of permethrin-treated uniforms in combination with DEET topical repellent for protection of French military troops in Cote d'Ivoire. J. Med. Entomol. 41, 914–921.

Diabate, A., Chandre, F., Rowland, M., N'Guessan, R., Duchon, S., Dabire, K.R., Hougard, J-M., 2006. The indoor use of plastic sheeting pre-impregnated with insecticide for control of malaria vectors. Trop. Med. Int. Hlth. 11, 597–603.

Diaz, J., 2016. Chemical and plant-based insect repellents: efficacy, safety, and toxicity. Wild Environ. Med. 27, 153–163.

Dolan, M.C., Jordan, R.A., Schultze, T.L., Schulze, C.J., Manning, M.C., Ruffolo, D., Schmidt, J.P., Piesman, J., Karchesy, J.J., 2009. Ability of two natural products, nootkatone and carvacrol to suppress *Ixodes scapularis* and *Amblyomma americanum* (Acari:Ixodidae) in a Lyme disease endemic area of New Jersey. J. Econ. Entomol. 102 (6), 2316–2324.

Dolan, M.C., Panella, N.A., 2011. A review of arthropod repellents. In: Paluch, G.E., Coats, J.R. (Eds.), Recent Developments in Invertebrate Repellents. Oxford University Press, Inc., Washington, DC, pp. 1–19.

Eamsila, C., Frances, S.P., Strickman, D., 1994. Evaluation of permethrin-treated military uniforms for personal protection against malaria in northeastern Thailand. J. Am. Mosq. Control Assoc. 10, 515–521.

Farrar, R.R., Kennedy, G.G., 1987. 2-Undecanone, a constituent of the glandular trichomes of *Lycopersicon hirsutum f. glabratum*: effects *on Heliothis zea* and *Manduca sexta* growth and survival. Entomol. Exp. Appl. 43, 17–23.

Faulde, M.K., Uedelhoven, W.M., Malerius, M., Robbins, R.G., 2006. Factory-based permethrin impregnation of uniforms: residual activity against *Aedes aegypti* and *Ixodes ricinus* in Battle Dress Uniforms worn under field conditions, and cross-contamination during the laundering and storage process. Mil. Med. 171, 472–477.

Faulde, M.K., Uedelhoven, W.M., Robbins, R.G., 2003. Contact toxicity and residual activity of different permethrin-based fabric impregnation methods for *Aedes aegypti* (Diptera:Culicidae), *Ixodes ricinus* (Acari:Ixodidae), and *Lepisma saccharina* (Thysanura: Lepismatidae). J. Med. Entomol. 40, 935–941.

Faulde, M., Uedelhoven, W., 2006. A new clothing impregnation method for personal protection against ticks and biting insects. Int. J. Med. Microbiol. 296, 225–229.

Frances, S.P., Auliff, A.M., Edstein, M.D., Cooper, R.D., 2003. Survey of personal protection measures against mosquitoes among Australian Defense Force personnel deployed to East Timor. Mil Med 168, 227–230.

Frances, S.P., Cooper, R.D., Sweeney, A.W., 1998. Laboratory and field evaluation of the repellents DEET, CIC-4 and

AI3-37220, against *Anopheles farauti* (Diptera: Culicidae) in Australia. J. Med. Entomol. 35, 690–693.
Frances, S.P., Cooper, R.D., 2007. Personal protective measures against mosquitoes: insecticide-treated uniforms, bednets and tents. ADF Health 8, 50–56.
Frances, S.P., Debboun, M., 2007. User acceptability: public perceptions of insect repellents. In: Debboun, M., Frances, S.P., Strickman, D. (Eds.), Insect Repellents, Principles, Methods and Uses. CRC Press, Boca Raton, FL, pp. 397–403.
Frances, S.P., Dung, N.V., Beebe, N.W., Debboun, M., 2002. Field evaluation of repellent formulations against daytime and nighttime biting mosquitoes in a tropical rainforest in Northern Australia. J. Med. Entomol. 39, 541–544.
Frances, S.P., Huggins, R.L., Cooper, R.D., 2008. Evaluation of the inhibition of egg laying, larvicidal effects and bloodfeeding success of *Aedes aegypti* exposed to permethrin and bifenthrin-treated military tent fabric. J. Am. Mosq. Control Assoc. 24, 598–600.
Frances, S.P., Rowcliffe, K.L., MacKenzie, D.O., 2020. Field evaluation of a metofluthrin fan-based emanator and DEET as repellents against *Aedes vigilax* (Diptera: Culicidae) in Southeast Queensland, Australia. J. Am. Mosq. Control Assoc 36, 120–122.
Frances, S.P., Wirtz, R.A., 2005. Repellents: past, present, and future. J. Am. Mosq. Control Assoc 21, 1–3.
Frances, S.P., 1987. Effectiveness of DEET and permethrin, alone, and in a soap formulation as skin and clothing protectants against mosquitoes in Australia. J. Am. Mosq. Control Assoc. 3, 648–650.
Frances, S.P., 2007. Efficacy and safety of repellents containing deet. In: Debboun, M., Frances, S.P., Strickman, D. (Eds.), Insect Repellents: Principles, Methods and Uses. CRC Press, Boca Raton, FL, pp. 311–325.
Frances, S.P., 2007. Picaridin. In: Debboun, M., Frances, S.P., Strickman, D. (Eds.), Insect Repellents: Principles, Methods and Uses. CRC Press, Boca Raton, FL, pp. 337–340.
Frances, S.P., 2007c. Evaluation of bifenthrin and permethrin as barrier treatments for military tents against mosquitoes in Queensland, Australia. J. Am. Mosq. Control Assoc. 23, 208–212.
Goodyer, L.I., Croft, A.M., Frances, S.P., Hill, N., Moore, S.J., Sangoro, P., Onyango, P., Debboun, M., 2010. Expert review of the evidence base for arthropod bite avoidance. J. Travel Med. 17, 182–192.
Goodyer, L., Behrens, R.H., 1998. Short report: the safety and toxicity of insect repellents. Am. J. Trop. Med. Hyg. 59, 323–324.
Goodyer, L., Grootveld, M., Deobhankar, K., Debboun, M., Philip, M., 2020. Characterisation of actions pf p-menthane-3,8,-diol repellent formulations against *Aedes aegypti* mosquitoes. Trans. R. Soc. Trop. Med. Hyg., 1–6. doi:10.1093/trstmh/traa045.
Graham, K., Mohammed, N., Rehman, H., Nazari, A., Ahmad, M., Skovmand, O., Guillet, P., Allan, R., Zaim, M., Yates, A., Lines, J., Rowland, M., 2002. Insecticide-treated plastic tarpaulins for control of malaria vectors in refugee camps. Med. Vet. Entomol. 16, 404–408.
Gupta, R.K., Rutledge, L.C., Reifenrath, W.G., Guterrez, G.A., Korte Jr., D.W., 1990. Resistance to weathering in fabrics treated for protection against mosquitoes. J. Med. Entomol 27, 494–500.
Gupta, R.K., Sweeney, A.W., Rutledge, L.C., Cooper, R.D., Frances, S.P., Westrom, D.R., 1987. Effectiveness of controlled-release personal-use arthropod repellents and permethrin-impregnated clothing in the field. J. Am. Mosq. Control Assoc. 3, 556–560.
Haleem, Z.M., Yadav, S., Cushion, M.L., Tanner, R.J., Carek, P.J., Mainous, A.G., 2020. Exposure to N,N-diethyl-meta-toluamide insect repellent and human health markers: population based estimates from the national health and nutrition examination survey. Am. J. Trop. Med. Hyg. 103 (2), 812–814.
Harbach, R.E., Tang, D.B., Wirtz, R.A., Gingrich, J.B., 1990. Relative repellency of two formulations of *N,N*-diethyl-3-methylbenzamide (DEET) and permethrin-treated clothing against *Culex sitiens* and *Aedes vigilax* in Thailand. J. Am. Mosq. Control Assoc. 6, 641–644.
Hewitt, S., Rowland, M., Muhammad, N., Kamal, M., Kemp, E., 1995. Pyrethroid-sprayed tents for malaria control: an entomological evaluation in Pakistan. Med. Vet. Entomol. 9, 344–352.
Hill, N., Lenglet, A., Arnez, A.M., Caniero, I., 2007. Plant based repellent and insecticide treated bed nets to protect against malaria in areas of early evening biting vectors: double blind randomized placebo controlled clinical trial in the Bolivian Amazon. Br. Med. J. 335, 1023–1027.
Hougard, J-M., Duchon, S., Zaim, M., Guillet, P., 2002. Bifenthrin: a useful pyrethroid insecticide for treatment of mosquito nets. J. Med. Entomol. 39, 526–533.
John, R., Ephraim, T., Andrew, A., 2008. Reduced susceptibility to pyrethroid insecticide treated nets by the malaria vector *Anopheles gambiae* s.l. in Western Uganda. Malar. J. 7, 92.
Jordan, R.A., Schulze, T.L., Dolan, M.C., 2012. Efficacy of plant-derived and synthetic compounds on clothing as repellents against *Ixodes scapularis* and *Amblyomma americanum* (Acari:Ixodidae). J. Med. Entomol. 49 (1), 101–106.
Joshi, R.M., Ghose, G., Som, T.K., Bala, S., 2003. Study of the impact of deltamethrin impregnated mosquito nets on malaria incidence at a malaria station. Med. J. Armed Forces India 5, 12–14.
Kawada, H., Maekawa, Y., Tsuda, Y., Takagi, M., 2004a. Laboratory and field evaluation of spatial repellency with metofluthrin-impregnated paper strip against mosquitoes in Lombok Island, Indonesia. J. Am. Mosq. Control Assoc 20, 292–298.

Kawada, H., Maekawa, Y., Tsuda, Y., Takagi, M., 2004b. Trial of spatial repellency of metofluthrin-impregnated paper strip against *Anopheles* and *Culex* in shelters without walls in Lombok, Indonesia. J. Am. Mosq. Control Assoc. 20, 434–437.

Kikankie, C.K., Brooke, B.D., Knols, B.G.J., Koekemoer, L.L., Farenhorst, M., Hunts, R.H, Thomas, M.B., Coetzee, M., 2010. The infectivity of the entomopathogenic fungus *Beauveria bassiana* to insecticide-resistant and susceptible *Anopheles arabiensis* mosquitoes at two different temperatures. Malar. J. 9, 71.

Killeen, G.F., Smith, T.A., 2007. Exploring the contributions of bednets, cattle, insecticides and excitorepellency to malaria control: a deterministic model of mosquito host-seeking behaviour and mortality. Trans. R. Soc. Trop. Med. Hyg. 101, 867–880.

Kimani, E.W., Vulule, J.M., Kuria, I.W., Mugisha, F., 2006. Use of insecticide-treated clothes for personal protection against malaria: a community trial. Malar. J. 5, 63–72.

Kroeger, A., Gerhardus, A., Kruger, G., Mancheno, M., Pisse, K., 1997. The contribution of repellent soap to malaria control. Am. J. Trop. Med. Hyg. 56, 580–584.

Lillie, T.H., Schreck, C.E., Rahe, A.J., 1988. Effectiveness of personal protection against mosquitoes in Alaska. J. Med. Entomol. 25, 475–478.

Lindsay, S.W., Gibson, M.E., 1988. Bednets revisited- Old idea, new angle. Parasitol. Today 4, 270–272.

Lines, J.D., Myamba, J., Curtis, C.F., 1987. Experimental hut trials of permethrin-impregnated mosquito nets and eave curtains against malaria vectors in Tanzania. Med. Vet. Entomol. 1, 37–51.

Liu, D., Zhang, A., Zhou, M., 2003. Preliminary selective studies on the long lasting effect repellent formulations. Chin. J. Vector Bio. Control. 14, 358.

Lucas, J.R., Shono, Y., Iwasaki, T., Ishiwatari, T., Spero, N., Benzon, G., 2007. U.S. laboratory and field trials of metofluthrin (SumiOne®) emanators for reducing mosquito biting outdoors. J. Am. Mosq. Control Assoc. 23, 47–54.

Maia, M., Sangaro, P., Thiele, M., Turner, E., Moore, S., 2012. Do topical repellents divert mosquitoes within a community? Malar. J. 11 (Suppl. 1), 120.

Mani, T.R., Reuben, R., Akiyama, J., 1991. Field efficacy of "Mosbar" repellent soap against vectors of Bancroftian filariasis and Japanese Encephalitis in southern India. J. Am. Mosq. Control Assoc. 7, 565–568.

McCready, R., KA Hamilton, K.A., Simpson, J.A., Cho, T., Luxemberger, C., Edwards, R., Looareesuwan, S., White, N.J., Nosten, F., Lindsay, S.W., 2001. Safety of the insect repellent *N,N*-Diethyl-*m*-toluamide (DEET) in pregnancy. Am. J. Trop. Med. Hyg. 65, 285–289.

McCulloch, R.N., 1946. Studies in the control of scrub typhus. Med. J. Aust. 1, 717–738.

McCulloch, R.N., 1947. The adaption of military scrub typhus mite control to civilian needs. Med. J. Aust. 1, 449–452.

McGinn, D., Frances, S.P., Sweeney, A.W., Brown, M.D., Cooper, R.D., 2008. Evaluation of Bistar 80SC (Bifenthrin) as a tent treatment for protection against mosquitoes in Northern Territory. Australia. J. Med. Entomol. 45, 1087–1091.

Miller, N.J., Rainone, E.E., Dyer, M.C., Gonzalez, M.L., Mather, T.N., 2011. Tick bite protection with permethrin-treated summer-weight clothing. J. Med. Entomol. 48, 327–333.

Miller, R.J., Wing, J., Cope, S.E., Klavons, J.A., Kline, D.L., 2004. Repellency of permethrin–treated battle–dress uniforms during Operation Tandem Thrust 2001. J. Am. Mosq. Control Assoc 20, 462–464.

Moore, S.J., Darling, S.T., Sihuincha, M., Padilla, N., Devine, G.J., 2007. A low-cost repellent for malaria vectors in the Americas: results of two field trials in Guatemala and Peru. Malar. J. 6, 101.

Naucke, T.J., Lorentz, S., Grunewald, H.W., 2006. Laboratory testing of the insect repellents IR3535 and DEET against *Phlebotomus mascitti* and *P. duboscqi* (Diptera:Psychodidae). Int. J. Med. Microbiol. 40S, 230–232.

Nentwig, G., Frohberger, R., Sonneck, R., 2017. Evaluation of clove oil, Icaridin, and transfluthrin for spatial repellent effects in three tests systems against the *Aedes aegypti* (Diptera: Culicidae). J. Med. Entomol. 54, 150–158.

N'Guessan, R., Rowland, M., Moumouni, T-L., Kesse, N.B., Carnevale, P., 2006. Evaluation of synthetic repellents on mosquito nets in experimental huts against insecticide-resistant *Anopheles gambiae* and *Culex quinquefasciatus* mosquitoes. Trans. R. Soc. Trop. Med. Hyg 100, 1091–1097.

Ogama, S.B., Moore, S.J., Maia, M.F., 2012. A systematic review of mosquito coils and passive emanators: defining recommendations for spatial repellency testing methodologies. Parasi. Vectors 5, 287.

Onyango, S.P., Moore, S.J., 2015. Evaluation of repellent efficacy in reducing disease incidence. In: Debboun, M., Frances, S.P., Strickman, D. (Eds.), Insect Repellents Handbook2nd ed. CRC Press, Boca Raton, FL, pp. 117–156.

Oo, W.H., Cutts, J.C., Agius, P.A., Aung, K.Z., Aung, P.P., Thi, A., Zaw, N.N., Thu, H.K., Htay, W.Y.M., Ayaide, R., O'Flaherty, K., Yawn, A.I, Soe, A.P., Beeson, J.G., Crabb, B., Patricha, N., Fowkes, F.J.I., 2018. Effectiveness of repellent delivered through village health volunteers on malaria incidence in villages in South-East Myanmar: a stepped-wedge cluster-randomised controlled trial protocol. BMC Infect. Dis. 18, 663.

Panella, N.A., Dolan, M.C., Karchesy, J.J., Xiang, Y., Peralta-Cuz, J., Khasaweneh, M., Montenieri, J.A., Maupin, G.O., 2005. Use of novel compounds for pest control: insecticidal and acaricidal activity of essential oil components from heartwood of Alaska yellow cedar. J. Med. Entomol. 42, 352–358.

Perich, M.J., Tidwell, M.A., Dobson, S.E., Sardelis, M.R., Zaglul, A., Williams, D.C., 1993. Barrier spraying to control the malaria vector *Anopheles albimanus*: laboratory and field evaluation in the Dominican Republic. Med. Vet. Entomol. 7, 363–368.

Philip, J.R., Sabin, A.B., 1944. Dimethyl phthalate as a repellent in control of *Phlebotomus* (*pappataci* or sandly) fever. War Med 6, 27–33.

Piesman, J., Eisen, L., 2008. Prevention of tick-borne diseases. Annu. Rev. Entomol. 53, 323–343.

Procacci, P.G., Lamizana, L., Kumlein, S., Habluetzel, A., Rotigliano, G., 1991. Permethrin-impregnated curtains in malaria control. Trans. R. Soc. Trop. Med. Hyg. 85, 181–185.

Puccetti, G., 2007. IR3535 (Ethyl Butylacetoaminopropionate) Strickman. In: Debboun, M., Frances, S.P. (Eds.), Insect Repellents: Principles, Methods and Uses. CRC Press, Boca Raton, FL, pp. 353–360.

Qualls, W.A., Smith, M.L., Muller, G.C., Zhao, T-Y., Xue, R-D., 2012. Field evaluation of a large-scale barrier application of bifenthrin on a golf course to control floodwater mosquitoes. J. Am. Mosq. Control Assoc. 28, 219–224.

Qualls, W.A., Xue, R-D., Holt, J.A., Smith, M.L., Moeller, J.J., 2011. Field evaluation of commercial repellents against the floodwater mosquito *Psorophora columbiae* (Diptera: Culicidae) in St. Johns County, Florida. J. Med. Entomol. 48 (6), 1247–1249.

Revay, E.E., Junnila, A., Xue, R-D., Kline, D.L., Bernier, U.R., Kravchenko, V.D., Qualls, W.A., Ghattas, N., Muller, G.C., 2013. Evaluation of commercial products for personal protection against mosquitoes. Acta Trop. 125, 226–230.

Rowland, M., Downey, G., Rab, A., Freeman, T., Mohammad, N., Rehman, H., Durrani, N., Curtis, C., Lines, J., Fayaz, M., 2004. DEET mosquito repellent provides personal protection against malaria: a household randomized trial in an Afghan refugee camp in Pakistan. Med. Vet. Entomol. 9, 335–342.

Rutledge, L.C., Sofield, R.K., Mussa, M.A., 1978. A bibliography of Diethyl toluamide. ESA Bull. 24, 431–439.

Schoenig, G.P., Neeper-Bradley, T.L., Fisher, L.C., Hartnagel Jr., R.E., 1994. Teratological evaluations of DEET in rats and rabbits. Fundam. Appl. Toxicol. 23, 63–69.

Scholte, E-J., Knols, B.G.J., Samson, R.A., Takken, W, 2004. Entomopathogenic fungi for mosquito control: a review. J. Insect Sci. 4, 19.

Schreck, C.E., Carlson, D.A., Weidhaas, D.E., Posey, K., Smith, D., 1980. Wear and aging tests with permethrin-treated cotton-polyester fabric. J. Econ. Entomol. 73, 451–453.

Schreck, C.E., Leonhardt, B., 1991. Efficacy assessment of Quwenling, a mosquito repellent from. China J. Am. Mosq. Control Assoc. 7, 43–436.

Schreck, C.E., Mount, G.A., Carlson, D.A., 1982. Wear and wash persistence of permethrin used as a clothing treatment for personal protection against the lone star tick (Acari: Ixodidae). J. Med. Entomol. 19, 143–146.

Schreck, C.E., Posey, K., Smith, D., 1978. Durability of permethrin as a potential clothing treatment to protect against blood-feeding arthropods. J. Econ. Entomol. 71, 397–400.

Schreck, C.E., 1991. Permethrin and dimethylphthalate as a tent fabric treatments against *Aedes aegypti*. J. Am. Mosq. Control Assoc. 7, 533–535.

Simmons, T.E., 1985. Insect-repellent and insecticidal soap composition. UK Patent No. 2160216A.

Sholdt, L.L., Holloway, M.L., Chandler, J.A., Fontaine, R.E., van Elsen, A., 1977. Dwelling space repellents: their use on military tentage against mosquitoes in Kenya, East Africa. J. Med. Entomol. 14, 252–253.

Smith, D.L., Battle, K.E., Hay, S.I., Barker, C.M., Scott, T.W., McKenzie, E., 2012. Ross, Macdonal, and a theory for the dynamics and control of mosquito-transmitted pathogens. Plos Pathol 8 (4), e1002588 1002510.1001371/journal.ppat.1002588.

Sochantha, T., van Bortel, W., Savannaroth, S., Marchotty, T., Speybroeck, N., Coosemans, M., 2010. Personal protection by long-lasting insecticidal hammocks against the bites of forest malaria vectors. Trop. Med. In.t Hlth. 15, 336–341.

Sota, J., Mendina, F., Dember, N., Berman, J., 1995. Efficacy of permethrin-impregnated uniforms in the prevention of malaria and leishmaniasis in Columbian soldiers. Clin. Infect. Dis 21, 599–602.

Sudakin, D.L., Wade, R., Trevathan, B.S., 2003. DEET: a review and update of safety and risk in the general population. J. Toxicol. Clin. Toxicol. 41, 831–839.

Syafruddin, D., Asih, P.B.S., Rozi, I.E., Permana, D.H., Hidayti, A.P.N., Syahrani, L., Zubaidah, S., Sidk, D., Bangs, M.J., Bpogh, C., Liu, F., Eugenio, E.C., Hendrickson, J., Burton, T., Baird, J.K., Collins, F., Grieco, J.P., Lobo, N.F., Achee, N.L., 2020. Efficacy of a spatial repellent for control of malaria in Indonesia: a cluster-randomized controlled trial. Am. J. Trop. Med. Hyg. 103, 344–358.

Thavara, U.A., Tawatsin, A., Chompoosri, J., Suwonkerd, W., Chansuang, U.R., Asavadachanukon, P., 2001. Laboratory and field evaluations of the insect repellent 3535 (ethyl butylacetylaminoproprionate) and DEET against mosquito vectors in Thailand. J. Am. Mosq. Control Assoc. 17, 190–195.

Thomson, M.C., Alessandro, U., Bennett, S., Connor, S.J., Langerock, P., Jawara, M., Todd, J., Greenwood, B.C., 1994. Malaria prevalence is inversely related to vector density in The Gambia, West Africa. Trans. R. Soc. Trop. Med. Hyg. 88, 638–643.

Tisgratog, R., Sukkanon, C., Grieco, J.P., Sanguanpong, U., Chauhan, K.R., Coats, J.R., Chareonviriyaphap, T., 2018. Evaluation of the constituents of Vetiver Oil against *Anopheles minimus* (Diptera: Culicidae), a malaria vector in Thailand. J. Med. Entomol. 55, 193–199.

Trigg, J.K., Hill, N., 1996. Laboratory evaluation of a eucalyptus-based repellent against four biting arthropods. Phytother. Res. 10, 10313–10316.

Trigg, J.K., 1996. Evaluation of a eucalyptus-based repellent against *Anopheles* spp. in Tanzania. J. Am. Mosq. Control Assoc. 12, 243–246.

Trout, R.T., Brown, G.C., Potter, M.F., Hubbard, J.L., 2007. Efficacy of two pyrethroid insecticides applied as barrier treatments for managing mosquito (Diptera: Culicidae) populations in suburban residential properties. J. Med. Entomol. 44, 470–477.

Witting-Bissinger, B.E., Stunpf, C.E., Donohue, K.V., Apperson, C.S., Roe, R.M., 2008. Novel arthropod repellent, BioUD, is an efficacious alternative to Deet. J. Med. Entomol. 45, 891–898.

World Health Organization., 2013. Guidelines for efficacy testing of spatial repellents. Pesticide Evaluation Scheme, Geneva.

World Health Organization., 2001. Report of the fifth WHOPES working group meeting. WHO/CDS/WHOPES/2001.4. World Health Organization, Geneva.

World Health Organization., 1975. Manual of Practical Entomology in malaria, Part II. Methods and techniques. World Health Organization, Geneva.

Xue, R-D., Qualls, W.A., Smith, M.L., Gaines, M.K., Weaver, J.H., Debboun, M., 2012. Field evaluation of the Off! Clip-on mosquito repellent (metofluthrin) against *Aedes abopictus* and *Aedes taeniorhynchus* (Diptera: Culicidae) in Northeastern Florida. J. Med. Entomol. 49, 652–655.

Yap, H.H., Jahangir, K., Zairi, J., 2000. Field efficacy of four insect repellents against vector mosquitoes in a tropical rainforest. J. Am. Mosq. Control Assoc. 16, 241–244.

Yap, H.H., Jahangir, S.C., Chong, C., Adanan, C.R., Chong, N.L., Malik, Y.A., Rohaizat, B., 1998. Field efficacy of a new repellent, KBR 3023, against *Aedes albopictus* (Skuse) and mosquitoes in a tropical rainforest. J. Vector Ecol. 23, 62–68.

Yap, H.H., 1986. Effectiveness of soap formulations containing DEET and permethrin as personal protection against outdoor mosquitoes in Malaysia. J. Am. Mosq. Control Assoc. 2, 63–67.

Index

Page numbers followed by "*f*" and "*t*" indicate, figures and tables respectively.

Printed in the United States
by Baker & Taylor Publisher Services